Encyclopedic Dictionary
of
Exploration Geophysics
Second Edition

Robert E. Sheriff

Society of Exploration Geophysicists

ISBN 0-931830-31-3
Library of Congress Catalog Card No. 84-051971

Society of Exploration Geophysicists
P.O. Box 702740, Tulsa, OK 74170-2740
© 1984 by the Society of Exploration Geophysicists.
Published 1984. Reprinted 1988, 1989

Table of Contents

Preface to the Second Edition

It is now 10 years since the *Encyclopedic Dictionary of Exploration Geophysics* was published. In this edition new terms have been added and definitions revised in order to keep up with the technological advances of this decade.

The first edition appeared when exploration geophysicists were starting to cope with data processing and help was needed with the processing vocabulary. The effects of processing improvements on the quality of data have been remarkable. Now an exploration geophysicist needs to extract more geologic significance from data. Accordingly, more geologic terms are included. The vocabulary of geophysics outside the exploration area has also been expanded. We see more exploration technology used to understand fundamentals of the earth, and better understanding of the entire earth also contributes to our ability to explore. We must narrow the gap between pure and applied geophysics and not lose touch with areas outside our specialties.

I use the following conventions in the Dictionary.

1) Entries usually begin with the heart of a definition, although occasionally a discussion is given first to provide a frame of reference. Restrictions on meanings sometimes are contained in a discussion which follows rather than being incorporated into the definition itself. Terms indicated as being synonomous are often used interchangeably even though they may not be identical in all respects. This is especially true of terms of well-log methods which involve the same principles but differ in details.

2) The numbering within an entry indicates different meanings, but the sequence does not indicate preference as to usage. Where different meanings are contradictory, this is stated explicitly and, in some cases, a preferred usage is indicated and an alternative suggested to avoid ambiguity. Letters subdivide an entry without implying differences in meaning. Only specialized meanings used in geophysics are included.

3) A definition in *italics* refers the reader to another entry which supplements the meaning. Cross-references are shown only where needed to complete the meaning. They are indicated by "see", "compare", or "q.v." (quo vide). Cross-references are also used to indicate preferences. For example, "*P*-wave" is preferred to other terms which mean exactly the same thing, so the other terms are referred to the *P*-wave entry. Likewise, "common midpoint" is preferred to "common-depth-point" or "common-reflection-point" because it expresses more accurately what it is that is common.

4) **Bold-face** within an entry indicates additional terms which are defined here (in effect).

5) Tradenames are included where they are in general use. Where such entries are used for a class of devices, the entry may begin with a lower-case letter even though the tradename may begin with a capital. Neither inclusion nor exclusion

of tradename credits should imply judgments about the merits of any devices or processes. Additional tradenames and company names are in the appendices.

6) References suggest a place to begin looking for further information. References are listed in the back. A single source which may be readily available is cited to provide the starting point for further search. The citation of a reference does not imply the original source nor the most complete or current reference. Readers who want more information will generally find additional sources suggested in the cited references.

7) Figures have been kept simple to illustrate the terminology and the important features of concepts without attempting to make them realistic or illustrate all features. Distinction between figures and tables has been eliminated, and both are now called "figures". The figure numbers are prefixed with the letter indicating the part of the book where they can be found.

Suggestions from the first edition and for the present edition have come from many sources. All have been considered and I thank all who assisted. Please continue sending suggestions so that errors may be corrected in future printings.

In addition to those explicitly cited in the preface of the first edition, we acknowledge the help of the following.

F. I. Ahmad	G. H. F. Gardner	H. Roice Nelson
Philip D. Antonelli	Lloyd P. Geldart	G. E. Parker
Edwin J. Ballantyne	Norman E. Goldstein	Layton M. Payne
Linda K. Barasch	Ronald Green	C. W. Racer
B. B. Bhattacharya	Stuart A. Hall	K. Ranto
Jules Braunstein	W. J. Hinze	Robert D. Regan
Alistair R. Brown	Mark Holzman	Carl H. Savit
S. J. Bullock	Glenda E. Johnson	F. Segesman
Kenneth E. Burg	W. C. Kellogg	Margaret S. Sheriff
John Butler	Richard L. Kirkpatrick	Richard K. Sheriff
R. E. Carlile	H. J. Koerner	Anne Simpson
D. A. Cavers	Thomas R. LaFehr	Al Singleton
Thomas C. Chen	Franklyn K. Levin	N. M. Soonawala
A. D. Christensen	Roy O. Lindseth	Nelson C. Steenland
Karel Cidlinsky	Keith W. Loucks	Kurt M. Strack
Jon F. Claerbout	Bill Lynch	Paul H. Taxil
Ph. Claude	Brian C. Mallick	Howard L. Taylor
J. D. Corbett	W. Harry Mayne	Richard E. Thayer
D. B. Daniel	L. A. McCarley	Raymond C. Todd
G. Clark Davenport	John A. McDonald	Keeva Vozoff
Leslie R. Denham	Neil McNaughton	Stanley H. Ward
T. L. Dobecki	James G. Morgan	Jack C. Weyand
Elmer Eisner	John F. Mueller	Randy A. Winney
E. E. Finklea	D. B. Murray	Kevin L. Woller
Wynn Gajkowski	Altan Necioglu	Howard J. Yorston

I also wish to thank the organizations for whom the people above work for their (sometimes unknown) support of this endeavor. I am especially grateful to those below who plowed through the final manuscript looking for errors or omissions:

Linda K. Barasch	Lloyd P. Geldart
J. D. Corbett	Margaret S. Sheriff
Leslie R. Denham	Richard K. Sheriff
Elmer Eisner	Kurt M. Strack

Appreciation is due to Stanley H. Ward for his efforts to make sure this work conforms to SEG standards so that it may be used as a guide by authors for GEOPHYSICS. We also appreciate the help of Samuel H. Mentmeir who revised Appendix H.

I especially wish to acknowledge the cooperation and support of the Geosciences Department and the Allied Geophysical Laboratories at the University of Houston on whose computer/word-processor this manuscript was prepared. In this connection, I especially thank Dr. John Butler, Dr. G. H. F. Gardner, Dr. John A. McDonald, Rudy Charest, and Teresa Weir (who spent many hours making entries and corrections on the word processor).

And finally, I wish to thank Jerry Henry and the SEG Publications staff for their efforts in publishing this edition, especially Jane R. Salas who accomplished the huge task of checking the multitudinous details required.

ROBERT E. SHERIFF
HOUSTON, 1984

Excerpts from the Preface to the First Edition

The "Glossary of Terms Used in Geophysical Exploration" which was published in GEOPHYSICS in February 1968 occasioned many expressions of appreciation for my attempt to help geophysical understanding and communication. Improvements and additions were suggested, including expansion into related fields. These suggestions resulted in the publication in April 1969 of an "Addendum" (with expanded coverage of mining geophysics) and the publication in December 1970 of the "Glossary of Terms Used in Well Logging."

The Glossary developed greater use as a reference than was anticipated. One well-received feature of the original glossary was the "extended entry," wherein a definition was expanded to clarify implications and relate it to other concepts. The expanded entries provided convenient access to basic equations and facts which are only used occasionally. Milton Dobrin observed that this work had developed into more than a mere glossary and suggested the name, "Encyclopedic Dictionary of Exploration Geophysics."

The *Encyclopedic Dictionary of Exploration Geophysics* was compiled for "practical" geophysicists rather than for researchers or other specialists. The novice surveyor can find the reason for surveying for a "walkaway." Common geologic terms are included because the end objective of most geophysical work is a geologic picture, although geologic time has been relegated to an appendix. Some "popular" information like sea states and earthquake intensity scales is included. Those who are only seismologists may find that their specialized vocabulary is also used in electrical exploration, and we hope they'll realize that geophysics includes many disciplines. I trust that browsing in this work may show the broad scope of geophysics.

The eclectic nature of this work inevitably involves inconsistencies in deciding which entries and how much information to include. The most likely reason why a particular entry may be missing is that I didn't happen to think of it. While I would have preferred to have a more balanced and complete work, this dictionary seems open-ended as if it could never be complete. To continue to refine it would only have delayed its availability. Therefore, despite faults, we have proceeded to press. It will be appreciated if readers who find errors or omissions or who may have other suggestions will inform me.

I do not claim to have invented the terms included here but merely to give the meanings in actual geophysical usage. Numerous references were consulted to assure reasonable conformity; sometimes this results in repeating someone else's phraseology.

Many people helped in this compilation, including many whom I have never met and some whose names I do not even know. I thank them all for their help even though I do not cite them specifically.

ROBERT E. SHERIFF
Houston, 1972

Excerpts from the Preface to the First Edition

A

A: See *A-type section*.

ABC method: A method of computing refractor depth based on refracted arrivals from sources near the surface. Especially used for weathering thickness determination from shots above the base of the weathering. See Figure A-1. The weathering time below *B*, t_w, is sometimes multiplied by a "*k*-factor" to give the vertical weathering time:

$$t_v = kt_w = t_w V_1 (V_1^2 - V_0^2)^{-1/2}.$$

AB electrodes: The current electrodes in resistivity surveying and well logging. Current is passed between the *A* and *B* electrodes while voltage is measured between the *M* and *N* electrodes. See Figures A-12, A-13, and E-7.

Abelian: *Commutative* (q.v.). Named for Niels Henrik Abel (1802-29), Norwegian mathematician.

ABEND: Acronym for *abnormal end* (q.v.).

abnormal end: Premature termination of a computer program due to a hardware or software error. Verb: **abort.**

abnormal events: Coherent events which are not reflections. Refractions, reflected refractions, diffractions, surface waves, and sometimes multiple reflections are included (although there is nothing "abnormal" about any of these).

abnormally high pressure: Formation fluid pressure which appreciably exceeds the **hydrostatic head**, the pressure produced by a column of water extending to the surface. **Hydrostatic pressure** is about 0.465 psi/ft (1.05 × 10^4 Pa/m, equivalent to 9.2 lb/gallon mud); **geostatic** or **lithostatic pressure** (the weight of the overlying section) is about 1 psi/ft (2.26 × 10^4 Pa/m or 20 lb/gallon mud). Pressures greater than about 0.7 psi/ft are considered "abnormal" and are associated with low seismic velocity and low density. The seismic effects of abnormal pressure are discussed in Sheriff and Geldart, v. 2 (1983, p. 13-15).

abort: To terminate without running to completion, as to "abort a computer run".

ab plane: See *axial surface*.

AB rectangular array: Gradient array; see *array*.

absorption: 1. A process whereby energy is converted into heat while passing through a medium. Absorption for seismic waves may be as large as 0.5 dB/cycle. See *Q* and Toksöz and Johnston (1982). **2.** The penetration of the molecules or ions of a substance into the interior of a solid or liquid.

absorption coefficient: If the amplitude *A* is expressed as

$$A = A_0 e^{-\alpha x}$$

where x = distance, α is the absorption coefficient or attenuation factor. Distinction may or may not be made as to the reason for attenuation (i.e., absorption or some other mechanism). See *Q*.

ac-bias recording: A system in which a modulating message biases a constant-amplitude carrier. See Figure M-10. Used in magnetic tape recording to avoid hysteresis distortion. The half-wave rectified portion is sometimes used as the information carrier.

acceleration factor: The velocity gradient factor *a* in the equation $V = V_0 + az$, where z = depth. In basins filled with clastic sediments, *a* is commonly about 0.6/s.

acceleration of gravity: See *gravitational constant*.

accelerometer: 1. A transducer whose output is proportional to acceleration. Accelerometers are used with shipboard gravimeter and inertial navigation systems. **2.** A seismometer or geophone whose output is proportional to the acceleration of earth particles. For example, a moving-coil geophone with velocity response proportional to frequency (as may be the case below the natural frequency) operates as an accelerometer.

acceptance criteria: Criteria which data must satisfy in order not to be rejected.

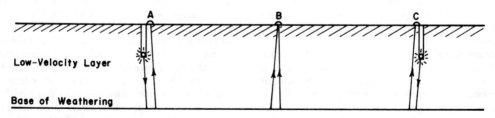

FIG. A-1. **ABC weathering method.** Weathering time t_w below B is
$$t_w = (T_{AB} + T_{BC} - T_{AC})/2,$$
where T_{AB} = surface-to-surface time from A to B (obtained by adding the uphole time to the first-break time), etc.

1

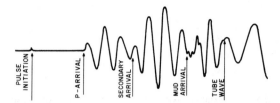

FIG. A-2. **Acoustic wave train** in a borehole (idealized). See also Figure C-2.

acceptor: A *p-type semiconductor* (q.v.).

access time: The time interval between when data are called for and when they are delivered. For example, the time it takes a computer to locate data on an I/O device (or in its memory) and transfer it to its arithmetic unit where computations are to be performed, or the time to transfer it back to the proper location in memory.

ac coupling: See *coupling*.

accumulate error: The result of a systematic cause of errors, so that errors tend to be of the same sign. Such errors have little tendency to cancel and may add up to a large cumulative error.

accuracy: 1. The degree of freedom from *error* (q.v.); the total error compared to the "true" value. Compare *precision, uncertainty*, and *sensitivity*. **2.** The capability of an instrument to follow a true value. **Inaccuracy** is the departure from the true value due to any instrument error such as lack of repeatability, drift, temperature effects, or other causes.

ac demagnetization: See *alternating-field demagnetization*.

acidic crust: See *sial*.

aclinic line: *Magnetic equator* (q.v.).

acoustic: Sonic; pertaining to sound. Usually refers to *P*-waves, sometimes is restricted to *P*-waves in fluids (liquids and gases).

acoustic basement: The deepest more-or-less continuous seismic reflector; often an unconformity below which seismic energy returns are poor or absent.

acoustic coupler: A device for transforming analog signals from acoustic form (as in a telephone handset) to electrical form (as in a modem), and vice versa.

acoustic emissions: Small-amplitude high-frequency transient elastic waves generated by deformation of a material.

acoustic impedance: Seismic velocity multiplied by density. Reflection coefficient at normal incidence depends on changes in acoustic impedance. **Specific acoustic impedance** is acoustic impedance divided by the acoustic impedance of water.

acoustic impedance section: A seismic display intended to represent acoustic impedance variations; the result of inversion of a reflectivity section. Compare *seismic log*.

acoustic log: 1. A generic term for well logs that display aspects of seismic-wave propagation. In some (sonic log, continuous-velocity log), the traveltime of *P*-waves

between two points is measured. In others (amplitude log), the amplitude of part of the wave train is measured. Other full-waveform acoustic logs (character log, 3-D log, VDL-log, microseismogram log, signature log) display part of the wave train in wiggle or variable-density form, and still others (cement-bond log, fracture log) are characterized by the objective of the measurements rather than their form. Borehole televiewer is also an acoustic log. **2.** Specifically, a *sonic log* (q.v.).

acoustic positioning: Determining location using sonar waves, as by (a) *Doppler-sonar* (q.v.) or (b) locating with respect to fixed sonar transponders.

acoustic velocity log: *Sonic log* (q.v.).

acoustic wave: 1. A *P*-wave, sometimes restricted to *P*-waves in fluids, but often including those in the solid earth. Synonyms: **sound wave, sonic wave. 2.** The wave train generated and detected by a sonic-logging sonde (see *acoustic log*). The wave train (Figure A-2) is a composite of various modes of energy transfer. The first arrival usually results from *P*-waves traveling in the formation; the **sonic log** measures the inverse of its velocity. A second arrival is sometimes identified as *S*-wave travel in the formation; it is sometimes a tube wave which travels at approximately the velocity of *S*-waves. Waves traveling through the mud usually have relatively high frequency content; they are sometimes called **fluid waves**. One or more modes of high-amplitude, low-frequency tube waves (sometimes called Stoneley waves) are usually distinct arrivals. **3.** More generally, any elastic wave or seismic wave.

Acoustilog: Acoustic-velocity log or *sonic log* (q.v.). Dresser-Atlas tradename.

activation logging: A well-logging technique in which the formation is irradiated with neutrons that transmute some nuclei into radioisotopes. Radiation from the radioisotopes is measured after a short time interval.

activation overvoltage: See *overvoltage*.

active: 1. A system or circuit which includes an energy source. Often pertains to electronic elements such as amplifiers and filters. Compare *passive*. **2.** A method which involves artificially induced signals. For example, see *controlled-source electromagnetics*. **3.** A positioning system which involves transmission from the mobile station which is to be located.

active beacon: See *passive*.

active margin: A plate margin where one plate is being subducted under another plate because of plate convergence. See Figure P-4. Also called a **Pacific** or **convergent margin**. Antonym: passive, Atlantic or trailing margin.

activity: The relative tendency of a substance to enter into a reaction. When shales adjacent to a reservoir act as perfect cationic membranes and the permeable bed is clean, the *electrochemical SP* (q.v.) can be found from the activities of the formation water and mud filtrate, which depend on the concentration of dissolved salts.

actuator: A vibrator designed for higher frequencies.

A/D: *Analog-to-digital* (q.v.).

Adachi formulas: Equations for solving the multilayer dipping refractor problem where the spread is perpendicular to the strike. See Adachi (1954) or Sheriff and Geldart, v. 1 (1982, p. 93-94).

adaptive deconvolution: A deconvolution where the inverse filter changes as the statistics of the data change.

adaptive processing: Data processing wherein processing parameters are varied with arrival time as measurements of data statistics change.

adder: A logic circuit device whose output represents the sum of its inputs. See *half adder*.

additive: A correction added to arrival times of seismic reflections measured from an arbitrary time origin. The additive normally adjusts to correspond to an arbitrary datum.

additive primary colors: See *primary colors*.

address: A label which identifies a specific location in a computer memory.

address space: The set of addresses available to a running program.

adiabatic: Involving no net gain or loss of heat in the system under consideration.

adiabatic approximation: Determination of changes in quantities by assuming that no heat is gained or lost from the system.

adjacent-bed effect: See *shoulder-bed effect* (q.v.).

adjoint: The inverse matrix times the determinant, also called **adjugate**. See *matrix*.

adjugate: See *adjoint*.

admittance: 1. The reciprocal of *impedance* (q.v.); the complex ratio of current to voltage. Unit of measure is the siemen or mho. **2.** The admittance S of a section of horizontal layers of thickness h, and resistivity ρ, overlying a basement of very high resistivity is:

$$S = \Sigma(h_i/\rho_i).$$

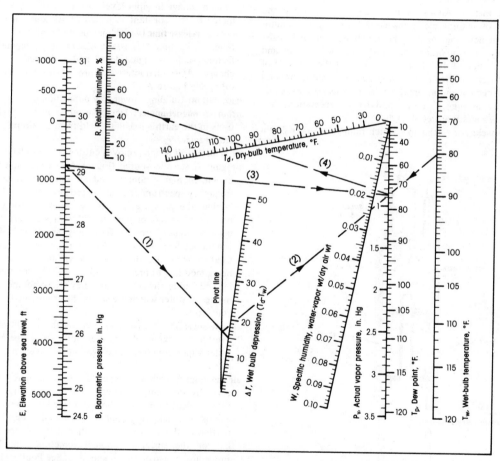

FIG. A-3. **Air pressure**/temperature/humidity relations. The nomograph involves the equations

$$P_v = P_w - B(T_d - T_w)/2700,$$
$$R = P_v/P_d,$$
$$W = P_v/1.61(B - P_v),$$
$$\Delta T = T_d - T_w,$$

where P_v = actual vapor pressure in inches of mercury (= vapor pressure at the dew point); P_w = vapor pressure at wet-bulb temperature T_w; P_d = vapor pressure at dry-bulb temperature T_d; B = barometric pressure; E = elevation in feet; W = specific humidity (= water-vapor weight/dry-air weight); R = relative humidity; ΔT = wet-bulb depression in °F; T_p = dew point. Example: At 750 ft elevation and dry-bulb temperature 95°F, the wet-bulb temperature is 80°F. (1) Align 750 ft on scale E with wet-bulb depression ΔT = 15; (2) align where line crosses pivot line with T_w = 80° and read P_v = 0.87 inches of mercury and dew point T_p = 74.5°; (3) align P_v = 0.87 with E = 750 ft and read W = 0.019 specific humidity; (4) align P_v = 0.87 with T_d = 95 and read relative humidity = 52.4 percent.

See also *S-rule*. **3.** The reciprocal of acoustic impedance. **4.** In the magnetotelluric method, the complex
ratio of electric field **E** to the perpendicular magnetic
field **H**, or the inverse of the impedance tensor Z. **5.**
Admittance is used in mechanical situations where an
analogy is made to an electrical circuit. Thus one might
refer to the "admittance of the ground" when discuss-
ing the transfer of energy from a source into seismic
wave energy.

adsorption: An electrostatic chemical process in which a
thin layer of molecules becomes fixed to the outer
surface of a solid. See *fixed layer*.

AEM: Airborne electromagnetic surveying system.

aeolotropy: *Anisotropy* (q.v.).

aerated layer: *Weathering* (q.v.) or near-surface low-
velocity layer.

aeromagnetic: Involving magnetic measurements made
from an aircraft.

Afmag method: Audio-frequency magnetic technique; the
use of natural electromagnetic noise in the audio-
frequency range to study lateral changes in earth resistivity. The quantities measured are the azimuth and
inclination of dip of the major axes of the ellipsoid of
polarization. See *polarization ellipse*. Used in mineral
prospecting and mapping faults and shear zones, especially in rugged terrain and heavy vegetation, See
sferics and Ward et al. (1968).

aftershock: An earthquake which follows a larger earth-

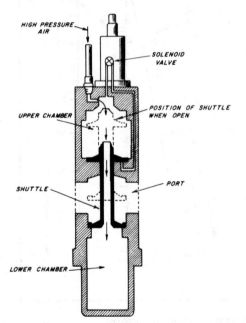

FIG. A-4. **Air gun.** High-pressure air flows continuously into the upper chamber and through the shuttle into the lower
chamber. Opening the solenoid valve puts high-pressure
air under the upper shuttle seat causing the shuttle to move
upward, opening the lower chamber and allowing its air to
flow out through ports to form a bubble of high-pressure air
in the water. The size of a gun is the size of its lower
chamber. (Courtesy Bolt Associates.)

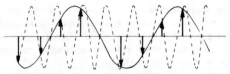

FIG. A-5. **Aliasing** of 200 Hz (dashed line) as 50 Hz (solid
line). Both 50 and 200 Hz waves give the same values
when sampled at 250 Hz (4 ms sampling).

quake or main shock and originates at or near the focus
of the larger earthquake. Major earthquakes are often
followed by a large number of aftershocks, decreasing
in frequency with increasing time.

AGC: *Automatic gain control* (q.v.).

AGC time constant: The time required for the voltage of a
system under automatic gain control to return to 63
percent (or $1 - 1/e$) of its final steady-state value after a
sudden change in input level. If the input change is an
increase, the constant is called **attack time**; if a decrease, **release time** (which may be different from attack
time). AGC time "constants" sometimes depend on
factors such as signal level and magnitude of the
change. AGC characteristics are usually specified in
dB/s. See Figure A-17.

aggradation: Building upward by deposition.

airborne magnetometer: Device used to measure variations in the earth's magnetic field from an aircraft. See
magnetometer.

air drill: A drill which removes cuttings by circulating air.

air gun: A seismic source (Figure A-4) which injects a
bubble of highly compressed air into the water. Its
frequency spectrum depends on the amount of air in the
bubble, the pressure, and the water depth (or water
pressure). Arrays of guns of different sizes are often
used so that a broader frequency spectrum will be
generated, See also *waveshape kit*. Air guns are also
used in boreholes or pushed down into marsh after
being modified to prevent mud, sand, etc., from entering and fouling the air gun, and they are sometimes used
in bags of water which are set on the ground surface for
land work.

air pressure: The pressure exerted by the weight of the
overlying column of air. Relations to temperature,
water vapor content, and elevation are given in Figure
A-3.

air shooting: A method of generating seismic energy in the
earth by detonating charges in the air. Charges are
usually placed on poles about a meter long so that the
explosive shock wavefront is distributed over a larger
portion of the ground surface than if the charge were
laid on the surface, although sometimes the latter
procedure is also called air shooting. See Poulter (1950).

air wave: 1. Energy which travels in the air at the velocity
of sound: $V \approx 1051 + 1.1F$ ft/s, where F = Fahrenheit
temperature, or $V \approx 331.5 + 0.607 C$ m/s, where C =
Celsius temperature. **2.** Audible sounds like distant
thunder associated with the arrival of *P*-waves from
nearby earthquakes.

Airy hypothesis: See *isostasy* and Figure I-6. Proposed by
George Biddell Airy (1809-1892), British astronomer

who determined the mean density of the earth from gravity measurements in mines.

Airy phase: A buildup in the amplitude of dispersed wave trains traveling by normal-mode propagation. The Airy phase is associated with a minimum in the curve of group-velocity versus frequency. The Airy phase is characterized by a constant-frequency wave train, often with a fairly abrupt termination at an arrival time corresponding to the minimum group velocity. See *channel waves* and Sheriff and Geldart, v. 1 (1982, p. 70-73).

albedo: Reflectivity of a free surface for electromagnetic radiation, especially for light; the fraction of the incident energy reflected.

alert: The time when a navigation satellite should pass within range so that a location fix can be obtained. See *satellite navigation*.

ALGOL: Acronym for Algorithmic Oriented Language, an algebraic and logic computer language.

algorithm: A step-by-step procedure for carrying out a numerical or algebraic operation. Compare *heuristic*.

alias: Frequency ambiguity resulting from the sampling process. Where there are fewer than two samples per cycle, an input signal at one frequency yields the same sample values as (and hence appears to be) another frequency at the output of the system; this is the **sampling theorem**. Half of the frequency of sampling is called the **folding** or **Nyquist frequency**, ν_N. The frequency $\nu_N + \Delta\nu$, appears to be the smaller frequency, $\nu_N - \Delta\nu$. The two frequencies, $\nu_N + \Delta\nu$ and $\nu_N - \Delta\nu$, are "aliases" of each other. See Figure A-5. To avoid aliasing, frequencies above the Nyquist frequency must be removed by an alias filter before the sampling. The passbands obtained by folding about the Nyquist frequency are called **alias bands**, **side lobes**, and **secondary lobes**. Aliasing is an inherent property of all sampling systems and it applies to sampling at discrete time intervals as with digital seismic recording, to the sampling which is done by the separate elements of geophone and shotpoint arrays (**spatial sampling**) and to sampling such as is done in gravity surveys where the potential field is measured only at discrete stations, etc. See Sheriff and Geldart, v. 2 (1983, p. 29-32 and 56-58).

alias bands: See *alias*.

alias filter: A filter used before sampling to remove undesired frequencies which the sampling process would otherwise *alias* (q.v.). An alias filter should have linear phase response and nearly flat amplitude response over the signal passband and should roll off rapidly above this band to provide high attenuation above the Nyquist frequency; see Figure F-7. Also called **antialias filter**.

alidade: A ruler, equipped with sights, which allows aligning a straight edge in the direction of an object. See Figure A-6. The telescope is usually equipped with vertical circle and stadia cross-hairs. The alidade is placed on a plane table and the object to be located is sighted through the telescope; a line drawn on the plane-table paper along the straight edge then indicates the direction to the object.

alkali-vapor magnetometer: See *optically pumped magnetometer*.

allocate: To assign a resource to a program for exclusive use. If the program is already running, the allocation is said to be dynamic.

allocthonous: Formed elsewhere than its present location.

alpha: 1. The ratio of pseudostatic self-potential (SP) to static SP. See **SSP. 2.** The ratio of array length to depth; see also *beta curve*. **3.** The symbol for *P*-wave velocity.

alpha centers: A conductivity inhomogeneity where con-

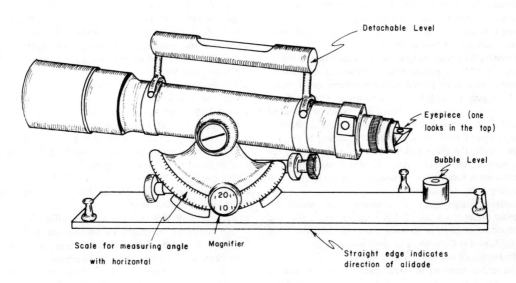

Detachable Level

Eyepiece (one looks in the top)

Bubble Level

Scale for measuring angle with horizontal

Magnifier

Straight edge indicates direction of alidade

FIG. A-6. **Alidade.**

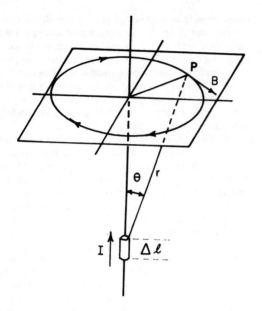

FIG. A-7. **Ampere's law.** A current I through a length of wire Δl creates at a point P a magnetic field $\Delta \mathbf{B}$ given by

$$\Delta \mathbf{B} = 10^{-7} \, I \sin \Theta \, \Delta l/r^2,$$

where $\Delta \mathbf{B}$ is in webers/m² when I is in amperes and r and Δl are in meters.

ductivity varies continuously with distance from a given point in the earth. Used to represent an ore body with gradational boundaries. Also applied to gravity or magnetics. See Edwards et al. (1978).

alpha configuration: See *array (electrical)*.

alphameric: See *alphanumeric*.

alphanumeric: Pertaining to a character set containing letters, numerals, and other characters.

alpine collision: A continent-to-continent plate collision. Also called an **A-type collision.**

alternating-field demagnetization: A method for determining the stable component of remanent magnetization along one axis by partial demagnetization and removal of components with low coercive force. The specimen is placed in a space with nulled field (such as produced with Helmholtz coils) and then subjected to an alternating magnetic field which is reduced gradually to zero by decreasing the current of the field coil or by pulling the specimen from the coil. Also called **ac demagnetization.**

alternator: A rotating electromechanical device for supplying alternating current.

altitude: 1. Height above a reference level, usually the geoid (mean sea level). 2. For a satellite or astronomic observation, angular distance above the horizon.

ALU: Acronym for *arithmetic logic unit* (q.v.).

AM: *Amplitude modulation* (q.v.).

ambient: Surrounding or background. Ambient noise is the pervasive noise associated with an environment, usually being a composite from both near and far sources.

ammeter: An instrument used to measure electrical current.

ammonium nitrate: A fertilizer which is used as an explosive when mixed with diesel fuel or other oxidizers. The mixture is confined by tamping and detonated by the explosion of a primer. It is water-soluble and will not detonate if wet. Ammonium nitrate is also an ingredient of some packaged explosives.

Ampere's law: A law giving the magnetic field due to a current. See Figure A-7. Also called the **Biot-Savart law.** The ampere, a unit of electric current, is named for André Marie Ampere (1775-1836), French physicist.

amplifier: A device which increases signal amplitude, voltage, or power. The output of a linear amplifier is the input multiplied by the amplifier gain. An amplifier is sometimes symbolized by a triangle in a circuit diagram, as shown in Figure A-8.

amplitude: The maximum departure of a wave from the average value. For "envelope amplitude," see *complex-trace analysis*.

amplitude anomaly: Local increase or decrease of seismic reflection amplitude, especially if attributable to a hydrocarbon accumulation. Amplitude anomalies may also be caused by geometric focusing, velocity focusing, interference, processing errors, or other reasons. Sometimes called **bright spot** (if an increase in amplitude) or **dim spot** (if a decrease). See *hydrocarbon indicators*.

amplitude distortion: See *distortion*.

amplitude equalization: A procedure by which the gain of each trace is adjusted to produce the same average amplitude for each trace.

amplitude log: A borehole log of the amplitude of a portion of the acoustic wave. See *cement-bond log* and *fracture log*.

amplitude modulation: Variations in the amplitude of a high-frequency carrier wave according to low-frequency information. Abbreviated **AM.** See Figure M-10.

amplitude of the envelope: See *complex trace analysis*.

amplitude recovery: Technique for recovering the amplitude of a seismic trace at any instant of time.

amplitude spectrum: Amplitude-versus-frequency relationship such as computed in a Fourier analysis. See *Fourier transform*.

AMT: *Audiomagnetotelluric method* (q.v.).

ANA: Prakla tradename for a radio-navigation system.

analog: 1. A continuous physical variable (such as voltage or rotation) which bears a direct relationship (usually linear) to another variable (such as motion of the earth) so that one is proportional to the other. 2. Continuous, as opposed to discrete or digital.

analog computer: See *computer*.

analog-digital converter: Device for converting analog signals into digital form.

analog modeling: A method of studying effects of subsurface bodies or structures by comparison with the response of models. Induced-polarization and resistivity surveys, for example, may be simulated by measurements in an electrolytic tank using conductive or polarizable shapes to represent subsurface bodies. **Physical modeling.**

analog tape formats: Analog information may be written

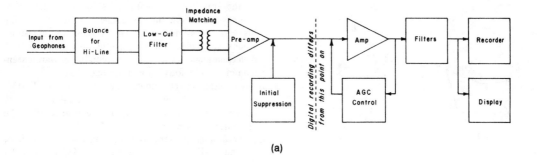

(a)

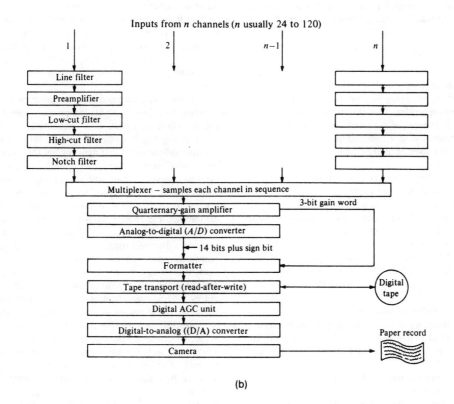

(b)

FIG. A-8. Seismic **amplifier**, (a) Schematic of a seismic analog amplifier. "Amp" is an amplification stage; there are usually several such stages. The AGC control produces negative feedback. Filters may be located at various positions. Usually there are many channels in parallel. (b) Block diagram of an IFP digital recording system. Each channel has its own components prior to the multiplex switch. The **line filter** reduces radio-frequency static picked up by the geophone cables. The **preamplifier** increases the signal level by a constant amount while providing impedance matching. The **low-cut filter** supplements geophone filtering by removing very low frequencies where ground roll is excessive. The **high-cut filter** prevents aliasing; its slope is typically 72 dB/octave. The **notch filter** reduces 50 or 60 Hz power-line pickup (or 16 2/3 Hz electric railroad pickup). The **multiplexer** connects each geophone sequentially to the **quaternary-gain amplifier** which automatically adjusts its gain in 4:1 steps until the amplitude falls within a prescribed range, after which a 3-bit word specifying the gain is sent to the formatter. The **A-D converter** measures the signal amplitude, one bit being output for polarity and fourteen bits for the magnitude. The **formatter** arranges the data for writing on magnetic tape by the **tape transport**. Separate **read heads** read the magnetic tape immediately after the data have been written. The output is amplified in the **digital AGC unit**, converted to analog form in the **D/A converter**, and written by a **camera** to give a monitor paper record. (From Sheriff and Geldart, v. 1, 1982, p. 173.)

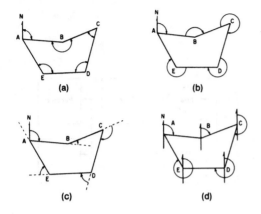

FIG. A-9. **Angle** measuring conventions. Specifying by **(a)**
interior angles; **(b) angles right**; **(c) deflection angles**;
(d) azimuth angles. The first leg of a loop is specified by
azimuth.

on magnetic tape in several forms. The magnetization
(less a constant bias) may be proportional to the input in
direct or **bias recording.** Information may be carried by
frequency variations of a carrier wave in FM or *frequency modulation* (q.v.) recording or by the width of
square-wave pulses in *pulse-width modulation* (q.v.).
See Figure M-10.

analog-to-digital: Digitizing; the conversion of analog data
to digital form. Abbreviated **A/D.**

analytic signal: See *Hilbert transform.*

anaseism: Initial earth movement away from the focus of
an earthquake. Antonym: **kataseism.**

anchor: See *charge anchor.*

andesite line: A closed line roughly ringing the Pacific
Ocean within which only basic rocks occur (andesite is
a basic rock). Surface waves are markedly attenuated at
the line.

AND gate: A circuit with multiple inputs which functions
only when signal is present at all inputs. If the inputs are
A and B, AND is denoted as $(A \cdot B)$, (AB), $(A \cap B)$
$(A \times B)$, or the "intersection of A and B." See **gate** and
Figure G-1.

anelasticity: Deviation from a linear proportionality between stress and strain.

AN/FO: A mixture of ammonium nitrate and fuel oil used
as an explosive.

angle of approach: The direction from which a wave
comes; the angle which a wavefront makes with a
surface.

angle of incidence: The acute angle which a raypath makes
with the normal to an interface. This is the same angle
an approaching wavefront makes with the interface in
an isotropic medium. In the anisotropic case, it is the
angle between the raypath and the normal, the raypath
not necessarily being perpendicular to the wavefront.
The angle of incidence may be complex.

angle right: See *angles.*

angles (surveying): The direction of a survey leg with
respect to the preceding leg of the survey traverse.

Several measuring conventions are used (Figure A-9).
Azimuth angles are measured to the right (clockwise)
with respect to north (either true north or magnetic
north), occasionally with respect to south. The first leg
of the traverse is usually specified by azimuth or
compass direction. **Interior angles** are the angles lying
inside a closed traverse. **Angles right** are measured
clockwise after backsighting on the previous station. A
deflection angle is the angle between the onward extension of the previous leg and the line ahead.

angular distance: The angle (measured at the Earth's
center) which is subtended by the great circle path
between two points, such as between an earthquake
epicenter and a receiver.

angular frequency: Repetition rate measured in radians/
second. Where ν = frequency in Hz, the angular
frequency ω, is $\omega = 2\pi\nu$.

angular unconformity: See *unconformity.*

anhysteretic remanent magnetization: The magnetic state
of a sample which has been subjected to a constant
magnetic field while a supplemental decaying alternating field has been progressively reduced to zero. This
procedure removes isothermal remanent magnetization.

anion: A negatively charged ion. Compare *cation.*

anisotropy: Variation of a physical property depending on
the direction in which it is measured. Geophysical
usage is sometimes qualified as **apparent anisotropy** or
effective anisotropy to distinguish from the "point property" such as possessed by crystals. (a) The general
elasticity tensor relating stress and strain in anisotropic
media contains 21 independent constants. If properties
are the same in two directions (**transversely isotropic
medium**), these reduce to five independent constants.
Isotropic media possess only two independent elastic
constants. In anisotropic media, the *S*- and *P*-modes of
body-wave propagation are not necessarily independent, wavefronts are not necessarily orthogonal to the
directions of wave propagation, and Snell's law requires modification. **Aeolotropy** is also used for anisotropy. (b) In seismic usage, a difference between velocity parallel to the bedding plane and velocity
perpendicular to the bedding (**transverse isotropicity**) for
a lithologic unit. Plate-like mineral grains and interstices tend to orient themselves parallel to the bedding.
This is called **microscopic anisotropy.** Velocity along the
bedding (as measured by refraction, for example) is
typically 10-15 percent higher than velocity measured
perpendicularly (as in a well). (c) Anisotropy is sometimes used to denote a difference between the velocity
parallel and perpendicular to the bedding for an entire
layered sequence, called **macroscopic anisotropy.** The
velocity parallel to the bedding appears greater because
the higher-velocity members carry the first energy,
whereas all members contribute in proportion to their
thickness in measurements perpendicular to the bedding. (d) The **resistivity anisotropy coefficient** is the
square root of the ratio of the resistivity measured
perpendicular to the bedding to that measured parallel
to the bedding. It usually has a value between 1 and 2.
Anisotropy of induced polarization in rocks is less than
anisotropy of resistivity. In foliated rocks the resistivity

parallel to foliation is less than that perpendicular to foliation.

anisotropy paradox: See *paradox of anisotropy*.

annulus: 1. The space between a drill pipe and the formations through which the returning drilling fluid (mud) returns to the surface. **2.** The space between tubing and casing or between casing and formation. **3.** A low-resistance ring about a borehole sometimes produced by invasion of mud filtrate into hydrocarbon-bearing beds. Because of their greater mobility, hydrocarbons may be displaced farther beyond the invaded zone than conductive formation water. See Figure I-5.

anode: An electrode at which oxidation occurs and electrons are produced (that is, are given up to the electrode). The positive terminal of an electrolytic cell or the negative terminal of a battery.

anomaly: 1. A deviation from uniformity in physical properties; a perturbation from a normal, uniform or predictable field. **2.** Observed minus theoretical value. **3.** A portion of a geophysical survey, such as magnetic or gravity survey, which is different in appearance from the survey in general. **4.** A gravity measurement which differs from the value predicted by some model, for example, a Bouguer or free-air anomaly. **5.** In seismic usage, generally synonymous with structure. Occasionally used for unexplained seismic events. **6.** Especially, a deviation which is of exploration interest; a feature which may be associated with petroleum accumulation or mineral deposits. **7.** An induced-polarization anomaly is usually positive and greater than background (or the **normal effect**) to be economically interesting. In the frequency domain, an anomalous region has a resistivity which decreases with frequency. An interesting resistivity anomaly is generally smaller than background.

ANSI: Acronym for American National Standards Institute.

antialias filter: *Alias filter* (q.v.).

anticipation function: A function which collapses a wave train into an impulse at the front end of the train. Involved with recursive filters.

anticline: A fold in stratified rocks in which the rocks dip in opposite directions from a crest; layers are convex upward. Antonym: **syncline.**

antiferromagnetism: Property of certain magnetic atoms which makes sublattices take an antiparallel ordering of spins (i.e., oriented opposite to each other), such that no net magnetization is observed. Compare *ferromagnetism* and *ferrimagnetism*.

antiparallel: Two vectors which point in opposite directions.

antiroots: Shallow high-density mantle material beneath thin portions of the relatively light crust of the earth. Characterized by shallow Moho. The effect is to bring topographically low areas (like ocean basins) into isostatic equilibrium. See Figure I-6.

antisymmetric: The property of a function which makes it change sign when its argument changes sign:

$$A(x) = -A(-x).$$

Also called **odd function.**

antithetic fault: 1. A secondary fault having throw in

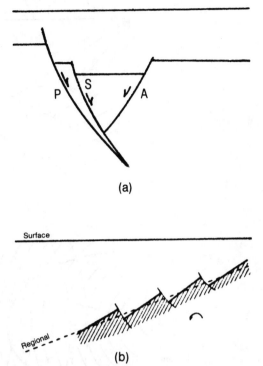

(a)

(b)

FIG. A-10. **Antithetic faulting.** (**a**) The secondary antithetic fault *A* has throw in the direction opposite to that of the primary fault *P*. *S* is a **synthetic fault.** (**b**) Faulting involving rotation so as to increase the throw on the faults.

opposite direction to the major **synthetic fault** with which it is associated. **2.** A fault associated with rotation so that the net slip is greater than it would have been without the rotation. See Figure A-10.

AP: 1. *Array processor* (q.v.). **2.** Attached processor (two CPUs); usually one of them has limited function (e.g., no I/O capabilities). Compare *MP*.

aperture: An opening, gate, or window that limits the information affecting a measuring device. The seismic spread length is sometimes considered the aperture of a seismic system. In processing, the spatial range of the data considered in the calculation (for example, in seismic migration) and/or the time range of the data considered (for example, in deconvolution).

aperture time: A designation of the location of the time interval which contains the data utilized in a calculation, usually indicated as starting and finishing time for the aperture (e.g., 1.5 to 2.5 s), sometimes indicated by specifying the center time and aperture width.

aperture width: The **effective aperture width** is the width of a boxcar with the same peak height and area.

API: 1. The American Petroleum Institute. **2.** The "proper" way to do a job: "strictly API."

API unit: 1. A unit of counting rate for the gamma-ray log. The difference between the high and low radioactivity sections in the API calibration pit is defined as 200 API

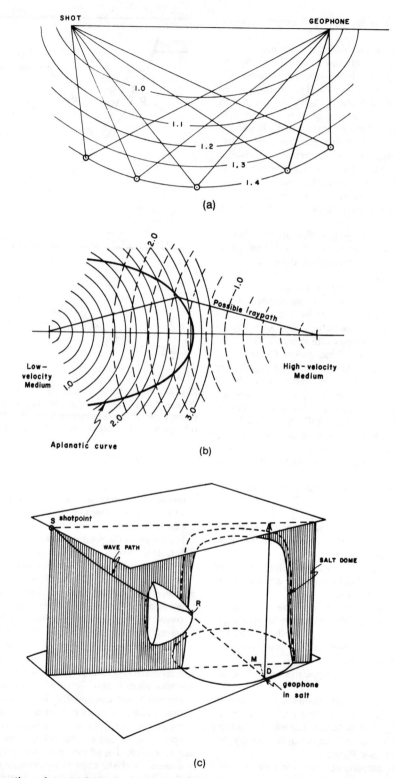

FIG. A-11. **Aplanatic surface**. (a) **Bathtub chart** for an offset geophone. Aplanatic surface can be found by the intersections of wavefronts about shotpoint and geophone. (b) To find the boundary between two media of different velocities where shotpoint and geophone are in different media, draw wavefronts about each and find intersections where the sum equals the traveltime (4.0 s for heavy curve). (c) Use of aplanatic surface in 3-D problem of locating salt dome flank by shooting into a geophone in a borehole in the salt. (From Gardner, 1949.)

units. **2.** A unit of counting rate for the neutron log. The reading in the Indiana limestone portion of the API neutron log calibration pit (which has 19 percent porosity and is saturated with fresh water) is defined as 1000 API units.

API well number: A unique number assigned by the American Petroleum Institute to each well drilled in the United States. A 12-digit decimal number which is broken down as follows:

Digits 1, 2: Code for state; numbers 1 to 49 are alphabetical for the states including the District of Columbia; Alaska and Hawaii are 50 and 51.

Digits 3-5: Code for county, parish, or offshore.

Digits 6-10: Code for the specific well.

Digits 11-12: Code for sidetrack, etc.

See API Bulletins D12 and D12A.

APL: Acronym for A Programming Language, a computer language designed for mathematical applications.

aplanatic surface: The surface between two media which is the locus for a given traveltime for wave energy reflected or refracted at a surface. Wavefronts on a wavefront chart are aplanatic curves for reflection times observed at the shotpoint; see Figure A-11a. The traveltime between two points via any point on the aplanatic surface is constant. Used (for example) in defining a salt-sediment interface (Figure A-11c). Each combination of shotpoint and geophone position defines one aplanatic surface. If the velocity and other assumptions are correct, the salt-sediment interface is the common tangent to all the aplanatic surfaces. See Gardner (1949) and Musgrave et al. (1967).

apodizing function: A weighting function used in truncating which reduces discontinuity effects. See *window*.

apogee: The point on a satellite's orbit which is farthest from the center of the Earth; see Figure E-9. The shortest distance is **perigee.**

apparent: 1. The value indicated by a measurement, as in "apparent velocity." **2.** The value of a property assuming the ground to be homogeneous, isotropic, and semiinfinite, as distinct from the "true" value. The subscript a is frequently used to indicate that a quantity is apparent, as with $(PFE)_a$, $(MF)_a$, etc.

apparent anistropy: See *anisotropy*.

apparent autocorrelation function: See *autocorrelation*.

apparent density: Density calculated from gravity measurements in a borehole.

apparent dip: 1. The angle which an emerging seismic wavefront makes with the surface; the angle whose tangent is the ratio of the vertical to horizontal components of displacement produced by the wavefront. Apparent dip can be related to the true dip of the reflector if the cross dip and velocity distribution are known. Compare *dip moveout* and *apparent velocity*. **2.** The angle from horizontal for a refracting horizon determined from the updip and downdip velocity of refracted waves; see Figure R-7. **3.** The dip of a rock layer as exposed in any section not at a right angle to the strike.

apparent polarity: A convention which relates a peak or trough of a seismic reflection to the sign of the reflection coefficient, assuming that the reflecting interface is an isolated one.

apparent resistivity: 1. The resistivity of homogeneous, isotropic ground which would give the same voltage-current relationship as measured. **Direct current apparent resistivity** ρ_a is an Ohm's-law ratio of measured voltage V to applied current I, multiplied by a geometric constant k which depends on the electrode array:

$$\rho_a = kV/I.$$

Usually has units of ohm-meters. See *resistivity*. **2.** With electromagnetic methods, quantities such as the moduli of the electric and magnetic field intensities (**E** and **H**) are measured for a certain frequency or time. If the subsurface were homogeneous and isotropic, these would yield the true resistivity via a certain equation. However, use of the same equation for a heterogeneous subsurface yields the "apparent" resistivity,

$$\rho_a = \rho_{HS}F(V)/F(V_{HS}),$$

where ρ_{HS} = resistivity of a homogeneous half-space, F is a function of V = observed voltage, and V_{HS} = voltage for a half-space. **3.** The resistivity recorded by an electrical log which differs from the true resistivity of the formation because of the presence of mud column, invaded zone, influence of adjacent beds, etc.

apparent-resistivity curve: A graph of apparent resistivity against electrode separation, frequency, or time. Apparent-resistivity curves are often plotted on logarithmic paper and compared with type curves (normalized theoretical curves) for interpreting the resistivity, thickness, and depth of subsurface layers.

apparent velocity: 1: The phase velocity which a wavefront appears to have along a line of geophones. If the wavefront makes the angle θ with the spread and the true velocity of the wavefront is V, then the apparent velocity is $V/\sin \theta$. See Figure W-2. **2.** The inverse of the slope of a refraction time-distance curve.

apparent-velocity filtering: Attenuating events based on their *apparent velocity* (q.v.). See *velocity filter*.

apparent wavelength: The distance between the correlative points on a wave train as seen by a geophone spread. Differs from actual wavelength if the wave direction makes an angle with the spread. See Figure W-2.

apparent wavenumber: See *wavenumber*.

applied geophysics: See *geophysical exploration*.

applied-potential method: See *equipotential line method*.

Aquapulse: *Sleeve exploder* (q.v.). Western Geophysical tradename.

Aquaseis: A marine seismic energy source in which a towed explosive cord is detonated. Imperial Chemical Industries tradename.

Arcer: A high-powered *sparker* (q.v.). Alpine Geophysical Systems tradename.

archaeological survey: A survey consisting of high-resolution subbottom profiler, magnetometer, side-scan sonar and echo sounder data required on U.S. offshore leases. Purpose is to determine if a cultural resource is present.

archeomagnetism: See *paleomagnetism*.

Archie's formulas: Empirical relationships between the **formation factor** F (sometimes F_R), porosity ϕ, water saturation S_W, and resistivities; in clean granular rocks,

$$F = R_0/R_W = a\phi^m,$$

$$R_0/R_t = S_w^n,$$

where m = cementation factor which varies between 1.3 and 3, a = proportionality constant varying from 0.6 to 1.5, R_0 = resistivity of the formation when 100 percent saturated with formation water, R_w = resistivity of the formation water, R_t = true resistivity of the formation, n = "saturation exponent"; often $n = 2$. **Archie's law** assumes that $m = 2$ and $a = 1$. The **Humble formula** assumes that $m = 2.15$ and $a = 0.65$.

architecture: Functional relationships between the parts of a computer or computer system.

arc shooting: *Fan shooting* (q.v.).

areal closure: See *closure*.

areal survey: A *three-dimensional survey* (q.v.).

Argo: A medium-frequency ($\sim$ 2 MHz) pulsed phase-measuring radio positioning system operable in either circular or hyperbolic mode. Cubic Western tradename.

argument: 1. Angle of a complex number. **2.** The parameter which determines the value of a function; e.g., ϕ is the argument for sin ϕ. **3.** Data passed to a subroutine; compare *parameter*.

Aries: The **first point of Aries** or the **vernal equinox** is the reference point on the celestial sphere from which right ascension and celestial longitude are measured. It is the intersection point of the celestial equator and the ecliptic, where the sun is located on the vernal equinox, March 21.

arithmetic logic unit: A computer hardware subsystem that performs arithmetic and logic functions.

arm: 1. A bow spring or lever connected to a logging sonde which presses against the borehole wall to centralize the tool, to push the tool to the opposite side of the borehole, or to hold a sensor pad to the borehole wall. **2.** To prepare a blaster or a perforating gun for firing.

arpent: Old French land measure, approximately one acre.

array: A *matrix* (q.v.).

array (computer): A high-level language data construction for accessing data elements in one or more dimensions. A specific element is accessed by an array name and subscripts (which represent position in the various dimensions of the array), e.g., "name (x, y)".

array (electrical): The arrangement of electrodes in resistivity prospecting, also called **configuration**. Several

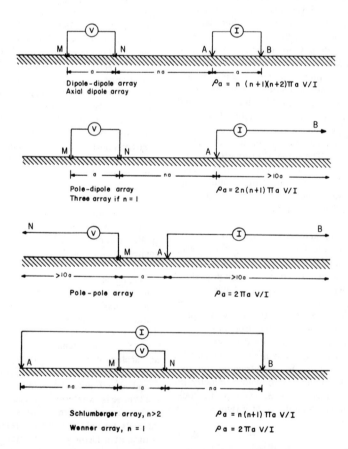

FIG. A-12. Collinear **arrays** used in resistivity surveying. The equation for apparent resistivity ρ_a is given for each array.

array types are shown in Figures A-12 and A-13. Resistivity array types include: (a) Azimuthal array; see *azimuthal survey*. (b) **Dipole-dipole array** or **axial-dipole array**, an array in which one dipole (a connected pair of electrodes) sends current into the ground and the other dipole serves as the potential-measuring pair. The separation between pairs is often comparable to (or only a few times greater than) the spacing within each pair, so the electrode pairs are not ideal dipoles; in deep resistivity sounding, the separation is larger. The dipole pairs are usually collinear (in line) but other orientations are used (Figure A-13). Resistivity and IP data from this array often are displayed as on Figure P-12. (c) **Gradient array** or **AB rectangular array**, an arrangement in which a pair of potential electrodes measure the voltage between points of a rectangular grid between two distant, fixed current electrodes. A variation of the Schlumberger array. (d) **Pole-dipole array**, a voltage-measuring pair of grounded potential electrodes separated successively from one current electrode (pole) while traversing a survey line. The second current electrode (the **infinite electrode**) is so far away that its location has negligible effect on the measurements. Data can be plotted below the midpoint between the current and the near potential electrode on a pseudosection. Called a **three array** if the electrodes are equally spaced. (e) **Pole-pole array** or **two array**, two electrodes (poles), a current and a potential, are traversed or successively expanded on a survey line. The other current and potential electrodes are located so far away that their location has negligible effect on the measurements. Data are plotted either at the potential electrode or halfway between the two poles. (f) **Radial array**; see *azimuthal survey*. (g) **Schlumberger array**, the inner voltage-measuring pair of potential electrodes is closer together than the outer current electrode pair, by a factor of about 6. (h) **Wenner array**, four equally spaced in-line electrodes; either the electrodes are moved along

a traverse or their interval is successively expanded. The usual or α **configuration** has the center two electrodes as the potential electrodes; the β **configuration** has the first two electrodes as potential electrodes; and the γ **configuration** alternates current and potential electrodes.

array (electromagnetic): An arrangement of antennas. Parasnis suggested for moving source methods the designation

$$T(a,b,c),\ L(d,e,f),\ R(g,h,i),\ r$$

to describe a configuration where the traverse direction is the x-axis, and the z-axis is in the vertical plane through the x-axis (the z-axis is north-south if the traverse is in a vertical borehole); a, b, and c are the direction cosines of the transmitter coil axis; d, e, and f those of the line L joining transmitter and receiver coils; g, h, and i those of the receiver coil axis; and r is the transmitter-receiver distance. See Parasnis (1970). Fixed source methods are described in terms of the type of source (long wire, large loop, small loop, etc.), orientation of the source, orientation of the receiver, and the relationship of the traverses to the source.

array (seismic): 1. A group of geophones connected to a single recording channel or a group of source points to be fired simultaneously. The records made from nearby source points when vertically stacked also effectively constitute use of a source array. Sometimes called a **pattern** or **patch. 2.** The arrangement or pattern of a group of geophones or source points (Figure A-14). Arrays discriminate against certain events on the basis of their stepout or apparent wavelength; see *directivity graph*. For a uniform array (see Figure D-12) of n geophones separated by distance d, the **array length** is nd and the first null response occurs when the apparent wavelength equals this. The half-width of the main lobe at 0.7 peak amplitude defines the **pass wavelengths**. For a nonuniform array, the **effective length** is the length of

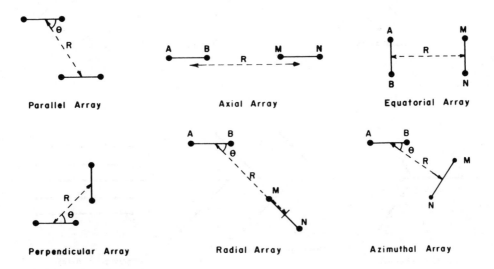

Parallel Array   Axial Array   Equatorial Array

Perpendicular Array   Radial Array   Azimuthal Array

FIG. A-13. Dipole-dipole **arrays** used in resistivity and IP exploration.

the uniform array which has the same pass wavelengths.

array factor: See *geometric factor.*

array length: See *array (seismic).*

array processor: A special-purpose processor utilized as a peripheral by a host computer to carry out special functions (such as matrix manipulations) more efficiently than the general purpose computer. See also *convolver.*

array station: Earthquake detection station which uses an array of seismometers to improve the detectability of weak signals. The **LASA** array (Figure L-1) uses 525 seismometers distributed over 200 km.

arrival: An **event**, a lineup of coherent energy signifying the arrival of a new wave train.

arrival time: 1. The time from energy release until an event arrives. **2.** Arrival time may allow for static and dynamic corrections.

arrow plot: A tadpole plot display of dipmeter or drift data; see Figure D-11.

artificial magnetic anomalies: *Cultural magnetic anomalies* (q.v.).

ascension: See *right ascension.*

asdic: *Sonar* (q.v.). The British acronym for their wartime Antisubmarine Detection Investigation Committee which developed its use.

aseismic: Free of natural earthquakes.

aspect ratio: 1. The ratio of vertical to horizontal scale. **2.** The ratio of shorter to longer axes for an ellipse or ellipsoid.

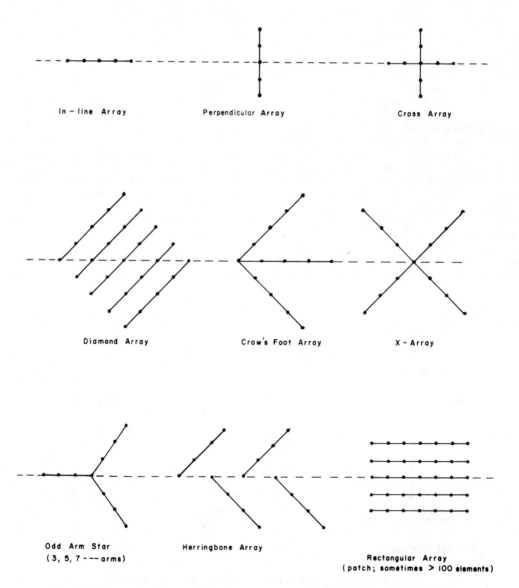

FIG. A-14. **Arrays** used for geophone or source groups. Dashed line indicates line direction.

assemble: To prepare a machine language program from a symbolic language program; to make an object program from a source program.

assembler: A program that translates symbolic assembler language code into binary code for execution by a computer.

assembler language: A source language that includes symbolic language statements in which there is a one-to-one correspondence between the instruction and data formats of the computer.

assembly program: A program written in *assembler language* (q.v.).

associative memory: Content-addressable memory capable of performing search and compare operations on all memory locations in one memory access. Looks for similar bit patterns in key words.

astatic: Having a negative restoring force which aids a deflecting force, thereby rendering the instrument more sensitive and/or less stable. The idea of astatic balance is illustrated in Figure A-15. Used in gravimeters and magnetometers.

asthenosphere: A yielding zone in the Earth's mantle thought to be involved in isostatic compensation and in plate-tectonic movements. The asthenosphere lies between the rigid lithosphere and above the mesosphere; it is approximately 200 km thick, has high attenuation of seismic energy (low Q), and little strength. See Figure E-1.

astronomic latitude: Latitude measured with respect to the vertical and the stars. Differs from "geodetic latitude" by a few seconds of arc where the mass distribution distorts the geoid equipotential surface, as near the roots of mountain ranges. See Figure G-2.

asymptote: The limit of the tangent to a curve as the point of contact approaches infinity.

asynchronous: Without a regular time relationship.

asynchronous I/O: A programming technique in which computation proceeds concurrently with input/output operations. The program subsequently interrogates the I/O to reestablish synchronization.

asynchronous protocol: A telecommunication protocol in which the transmitting station must indicate (by special sequences of transitions of state) the beginning and end of data transmission. Compare *binary synchronous* and *synchronous data link control* (SDLC).

asynchronous system: A system in which the components operate at independent speeds, requiring a handshake or interlock sequence for intercomponent communication.

atmospheric electricity: See *geophysics* and Figure A-16.

atomic capture cross-section: See *capture cross-section*.

atomic clock: A clock which determines time by counting atomic oscillations. The standard is the atomic second, 9 129 631 770 oscillations of a cesium-133 atom.

attack time: See *AGC time constant*.

attenuation: A reduction in amplitude or energy caused by the physical characteristics of the transmitting media or system. Usually includes geometric effects such as the decrease in amplitude of a wave with increasing distance from a source. Also used for instrumental reduction effects such as might be produced by passage through a filter.

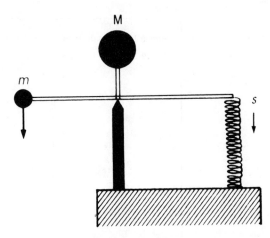

FIG. A-15. **Astatic balance** principle. When the force on *m* is balanced by the spring tension *s*, the large weight *M* exerts no net force, but once unbalance occurs, *M* throws the system farther out of balance.

attenuation factor: If the amplitude of a plane wave is reduced by the factor $e^{-\alpha x}$ in traveling a distance of x meters, the attenuation factor is α.

attenuator: An adjustable passive device for reducing the amplitude of a signal.

attitude: The relation of a feature to horizontal. The strike and dip for a bed or other planar feature; the trend direction and plunge for a linear feature such as an anticline.

attribute: A measurement based on seismic data, such as envelope amplitude ("reflection strength"), instantaneous phase, instantaneous frequency, polarity, velocity, etc. See *complex-trace analysis* and *hydrocarbon indicator*.

A-type collison: *Alpine collision* (q.v.).

A-type section: A three-layer model in which resistivites increase with depth. See Figure T-5.

audio: Pertaining to the frequencies corresponding to normal voice communication, i.e., 15 Hz to 20 kHz.

audio-frequency magnetic method: *Afmag method* (q.v.).

audiomagnetotelluric method: A magnetotelluric method involving measurement of natural plane-wave electromagnetic signals, mainly sferic energy, in the 10 to 10^4 Hz range to determine subsurface resistivity. Abbreviated **AMT**. See also *magnetotelluric method*, which involves the 10^{-3} to 10 Hz range.

auger: A drilling tool designed so that the cuttings are carried to the top of the hole continuously during the drilling operation by helical grooves on a rotating drill pipe. In the **wet auger,** fluid is injected at the bit to assist in the removal of cuttings.

aulacogen: *Failed arm* (q.v.).

Autocarta: An on-board computer and plotting system for marine processing of positioning and/or water-depth data. Racal Decca tradename.

autoconvolution: Convolution of a function with itself. See *retrocorrelation*.

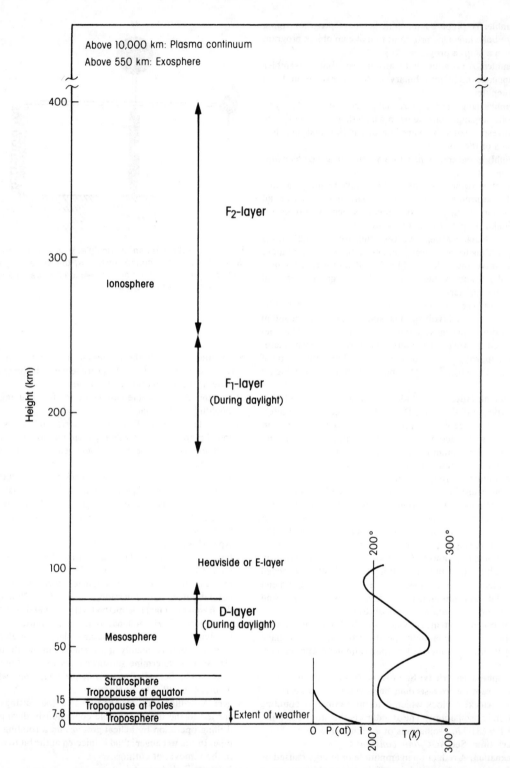

FIG. A-16. **Atmospheric layering.** Weather is mainly confined to the troposphere. The jet stream at about 11 km involves a wave-shapped course at midlatitudes. A circumpolar vortex extends sometimes to 60 km. Ionized layers in the ionosphere are important in radio transmissions. Atmospheric pressure falls off rapidly above the surface and is very small above 20 km. Temperature falls rapidly in the troposphere, is about 210-220° in the stratosphere and then rises to about 275° at 50 km because of the absorption of 2100-2900 Å radiation by ozone.

autocorrelation: Correlation of a waveform with itself. The normalized autocorrelation function $\phi_{11}(\tau)$ for a continuous stationary waveform is:

$$\phi_{11}(\tau) = \lim_{T \to \infty} \frac{\displaystyle\int_{-T/2}^{T/2} f(t)f(t+\tau)\,dt}{\displaystyle\int_{-T/2}^{T/2} f^2(t)\,dt},$$

where $f(t)$ represents a waveform (or seismic trace) and τ is the time shift or lag. Where the integrals are taken only between certain limits, the **apparent autocorrelation function** results. For equally sampled (digital) data this becomes:

$$\phi_{11}(\tau) = \lim_{N \to \infty} \frac{\displaystyle\sum_{k=-N}^{N} f_k f_{k+\tau}}{\displaystyle\sum_{k=-N}^{N} f_k^2}.$$

The time limits (as from t_1 to t_2) specify the **gate** or **window**. The denominators in the preceding equations are the **normalizing factors** and sometimes are not included. The autocorrelation function is a measure of the statistical dependence of the waveform at a later time τ on the present value, or the extent to which future values can be predicted from past values. The autocorrelation function contains all of the amplitude-frequency information in the original waveform but none of the phase information. An autocorrelation function is symmetrical about zero shift, that is, it is **zero phase**. Deconvolution operators are often based on autocorrelations; see Sheriff and Geldart, v. 2 (1983, p. 43). Autocorrelation is equivalent to passing a waveform through its matched filter; see Anstey (1964).

autocorrelation pulse: The autocorrelation of a sweep signal, Sosie sequence, or the like. See *Klauder wavelet* and *Vibroseis*.

autocorrelogram: A display of half of the autocorrelation function (the half for positive time shifts) of seismic traces, usually in record-section format.

autocovariance: Similar to an autocorrelation except that the mean value $\bar{f}$ is subtracted before the integration, and normalization is not done:

$$\int [f(t) - \bar{f}][f(t+\tau) - \bar{f}]\,dt.$$

For functions which have a zero mean, autocovariance is the same as an autocorrelation function which is not normalized.

automatic gain control: AGC, a system in which the output amplitude is used for automatic control of the gain of an amplifier. Seismic amplifiers used to have individual AGC for each channel, although multichannel control was sometimes used. See Figure A-17 and *gain control*.

autoregressive series: A time series generated from another time series as the solution of a linear-difference equation. Usually previous values of the output enter into the determination of a current value.

Autotape: A short-range (~100 km) radio-positioning system operating in the 3 GHz range. Cubic Western tradename.

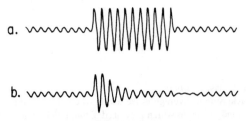

FIG. A-17. **Automatic-gain-control** (AGC) action. (**a**) Input; (**b**) output. The AGC restores the output to the same level after an increase or decrease in input amplitude.

auxiliary storage: Any storage not directly addressable by the processor.

AVC: Automatic volume control or *automatic gain control* (q.v.).

average: The arithmetic mean over some ensemble, population, etc.

average velocity: The distance traversed by a seismic wavelet divided by the time required, both with respect to some particular travel path and to a certain datum. For reflections, often refers to a ray reflected at normal incidence, sometimes to a vertical travel path. See *velocity*.

axial-dipole array: See *array (electrical)*.

axial surface: The surface about which folded beds are more or less symmetrical. Sometimes called the **axial plane**, though not geometrically "plane." Also called **ab plane**. See Figure F-13.

azimuth: The horizontal angle specified clockwise from true north. Occasionally azimuth will be referenced to south or to magnetic rather than true north. Also called **true bearing**.

azimuthal array: See Figure A-13 and *azimuthal survey*.

azimuthal projection: A map projection in which a spherical surface is developed on a tangent plane. Azimuths to any point from the point of tangency are correctly represented.

azimuthal survey: 1. A survey in which current electrodes on the ground surface at specific azimuths from a drill hole are used with one or both potential electrodes in the hole. The electrodes in the borehole may be successively raised to develop a log with the objective of determining the direction toward better mineralization. **2.** A survey method in which potential electrodes are moved along radii about a drill hole which contains a fixed current electrode. The second current electrode ("infinite electrode") is a great distance away. Also called **radial survey**. **3.** Azimuthal and radial "arrays" usually refer to dipole-dipole surveys where all electrodes are on the surface, in contrast to an azimuthal "survey" involving a drill hole.

azimuth bar: An established survey point near (200 to 2000 ft) a triangulation station, used to aid in orienting a transit.

B

back bias: See *bias*.

background: Average noise level, whether systematic or random, upon which a desired signal (such as a reflection) is superimposed. Usually refers to the total noise independent of the presence of the signal.

background polarization: The relatively weak IP response exhibited by unmineralized rocks, particularly those containing abundant clay minerals or layered or fibrous minerals. Also broad-scale, pervasive mineralization which is not of economic interest. **Normal effect.**

backplane: The component of an electronic system that physically holds printed circuit boards and provides the interconnections between them.

backsight: 1. A sight on a previously established survey point with the objective of determining the position and elevation of the survey instrument. The closing sight of a traverse or level-line loop is not considered a backsight. **2.** In plane-table traversing, orientation of the table on aligning the alidade on an established mapped point.

back-to-back: Processes which follow each other sequentially without any judgment being exercised in intermediate stages.

Backus filter: An inverse filter which removes the effects of reverberation involving a simple water bottom. The filter's z-transform expression is

$$1 + 2kz^q + k^2 z^{2q},$$

where k is the water-bottom reflection coefficient, and qt_s is the two-way traveltime through the water layer if t_s is the sample interval. See Backus (1959).

backward branch: The part of a diffraction event which lies under the reflection event, as opposed to the **forward branch** which tends to carry the reflection on beyond its termination. See Figure D-8. Compare *reverse branch*.

backward prediction: The use of future values to predict past values.

backward crossover: Opposite of *proper crossover* (q.v.).

baked test: Remanent magnetism is destroyed by heating to high temperatures. Therefore, the heated country rock adjacent to an intrusive should have the same remanent direction as the intrusive (appropriate to that of the time of the intrusion) but different from the unheated country rock (which should be indicative of the time of formation of the country rock). Checking that this is so is called the "baked test".

balanced input: A symmetrical input circuit having equal impedance from both input terminals to ground.

balanced section: A structural section which accounts for conservation of mass and bed length during structural deformation.

balancing a survey: Distributing cumulative errors among the legs of a survey.

band: 1. A range of frequencies such as those passed (**band-pass**) or rejected (**band-reject**) by a filter. Mea-surements are usually made between points where the amplitude is down by 3 dB (or 70 percent) from the peak value. **2.** The names given to the frequency ranges of the electromagnetic spectrum, as shown in Figure E-8. **3.** Landsat data channels; see *Landsat image*. **4.** A track on magnetic tape.

band-limited function: A function whose Fourier transform vanishes (or is very small) outside some finite band of frequencies.

band-pass: See *band*.

band-pass filter: See *filter*.

band-reject filter: A *filter* (q.v.) which attenuates a range of frequencies; the inverse of a band-pass filter.

bandwidth: 1. The range of frequencies over which a given device is designed to operate within specified limits. **2.** The differences between half-power points, i.e., the frequencies at which the power drops to half the peak power (3 dB). **3.** The **effective bandwidth** is

$$\int P(\nu)\, d\nu / P_{max},$$

where $P(\nu)$ is the power at the frequency ν and P_{max} is the maximum power. It is the width of a boxcar with the same total power and the same peak power. **4.** The rate at which a computer resource can carry (accept or deliver) data. Usually expressed in bytes per second or bits per second.

Banta method: A curved-raypath correction method that assumes that successively greater refraction times have penetrated to greater depths. See also *diving waves*.

bar: A unit of pressure, 10^5 pascals or 10^5 N/m^2; approximately one atmosphere or 14.5 psi.

barite: Barytes or barium sulphate, used to make drilling mud heavier.

barium titanate: A ceramic having piezoelectric properties. Used in transducers such as hydrophones.

barker word: The distinctive word in the message from a navigation satellite which indicates the beginning of the 2-minute transmission cycle.

Barkhausen noise: Noise introduced by the discreteness of magnetic structure so that magnetization occurs as a series of small steps rather than continuously.

barn: A unit for measuring capture cross-section; 10^{-28} m^2.

barrel: A volume of 42 U.S. gallons or 144 liters. Actual drums usually hold 55 gallons.

barrels of oil equivalent: A unit of energy equivalency; 5604 ft^3 natural gas; 5.8×10^6 BTU; 1700 kWh; 0.22 ton bituminous coal.

Barry's method: A refraction interpretation method utilizing delay times. See Barry (1967) or Sheriff and Geldart, v. 1 (1982, p. 220-221).

Barthelmes method: A refraction interpretation method involving continuous profiling. See Barthelmes (1946).

barytes: *Barite* (q.v.).

base: 1. The reference integer in a number system. Also called **radix. 2.** The transistor element that corresponds to the grid of a vacuum tube. **3.** *Base station* (q.v.).

baselap: Onlap or downlap *reflection configuration* (q.v.). See Figure R-8.

baseline: 1. A line used as a reference for measurements. See *shale baseline*. **2.** The line between two radio-positioning base stations whose transmissions are synchronized.

baseline extension: The straight-line extension of a baseline beyond the base stations.

baseline shift: A change in the location of the shale baseline on an *SP*-curve. A shift may occur when waters of different salinities are separated by shale beds which do not act as perfect cationic membranes, when the formation water salinity changes within a permeable bed, or when the resistivity of the mud in the borehole changes.

base map: A map showing location data, which can be used to post and map other data.

basement: 1. Geologic basement is the surface beneath which sedimentary rocks are not found; the igneous, metamorphic, granitized, or highly folded rock underlying sedimentary rocks. **2.** Petroleum **economic basement** is the surface below which there is no current exploration interest, even though some sedimentary units may lie deeper. **3. Magnetic basement** is the upper surface of igneous or metamorphic rocks which are magnetized much more than sedimentary rocks. **4. Electric basement** is the surface below which resistivity is very high and hence variations below this surface do not affect electrical-survey results significantly. **5. Acoustic basement** is the deepest more-or-less continuous reflection. **6. Gravity** or **density basement** is where a large density contrast exists, so large that the anomalies resulting from deeper contrasts are effectively lost in the noise. **7. Hydrologic basement** is the deepest point where significant porosity exists. The different types of basement may not coincide.

base of low-velocity layer (LVL): See *base of weathering*.

base of weathering: The boundary between the surface layer of low seismic velocity and an underlying layer of appreciably higher velocity. It may or may not correspond to the geologic weathering or to the water table. The boundary is involved in deriving static time corrections for seismic records. Also called **base of LVL (low velocity layer)**. See also *weathering*.

base station: 1. A reference station that is used to establish additional stations. Quantities under investigation have values at the base station that are known (or assumed to be known) accurately. Data from a base station may be used to normalize data from other stations, as in the telluric-current method. **2.** Accurately located fixed station for radio positioning.

base temperature: The temperature in the region of uniform temperature normally found in the lower part of a convecting system.

Basic: Beginner's All-purpose Symbolic Instruction Code, a conversational computer programming language which permits the use of simple English words, abbreviations, and familiar mathematical symbols to perform logical and arithmetic operations. Symbolic instructions are interpreted "on-the-fly" as opposed to languages which are compiled prior to execution.

basic crust: See *sima*.

basic functions: Functions that form the basis for approximate methods used in numerical modeling for interpolating, approximating a function, or numerical integration. The functions may be polynomials, splines, trigonometric functions, sinc functions, etc.

basic wavelet: *Embedded wavelet* (q.v.).

basin: A depressed, sediment-filled area. Sometimes roughly circular or elliptical in shape, sometimes very elongate.

basis functions: Algorithms that form the basis for numerical modeling and for methods of approximating.

batch processing: 1. Processing in which the entire job (usually large) is completed without further instruction by the inputter. Compare *interactive processing*. **2.** A procedure in which computer processing work that is similar is accumulated and submitted together to increase efficiency.

bathtub chart: A chart of the loci of reflection times for an offset geophone; see Figure A-11a. The loci are called aplanatic surfaces. A reflector tangent to one of the curves at any point satisfies the reflection time appropriate to that curve. A bathtub chart is simply a wavefront chart for nonzero offset.

baud: 1. A measure of the ability of a transmission medium to change states. **2.** The speed at which a channel transmits information (somewhat lower due to protocol overhead). One pulse (bit) per second.

Baumgarte ray-stretching method: 1. A graphical reflection interpretation method in which the positions of successive layers are constructed as surfaces tangent to fictitious wavefronts which are projected backward from the observing stations. **2.** A graphical refraction interpretation method. See Baumgarte (1955).

b-axis: Beta axis, the longitudinal axis of a fold structure. Defined on a Schmidt net by the intersection of great circles which represent foliation surfaces. See Figure F-13.

bay: A transient magnetic disturbance having a period of about an hour or so and the appearance of "a bay along the sea coast" on an otherwise undisturbed magnetic record. Other transient magnetic disturbances include micropulsations (small rapid variations), giant pulsations, and magnetic storms. The onset of bay is usually accompanied by a micropulsation burst.

bay cable: A marine seismic cable which is laid in place on the water bottom, as opposed to a drag cable or a streamer which are towed into place.

BCD: *Binary-coded decimal* (q.v.).

b/d: Barrels-per-day. Sometimes written bpd or bpcd (barrels-per-calendar day), bcd, bcpd (barrel condensate per day).

beacon: 1. A fixed navigation aid. Beacons may be either passive or active transponders. **2.** A radar reflector (such as a corner reflector mounted on a buoy) used as a navigation or positioning aid or to locate the tail of a seismic streamer.

Beaman arc: A unit of measure of the quantity $50(1 + \sin 2\alpha)$ used in calculating vertical displacement of a stadia rod with respect to the transit, where α is the angle which the line of sight makes with the horizontal. See also *stadia tables*. Named for William M. Beaman (1867-1937), American engineer.

Beaufort wind scale	Wind speed		Weather bureau terms	Observed effects	Sea description	Wave heights, ft	Douglas sea-state
	knots	km/hr					
0	<1	<1	Calm	Sea like a mirror	Calm	0	0
1	1-3	1-5	Light	Ripples; smoke drifts	Smooth	<1	1
2	4-6	6-11	Light breeze	Small wavelets; breeze felt	Slight	1-3	2
3	7-10	12-19	Gentle	Waves begin to break; leaves in constant motion	Moderate	3-5	3
4	11-16	20-28	Moderate	Numerous whitecaps; dust and leaves blow	Rough	5-8	4
5	17-21	29-38	Fresh	Some spray; small trees sway			
6	22-27	39-49	Strong wind	Large waves; white foam crests; Large branches in motion			
7	28-33	50-61	Stiff wind	White foam blown downwind	Very rough	8-12	5
8	34-40	62-74	Stormy wind	Small branches broken			
9	41-47	75-88	Strong gale	Slight structural damage	High	12-20	6
10	48-55	89-102	Whole gale		Very high	20-40	7
11	56-63	103-117	Storm		Mountainous	40	8
12	>64	>118	Hurricane		Confused		9

FIG. B-1. **Beaufort wind scale** and Douglas sea-state scale.

beam pointing: *Apparent velocity filtering* (q.v.).

beam steering: A method for emphasizing energy from a particular direction by delaying successive channels so that events of a certain dip moveout (or apparent velocity) occur at the same time, and then summing them. This procedure can be repeated for a succession of different dip moveouts to steer for other dips. See also *sonogram*, *Rieberize*, and *tau-p mapping*.

beamwidth: The angular width of the beam of a directional transducer or array of transducers, typically measured between the 3 dB points. The beamwidth is sometimes the combined effective beamwidths of transmission and reception.

bearing: The horizontal direction of one point with respect to another, usually measured as a clockwise angle. **True bearing** is the same as azimuth with respect to true north.

beat: The periodic increase and decrease in wave amplitude caused by the interference of two waves of nearly equal frequencies.

Beaufort number: A numerical scale indicating wind speed. See Figure B-1. Named for Admiral Sir Francis Beaufort (1774-1857).

bed of nails: A two-dimensional comb or the 2-D sampling function. An impulse is located at the intersection of all integral coordinate values.

bedrock: Any solid rock, such as may be exposed at the surface or overlain by unconsolidated material.

bell-shaped distribution: Normal or *Gaussian distribution* (q.v.).

benchmark: 1. A relatively permanent metal tablet or other marker firmly embedded in a fixed object indicating a precisely determined elevation and bearing identifying information. Used as a reference in topographic surveys. **2.** A test which can be used to evaluate performance. **3.** A standard set of jobs used to measure computer performance.

bender: A type of piezoelectric transducer used in hydrophones. Two thin plates of piezoelectric material with metallic film on opposite surfaces are bonded onto a brass block so that only the ends of the plates are supported. Pressure bends the plates, producing a voltage across each plate. See Figure H-7.

Benioff zone: A dipping zone containing earthquake hypocenters lying along the top of a subducting plate; see Figure P-4. Where plates of the earth's lithosphere converge so that one plate moves downward beneath another, the epicenters of earthquakes resulting from their movements lie near the upper boundary of the subducting plate. This type of plate collision is called a **Benioff-type** or **B-type collision.** Named for Hugo Benioff (1899-1968). American geophysicist and seismologist.

bentonite: A highly plastic, colloidal clay which increases its volume upon addition of water. Used in drilling mud. Largely made up of the mineral montmorillonite.

beta: See *beta curve.*

beta axis: See *b-axis.*

beta configuration: See *array (electrical).*

beta curve: A type of curve used in interpreting pulse IP data, also called **pulse curve. Beta**, the ratio of observed apparent chargeability to the true chargeability of the lower medium (a bilogarithmic weighting function developed from IP theory), is proportional to IP phase angle. A beta curve for a simple single horizontal layer where only the lower material is polarizable shows the resistivity contrast factor as a function of the ratio of array interval to depth (**alpha**) and beta. See Seigel (1959, chapter 2).

beta diagram: A *cyclographic diagram* (q.v.).

Betsy: A relatively weak seismic source which involves detonating a shotgun shell. Mapco tradename.

BHC: Borehole-compensated sonic log; see *sonic log* and *compensated log*. BHC is a Schlumberger tradename; BHC Acoustilog is a Dresser Atlas tradename.

BHT: Bottom-hole temperature.

BHTV: *Borehole televiewer* (q.v.).

bias: 1. The amount by which the average of a set of values departs from a reference value. **2.** Superposing an additional magnetic field upon the magnetic field associated with the signal during magnetic tape recording. Magnetic biasing is used to obtain a linear relationship between the amplitude of the signal and the remanent flux density in the recording medium. See *ac-bias recording*. **3.** A voltage which is maintained at a point in a circuit so that the device will operate with desired characteristics. **4.** A diode is said to have **forward-bias** when the voltage across it is such that current flows through it, and **back-bias** when the opposite polarity is maintained so that no current flows.

bias recording: See *analog tape formats*.

bilinear interpolation: A two-dimensional interpolation method in which values are first interpolated in one direction and then in the orthogonal direction. Computer programs often interpolate stacking velocity in this

way, first interpolating in time between picks at velocity analysis points and then spatially between velocity analysis locations.

bimodal: A set of observations which includes members of two populations that have different distributions.

bin: One of a set of discrete areas into which a survey area is divided. Data are sorted among bins according to midpoint locations for unmigrated data or according to reflecting points for migrated data. After sorting, the data elements within each bin are summed (stacked) and divided by the number of elements to obtain the output trace for the particular bin.

binary: Composed of only two elements. A number system in which two digits, 0 and 1, are used, the position of the digits representing powers of two; e.g., 11011 represents $2^4 + 2^3 + 2^1 + 2^0 = 16 + 8 + 2 + 1$ or the decimal number 27. See Figure N-4. Negative numbers may be represented by a minus sign or by codes such as the one's-complement or two's-complement codes.

binary-coded decimal: BCD; a number system code in which decimal digits are represented by four binary digits; see Figure N-4. For example, in the BCD number 00100111, the 0010 represents $2^1 \times 10 = 20$ and the 0111 represents $2^2 + 2^1 + 2^0 = 7$, giving the decimal number 27. The **excess-three code** (BCD×S3) is the BCD number plus 0011 (three).

binary digit: Bit; a mark used to represent 0 or 1, as used in the binary system.

binary gain: A gain-control system in which amplification

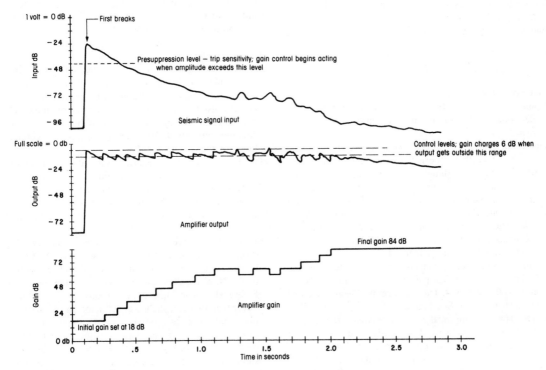

FIG. B-2. **Binary-gain** action in seismic amplifier. (Courtesy Seismic Data Service.)

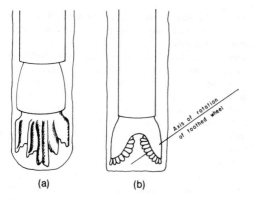

(a)   (b)

FIG. B-3. Drill **bits**. (**a**) Drag bit or fishtail bit; the teeth on drag bits tear into soft formations like sand and clay as the drill stem is rotated. (**b**) Rock bit or roller bit; the teeth on rock bits are on wheels which turn as the drill stem is rotated, so that they alternately put pressure on the rock and relieve the pressure, which causes rock pieces to flake off.

is changed only in discrete steps by factors of 2. The action of binary-gain control in a seismic amplifier is shown in Figure B-2. The times at which the gain steps occur is recorded so that the amplitude can be recovered later. The gain code represents amplification applied to the analog data for presentation to the analog-to-digital converter. Thus, samples with large gain codes represent small values. Compare *floating point*.

binary synchronous communication: A *synchronous protocol* (q.v.) that supports the transmission of binary data as well as character data; **bisync.**

binate: Retaining every other sample of sampled data. Also called "decimate".

binomial expansion: If $|y| < |x|$,

$$(x \pm y)^n = x^n \pm nx^{n-1}y + n(n-1) x^{n-2}y^2/2!$$
$$\pm n(n-1)(n-2)x^{n-3}y^3/3! + ..., + (-1)^n y^n.$$

Biot-Savart law: *Ampere's law* (q.v.).

bipole: A dipole electrode arrangement in which the electrodes of the dipole are an appreciable distance apart.

bipole-dipole array: A direct-current resistivity array in which the earth is energized using a pair of widely spaced electrodes (**bipole**), and the resultant electric field is mapped at numerous locations up to several miles from the bipole using orthogonal pairs of dipoles to determine the magnitude and direction of the electric field. The bipole-dipole method is generally used for reconnaissance because of low cost per station compared to other resistivity methods.

biquinary: A number-system code in which a decimal digit is represented by a digit pair, $5A + B$, where A is 0 or 1, and B is 0, 1,2,3, or 4. Thus the decimal 7 is represented as 12 ($1 \times 5 + 2$). See Figure N-4.

Birch's law: An empirical relation between density ρ in g/cm^3, *P*-wave velocity V in km/s, and m = mean atomic number. For nonporous rocks, $\rho = a(m) + 3.05$ V. For $m = 21$, $a(m) = -137$. Also sometimes written $V = A(m) \rho^{1.5}$. Both forms give about equally good fits.

bird: 1. A sensor (such as a magnetometer) suspended

from an aircraft by a cable to make geophysical measurements. **2.** A *depth controller* (q.v.).

bird-dog: 1. To pay close attention to a job or to follow up a job until it is finished. **2.** The one to whom such surveillance is assigned. **3.** The representative of the company who hires a geophysical crew, responsible for the work of the crew. Usually a staff or liaison position not carrying line responsibility.

bisync: *Binary synchronous communication* (q.v.).

bit: 1. A binary digit, the smallest unit of information; the necessary and sufficient information to distinguish between two choices. A bit may represent zero or one, yes or no, on or off, etc. **2.** A magnetized spot on a digital magnetic tape conveying a binary digit. **3.** The element on the end of a drill pipe which actually does the cutting (Figure B-3).

bit density: Packing density; the number of bits per unit length of magnetic tape, measured in bits per inch per track (bpi). A 9-track tape at 800 bpi actually contains 7200 bits of information in one inch of tape (6400 bits of data and 800 bits of parity). Bit density normally implies linear density as opposed to areal density.

bit recording: See *sign-bit recording*.

bit-shift: Multiplying or dividing a binary number by a power of two by shifting the radix point (decimal place).

bit slice processor: A processor composed of a number of identical chips operating in parallel, each chip providing full processor functionality for a limited number of bits. For example, a 16-bit processor might be composed of 4 chips each comprising a 4-bit processor.

black box: A unit or device whose basic function is specified but whose method of operation is not specified. Sometimes used in a derogatory manner for an untried or unproven method, especially for one whose method of operation is not understood.

blank: To set equal to zero; to *mute* (q.v.).

blast: An explosion.

blaster: Device used (a) to detonate an explosive charge by sending an electric current through a blasting cap and (b) to transmit the time-break to the recording unit. Usually also includes (c) a current-limited ohmmeter for checking the blasting circuit, (d) phone or radio for communication with the recording unit, and (e) connections for the uphole geophone.

blastphone: A device to record an air wave, used in early seismic work to determine the source instant when the distance was known, or the distance when the source instant was known (by radio transmission).

blind hole: A borehole characterized by lost circulation of the drilling fluid.

blind zone: 1. A layer with lower velocity than overlying layers so that it does not carry a head wave. **2.** A layer which cannot be detected by first-break refraction methods, also called **hidden layer** or **shadow zone.** See Figure H-4. The blind zone (a) may have a velocity lower than that of a shallower refractor, in which case it may lead to an overestimate of the depth of deeper refractors, or (b) it may have a velocity intermediate between those of layers above and below but not have sufficient velocity difference or thickness to produce first arrivals; in this case it is apt to cause an underestimate of the depth of deeper refractors. **3.** A zone from which reflections do not occur. **4.** A zone from which no drill cuttings are returned to the surface. **5.** A portion of a formation in which a logging tool response is too low. A blind zone occurs because of the finite size or configuration of the logging tool. For the lateral curve a blind zone (abnormally low reading) is recorded when a bed which is highly resistive compared to the overlying and underlying formations is present between current and measuring electrodes. **6.** A layer which cannot be detected by electrical methods because its resistivity is not sufficiently different from the resistivity of other layers or because it is too thin.

BLM: Bureau of Land Management, nonregulatory group advisory to the U. S. Geological Survey.

block: A group of words or files considered as a unit.

block diagram: 1. Diagram showing the functions of a system, processes, or devices and how they are interrelated, without showing construction details. **2.** Diagram showing the component operations of a computer program.

Blondeau method: A method of determining vertical time to a predetermined depth based on first-break data and the assumption that the instantaneous velocity is proportional to a power of the depth. The modified Blondeau method assumes that the constant of proportionality and the exponent can vary with depth. See Musgrave and Bratton (1967).

blowup: 1. To become unstable; to fail to converge. **2.** To enlarge, such as by photographic means.

body waves: $P-$ and S-waves which travel through the body of a medium, as opposed to surface waves. See *P-wave* and *S-wave*.

BOE: Barrels of oil equivalent.

bomb: 1. Explosive charge used as a seismic energy source. **2.** To fail to execute on the computer, e.g., a data processing operation "bombs".

Boolean algebra: An algebra of elementary logical properties of statements; a system involving yes or no decisions used in computer design and programming. The rules of Boolean algebra are often shown in **truth tables** where 0 = no and 1 = yes, such as the following:

A	B	$\overline{A}$	AND	OR	EXCEPT	NOR	NAND
0	0	1	0	0	0	1	1
1	0	0	0	1	1	0	1
0	1	1	0	1	1	0	1
1	1	0	1	1	0	0	0

Symbolic representations of "A and B" are **A.B, AB, A×B,** and **A∩B**: representations of "A or B" are **A+B** and **A∪B**. Inversion is often shown by a superscript line: $\overline{A}$ = not **A**. The laws of Boolean algebra include:

$$A.0 = 0 \qquad A + 0 = A$$
$$A.1 = A \qquad A + 1 = 1$$
$$A.A = A \qquad A + A = 0$$
$$A.\overline{A} = 0 \qquad A + \overline{A} = 1$$

$$\left.\begin{array}{l} \overline{A.B} = \overline{A} + \overline{B} \\ \overline{A.B} = \overline{A} + B \end{array}\right] \quad \textbf{deMorgan's theorems.}$$

EXCEPT (exclusive OR) can be shown as $\overline{A}.B + \overline{AB}$, NAND by $\overline{A.B}$, and NOR by $\overline{A + B}$. See also Figure G-1. Named for George Boole (1815-1864), English mathematician.

Boomer: 1. A marine seismic-energy source in which capacitors are charged to high voltage and then discharged through a transducer in the water. The transducer consists of a flat coil with a spring-loaded aluminum plate (or plates). Eddy currents force the plates to separate sharply, producing a low-pressure region between the plates into which the water rushes, generating a pressure wave by implosion. Frequency 300-1500 Hz. EG & G tradename. **2.** A very strong, usually low-frequency reflection event identified with a distinctive massive reflector.

booster: An intermediate explosive which has to be detonated by another explosive (often a blasting cap), the purpose of which is to detonate the main explosive charge.

boot: 1. Protective shield placed around a hydrophone, cable connector, sleeve-exploder unit, etc. **2.** To cause a computer to exercise a bootstrap program (see *bootstrapping*).

bootstrapping: A system bringing itself into a desired state by its own actions. The execution of a **bootstrap program,** a set of permanently stored instructions; may involve loading (from external storage) and executing a set of instructions to bring a computer to some desired state of initialization.

borehole effect: A distortion of a well log because of the size and influence of the borehole or (sometimes) the invaded zone.

borehole gravimeter: A remote reading gravimeter which can be lowered through a borehole while a gravity value is measured. The difference between the gravity readings at two different depths gives the apparent density ρ between the depths:

$$\rho = 3.686 - 128.5 \, \delta g/\delta h_1$$
$$= 3.686 - 39.18 \, \delta g/\delta h_2$$

where δg is the gravity difference in mGal, δh_1 is the depth difference in meters, and δh_2 the depth difference in feet.

borehole log: *Well log* (q.v.).

borehole televiewer: A well log wherein a pulsed, narrow acoustic (sonar) beam scans the borehole wall in a tight helix as the tool moves up the borehole. A display of the amplitude of the reflected wave on a cathode ray tube (television screen) is photographed yielding a picture of the borehole wall to reveal fractures, vugs, etc; see Figure B-4. BHTV is a Mobil Oil tradename. See Zemanek et al. (1970).

bottom lock: Situation where Doppler-sonar measurements are based on reflections from the sea bottom (the normal operational mode in less than 400-1000 ft of water). As opposed to the water-scatter mode which occurs in deeper water.

Bouguer anomaly: 1. The value obtained after latitude correction, elevation correction (including both free-air and Bouguer corrections), and (usually) terrain correction have been applied to gravity data. Pronounced bō gâr' or boo gâr' sometimes boo gay' or boo zhay'. The Bouguer gravity field is not the same as the field which would have been observed at the datum elevation, because the shape of anomalies due to remaining density irregularities still are appropriate to the elevation of measurement rather than to those of the datum elevation. **2.** Sometimes a departure from smoothness in the contours on a map showing Bouguer gravity values (i.e., an anomaly in Bouguer anomaly values). Named for Pierre Bouguer (1698-1758), French mathematician who tried to determine the figure of the earth.

Bouguer correction: 1. A correction to gravity data because of the attraction of the rock between the station and the elevation of the datum (often sea level) or, in the case of stations below the datum elevation, for rock that is missing between the station and datum. The Bouguer correction is 0.04185 ρh mGal, where ρ is the specific gravity of the intervening rock and h is the difference between the station and datum elevations in meters (or 0.01276 ρh mGal if h is in feet). **2.** In surface ship gravity data, the Bouguer correction replaces the sea water with rock, and ρ in the preceding expression is the difference in specific gravity of the replacement rock and that of sea water (usually taken as 1.03).

Bouguer plate: An infinite slab of finite thickness h and density ρ (g/cm^3); its gravity effect is 0.04185 ρh mGal if h is in meters.

boundary condition. A constraint that a function must satisfy along a boundary.

boundary-value problem: A differential equation together with boundary conditions.

boundary wave: A mode of wave propagation along the interface between media of different properties. Also called **surface wave**.

bound layer: *Fixed layer* (q.v.).

bound water: Water absorbed in or chemically combined

with shales, gypsum, or other material and which is not free to flow under natural conditions.

bow tie: The appearance of a buried focus on a seismic record section. See Figure B-8.

boxcar: A rectangular window function:

$$\text{box}_a(t) = 0,\ t < -a/2 \text{ or } t > a/2$$
$$= 1,\ -a/2 < t < a/2.$$

Its transform is

$$\text{box}_a(t) \leftrightarrow a\ \text{sinc}(\pi a/t) = a\ \sin(\pi a/t)/(\pi a/t).$$

Sometimes called a **gate**.

box corer: See *corer*.

BP: Before the present. Used in designating geologic age; see Appendix L.

bpi: Bits per inch, refers to the linear bit density, the inverse spacing of bits along a single track on magnetic tape in the direction of motion.

brachistochrone: 1. Least-time path. Any raypath (in the usual sense) is a brachistochrone. See *Fermat's principle*. **2.** A table of reflection time versus depth.

branch: 1. An instruction which can cause selection of the computer's next instruction to be from a location other than the next sequential location. A branch can be unconditional or conditional based on the magnitude or state of some value. Synonym: jump. **2.** One set of values for a multivalued function. **3.** One reflection event from a given reflector where other reflections from the same reflector are also present, as where the reflector is concavely curved or discontinuous. See *buried focus*. Multiple reflection branches can also result from large velocity gradients; see *diving waves*. **4.** More than one refraction event may be observed at a given point because of the configuration of the refractor.

branch point: 1. A point where a decision between alterna-

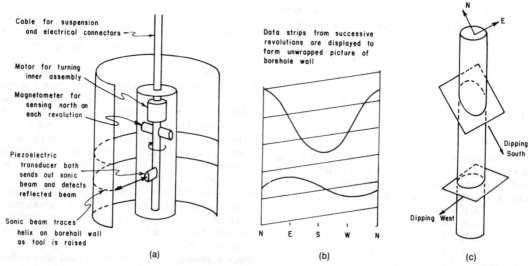

FIG. B-4. **Borehole televiewer.** (a) Schematic of the sonde in the borehole. (b) Schematic appearance of two plane fractures, one with steep south dip and one with gentle west dip as shown in (c). (From Zemanek et al. 1970, p. 255 and 257.)

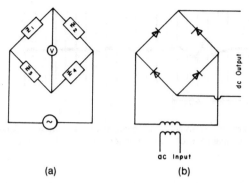

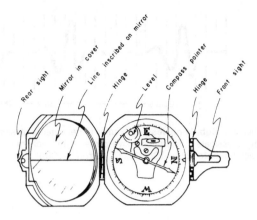

(a) (b)

FIG. B-5. (**a**) **Bridge** circuit. When the bridge is balanced no voltage is measured by V and $Z_1/Z_2 = Z_3/Z_4$. Used to measure one unknown Z in terms of the other three known Z. (**b**) Bridge rectifier for full-wave rectification.

FIG. B-6. **Brunton compass**.

tives must be made; a **node. 2.** The point of a Riemann surface at which two or more branches of a multivalued analytic function come together.

breadboard: An electrical circuit connected temporarily, such as to test out a new circuit before the design is finalized.

break: 1. Onset of an event, especially the first break. A burst of energy indicating the arrival of new energy. **2.** See *time break* and *cable break*.

breakpoint: A location in a program at which execution of that program can be halted to permit visual checking, printing out, or other performance analysis.

Bremsstrahlung: Radiation which results when a charged particle experiences an acceleration while in the field of another charged particle. An electron decelerating as it approaches a nucleus will be deflected and low energy γ-rays and x-rays will be emitted. The radiation is a continuum from zero energy to the maximum kinetic energy of the electron. From the German for braking radiation.

bridge: 1. An obstruction in a drill hole above the bottom of the hole usually formed by caving. Bridges block the passage of drilling tools or an explosive charge. A **bridge plug** may be set deliberately. **2.** An electrical network usually having four or more arms connected in a diamond arrangement and containing one branch (the "bridge") which connects two points of equal potential when the circuit is properly balanced; see Figure B-5. Used for measuring electrical impedance. **3.** A jumper or wire used to short-circuit around parts of an electrical circuit.

bridge plug: See *bridge*.

bridge rectifier: A full-wave rectifier circuit in which there are four arms, each containing a rectifier. See Figure B-5.

bridle: 1. To connect in parallel a group of amplifiers to a common input. **2.** A seismic record produced with the amplifiers bridled; see *parallel record*. **3.** An arrangement for towing a seismic streamer. **4.** The insulation-

covered lower portion of the cable to which the logging tool is connected.

bright spot: 1. A local increase of amplitude on a seismic section for any reason. **2.** An increase of amplitude assumed to be caused by hydrocarbon accumulation; an **amplitude anomaly.** See *hydrocarbon indicator*. Also **brite spot.** Opposite of **dim spot.**

brite spot: See *bright spot*.

broadside: 1. A reflection shooting arrangement in which the source point is appreciably (more than say 200 ft) outside the line of the spread. Also called **L-spread** or **T-spread**, depending on whether the source point is opposite one end or the center of the spread. See Figure S-17. **2.** A refraction technique in which the spread is perpendicular to the line connecting it with the source point. The source-to-spread distance is usually kept nearly constant. **3.** Electromagnetic-surveying procedure in which the transmitter coil is moved along one line while the receiver coil is moved along a parallel line. Compare *in-line*.

broomstick charge: A directional charge consisting mainly of a long helical coil of detonating cord wound around a dowel (broomstick). The pitch of the helix is such that the speed of the explosion front along the axis of the helix is approximately equal to the seismic velocity in the surrounding medium. Used to increase the sharpness of the downward-traveling wave and to reduce ghosting.

Brunton: A small magnetic pocket compass with levels, sights, and a mirror reflector, used for measuring angles as well as for determining directions. See Figure B-6. Various arrangements of the sights and mirror permit one to sight on objects and simultaneously read the scale and level. The Brunton is often used as a protractor. Named for David W. Brunton (1849-1927), American engineer.

brute stack: A *common-midpoint stack* (q.v.) where final static and normal-moveout corrections have not yet been applied.

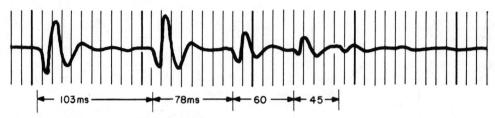

FIG. B-7. **Bubble pulses** from a small underwater explosion. The time between successive implosions decreases as energy is dissipated. The successive pulses effectively generate additional records superimposed on the first; the result is called **bubble noise**. (Courtesy Chevron Oil Co.)

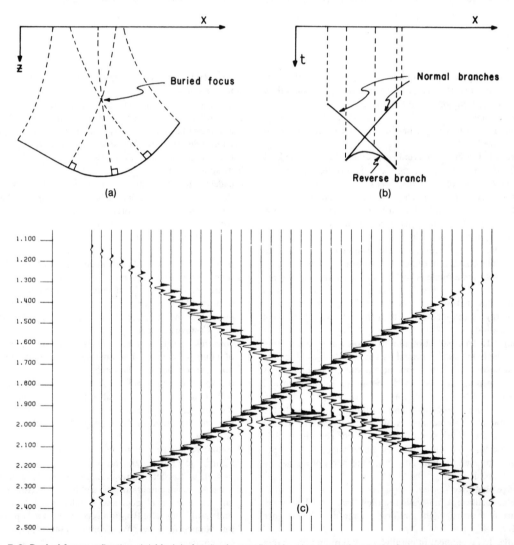

FIG. B-8. **Buried focus** reflection. (a) Model of a simple syncline showing several raypaths for common source-detector traces. (b) Schematic time section corresponding to model. (c) Record section for a cylindrical model similar to that in (a), showing three branches. This pattern is sometimes called a **bow-tie**. Note the 90-degree phase shift in the reverse branch. (Courtesy Chevron Oil Co.)

Btu: British thermal unit = 1054.5 joule.

bubble. An oscillating bubble of gases, such as formed by an explosion in water. Most marine seismic sources generate bubbles; the bubble oscillation frequency is given by the Rayleigh-Willis relation (see Figure R-4). Various arrangements such as Flexotir or waveshape kits are used to attenuate bubble effects or (as with Maxipulse) to correct for them. Bubble oscillations can be prevented if the source is so close to the surface that the gases vent before the bubble collapses. This occurs if an explosive of w pounds of dynamite is fired shallower than d ft, where $d = 3.8\ w^{1/3}$.

bubble effect: The result of *bubble* (q.v.) oscillations, often called **bubble noise.** The **bubble pulses** (repeated collapses of a bubble) for a large bubble (both physically large and involving considerable energy) show on a seismic record as repeats of the first arrivals and all other source-generated events. See Figure B-7. Bubble effects are seen occasionally in marsh or land shooting.

buck: To *bias* (q.v.).

bucking electrodes: *Guard electrodes* (q.v.).

buffer: 1. An intermediate storage device which accommodates differences between the rate at which information is fed into a computer and the rate at which the computer can receive it, or which performs the same function for information output by the computer or between component parts of a computer. **2.** An isolating circuit to prevent a reaction of the circuit that is driven from affecting the circuit which is doing the driving. A **buffer amplifier** may follow a critical stage to isolate subsequent stages from load impedance variations.

buffer amplifier: See *buffer.*

bug: 1. Error or a malfunction in equipment, computer program, etc. **2.** A *geophone* (q.v.), especially the *uphole geophone* (q.v.).

bug time: *Uphole time* (q.v.).

bulk modulus: See *elastic constants.*

bullet: 1. A device for obtaining sidewall cores. **2.** A device for perforating.

bump: To increase a count by one.

buried-focus effect: The effect of energy passing through a focal region before it reaches the recording plane. A buried-focus situation commonly causes several branches (usually three) of a reflection to be observable at the same surface location (i.e., the law of reflection is satisfied for several points on the reflector); see Figure B-8. The portion for which raypaths pass through the focus is called the **reverse branch.** There is a phase shift in the reverse branch; see Sheriff and Geldart, v. 1 (1982, p. 112-117). For zero offset and constant velocity, a buried focus occurs if a reflector's center of curvature lies beneath the recording plane. Less curvature is required to produce a buried focus for offset traces than for normal-incidence traces and, hence, buried-focus effects are more likely on long-offset traces. They are also more likely deeper in the section. Velocity gradients and curvature of isovelocity surfaces (which are apt to occur in structural areas) affect buried-focus effects. Curvature of the reflector out of the plane of the seismic line can also cause multiple branches. Buried-focus effects can also be produced by lateral velocity variations which bring about focusing conditions because of lens effects.

burn-in: 1. To implant a circuit pattern permanently on a chip. **2.** To perform an endurance test, sometimes under adverse conditions.

burst out: A large sudden increase in amplitude which threatens to exceed the linear capabilities of a system.

burst rate: The rate at which a device transfers data once the data have been accessed, as opposed to the effective data rate which also involves access time.

bus: 1. Common connector for electrical power or computer data. **2.** A circuit or group of circuits which provides a communication path between two or more devices, such as between central processor, memories, and peripherals.

bust: A failure to tie a survey loop within acceptable standards.

butterfly filter: A velocity filter that rejects a prescribed band of apparent velocities or moveouts. As opposed to a **pie-slice** filter which passes a band of moveouts. Texas Instruments tradename.

Butterworth filter: A type of frequency filter characterized by a very flat passband, often used as an alias filter. Sometimes called a **maximally flat filter.** See Sheriff and Geldart, v. 2 (1983, p. 188-189).

button: The small circular electrode on a microresistivity sonde.

byte: A small group of adjacent binary digits (as those across the width of a tape) which are manipulated as a unit. See also *character.*

C

cable: 1. The assembly of electrical conductors used to connect geophone or hydrophone groups to the recording truck. See also *streamer*. **2.** The assembly of electrical conductors and tensile members used to support a logging sonde, well geophone, or bird.

cablebreak: An arrival in a well-velocity survey caused by energy travel in the cable which supports the well geophone.

cable drilling: A method of making a hole in hard rock by alternately lifting and dropping the tool.

cable strum: Vibration of a marine streamer produced by occasional sudden tension such as might be caused by pitching of a towing ship or jerks from a tail buoy. A source of noise in marine seismic data.

cable tools: Equipment for *cable drilling* (q.v.). Occasionally used for seismic shothole drilling in areas of extremely hard surface rock. See *spudder*.

cache memory: A buffer memory which temporarily holds data and memory until they can be transferred into the computer's rapid-access memory.

cadastral survey: A survey to determine boundary lines.

cage: Perforated steel which surrounds a marine energy source (small explosive or air gun) to attenuate the *bubble effect* (q.v.) by dissipating energy in the turbulent flow of water in and out of the cage. See Figure F-10b.

cage shooting: *Flexotir* (q.v.).

Cagniard apparent resistivity: 1. A resistivity calculated simply from orthogonal electric and magnetic field measurements. Valid for magnetotelluric measurements in a layered earth. For more complicated structure, the full tensor impedance must be considered. **2.** In the magnetotelluric method, an apparent resistivity in ohm-m at period T obtained from $0.2T|Z|^2$, where Z is the *Cagniard impedance* (q.v.).

Cagniard impedance: In the magnetotelluric method, the ratio of horizontal electric field component in some direction E_x to the magnetic field H_y in a perpendicular direction. Defined in the specific units of E in nV/km and H in gammas. For a horizontally layered earth, it is independent of the choice of x-direction.

cake: *Mud cake* (q.v.).

calibrate (a sonic log): To adjust values so that the integrated traveltime agrees with data from surface shots into a well seis in the borehole.

calibration: 1. Determination of the number of units of a quantity being measured per scale division of the read-out device. **2.** A check of equipment readings made by measuring a known standard. **3.** Adjusting an apparatus so that it reads values correctly.

calibration resistor: A pure resistance of known value used to calibrate a frequency-domain transmitter and receiver.

calibration tails: Calibration records run before and/or after a log run and attached to the logs.

caliper log: A well log which measures hole diameter. Open hole caliper logging tools sometimes have four or more arms. Also called **section gauge.** See Figures M-6 and S-12. Tools for studying the corrosion of casing or tubing use many "fingers."

calculus of variations: The mathematics of finding a function which will maximize (or minimize) a definite integral.

camera: A recording oscillograph used to produce a visible pattern representing electrical signals or to make a visible seismic record on photosensitive paper or film or by xerography. See also *plotter*.

canonical transformation: A transformation from one set of coordinates and momenta to another set in which the equations of motion are preserved.

cap: A small explosive designed to be detonated by an electrical current and which in turn detonates another explosive. **Seismic caps** are designed to detonate with little uncertainty associated with their time of detonation.

capacitance: The ratio of charge (Q in coulombs) on a capacitor to the potential across it (V in volts) is the capacitance (C in farads):

$$C = Q/V.$$

capacitive coupling: See *coupling*.

capacitive reactance: Electrical impedance X_c due to capacitance:

$$X_c = 1/(2\pi\nu C) \text{ ohms},$$

where ν is frequency in hertz and C is capacitance in farads.

capacitivity: Permittivity, the property of a material which enables it to store electrical charge. Measured in farads per meter. The ratio of the capacitivity of a material to that of free space is the **dielectric constant** of the material. Free space has a capacitivity of 8.854 × 10^{-12}/m. See also *electric susceptibility*.

capacitor: A device which adds capacitance to an electrical circuit. The standard color code for capacitors is shown in Figure C-1.

caprock effect: A sharp positive gravity anomaly produced by the dense caprock of a salt dome, superimposed on a broader negative caused by the salt dome. Usually shallow salt is itself denser than surrounding sediments, so caprock is not essential to produce this effect.

capture cross-section: 1. Atomic capture cross-section for neutrons is the effective area within which a neutron has to pass in order to be captured by an atomic nucleus. It is a probabilistic value dependent upon the nature and energy of the particle as well as the nature of the capturing nucleus. Atomic capture cross-section is often measured in **barns** (1 barn = 10^{-28} m²). **2. Macroscopic capture cross-section** Σ is the effective cross-sectional area per unit volume of material for capture of neutrons; hence, it depends on the number of atoms present as well as their atomic capture cross-sections. The unit of measure for Σ is $cm^2/cm^3 = 10^3$ cu, where cu = *capture unit* (q.v.). **3.** The rate of absorption of thermal neutrons with a velocity V is $V \Sigma$.

capture unit: A unit of measure of macroscopic capture cross-section; 0.1 m^2/m^3. Also called sigma unit and abbreviated cu or su.

card: 1. A plug-in printed-circuit module. **2.** A punched card used for computer input or output.

cardinal theorem: *Sampling theorem* (q.v.).

Carpenter electrode array: Four collinear equispaced electrodes used in electrical surveying. In the **Carpenter-1** (or Wenner) array, the two central electrodes are the potential electrodes; in the **Carpenter-2** array, one end pair is the potential electrode; in the **Carpenter-3** array, the current and potential electrodes alternate.

carry: 1. In arithmetic operations, the transfer of a value from a lower order position to the next higher order position as a result of the lower order having equaled or exceeded the base of the number system. For example, if subscripts indicate the base,

$$8_{10} + 2_{10} = 0 + \text{carry } 10_{10};$$
$$1_2 + 1_2 = 0 + \text{carry } 2_2.$$

2. A computer status bit that indicates whether or not the last arithmetic operation resulted in a carry out of the high-order bit of the ALU result register.

cascade: To arrange in series.

casing: Tubes or pipes used to keep a borehole from caving in. Shothole casing is usually made in 10-ft lengths which screw together.

casing-collar: The coupling between joints of casing.

casing-collar locator: A magnetic or scratcher device for locating casing collars, used to correlate cased-hole logs and serve as reference depths in completion operations.

casing point: The lowest depth in a well where casing is set and cemented.

casing shoe: The lowest point on a string of casing; the casing point.

catalog: 1. Collection of master curves or *type curves* (q.v.) for interpreting magnetic, electrical, and electromagnetic survey measurements. **2.** A list of computer programs; a **menu** or **library.**

catcher: 1. A device to prevent cap wires from being blown out of the hole by the shot and endangering personnel by making contact with high-voltage power lines. **2.** A device to catch a Dinoseis gas gun before it falls back to the ground after an explosion, thus preventing a second sharp impact which would complicate the waveform. **3.** A device used to retain a core in a core barrel.

cathode: 1. The electrode at which reduction occurs and electrons are taken up. The negative terminal of an electrolytic cell or the positive terminal of a battery. **2.** The source electrode for electrons inside a vacuum tube or semiconductor diode device that converts information into an electron beam (and subsequently to light energy to provide a visual display).

cathode-ray tube: A device for viewing waveforms (such

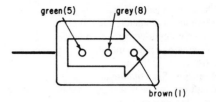

A three-dot 580 μμF mica capacitor

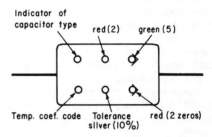

A six-dot 2500 μμF mica capacitor

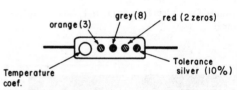

A 3800 μμF ceramic capacitor

FIG. C-1. **Capacitor** identification codes. The color code is brown = 1, red = 2, orange = 3, yellow = 4, green = 5, blue = 6, violet = 7, gray = 8, white = 9, black = 0, gold = 5% tolerance, silver = 10% tolerance.

as voltage as a function of time) and other information. Abbreviated **CRT** or **CRO.**

cathodic protection: Protection for buried pipelines and other metallic materials subject to electrochemical corrosion. The pipeline is made electrically negative with respect to ground.

cation: A positively charged ion. Compare *anion.*

cationic membrane: A membrane which permits the passage of cations but not of anions. Shale acts as such a membrane, allowing sodium ions to pass but not chloride ions. Important in generating *electrochemical SP* (q.v.).

causal: Not existing before some finite starting time and having a finite total energy.

caved portions: Parts of a borehole where the hole diameter becomes large.

cavitation: The situation where the pressure in a liquid becomes lower than the hydrostatic pressure; it often reaches the vapor pressure of the liquid. The collapse of liquid into the region produces very large pressures and generates a shock wave by implosion. The outward momentum of water and gases from an underwater explosion (and other marine energy sources) tends to leave behind such a low-pressure region, a collapse of which produces a bubble effect.

CBL: *Cement-bond log* (q.v.).

CCD: Charge-coupled device.

CDM: Continuous dipmeter; see *dipmeter*.

CDP: Common-depth-point; see *common midpoint*.

CDPS: Common-depth-point stack; see *common-midpoint stack*.

celestial equator: See *equinoctial*.

celestial latitude: Angular distance north or south of the ecliptic. Differs from declination.

celestial longitude: Angular distance east of the vernal equinox measured along the ecliptic. Different from right ascension.

celestial navigation: See *positioning*.

celestial pole: The projection of the Earth's axis onto the celestial sphere.

celestial sphere: An imaginary sphere of infinite radius concentric with the Earth on which all celestial bodies are imagined to be projected.

Celsius: A temperature scale where water freezes at 0° and boils at 100°; formerly called **centigrade.** If T_C, T_K, and T_F are Celsius, Kelvin and Fahrenheit temperatures, respectively,

$$T_C = T_K - 293.15 = (5/9)(T_F - 32).$$

Named for Anders Celsius (1701-1744), Swedish astronomer.

CEM: Crone electromagnetic method; see *shootback method*.

cementation factor: The exponent m in the *Archie formula* (q.v.).

cement-bond log: A well log of the amplitude of the acoustic wave which indicates the degree of bonding of cement to casing and formations. See Figure C-2. The log may consist of (a) **an amplitude log (CBL)** which represents the amplitude of a portion of the longitudinal acoustic wave train, or (b) a display of the wave train such as the character log, 3-D, microseismogram, VDL, or acoustic signature log.

centigrade: *Celsius* (q.v.).

centipoise: A unit of viscosity, equal to 10^{-3}kg/s.m. The viscosity of water at 20°C is 1.005 centipoise.

central induction sounding: An electromagnetic technique in which the vertical magnetic field is measured as a function of frequency at the center of a large horizontal transmitting loop.

central processing unit: The heart of a computer which controls operations and interprets and executes programs. Abbreviated **CPU.** Typically includes the arithmetic logic unit and an instruction processing unit.

CEP: Circular error probability; the radius of a circle such that half of the measurements fall within the circle. A

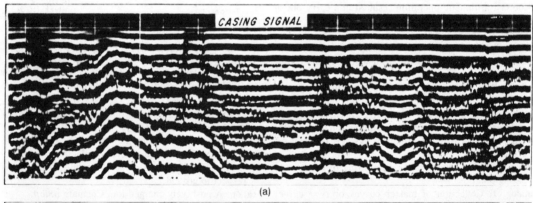
(a)

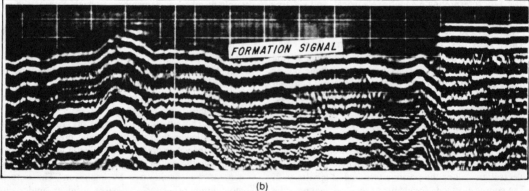
(b)

FIG. C-2. **Cement-bond** application of microseismogram log. (**a**) Before squeeze; (**b**) after squeeze. If the casing is poorly cemented, energy which travels through the steel casing is strong and little energy travels in the formation; if the casing is well cemented, the casing signal nearly disappears and the formation signal is strong. Time increases from top to bottom, depth increases from right to left. (Courtesy Welex.)

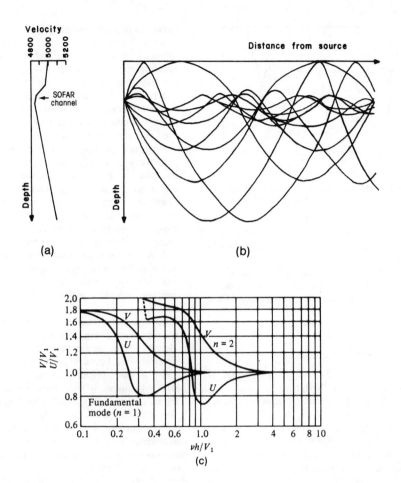

(a)

(b)

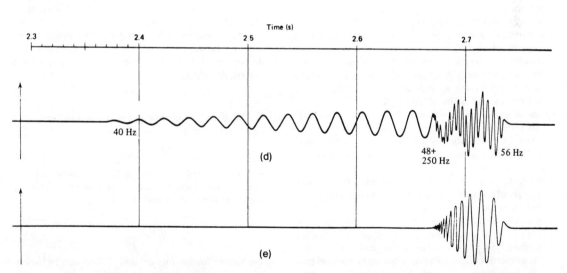

(c)

(d)

(e)

FIG. C-3. **Channel waves**. (a) The **SOFAR channel** is formed by velocity inversion. (b) Energy from a source in the channel is repeatedly refracted back toward the velocity minimum and so undergoes less divergence than normal. (From Ewing et al, 1948.) (c) Phase and group velocity versus normalized frequency for a liquid layer on an elastic substratum. (From Ewing et al, 1957.) (d) First-mode wave train from a source 4 km distant where the ocean constitutes the channel. (e) High-frequency portion of (d), called the **water wave**; its arrival is used in refraction work to determine the range. (From Clay and Medwin, 1977.)

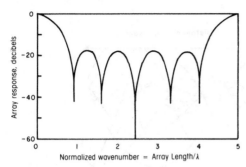

FIG. C-4. Chebychev array response has all minor lobes of equal height. Response shown is for 6-element array.

circle containing 90 percent of the measurements would be **90% CEP.**

cepstrum: The Fourier-transform of a log frequency distribution. If $\leftrightarrow$ indicates a Fourier transform operation and $\hat{g}(\delta)$ the cepstrum as a function of the saphe δ for a function $g(t)$,

$$\hat{g}(\delta) \leftrightarrow \log G(\omega)$$

$$G(\omega) \leftrightarrow g(t).$$

See Sheriff and Geldart, v. 2 (1983, p. 44, 184-185). A permutation of the letters in "spectrum".

cesium magnetometer: A type of *optically pumped magnetometer* (q.v.).

chain: 1. A unit of length equal to 66 ft. 2. To measure distances directly with a steel tape (chain) as opposed to measuring by stadia, electronic distance measurement or triangulation. Accuracy of the order of 1:5000 can be achieved with ordinary care. See *chainman*. 3. An ordered group of computer records.

chaining: 1. Measuring distances with a chain. 2. A system of storing data records in which each record belongs to a specific group or chain.

chainman: A surveyor's assistant, who helps to measure distances with a chain. Chainmen often work in pairs to measure out seismic spreads using a steel tape or wire which is as long as the group length.

Chandler wobble: A precession of the earth's rotational axis about its axis of greatest moment of inertia. Also called the **free nutation of the earth.** The wobble period is about 435 days and the amplitude is about 0.14 s of arc.

channel: 1. A single series of interconnected devices through which data can flow from source to recorder. Most seismic systems are 96 channel, allowing the simultaneous recording of energy from 96 groups of geophones. 2. A localized elongated geologic feature resulting from present or past drainage or water action. The properties of the material in-filling the channel may differ from those of the material into which the channel is cut, creating various geophysical effects. 3. An allocated portion of the radio-frequency spectrum. 4. A time gate during which measurements are made in time-domain electromagnetic surveying. Measurements made during several time gates following a source pulse yield several channels of data. 5. A layer whose veloci-

ty is such that seismic energy gets trapped; see *channel wave*. 6. A gate with upper and lower energy thresholds followed by a count rate meter, used in gamma ray spectrometers. 7. A component of a mainframe processor that supports and controls I/O.

channel wave: An elastic wave propagated in a layer where the energy is trapped. The layer may (a) have lower velocity than those on either side of it (so that total reflection can occur at the boundaries) or (b) a layer boundary may be a free surface (so that the reflectivity is nearly one). Instead of having sharp interfaces as boundaries, channels may also be produced by an increasing velocity gradient in either direction. Energy is largely prevented from escaping the channel because of repeated total reflection at the channel boundaries or because rays which tend to escape are bent back toward the channel. See Figure C-3. A channel is also called a **wave guide.** Coal seams and a surface water layer often carry channel waves. The *SOFAR channel* (q.v.) in the deep oceans is another example. See Sheriff and Geldart, v. 1 (1982, p. 70-73).

character: 1. The recognizable aspect of a seismic event; the waveform which distinguishes it from other events. Usually a frequency or phasing effect, often not defined precisely and hence dependent upon subjective judgment. 2. The recognizable aspect of a graph or picture which identifies it with some situation. 3. A single letter, numeral, or special symbol in a computer system. See also *byte*, with which it is sometimes used interchangeably. 4. A precocious offspring.

characteristic root: Eigenvalue; see *eigenfunction*.

character log: A *full waveform log* (q.v.).

characters per inch: cpi.

chargeability: One of several units of induced polarization in the time domain. Symbol: M. 1. The ratio of initial decay voltage (or secondary voltage) to primary voltage. 2. The dimensionless, induced-polarization parameter of a material in which there is an induced-current dipole moment per unit volume P energized by a current density J:

$$P = -MJ.$$

3. The fractional change in resistance measured on a decay curve, as a function of time:

$$M = \delta\rho(t)/\rho$$

4. The integrated area under an IP decay curve between times t_1 and t_2, normalized by the primary voltage V_p. Units are millivolt-seconds per volt.

$$M = (1/V_p) \int_{t_1}^{t_2} V(t) \, dt.$$

For standarization, on-time and off-time may be indicated by subscripts; that is $_{33}M_1$ or $M_{(331)}$ means "current on for three seconds and decay measured for the first second of a three-seconds off time." Field measurements of chargeability are usually calibrated to the $M_{(331)}$ standard, which differs by a factor of about 1000 from the value of M given in definitions 1 and 2 above. 5. The quantity described above multiplied by the conductivity, often expressed in millifarads/unit length. Called **specific capacity** when measured in farads/m. **6.** Chargeability can be related to *frequency effect* (q.v.).

charge anchor: A device fastened to an explosive charge to hold the charge in a fixed position in the shothole and prevent it from floating or moving.

charge-coupled device: An electronic medium in which data are encoded as the presence of positive or negative charges. **CCD.**

Chebychev array: A frequency filter or uniformly spaced linear array in which elements are weighted according to Chebychev polynomials. Such weighting equalizes the amplitude of minor lobes and gives a sharp cutoff. Called **equal-ripple filter.** See Figure C-4. Also spelled **Tchebyscheff array.** Named for Fafnutiy Lvovich Chebychev (1821-1894), Russian mathematician.

check: A test of the reliability of data. **Check problems** with known results are run through computers to verify correct functioning. (a) A **summation check** consists of adding a set of figures and using the sum (**check sum**) to verify accuracy and completeness. (b) A **duplication check** requires that an operation be repeated and give identical results when repeated. (c) An **echo check** is used to verify transmitted data; the data are sent back (**echoed**) to the source station for comparison with the original data. (d) **Error-detecting checks** search data for forbidden combinations. A **parity check** is such a check; the number of one bits plus a parity bit must add up to an odd number for odd parity or to an even number for even parity. A **longitudinal parity check** (LPC), written at the end of a record, does for each track what the parity bit does for each byte. (e) A **validity check** is a verification that a figure lies within a certain permissible range.

check bit: Parity bit; see *check.*

check shots: Shots into a well seismometer (see *well shooting*) to check the results of integrating a continuous velocity or sonic log.

check sum: A summation check; see *check.*

chemical remanent magnetism: (CRM) See *remanent magnetism.*

chemisorption: Adsorption due to chemical rather than simply electrostatic causes.

chip: A solid-state circuit or circuit element; an integrated circuit or IC.

chirp: Signal of continuously varying frequency. Often implies a linear change of frequency with time.

chi-square: A quantity distributed as

$$x_1^2 + x_2^2 + \cdots + x_k^2,$$

where $x_1, x_2, \cdots, x_k$ are independent, Gaussian, with zero mean and with unit variance. The chi-square test is a statistical test of data distribution. If f_i are observed frequencies of occurrence, and e_i are the expected number of occurrences, the **goodness of fit** y is

$$y = \sum \frac{(f_i - e_i)^2}{e_i}.$$

The goodness of fit can be related (by the use of tables) to the probability that the observed data fit the model with the assumed number of degrees of freedom.

chlorine log: A log based on the counting rate of gamma rays produced by capture of thermal neutrons by chlorine in the formation. By limiting the count to a certain energy range, the tool is made more sensitive to chlorine and relatively insensitive to formation porosity. The chlorine log has been mainly replaced by the neutron-lifetime log and thermal-decay-time log.

chopper: 1. An electrical switching device sometimes including an oscillator, used to interrupt a dc or low-frequency ac voltage so that it can be measured by an ac voltmeter or amplified by an ac amplifier. **2.** A helicopter.

chromatograph: See *partition gas chromatograph.*

chronographic chart: A diagram which summarizes conclusions from seismic-sequence analysis. Geologic time increases upward and the geographic distribution of units is plotted horizontally. See Figure C-5. Also called **chronostratigraphic chart.**

chronostratigraphic chart: See *chronographic chart.*

CI: *Contour interval* (q.v.).

Circular error probability: *CEP* (q.v.).

circulation: The movement of drilling fluid from a mud pit through pump, drill pipe, annular space in the hole, and back into the mud pit.

circulation loss: Loss of drilling fluid into a porous (or "thief") formation.

circumferential wave: Seismic wave that travels parallel to the earth's surface.

cis θ: Euler's identity:

$$\text{cis } \theta = \cos \theta + i \sin \theta = e^{i\theta}.$$

Clarke ellipsoid: The basis for the North American geodetic datum, the reference datum in most of the Western Hemisphere; the Clarke 1866 ellipsoid. (There is also a Clarke 1880 ellipsoid which is used in Africa.) See Figure G-3. Named for Alexander Ross Clarke (1828-1914), British geodesist.

clastic rock: A rock composed of fragments derived from other rocks.

clean: 1. Containing no appreciable amount of clay or shale. Applied to sandstones and carbonates. Compare *dirty.* **2.** To remove soft magnetization so that hard remanent magnetization may be studied. See *degaussing.*

clipped: Distorted because amplitude exceeded a maximum permitted amount. Clipping in analog systems usually occurs because of saturation of some element of the system, resulting in distortion of the waveform; see Figure C-6. Digital clipping (loss of the first bit) may have variable effects depending on what the lost bit represents (i.e., whether it is the sign, the most significant bit, a complement code, etc.); the clipped signal sometimes bears little resemblance to the unclipped signal.

clock: A generator of the basic timing signal pulses to which all system operations are synchronized.

closure: 1. The property of a structure whereby it has a closing contour. **Vertical closure** is the vertical distance from the apex to the lowest closing contour; **areal closure** is the area contained within the lowest closing contour. Compare *trap.* **2.** The cumulative error around a survey loop; *mis-tie* (q.v.).

cluster analysis: Analysis to see if subsets of a data set sort

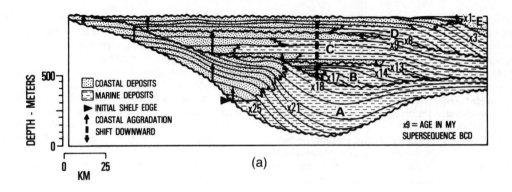

(a)

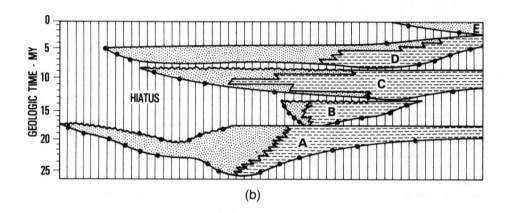

(b)

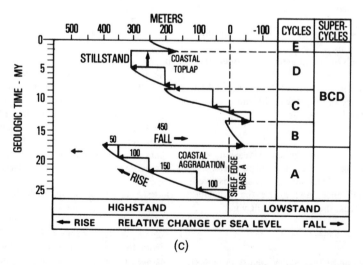

(c)

FIG. C-5. **Chronographic chart**. (a) Cross-section interpreted from seismic line with unconformities interpreted as sequence boundaries. (b) Chronographic chart corresponding to (a). (c) **Relative sea-level chart** interpreted from (a). (After Vail et al, 1977.)

out into separate regions defined by combinations of variables.

clutter: 1. Noise on a radar screen resulting from unwanted echoes or scattering such as from a rough sea or from rain. **2.** Coherent interference, in contrast to incoherent nonrepeatable interference.

cmos: Complementary metallic oxide semiconductor (MOS). A technology that employs both negative and positive MOS to minimize power and cooling requirements.

coagulation: Grouping together data with values sufficiently close to each other. Used especially prior to a color display in which different groups are assigned different colors.

coastline effect: Distortions in electromagnetic ground waves and natural magnetic fields produced by the contrast in electrical conductivity between land and highly conductive ocean water.

COBOL: Common Business-Oriented Language, one of the first languages designed for commercial data processing, utilizing commonly used English nouns, verbs, and connectives.

COCORP: Consortium for Continental Reflection Profiling, a program of seismic work to study the earth's deep crust and upper mantle by common-midpoint techniques.

code: A system of characters and rules for representing information in a language capable of being understood by a computer. See *source program, object program.*

coefficient of coherence: See *coherence.*

coefficient of variation: Ratio of standard deviation to the mean. See *statistical measures.*

coercive force: The magnetic field intensity required to reduce the magnetization of a sample to zero. Generally associated with remanent magnetization. See *hysteresis* and *coercivity.*

coercivity: The demagnetizing field intensity required to reduce the induction of a magnetic material from saturation to zero. See Figure H-8. Coercivity is used as a figure of merit for magnetic hardness of a material, particularly in reference to the distribution of coercive forces of a magnetic system's components. The latter might be different mineral phases, grain sizes, etc.

cofactor: See *matrix.*

coherence: 1. The property of two wave trains having a well-defined phase relationship, i.e., being in-phase. **2.** A measure of the similarity of two functions or portions of functions. If the functions have power spectra P_{ii} and P_{jj} and cross-power spectra P_{ij} (which may be complex), their coherence is

$$P_{ij}/(P_{ii}P_{jj})^{1/2}.$$

Also called **coefficient of coherence.** Coherence is a frequency-domain concept analogous to correlation in the time domain. **3.** Measures of the similarity among more than two functions. For example, seismic reflection events are coherent in a linear way with respect to dip, in a hyperbolic way with respect to normal moveout, in a nonanalytic though systematic way with respect to geophone locations. The principal evidence for a separate seismic event is coherence among the members of a set of seismic traces over a short time interval

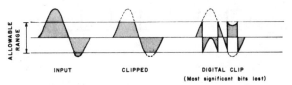

FIG. C-6. **Clipping.** Loss of a digital bit can cause various effects depending on how the data are formatted (e.g., the sign bit might be lost).

(of the order of a $1\frac{1}{2}$ or so cycles of the dominant frequency) compared with less coherence elsewhere. Coherence is used qualitatively in record picking but quantitative measures of coherence are used in automatic picking schemes. See *semblance* and Sheriff and Geldart, v. 2 (1983, p. 36, 39).

coherence filtering: 1. A multichannel filter that emphasizes coherent events. **2.** A method of noise suppression that filters out the coherent portion of two signals.

coherent detection: *Coherence filtering* (q.v.).

coherent noise: Noise wave trains which bear a systematic phase relation (coherence) between adjacent traces. Most source-generated seismic noise (ground roll, shallow refractions, multiples, etc.) is coherent. The distinction between random and coherent is sometimes a matter of sampling (trace spacing). See Sheriff and Geldart, v. 1 (1982, p. 125-126).

colatitude: Angular distance from the North Pole, the complement of the latitude or 90 degrees minus the latitude in the Northern Hemisphere, or 90 degrees plus the latitude in the Southern Hemisphere.

Cole-Cole plot: A plotting convention of in-phase versus quadrature measurements in which frequency relaxation appears as a semicircle. Useful for dielectric relaxation, seismic velocity, and induced-polarization measurements.

collar log: *Casing-collar locator* (q.v.).

collimated: Nondivergent, parallel.

cologarithm: The logarithm of the reciprocal of a number prior to loading and running.

color display: A display of measurements which have been color-encoded. See *attributes.*

colored: Having an arbitrarily shaped amplitude spectrum. Compare *white.*

color mimicry: A technique for correlating the response to different types of measurement for features in two dimensions. A set of photos, maps or other display are projected through primary color filters and the images superimposed. See Grossling (1969).

color processing: Encoding a set of measurements as a set of colors.

colored sweep: A Vibroseis sweep which is not linear with frequency so that its frequency spectrum is not flat.

column vector: See *matrix.*

comb: An infinite sequence of impulses $\delta(t - n\,\Delta t)$ occuring at time intervals Δt:

$$\text{comb}(t) = \sum_{n=-\infty}^{\infty} \delta(t - n\Delta t).$$

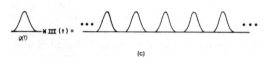

FIG. C-7. (a) The **comb**. (b) Multiplication of a function (dashed) by a comb samples the function. (c) Convolution of a function $g(t)$ with a comb replicates the function.

Called the **sampling function** because multiplication of a function and a comb gives the sample values at the comb interval, and the **replicating function** because convolution with a waveform reproduces the waveform at the position of each spike. Also called a **shah**. The Fourier transform of a comb is also a comb:

$$\text{comb}(t) \leftrightarrow \text{comb}(\nu)$$

where frequency $\nu = 1/t$ if t is time, and $\leftrightarrow$ indicates a Fourier transform operation. See Figure C-7 and F-17. If the impulses are spaced T apart,

$$\text{comb}(t) \leftrightarrow t\,\text{comb}(T\nu).$$

Combisweep: A **Vibroseis** sweep technique using several sequential linear sweeps separated by short listening periods, used to reduce correlation ghosts. A Prakla-Seismos tradename.

common-depth-point: See *common midpoint*. The use of common-depth-point is discouraged because there is no common point at depth if reflectors dip. **CDP.**

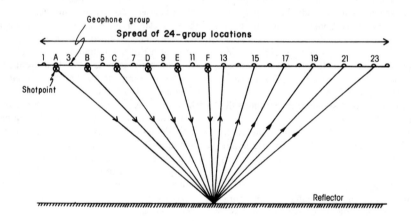

(a)

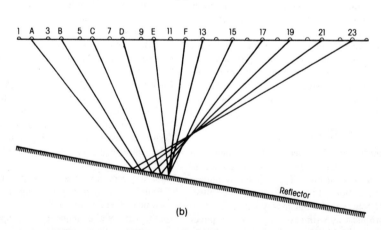

(b)

FIG. C-8. **Common-midpoint** shooting. (a) In six-fold shooting with 24-geophone groups, the shotpoint is moved two group intervals between successive shots; hence the same subsurface is sampled six times (A → 23, B → 21, C → 19, D → 17, E → 15, F → 13). (b) If the reflector dips, there is no longer a common reflecting point and common-midpoint stacking will result in **smearing**.

common-depth-point gather: A *common-midpoint gather* (q.v.).

common-depth-point stack: A *common-midpoint stack* (q.v.).

common-geophone gather: A set of seismic traces having the same geophone.

common midpoint: Having the same midpoint between source and detector. Also called **common-depth-point** and **common-reflection-point**. See Figure C-8.

common-midpoint gather: The set of traces which have a *common midpoint* (q.v.).

common-midpoint stack: A sum of the traces corresponding to the same midpoint; see Figure C-8. The traces from different profiles having different offset distances are gathered together, corrected for statics and normal moveout, and then summed (or stacked). The objective is to attenuate random effects and events whose dependence on offset is different from that of primary reflections. Also called **roll-along** and **horizontal stacking. CMPS.**

common mode: Having signals that are identical in phase.

common-mode rejection: A differential amplifier which ignores a signal which appears simultaneously at both input terminals.

common-offset gather: A side-by-side display of traces which have the same shot-to-geophone distance (**offset**).

common-offset stack: A stack of traces which have the same offset and which are located within a limited range of midpoint locations.

common-range gather: *Common-offset gather* (q.v.).

common-reflection point: *Common midpoint* (q.v.). The use of common reflection point is misleading because reflecting points are not common if there is dip.

communication: 1. A connection path between formations so that fluids from one can flow into the other. 2. The ability to interchange data, as when two computers "have communication with each other."

commutate: 1. To reverse the direction of a unidirectional electric current (for example, by periodically changing connections), or to reverse every other cycle of an alternating current to form a unidirectional current. 2. To change digital seismic data from a trace-sequential to a time-sequential sequence, i.e., to multiplex.

commutative: Operations which yield the same results regardless of the sequence in which they are performed.

compaction: Loss of porosity with pressure, usually in a nonelastic way, that is, by grain deformation, repacking, recrystallization, etc.

compaction correction: An empirical correction applied to porosity values derived from the sonic log in uncompacted formations. Undercompaction is indicated by low velocities in adjacent shales. See also *differential compaction*.

comparative interpretation: The comparison of survey data with type curves which have been calculated for bodies of assumed contrasts and geometry.

comparator: A circuit which compares two signals and indicates the result of the comparison.

compensated log: A well log made with a sonde designed to correct unwanted effects. The **compensated density log** (FDC) uses the signal from a secondary detector to correct for the effect of mud cake and small irregular-

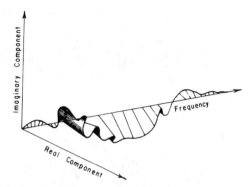

FIG. C-9. A **complex function** (such as a frequency spectrum) requires three-dimensional representation. The distance of a point from the frequency axis is the **modulus**, and the angle with the real plane is the **phase**.

ities in the borehole wall. The **compensated sonic log** (BHC) uses a special arrangement of the transducers to correct for irregularities in borehole size and sonde tilt.

competent: A bed which retains its stratigraphic thickness under stress. It folds or breaks under stress, in comparison with adjacent incompetent beds which tend to flow.

compiler: A program for converting a source program in a high-level language to an object program in machine language prior to loading and running. An **interpreter** involves conversion concurrently with running the program.

complement: The difference between a particular value and full scale. Thus in the decimal system, the complement of x is $(10^\mu - x)$, where μ is a fixed number; and in binary the complement of x is $(2^\nu - x)$. Adding the complement of x is equivalent to subtracting x. Computers often find it easier to generate the complement and add than to subtract. See also *one's complement* and *two's complement*.

complement of chargeability: An IP time-domain measurement of the area under the decay curve, integrating over the interval between 0.45 and 1.75 s measured on a Newmont-type receiver.

complex frequency: A damped wave can be expressed as the product of an absorption factor $e^{-\alpha t}$ and a periodic factor $e^{j\omega t}$:

$$Ae^{-\alpha t}\,e^{j\omega t} = Ae^{j(\omega + j\alpha)t},$$

where $(\omega + j\alpha)$ is the complex frequency.

complex number: A number with both real and imaginary parts, such as

$$z = x + jy = Ae^{j\theta},$$

where $j = (-1)^{1/2}$. [The symbol i is also used to indicate $(-1)^{1/2}$.] The **modulus** or magnitude of the above complex number is $A = (x^2 + y^2)^{1/2}$ and the angle indicating its direction with respect to the real axis is $\theta = \tan^{-1}(y/x)$. A graph of a complex function or quantity (such as a frequency spectrum) is shown in Figure C-9.

complex ratio: See *Turam method*.

complex resistance: *Impedance* (q.v.).

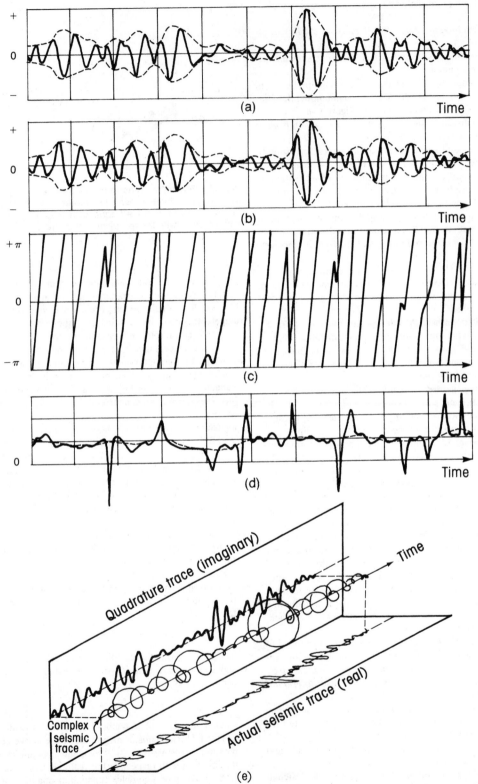

FIG. C-10. **Complex trace analysis**. Real (**a**) and quadrature (**b**) traces for a portion of a seismic trace. The envelope is shown as the dotted curve in (a), (b). Instantaneous phase is plotted in (**c**), instantaneous frequency in (**d**), and weighted average frequency as the dotted curve in (d). (**e**) Isometric diagram of complex trace. (From Taner et al, 1979.)

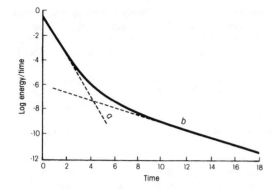

FIG. C-11. **Composite decay curve** for material containing two radio nucleides. The decay curve can be decomposed into the two component curves. The slope of **a** and **b** give their respective half-lives.

complex resistivity: Representation of apparent resistivity as having real and imaginary parts. Complex resistivity is the proportionality between voltage and current where the two are not in-phase. It is used to accommodate variations in resistivity with frequency as observed in induced-polarization surveys.

complex spectrum: See *Fourier transform*.

complex-trace analysis: Finding a complex number representation $F(t)$ of a real time-series $f(t)$:

$$F(t) = f(t) + jf_\perp(t) = A(t)\,e^{j\gamma(t)};$$

$f_\perp(t)$ is the **quadrature series,** $A(t)$ is the **amplitude of the envelope** of the trace (also called **reflection strength**), and $\gamma(t)$ is the **instantaneous phase. Instantaneous frequency** is $d\gamma(t)/dt$. Used to determine attributes of seismic data. See Figure C-10 and Taner et al. (1979). Often involves Hilbert transform.

compliance: The relationship of strain to stress. For an elastic substance, the reciprocal of the elastic modulus; see *elastic constants*. The mechanical or acoustical equivalent of electrical capacitance.

composite: 1. Made up from two or more different elements, as a *composite decay curve* or *composite time-distance curve* (q.v.). 2. To mix or combine the energy of different seismic channels without first applying static and normal-moveout corrections.

composite decay curve: A time-domain decay curve containing more than one component, often with different time constants or even a combination of positive and negative decay curves. See Figure C-11.

composite reflection: A wave train composed of two or more overlapping reflections.

composite time-distance curve: A refraction arrival time versus offset-distance graph synthesized from a number of shots and shorter spreads at various locations, as opposed to the graph which would have been obtained from a single shot into an extensive spread of geophones. See Figure T-7.

compositor: A device for *mixing* (q.v.).

compressed section: A record section with the horizontal scale compressed so that a large distance is represented, used to present regional aspects of seismic data. Associated with considerable vertical exaggeration. Also called **squash plot**.

compressibility: Change of density with pressure; reciprocal of bulk modulus. See *elastic constants*.

compressional wave: A *P-wave* (q.v.).

Compton scattering: The principal interaction mechanism for gamma photons in the 0.4 to 3.0 MeV range for media in the intermediate atomic number range. The incident photon changes direction and is available for repeated scattering, so that the probability of total dissipation increases with the volume of the scatterer. Named for Arthur Holly Compton (1892-1962), American physicist.

computer: 1. One who computes corrections for geophysical data; a **computor**. 2. A machine capable of accepting information, applying prescribed processes to the data (a sequence of arithmetic and logical operations as requested by program instructions), and supplying the results to some output device. A simple computer program is shown in Figure F-11. 3. An **analog computer** uses a physical analogy of position, electric current, flow, temperature, etc., to solve relationships. 4. A **digital computer** applies numerical processes to sets of discrete numbers.

computer generations: Historical levels of computer hardware technology: **first generation**, vacuum tubes; **second generation**, transistors; **third generation**, integrated circuits; **fourth generation**, very large scale integration (VLSI); **fifth generation**, massively parallel processors.

computer graphics: The visual display of data stored in a computer. Usually refers to a display on a cathode ray tube (CRT).

computer language: The form in which program instructions can be supplied to a computer for translation into a machine language program which can then be executed. Such languages include Algol, Cobol, Fortran, PL-1, APL, Pascal, C, Basic, and others.

computer modeling: See *numerical modeling*.

concatenate: To unite in a series; to connect together; to chain. The action by a computer program of relating data in some organized manner so as to treat multiple data sets or files as one.

concentration cell: See *electrochemical SP*.

concentration overvoltage: See *overvoltage*.

concentric fold: A flexural-slip fold. See *folding*.

concordance: Parallelism of reflections to seismic sequence boundaries.

condensate well: See *GOR*.

condition: 1. To circulate mud through a borehole to make the mud uniform throughout the system. 2. A restriction or constraint, as in *initial condition* or *boundary condition* (q.v.).

conditional jump: An instruction causing a transfer to an instruction location other than the next sequential instruction only if a specific condition is satisfied. The next sequential instruction is executed if the condition is not satisfied. See *branch*.

conductance: With direct current, the reciprocal of resistance. With alternating current, the resistance divided by the impedance squared; the real part of admittance.

General form for conic sections:

$Ax^2 + Bxy + Cy^2 + Dx + Ey + F = 0$, or
$x^2 + y^2 = e^2(x + a)^2$.

For a parabola, $e = 1$, or $B^2 = 4AC$
For a hyperbola, $e > 1$, or $B^2 > 4AC$
For an ellipse, $e < 1$, or $B^2 < 4AC$.

Common forms for conic sections:

Circle of radius r centered at (a, b):
$(x - a)^2 + (y - b)^2 = r^2$

Circle with origin on circumference centered at
(a, b): $r^2 = 2b^2 \cos(\theta - a)$

Ellipse where a, b = semimajor, semiminor
axes: $(x/a)^2 + (y/b)^2 = 1$.

Ellipse centered at the origin:
$r^2 = a^2 b^2 / (a^2 \sin^2\theta + b^2 \cos^2\theta)$.

Ellipse centered at one focus in polar form
where e = eccentricity (see Figure E-9) and
h = distance from a directorix straight line:
$\rho = eh / (1 - e \cos \theta)$.

Parabola with origin at vertex and focus at
$(0, \rho/2)$: $y^2 = 2\rho x$.

Parabola with origin at focus: $r = \rho / (1 - \cos \theta)$.

Hyperbola with origin at center: $(x/a)^2 - (y/b)^2$
$= 1$ or $r^2 = a^2 b^2 / (a^2 \sin^2\theta - b^2 \cos^2\theta)$.

FIG C-12. Equations of **conic sections**.

Measured in siemen (= mho = inverse ohm).

conduction angle: The number of degrees in a half-cycle ac wave during which a silicon controlled rectifier is turned on. If ϕ is the phase control angle, the conduction angle is $\pi - \phi$.

conduction contact: See *galvanic contact*.

conduction current: Electrical current resulting from the motion of free charges under the influence of an electric field. The density of conduction current **J** at a point in an isotropic medium is

$$\mathbf{J} = \sigma \mathbf{E},$$

where σ is the conductivity and **E** is the electric field. Conduction currents usually are more important than displacement currents in electromagnetic prospecting, depending on the frequency range.

conductivity: The ability of a material to conduct electrical current. In isotropic material, the reciprocal of resistivity. Sometimes called **specific conductance**. Units are siemen/m.

conductivity log: *Induction log* (q.v.).

conductor: A body within which electrical current can flow readily. Often, the "target" of an electromagnetic survey. An **electronic conductor** conducts electricity primarily by electron mobility. An **ionic conductor** conducts electricity primarily by means of ion mobility. Electrolytes are ionic conductors.

conductor casing: A second casing string set to a depth of 500-1000 ft with the annular space filled with cement to protect fresh-water sands.

confidence level: The probability that an interval contains an element with given characteristics; the limits between which a specified percentage of measurements are expected to lie, as in "90 percent confidence limits."

configuration: 1. Arrangement, as of geophones in a group; *array* (q.v.). 2. The hardware and/or software making up a computer system, and how it is put together.

conformability: See *conformal mapping* and *map projection*.

conformable: Two adjacent parallel beds separated by a surface of original deposition, where no disturbance or denudation occurred during their deposition.

conformal mapping: One area can be mapped into another area conformally if there is a continuous one-to-one correspondence of points and if angles are preserved.

conic: A curve defined by a quadratic equation. See Figure C-12.

conical wave: *Head wave* (q.v.).

conjugate: The conjugate of a complex number is the number with the sign of the imaginary part reversed. Often designated by a superscript asterisk or superscript bar. If $Z = a + jb = Ae^{j\theta}$, then $Z^* = \overline{Z} = a - jb = Ae^{-j\theta}$.

conjugate points: Pairs of object and image points which correspond to each other.

conjunction: The condition for which an AND gate is used; **intersection**. Each of two (or more) situations must occur. Often written A∩B or A.B (read as "A and B"). See Figure G-1.

connate water: Water trapped in the interstices of the sediments at the time of deposition, as opposed to water which migrated into the formations after deposition. See *interstitial water*.

Conrad discontinuity: A sharp increase in the *P*-wave velocity in the crust, commonly at a depth of 17–20 km. The velocity below the Conrad discontinuity is of the order of 5.5–6.7 (often 6.0–6.7) km/s (compared to around 8.1 km/s for the upper mantle immediately below the Moho). The Conrad discontinuity is not observed everywhere.

console: The computer operator's control panel. Generally includes start-stop keys, keyboard for entering instructions or data, and display.

constructive interference: See *interference*.

contact log: A *microresistivity log* (q.v.) in which the sonde is held against the borehole wall.

contact resistance: The resistance observed (a) between a grounded electrode and the ground, (b) between an electrode and a rock specimen, or (c) between electrical contacts.

continuation: Determining a field over one surface from measurements of the field over another surface (specifically, at another elevation). The field at the elevation z, $F(x,y,z)$ can be found from the field on the surface, $F(x',y',0)$. Where the surfaces are horizontal and no sources intervene, the **upward continuation** relation (an application of Green's theorem) is:

$$F(x,y,z) = \frac{|z|}{2\pi} \iint \frac{F(x',y',0)}{R^3} \, dx' \, dy',$$

where
$$R = [(x - x')^2 + (y - y')^2 + z^2]^{1/2}.$$

Interchange of the two fields in this equation gives the **downward continuation** relation. See Peters (1949).

continuity: Condition of an unbroken electric circuit.

continuous profiling: A seismic method in which geophone groups are placed uniformly along the length of the line and shot from holes so spaced that continuous (usually 100 percent) subsurface coverage is obtained along the line. Continuous profiling can be accomplished with a variety of spread types. Refraction continuous profiling requires continuous control on the refractor being mapped which may require irregular surface layouts.

continuous sweep Vibroseis: Conventional *Vibroseis* (q.v.) in which the vibrator frequency is varied continuously during a long sweep duration.

continuous-velocity log: A *sonic log* (q.v.). Abbreviated **CVL**.

contour: A line separating points of value higher than the contour value from points lower, representing the locus of a constant value on a map or diagram.

contour interval: The difference in value between two adjacent contour lines. Abbreviated **ci** or **CI**.

control: 1. Accurately known data which can be used to check the validity or accuracy of a series of measurements. 2. The data on which a map or section is based. Posting the control on a map or section allows one to evaluate interpretation as to what happens between control points. 3. The section of computer code that is currently executing.

control character: A character whose occurrence in a particular context initiates, modifies, or halts operation.

controlled mosaic: A composite aerial photograph made by rephotographing component vertical photographs to compensate for scale variations resulting from tilt and variations in flight altitude.

controlled rectifier: An electronic circuit element consisting of a controlled diode or solid-state switch. The diode is usually turned "on" by a small voltage from an external circuit and turned "off" when the voltage is reversed. Used to switch large currents in IP transmitters. When the semiconductor is silicon, a controlled rectifier is called **SCR**. A **GTO** (gate turn-off switch) controlled rectifier can be turned on and off independently of the current through the diode.

controlled-source electromagnetics: Any electromagnetic sounding or prospecting system which uses artificially generated fields with prescribed characteristics rather than natural fields. See *active*.

controller: 1. See *depth controller*. 2. A computer peripheral that handles multiple devices of the same kind (e.g., a tape controller or a disk controller).

control station: A point whose position (horizontally and/or vertically) is used as a base for a dependent survey or as control for adjusting survey errors.

convergence: 1. The effect of computing a survey on a curved surface as if the surface were plane. The correction from assumed rectangular coordinates to geodetic coordinates allowing for earth curvature is the **convergence correction**. 2. In iterative operations such as modeling, the condition when calculated values are sufficiently close to observed values. 3. The condition when calculated values approach finite limits as the number of terms used increases.

convergence correction: See *convergence*.

convergent margin: *Active margin* (q.v.).

conversational mode: An interactive procedure in which each entry from a terminal elicits a response from the computer and vice versa.

converted wave: Seismic energy which has traveled partly as a *P*-wave and partly as an *S*-wave, being converted from one to the other upon reflection or refraction at oblique incidence on an interface. Since mode conversion is small for small incident angles, converted waves become more prominent as offset increases.

converter: A device to perform digital-to-analog (D/A) or analog-to-digital (A/D) conversion.

convolution: Change in waveshape as a result of passing through a linear filter. 1. A mathematical operation between two functions, $g(t)$ and $f(t)$, often symbolized by an asterisk:

$$g(t)*f(t) = \int_{-\infty}^{\infty} g(\tau) f(t - \tau) \, d\tau.$$

Convolution is not restricted to one dimension. For example:

$$g(x, y)*f(x, y) = \int_{-\infty}^{\infty} \int_{-\infty}^{\infty} g(\alpha, \beta) \cdot$$
$$\cdot f(x - \alpha, y - \beta) \, d\alpha \, d\beta.$$

2. **Linear filtering.** If a waveform $g(t)$ is passed into a linear filter with the impulse response $f(t)$, then the output is given by the convolution of g with f. In discrete form where the input can be thought of as a series of impulses of varying size, each will generate an $f(t)$ of proportional amplitude and the output will be the superposition of these. This can be expressed as:

$$g_t*f_t = \sum_{k=0}^{L} g_k f_{t-k}.$$

This expresses that the output of a linear filter at the instant t is a weighted linear combination of the inputs. L is the convolution **operator length** in time divided by the same interval and $(L + 1)$ is the number of **points** in the operator. (A simple computer program is shown in Figure F-11.) The frequency-domain operation equivalent to time-domain convolution consists of multiplying frequency-amplitude curves and adding frequency-phase curves. Convolution is sometimes done by (a) **replacement** where each spike of the input is replaced by a proportionally scaled version of the impulse response and superposition forms the output; (b) **folding** where the impulse response of the filter is reversed in time and slid past the input, the output for each position of the impulse response being the sum of the products of input and impulse response for corresponding points; (c) **multiplying z-transforms** of the input and the impulse response to give the *z*-transform of the output; or (d) **multiplying Fourier or Laplace transforms** to give the Fourier or Laplace transform of the output. See Sheriff and Geldart, v. 2 (1983, p. 28-29, 167-169, and 178-179). **3.** Convolution in two dimensions is used with gravity, magnetic, and other data to produce grid residual, second derivative, continuation maps, etc.; see Fuller (1967).

convolutional model: The model in which a seismic trace $f(t)$ is given by the convolution of an embedded (equivalent) wavelet $w(t)$ with a reflectivity function $r(t)$ plus random noise $n(t)$:

$$f(t) = w(t) * r(t) + n(t)$$

The convolutional model is implied in most seismic processing and interpretation.

convolution theorem: The Fourier transform of the convolution of two functions is equal to the product of their individual transforms (or multiplying their amplitude spectra and summing their phase spectra).

convolver: A subcomputer under the control of a main computer performing operations which involve convolution or correlation. Sometimes called a **digital filter box.** Superseded by array processors.

cookbook: A method which is prescribed step-by-step.

Cooley-Tukey method: A Fourier-analysis algorithm . which considerably reduces computing time; see *fast Fourier transform* and Cooley and Tukey (1965).

coordinate transform: See Figure C-13.

coordinatograph: *X-Y reader* (q.v.).

copy: The degree to which events or traces look alike.

core: 1. A rock sample cut from a borehole or retrieved from the seafloor. See *core analysis*. **2.** The Earth's core is the central portion bounded by the Gutenberg discontinuity ($\sim$ 2900 km deep) which separates it from the mantle. The core's radius is $\sim$ 3500 km. See Figure E-1. Divided into **outer core** which will not transmit S-waves and **inner core,** the radius of the inner core being $\sim$ 1300

km. **3.** A *magnetic core* (q.v.) is a device used in rapid-access memories. **4.** A material of high magnetic permeability placed in the center of a coil of wire. Used in the flux-gate magnetometer for measuring magnetic fields.

core analysis: Cores from boreholes are analyzed for porosity, permeability, fluid content, and identification (water and/or oil saturation), lithology, and structure (fractures, cross bedding, etc.). Results are often illustrated on a log or graphed against depth.

corer: A device for obtaining a solid sample of rock from a borehole or from the ocean bottom. A **box corer** usually penetrates less than 3 ft into the sea floor and, has a spade-like device that retains a sample. A **gravity corer** penetrates the ocean floor solely by its own weight. The piston in a **piston corer** retracts as the cylinder penetrates the sediments. The jaws of a **grab sampler** seizes a portion of the sea bottom for retrieval. A **sidewall corer** obtains a sample from the borehole wall by firing a hollow cylindrical bullet from a tool suspended in the borehole. Cores are also obtained from the bottom of a borehole with a **wireline corer,** the core barrel being retrievable without having to trip out of the hole. A **core barrel** is a hollow cylinder attached to a special bit, used to obtain a continuous core section from the bottom of a borehole.

core slicer: A device using diamond-edged blades which cuts a triangular core about 1 inch on a side and up to 3 ft long from the side of a smooth borehole.

Coriband: A computer-produced log of secondary porosity, grain density, water saturation, porosity, fluid analysis, and formation analysis, calculated from resistivity, density, neutron, and sonic log data. Schlumberger trademark.

Coriolis acceleration: The acceleration of a body in motion with respect to the Earth resulting from the rotation of

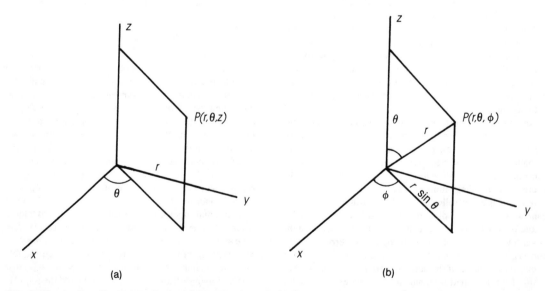

FIG. C-13 a, b. **Coordinate transform**. (a) Rectangular-cylindrical conversion; and (b) rectangular-spherical conversion. Unit vectors in x, y, z, Θ, Φ, directions are indicated by **i, j, k,** Θ, Φ.

	Rectangular coordinates	Cylindrical coordinates (a)	Spherical coordinates (b)
Conversion to rectangular coordinates		$x = r\cos\theta,\quad y = r\sin\theta,\quad z = z$	$x = r\cos\varphi\sin\theta,\quad y = r\sin\varphi\sin\theta,\quad z = r\cos\theta$
Gradient of ψ	$\nabla\psi = \frac{\partial\psi}{\partial x}\mathbf{i} + \frac{\partial\psi}{\partial y}\mathbf{j} + \frac{\partial\psi}{\partial z}\mathbf{k}$	$\nabla\psi = \frac{\partial\psi}{\partial r}\mathbf{r} + \frac{1}{r}\frac{\partial\psi}{\partial\theta}\boldsymbol{\theta} + \frac{\partial\psi}{\partial z}\mathbf{k}$	$\nabla\psi = \frac{\partial\psi}{\partial r}\mathbf{r} + \frac{1}{r}\frac{\partial\psi}{\partial\theta}\boldsymbol{\theta} + \frac{1}{r\sin\theta}\frac{\partial\psi}{\partial\varphi}\boldsymbol{\phi}$
Divergence of $\mathbf{A}$	$\nabla\cdot\mathbf{A} = \frac{\partial A_x}{\partial x} + \frac{\partial A_y}{\partial y} + \frac{\partial A_z}{\partial z}$	$\nabla\cdot\mathbf{A} = \frac{1}{r}\frac{\partial(rA_r)}{\partial r} + \frac{1}{r}\frac{\partial A_\theta}{\partial\theta} + \frac{\partial A_z}{\partial z}$	$\nabla\cdot\mathbf{A} = \frac{1}{r^2}\frac{\partial(r^2 A_r)}{\partial r} + \frac{1}{r\sin\theta}\frac{\partial(A_\theta\sin\theta)}{\partial\theta} + \frac{1}{r\sin\theta}\frac{\partial A_\varphi}{\partial\varphi}$
Curl of $\mathbf{A}$	$\nabla\times\mathbf{A} = \begin{vmatrix} \mathbf{i} & \mathbf{j} & \mathbf{k} \\ \frac{\partial}{\partial x} & \frac{\partial}{\partial y} & \frac{\partial}{\partial z} \\ A_z & A_y & A_z \end{vmatrix}$	$\nabla\times\mathbf{A} = \begin{vmatrix} \dfrac{\mathbf{r}}{r} & \boldsymbol{\theta} & \dfrac{\mathbf{k}}{r} \\ \dfrac{\partial}{\partial r} & \dfrac{\partial}{\partial\theta} & \dfrac{\partial}{\partial z} \\ A_r & rA_\theta & A_z \end{vmatrix}$	$\nabla\times\mathbf{A} = \begin{vmatrix} \dfrac{\mathbf{r}}{r^2\sin\theta} & \dfrac{\boldsymbol{\theta}}{r\sin\theta} & \dfrac{\boldsymbol{\phi}}{r} \\ \dfrac{\partial}{\partial r} & \dfrac{\partial}{\partial\theta} & \dfrac{\partial}{\partial\varphi} \\ A_r & rA_\theta & rA_\varphi\sin\theta \end{vmatrix}$
Laplacian of ψ	$\nabla^2\psi = \frac{\partial^2\psi}{\partial x^2} + \frac{\partial^2\psi}{\partial y^2} + \frac{\partial^2\psi}{\partial z^2}$	$\nabla^2\psi = \frac{1}{r}\frac{\partial}{\partial r}\left(r\frac{\partial\psi}{\partial r}\right) + \frac{1}{r^2}\frac{\partial^2\psi}{\partial\theta^2} + \frac{\partial^2\psi}{\partial z^2}$	$\nabla^2\psi = \frac{1}{r^2}\frac{\partial}{\partial r}\left(r^2\frac{\partial\psi}{\partial r}\right) + \frac{1}{r^2\sin\theta}\frac{\partial}{\partial\theta}\left(\sin\theta\frac{\partial\psi}{\partial\theta}\right) + \frac{1}{r^2\sin^2\theta}\frac{\partial^2\psi}{\partial\varphi^2}$

(c)

FIG. C-13 c. **Coordinate transform** (cont.). (c) Vector operations in rectangular, cylindrical, and spherical coordinates. Unit vectors in x, y, z, Θ, Φ, directions are indicated by **i, j, k,** Θ, Φ.

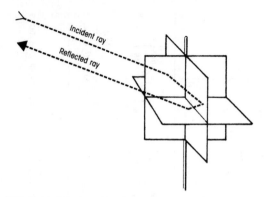

FIG. C-14. **Corner reflectors**. After three reflections, a ray emerges parallel to the incident ray regardless of the approach direction of the incident ray.

the Earth, as seen by an observer on the Earth. The Coriolis acceleration on a body moving on the surface of the Earth with a velocity V is $2\omega V \sin \phi$ where $\omega =$ angular rotational velocity of the Earth and $\phi =$ latitude. A Coriolis acceleration of a moving gravimeter is involved in the *Eötvös effect* (q.v.). Named for Gustave Gaspard Coriolis (1792-1843), French mathematician.

corner reflector: A radar reflector made of sheets of metal or metal screen at right angles to each other. It reflects like a mirror at normal incidence no matter from which direction it is viewed. See Figure C-14.

corrected tape: A magnetic tape copied from a field tape after making certain corrections.

correction: A quantity that is applied to a measured quantity to negate known effects, that is, to reduce a measurement to some arbitrary standard.

correlation: 1. Identifying a phase of a seismic trace (or record) as representing the same phase on another trace (or record), indicating that the events are reflections from the same stratigraphic sequence or refractions from the same marker. **2.** The degree of linear relationship between a pair of traces; a measure of how much two traces look alike or the extent to which one can be considered a linear function of the other. The time-domain concept analogous to coherence in the frequency-domain. See *autocorrelation* and *crosscorrelation function*. **3.** Determination of equivalence in stratigraphic position of formations; for example, in different wells based upon similarities in well-log character. **4.** The matching of different well logs and other well data either in the same well or in different wells.

correlation coefficient: A measure of the goodness of fit of one function to another. A normalized crosscorrelation; see *crosscorrelation function*.

correlation filter: A *matched filter* (q.v.).

correlation ghost: Nonlinearities in generating a Vibroseis signal introduce second harmonics. The correlation of reflected second harmonics with the generated frequency produces correlation ghosts which follow the signal for a downsweep but anticipate it for an upsweep.

correlation method: A seismic method of shooting isolated profiles and correlating events to learn the relative

structural positions of reflection horizons. The correlation is often based on similarities in the character of events and in the intervals between events.

correlation shooting: See *correlation method*.

correlator: 1. That which one correlates with; either a function or a device. **2.** *Matched filter* (q.v.).

correlogram: Graph of the autocorrelation function for positive time shifts.

COS: *Common-offset stack* (q.v.).

cosine law: In any plane triangle with sides a, b, and c and the angle C between sides a and b,

$$c^2 = a^2 + b^2 - 2\,ab \cos C.$$

If $C > 90$ degrees, the cosine is negative and the last term becomes additive. See Figure S-9.

cosine transform: The *Fourier transform* (q.v.) of the even or symmetrical part of a function. The **sine transform** involves the odd or antisymmetrical part.

cosmetics: The appearance of a record section apart from its information content. Cosmetic procedures are designed to improve the appearance of the section and ease of interpretation rather than signal/noise.

cospectrum: See *cross-spectrum*.

Coulomb's law: 1. A force F exists between electrical charges Q_1 and Q_2 which are separated by the distance r. The force is attractive for charges of unlike sign and repulsive for charges of like sign:

$$F = kQ_1Q_2/r^2;$$

k is 9×10^9 newton m^2/coulomb^2. **2.** A similar relationship between magnetic poles Q_1 and Q_2; k is 10^{-7} webers/amp m, where Q is in ampere seconds and r in meters. The coulomb is a unit of electric charge named for Charles Augustin Coulomb (1736-1806), French physicist.

coupled wave: A mode of wave propagation which involves the transfer of energy back and forth between two different wave-propagation modes with the same apparent phase velocity; a **C-wave**.

coupler: 1. The telephone cradle used in connecting a computer or teletype unit with a telephone line, such as used in time-share computer connections. See *acoustic coupler*. **2.** A device for connecting charges together to make a larger explosive.

couplet: *Doublet* (q.v.).

coupling: Interaction between systems. **1.** A device for fastening together, as the plugs for connecting electrical cables. **2.** Aspects which affect energy transfer. Thus the "coupling of a geophone to the ground" involves the quality of the plant (how firmly the two are in contact) and also considerations of the geophone's weight and base area, because the geophone-ground coupling system has natural resonances and introduces a filtering action. **3.** The type of mutual electrical relationship between two closely related circuits. **Ac coupling** would exclude dc voltages by employing a series capacitive element. **Dc** or **direct coupling** may exclude higher frequency signals by using a capacitive element shunted across the inputs, or it may allow all components to pass. **Capacitive coupling** may occur because of mutual capacitive impedance, as between the wires in IP circuits or between a wire and ground. **Inductive coupling** occurs because of mutual inductive impedance, such as between grounded IP transmitter and receiver circuits, especially at higher frequencies, greater distances, or lower earth resistivity. This may give rise to false IP anomalies. Also called **electromagnetic** or **EM coupling**. **Resistive coupling** in IP surveying is due to leakage between wires, between a wire and ground, or through the resistance of the ground itself between two grounded circuits.

covariance: A measure of the common variance between two quantities; a crosscorrelation function which is not normalized. A mean of zero is implied.

cpi: Characters per inch.

CP/M: Acronym for control program for microprocessors. An operating system used on many microprocessors.

cps: Cycles per second; hertz.

CPU: *Central processing unit* (q.v.).

crab: To maintain an angle between a ship's heading and the desired course, such as to compensate for a crosswind or cross sea. Compare **yaw** which is oscillation of the ship's heading.

Cramer's rule: The solution to a set of linear simultaneous equations,

$$a_{11}x_1 + a_{12}x_2 + \cdots + a_{1n}x_n = b_1$$
$$a_{21}x_1 + a_{22}x_2 + \cdots + a_{2n}x_n = b_2$$
$$\cdots \qquad \cdots$$

is the determinant of the coefficients obtained by replacing a row with the b_i, divided by the determinant of the coefficients (sometimes called "Δ").

$$x_1 = (1/\Delta) \begin{vmatrix} b_1 \, a_{12} \cdots a_{1n} \\ b_2 \, a_{22} \cdots a_{2n} \\ \cdots \qquad \cdots \end{vmatrix}$$

$$x_2 = (1/\Delta) \begin{vmatrix} a_{11} \, b_1 \cdots a_{1n} \\ a_{21} \, b_1 \cdots a_{2n} \\ \cdots \qquad \cdots \end{vmatrix}$$

$$\Delta = \begin{vmatrix} a_{11} \, a_{12} \cdots a_{1n} \\ a_{21} \, a_{22} \cdots a_{2n} \\ \cdots \qquad \cdots \end{vmatrix}$$

Method enunciated by Gabriel Cramer about 1750. This

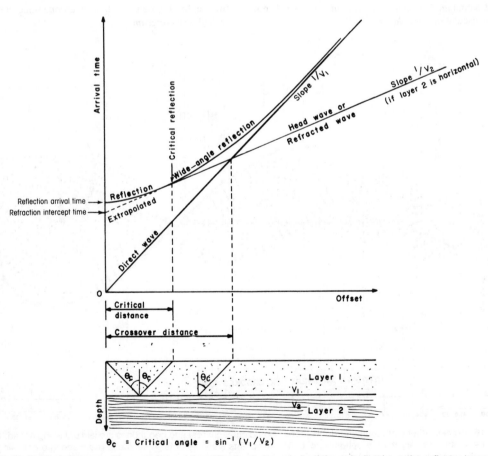

FIG. C-15. Critical distance. The curvature of the reflection is hyperbolic if the velocity above the reflector is constant. The reflection amplitude may be exceptionally large in the vicinity of the critical reflection.

is usually not an economical way for computers to solve simultaneous equations.

crater: A large funnel-shaped cavity at the top of a borehole, usually resulting from the detonation of a shot in the borehole.

creep: Gradually increasing deformation of a body under a continuous load. Usually associated with high temperature and pressure. Creep often becomes important at 40–50 percent of the temperature at which a phase-change occurs.

crest: 1. The highest point on a structure. **2.** The **peak** of a seismic (or other) wave.

crew: *Party* (q.v.).

critical angle: Angle of incidence θ_c for which the refracted ray grazes the surface of contact between two media (of velocities V_1 and V_2):

$$\sin \theta_c = V_1/V_2.$$

See Figure C-15. Has meaning only where $V_2 > V_1$. In general, four critical angles can be defined for the ratios of *P*- and *S*-wave velocities in the two media:

$$V_{P1}/V_{P2}, \ V_{S1}/V_{S2}, \ V_{S1}/V_{P2}, \ V_{P1}/V_{S2}.$$

Usually the first is intended unless otherwise specified.

critical damping: The minimum damping which will not allow oscillation. See *damping*.

critical dip: 1. Dip in the direction opposite to the regional attitude, indicating a closure. **2.** Dip at the most critical point which is necessary to establish closure.

critical distance: 1. The offset at which the reflection time equals the refraction time, that is, the offset for which reflection occurs at the critical angle; see Figure C-15. **2.** Sometimes incorrectly used for *crossover distance* (q.v.), the offset at which a refracted event becomes the first break.

critical reflection: A reflection at the critical angle. Amplitude may be exceptionally large in this vicinity. Reflection at angles in this vicinity is called **wide-angle reflection**. See Figure C-15.

CRM: *Chemical remanent magnetization* (q.v.).

CRO: *Cathode-ray oscilloscope* (q.v.).

Crone shootback: See *shootback method.*

crooked line: A seismic line which is not straight by a significant amount, that is, (a) where the offset (source-to-geophone) distance is sufficiently different from distances measured along the line that normal moveout corrections would be significantly in error; or (b) where cross-dip is large enough to confuse in-line dip measurements and cause significant error in dip calculations. Lines are sometimes shot deliberately in a crooked manner so that cross-dip can be determined as well as in-line dip. Figure C-16 shows a crooked-line plot, also called a **scattergram**.

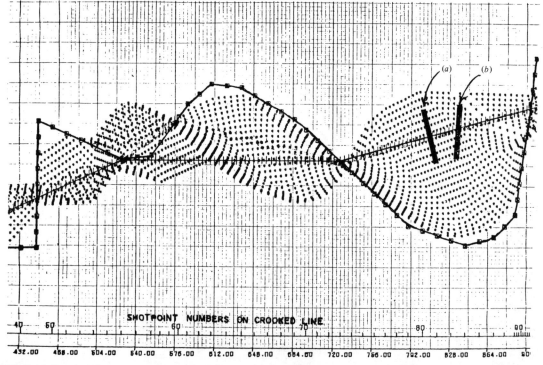

FIG. C-16. **Crooked-line** plot (computer drawn). The shotpoints (squares) and geophones are laid out along a road, there being one shotpoint every third geophone group. The midpoints show as dots. A synthetic line made in processing has cross-dashes showing the output trace spacing. The boxes show the areas of midpoints which might be combined to make a single trace by projecting (a) perpendicular to the line or (b) along the strike. (Courtesy Seiscom Delta United.)

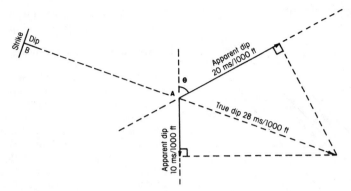

FIG. C-17. **Cross information**. Resolution of data from two nonorthogonal lines. The apparent dip seen on each line is the component of dip in the direction of the line. For observation point A, the reflecting point B is updip as indicated by the dip-strike symbol.

cross: *Cross-spread* (q.v.).

crosscorrelation filter: *Matched filter* (q.v.).

crosscorrelation function: A measure of the similarity of two waveforms, of the degree of linear relationship between them, or of the extent to which one is a linear function of the other. For two waveforms $G(t)$ and $H(t)$, the normalized crosscorrelation function $\phi_{GH}(\tau)$ is given as a function of the time shift τ between the functions by

$$\phi_{GH}(\tau) = \lim_{T \to \infty} \frac{\int_{-1/2T}^{1/2T} G(t)\,H(t+\tau)\,dt}{\left[\int_{-1/2T}^{1/2T} G^2(t)\,dt \int_{-1/2T}^{1/2T} H^2(t)\,dt\right]^{1/2}}.$$

For digital data, this becomes

$$\phi_{GH}(\tau) = \lim_{N \to \infty} \frac{\sum_{k=-N}^{N} G_k H_{k+\tau}}{\left[\sum_{k=-N}^{N} G^2_k \sum_{k=-N}^{N} H^2_k\right]^{1/2}}.$$

The denominator in the above two expressions is the **normalizing factor** and is sometimes omitted (as in Wiener filtering). When normalized, a crosscorrelation of 1 indicates perfect match, values near zero indicate very little correlation, negative values indicate a degree of match if one of the phases is inverted. Normalized crosscorrelation is also called **correlation coefficient.** See also *autocorrelation*. Unnormalized crosscorrelation can be accomplished by reversing one function in time and convolving:

$$\phi_{ab} = a(t) * b(-t).$$

The equivalent operation in the frequency domain involves multiplying the amplitudes of common frequencies and subtracting phase-reponse curves. See Sheriff and Geldart, v. 2 (1983, p. 33-35 and 168-169).

crosscorrelation theorem: The Fourier transform of the crosscorrelation of $g_1(t)$ and $g_2(t)$ is

$$\phi_{12}(t) \leftrightarrow \overline{G_1(\nu)}\,G_2(\nu) = \Phi_{12}(\nu),$$

where $G_1(\nu)$, $G_2(\nu)$ are the Fourier transforms of $g_1(t)$, $g_2(t)$, and the superscribed bar indicates a complex conjugate. $\Phi_{12}(\nu)$ is called the **cross-energy spectrum**. See Sheriff and Geldart, v. 2 (1983, p. 168-169).

crosscoupling effect: The effect in shipboard gravity measurements produced by simultaneous accelerations in two different directions.

cross dip: The component of dip in the direction perpendicular to a seismic line.

cross-energy spectrum: The Fourier transform of a crosscorrelation; see *crosscorrelation theorem*.

cross equalization: Filtering one channel to match the frequency spectrum of adjacent channels. The matching involves a phase shift as well as adjustment of the amplitude of frequency components. Tends to align coherent events better but may increase short-period reverberations.

crossfeed: *Crosstalk* (q.v.).

cross-hole method: 1. Technique for measuring in-situ *P*- and/or *S*-wave velocities by recording transit times from a source within one borehole to geophones in other boreholes. Sources may be explosive or directional to enhance either *P*- or *S*-waves, and geophones are often 3-component types. 2. Technique for resistivity or electromagnetic measurements between boreholes, used in cavity detection.

cross information: Information about the direction from which an event approaches the spread, especially the component outside the plane of the section (i.e., outside of a vertical plane which includes the line). The objective is to determine the orientation of the reflector in space; see Figure C-17. Cross information is obtained with cross-spreads, from intersecting seismic lines, from crooked-line data, or in other ways.

crossover: 1. The reversal of electromagnetic-dip direction over the apex of a conductor. The undisturbed electromagnetic field of stations in the plane of a vertical source loop is horizontal. In the presence of a subsurface conductor, the field will be horizontal over the apex but will have vertical components in opposite

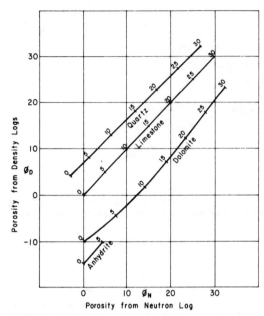

FIG. C-18. **Crossplot**. Plot of two different measurements on the same samples. In the example, porosity measured from the neutron log is plotted against porosity measured from the density log, showing the responses of different rock types. (Courtesy Schlumberger.)

directions on either side of the conductor. **2.** The intersection of two curves.

crossover distance: The source-to-receiver distance at which refracted waves following a deep high-speed marker overtake direct waves or refracted waves following shallower markers. Sometimes (erroneously) called "critical distance." See Figure C-15.

crossplot: A graph used to determine the relationship between two different measurements. For example, a crossplot of porosity measured from one type of log against porosity from another type log (Figure C-18) is used to show secondary porosity which affects the two logs differently.

crosspower spectrum: The Fourier transform of the cross-correlation function.

cross product: 1. A type of vector multiplication. If **i**, **j**, and **k** are mutually orthogonal unit vectors so that two vectors **A** and **B** may be expressed in terms of components in these directions:

$$\mathbf{A} = a_1\mathbf{i} + a_2\mathbf{j} + a_3\mathbf{k}$$

and

$$\mathbf{B} = b_1\mathbf{i} + b_2\mathbf{j} + b_3\mathbf{k},$$

then the cross product $\mathbf{A} \times \mathbf{B}$ is orthogonal to both **A** and **B**:

$$\mathbf{A} \times \mathbf{B} = (a_2b_3 - a_3b_2)\mathbf{i} + (a_3b_1 - a_1b_3)\mathbf{j}$$
$$+ (a_1b_2 - a_2b_1)\mathbf{k}.$$

Also called **outer product. 2.** The terms in an algebraic multiplication which involve elements of different kind;

e.g., $2ab$ is the **cross product term** in
$$(a + b)^2 = a^2 + 2ab + b^2.$$

cross-section: 1. A diagram showing the spatial relation of elements in a vertical plane. **2.** A geologic diagram showing the formations and structures cut by a vertical plane. **3.** A plot of seismic reflection events along a seismic line. Events are usually (but not always) migrated and the vertical scale is usually depth (but occasionally time). See *plotted section*. **4.** A concept to represent the probability of collision between particles. A particle has to pass within a certain distance of another particle for the two to interact, the effective distance depending on the type of interaction. See *capture cross-section*.

cross-spectrum: The expression of the mutual frequency properties of two time functions or series. The cross-spectrum is in general a complex-valued function and hence involves a pair of real relationships, such as the amplitude and phase as functions of frequency. The real part of the cross-spectrum is also called the **cospectrum** and the imaginary part the **quadrature spectrum**.

cross-spread: 1. A spread which makes a large angle (often a right angle) with the line of traverse. The objective is to obtain cross information, i.e., information which will allow one to determine the true direction from which the energy reaches the spread so that the true position of the reflector in space can be determined. See Figure C-17. **2.** A spread in the shape of a cross; for example, an equal number of groups laid out in-line and perpendicularly. See Figure S-17.

crosstalk: 1. **Crossfeed**, interference resulting from the unintentional pickup of one channel of information or noise on another channel. **2.** Specifically, interference between the two sides of an acoustic system such as side-scan sonar.

cross-track: Perpendicular to a seismic line.

CRP: Common reflection point. See *common midpoint*. Petty-Ray tradename.

CRT: Cathode-ray tube; also **CRO** (cathode ray oscilloscope).

crust: The outermost shell of the Earth; the portion above the Moho. The crust has a *P*-wave velocity which is usually <7 km/s and a mean density of 2.8 to 2.9 g/cm³. Often used in an imprecise way, sometimes for the lithosphere thus including the upper mantle, sometimes for only continental crust. Continental crust (**acidic crust** or **sial**) is granitic to gabbroic; oceanic crust (**basic crust** or **sima**) is basaltic. See Figure E-1.

cryogenic magnetometer: A magnetometer which operates at the temperature of liquid nitrogen. See *Squid magnetometer*.

crystal clock: A clock which uses a crystal oscillator as a reference frequency.

cu: *Capture unit* (q.v.).

cubic packing: A three-dimensional arrangement of particles. Described in rectangular coordinates, particles are centered at each location ($n\Delta$, $m\Delta$, $p\Delta$) and only at such locations, where Δ = a constant and n,m,p are integers. Cubic packing is not gravitationally stable.

cultural magnetic anomalies: Local magnetic fields caused by man-made features such as transmission and tele-graph lines, electric railways, steel drill casing, pipe-lines, tanks, etc. Also called **artificial magnetic anoma-lies**.

cultural noise: Man-made noise.

Curie: A unit of radiation equal to 3.7×10^{10} disintegra-tions per second.

Curie depth: The depth in the Earth at which the *Curie point* (q.v.) is reached, of the order of 40 km. Named for Pierre Curie (1859-1906), French physicist.

Curie point: The temperature at which a material loses its ability to retain magnetism, that is, where it changes from ferromagnetic to paramagnetic behavior. Below this temperature atoms interact so that their magnetic moments couple and behave collectively. At the **Curie temperature** the atom's thermal energy equals the cou-pling energy, and above this temperature the atomic magnetic moments are not coupled and the substance behaves paramagnetically. The Curie temperature of most rocks is approximately 550°C which is usually reached at depths of about 40 km. The analogous point with antiferromagnetic materials is the **Neel point**.

Curie's law: Magnetic susceptibility is inversely propor-tional to the absolute temperature. This law applies where dipoles are far enough apart that their interaction is small, as in solutions of paramagnetic salts. In paramagnetic solids the susceptibility is inversely pro-portional to the difference between the temperature and the Curie point, this latter fact being called the **Curie-Weiss law**.

Curie-Weiss law: See *Curie's law*.

curl: The curl of the vector **A** is given by the vector operation:

$$\text{curl } \mathbf{A} = \nabla \times \mathbf{A},$$

where ∇ is the operator *del* (q.v.). Curl **A** is expressed in rectangular, cylindrical, and spherical coordinates in Figure C-13.

current density: Current per unit cross-sectional area, determined by the velocity and density of charge carri-ers. Current density is a vector quantity, measured in amperes per square meter.

current electrode: The *A* or *B* electrode in electrical logging, resistivity or IP surveying; see Figures A-12 and E-7. Low electrical resistance of such contacts is desirable to maximize the current into the ground.

curvature: The rate of change of direction of a curve or surface; the reciprocal of the radius of curvature ρ.

$$\text{Curvature} = 1/\rho = \frac{d^2y}{dx^2}\left[1 + \left(\frac{dy}{dx}\right)^2\right]^{-3/2}.$$

See Figure C-19.

curvature of gravity: A vector calculated from torsion-balance data indicating the shape of an equipotential surface. It points in the direction of the longer radius of curvature.

curved path: A seismic raypath which is curved because refraction changes the direction of the ray as the velocity changes with depth. Increase in velocity with depth makes the raypath concave upward.

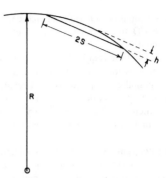

FIG. C-19. **Curvature** is the reciprocal of the radius of cur-vature. For small arcs, curvature is approximately $2h/S^2$.

curve fitting: Finding an analytic equation which approxi-mates a set of data. The most common curve-fitting technique is least-squares but other methods (such as a quadratic spline) are also used.

curve matching: An interpretation method whereby obser-vations are compared with master or *type curves* (q.v.). Achieving a close fit implies that the actual situation is similar to the model which the type curve represents; this is not necessarily true because of inherent ambigu-ity.

curve of maximum convexity: A *diffraction curve* (q.v.).

cut: To dilute, as may happen to drilling mud if formation water or gas enters the hole.

CUT: Coordinated universal time, same as Greenwich time.

cutoff: The frequency at which a filter response is down by a predetermined amount; usually 3 dB. The cutoff points designate the filter; e.g., an 18–57 filter has a low-frequency cutoff at 18 Hz and a high-frequency cutoff at 57 Hz.

cuttings: Material drilled out of shotholes or drill holes.

CVL: A continuous-velocity log or *sonic log* (q.v.). CVL is a Birdwell tradename.

CW: Continuous wave.

C-wave: *Coupled wave* (q.v.).

cycle: The change before a function or series repeats itself. Where the variable is time, a cycle is one **period**; where distance, a cycle is one **wavelength**. See Figure W-2

cycle breadth: *Period* (q.v.).

cycle redundancy check: An error-detection scheme, usu-ally hardware implemented, in which a check character is generated by the remainder after dividing the sum of all the bits in a block of data by a predetermined number. The remainder is recalculated later to verify that data have not been lost.

cycle skip: 1. Jumping a leg in correlating events, as may occur in matching noncorresponding peaks in automatic statics programs. **2.** In sonic logging, the first arrival may be strong enough to trigger one receiver but not the other receiver, which may then be triggered by a later cycle. The consequence is an abnormally high calculat-ed transit time.

cycle stealing: A characteristic of direct memory access devices. An input/output (I/O) device can delay CPU use of the I/O bus for one or more cycles while it accesses memory.

cycle time: 1. The time required by a computer to cycle a resource such as the arithmetic logic unit or memory. The fundamental clock period of that resource. **2.** The time required by a computer to read from or write into the system memory. If system memory is core, the read cycle time includes a write-after-read (restore) subcycle.

cyclographic diagram: A sterographic projection showing planes as great-circle intersections of a sphere. Used in three-dimensional structural representation. Also called **beta diagram**. Compare *pole diagram*.

cylindrical: 1. Having symmetry so that measurements do not depend on azimuth angle. **2.** *Two-dimensional* (q.v.).

cylindrical divergence: Decrease in the amplitude of a wave with distance because of geometrical spreading. The energy spreads out as a wavefront expands in a larger circle and hence the energy density varies inversely as the distance. Surface waves undergo cylindrical divergence whereas body waves undergo spherical divergence.

cylindrical hydrophone: A voltage is generated between the outside and inside of a hollow cylinder of piezoelectric material when subjected to radial pressure. Such hydrophones are very stable and durable and their sensitivity is independent of operating depth. Most streamer hydrophones are of this type. See Figure H-7.

D

D/A: *Digital-to-analog* (q.v.).

daisy chain: A method of propagating signals along a bus, often used in applications in which devices not requesting a signal respond by passing the signal on. The first device requesting the signal breaks the daisy-chain continuity. A daisy-chain scheme assigns priorities based on the electrical position of a device along the bus.

daisy wheel: A wheel with images of letters, numbers, and symbols used in printing the output from a computer or word processor.

damping: 1. A slowing down or opposition to oscillation due to dissipation of the oscillation energy. (a) **Critical damping** μ_c is the minimum damping which will prevent oscillation from taking place. (b) The **damping factor** μ is the ratio of the system friction to that necessary for critical damping, or the quotient of the logarithm of the ratio of two successive oscillations if the system is underdamped. The damping factor is one for critical damping, less than one for an **underdamped** system (which will oscillate), and greater than one for an **overdamped** system. See Figure D-1. (c) Most geophones are slightly underdamped, often having **optimum damping** which is 0.66 μ_c. 2. Damping is also used with reference to a site, often one containing structures, where it is concerned with natural resonances and the response to standing waves. Measuring the damping and Q generally requires the use of controlled vibrators. Site damping can be thought of as either the rate of amplitude decrease after cessation of excitation or as the decrease in amplitude response as the excitation frequency differs from the resonant frequency.

darcy: A unit of permeability; the permeability which will allow a fluid flow of 1 milliliter per second of 1 centipoise viscosity through 1 square centimeter under a pressure gradient of 1 atmosphere per centimeter. The commonly used unit is the millidarcy.

Darcy's law: A relationship for the fluid flow rate q:

$$q = (kA/\mu)\, \Delta p/\Delta x,$$

where k = permeability, A = cross-sectional area, μ = viscosity, and Δp = pressure differential across the thickness Δx.

dar Zarrouk: The name given by Maillet to resistivity parameters or curves that deal with layered anisotropic materials. The **dar Zarrouk variable** is the transversal unit resistance (ρ_T), the depth integral of the transverse resistivity perpendicular to the strata; the **dar Zarrouk function** is the longitudinal unit conductance ($1/\rho_L$), the depth integral of the conductivity parallel to the strata; and the **dar Zarrouk curve** is a plot of the mean resistivity (ρ_T) of the formation down to the depth Z, plotted versus the anisotropy coefficient times $Z(\rho_T\rho_L)^{1/2}$. See Maillet (1947).

data base: A collection of data organized and managed by a central facility.

data-base management system: A centralized computer facility to manage and provide consistent and secure access to a data base; **DBMS.**

data dictionary: A facility (usually within a DBMS) used to permit access to data by a common name.

data reduction: Transforming experimental data into a useful, ordered form, e.g., by correcting for known effects such as elevation differences, measurement system characteristics, etc.

data set: 1. A set of data. 2. A device for converting signals from a terminal into a form suitable for transmission and vice versa. 3. A named collection of data on a computer storage medium.

datum: 1. The reference value to which other measurments are referred; the arbitrary reference level to

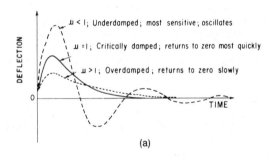

(a)

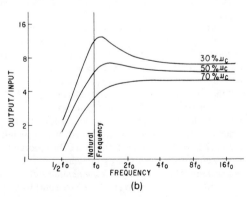

(b)

FIG. D-1. **Damping.** (a) Effect of damping on an impulsive input. (b) Effect of damping on the amplitude of a periodic input.

which measurements are corrected. **2. Elevation datum,** the reference level for elevation measurements, often sea level. **3. Seismic datum,** an arbitrary reference surface, reduction to which minimizes local topographic and near-surface effects. Seismic times and velocity determinations are referred to the datum plane as if sources and geophones had been located on the datum plane and as if no low-velocity layer existed. **4. Paleodatum,** sometimes used on stratigraphic sections to restore strata to the structural position they held at an ancient time.

datum correction: A calculation of the time required for a seismic pulse to travel from the source to the datum plane and from the datum plane to the geophone. This value is then subtracted from observed reflection times to give the arrival time corrected to datum. This is one part of the *static correction* (q.v.), the purpose being to remove the time-delay effects of the weathered layer and elevation differences.

datum elevation: A reference elevation used in mapping, usually sea level. In seismic mapping an elevation near the surface is generally used.

datuming: The arbitrary flattening and straightening of a particular reflection for use as a reference on a cross-section. The result emphasizes differences between this and other reflections.

datumized section: See *flattened section.*

datum plane: See *datum.*

datum velocity: The velocity assumed beneath the datum surface, often the subweathering velocity.

daughter: **1.** An isotope formed by radioactive decay of a parent isotope. **2.** A precocious female offspring.

dB: *Decibel* (q.v.).

dBm: Decibels less than 1 milliwatt of power. Used, for example, in specifying sensitivity for certain input impedance, such as "50 dBm at 5000 ohms."

DBMS: Acronym for *data-base management system* (q.v.).

dB/octave: Unit for expressing the slopes of filter curves.

dc: Direct current.

dc coupling: See *coupling.*

dc-pulse method: See *pulse method.*

dead: **1.** Not electrically connected, as a geophone whose connection to the cable has pulled loose. **2.** Having no signal, as a dead trace.

dead reckoning: Determining position by extrapolation of the track and direction from a previously known point. Inertial navigation and Doppler sonar are sophisticated versions of dead reckoning. See *positioning.*

debug: To search for and remedy malfunctions or errors, as with instruments or computer programs.

decade: A factor of 10 (or 1/10) in frequency.

decade-normalized PFE: See *percent frequency effect.*

decatrack: A system employing microheads which permits several channels of analog information to be recorded in the space normally occupied by one channel.

decay constant: The time for an exponentially changing voltage to vary by 1/*e* (or to change 63 percent) from its initial value. Also called **time constant.**

decay curve: A graph of the decay of a quantity as a function of time. An IP voltage decay curve may be characteristic of a particular material. In theory it can be transformed to a resistivity spectrum.

decay lifetime: See *pulsed neutron-capture log.*

Decca: One of several radio positioning systems available from Decca Survey Ltd. The **Decca Navigator** is a high-power fixed installation with a range of about 300 miles which employs four frequencies in the ratio 5:6:8:9. **Survey Decca** is a similar system used for temporary installations, with a range of about 100 miles. **Two-range Decca** is a variation of Survey Decca in which the master transmitter is at the mobile station. Decca **Hi-fix** is a short-range system which involves the broadcasting of a frequency between 1700 and 2000 kHz in a time-sharing arrangement between three fixed stations. **Dectra** and **Delrac** are long-range systems which operate at 100 kHz and at 10 to 14 kHz, respectively. See also *lambda.*

decibel: A unit used in expressing power or intensity ratios: 10 $\log_{10}$ of the power ratio. Symbol: **dB.** An amplitude ratio of 2 (which represents a power ratio of 4) is equivalent to 6 dB. Also expressed as 20 $\log_{10}$ of the amplitude ratio. See also Appendix G. 1 dB = 0.1151 neper. Named for Alexander Graham Bell (1847–1922), American inventor.

decibel/octave: The change in response between frequencies an **octave** apart, i.e., between frequencies having the ratio either 1/2 or 2. Used to describe filter slopes.

decimate: To resample eliminating certain sample values. See *subsample.*

declination: **1.** The angle between geographic north and magnetic north. **2.** The angle between the celestial equator and a celestial body. Differs from celestial latitude.

decoder: A logic device which converts data from one number system to another (e.g., an octal-to-decimal decoder). Decoders are also used to recognize unique addresses (such as a device address) and bit patterns.

décollement: A detachment surface involved in structural deformation across which deformation styles differ. Usually involves bedding-plane slippage (faulting) and/or plastic flow.

decomposition: **1.** Separating effects of different kinds or attributable to different causes. **2.** Separating a potential (e.g., gravity) map into regional and residual; *residualizing* (q.v.). **3.** Removing the effects of filtering from a filtered waveform; *deconvolution* (q.v.). **4.** Deriving resistivity stratification from a kernel function.

deconvolution: **1.** A process designed to restore a waveshape to the form it had before it underwent a linear filtering action (convolution); **inverse filtering.** A data processing technique applied to seismic reflection and other data for the purpose of improving the recognizability and resolution of reflected events. The objective of deconvolution is to nullify objectionable effects of an earlier filter action. Deconvolution may mean any of a number of operations designed to remove the effects of various types of previous convolutions. Specifically, it may mean (a) **system deconvolution** to remove the filtering effect of the recording system; (b) **dereverberation** or **deringing** to remove the filtering action of a water layer; see also *Backus filter* and *gapped deconvo-*

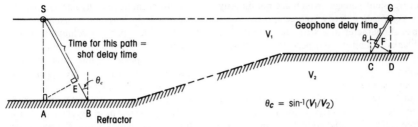

FIG. D-2. **Delay time: shot delay time** $= SB/V_1 - AB/V_2 = SE/V_1$; **geophone delay time** $= CG/V_1 - CD/V_2 = FG/V_1$. Shot delay time + geophone delay time = intercept time.

lution; (c) **predictive deconvolution** to attenuate multiples which involve the surface or near-surface reflectors. (d) **deghosting** to remove the effects of energy which leaves the source in the upward direction; (e) **whitening** or **equalizing** to make all frequency components within a band-pass equal in amplitude; (f) **shaping** the amplitude-frequency and/or phase response to match that of adjacent channels; or (g) *wavelet processing* (q.v.). Deconvolution results may vary markedly with different phase assumptions, gate location or widths, or operator lengths. See Sheriff and Geldart, v. 2 (1983, p. 32–33, 40–47). **2.** Potential maps and other data sets besides time series may be deconvolved.

Dectra: See *Decca.*

dedicated: Devoted exclusively to; for example, a "dedicated" power generator might supply the power for a seismic recording system only, another dedicated generator might power the navigation equipment, both being separate from a ship's normal power supply.

deep seismic sounding: DSS, a seismic profile (usually refraction) which has the objective of studying the crust, Moho, and upper mantle.

deflection angle: See *angles (surveying)* and Figure A-8.

deflection of the vertical: The angular difference between a plumb line (the vertical) and a perpendicular to the geodetic ellipsoid. Produced by irregularities in the Earth's mass distribution. See Figure G-2.

degaussing: Demagnetizing; removing the magnetization of a sample by placing the specimen in an oscillating magnetic field, the magnitude of which is gradually reduced to zero. The operation has to be done in a nulled magnetic field (such as may be produced with Helmholtz coils) to effect complete demagnetization. This procedure is used to **clean** magnetic tape (remove the data stored on it) so that the tape can be reused; the tape is often rotated during demagnetization to remove the effect of the Earth's magnetic field.

degeneracy: The situation where more than one eigenfunction is associated with the same eigenvalue, as where two vibration modes have the same frequency. *S*-waves have a degeneracy of 2 (*SH*- and *SV*-waves) in isotropic media.

deghosting: See *deconvolution.*

degree: The highest power which a variable assumes. For a differential equation, the power of the highest derivative. **Linear** implies that the degree is one; **quadratic,** that it is two; **cubic** three; **quartic** four, etc.

degrees of freedom: The minimum number of independent variables which must be specified to define a system.

del: The vector gradient operator; symbol is ∇. In rectangular coordinates:

$$\nabla = \mathbf{i}\,\frac{\partial}{\partial x} + \mathbf{j}\,\frac{\partial}{\partial y} + \mathbf{k}\,\frac{\partial}{\partial z},$$

where $\mathbf{i}, \mathbf{j}, \mathbf{k}$ are unit vectors in the x, y, z directions. ∇U is the gradient of the scalar potential field U. The operator ∇^2, sometimes called the **Laplacian,** appears frequently:

$$\nabla^2 = \nabla \cdot \nabla = \frac{\partial^2}{\partial x^2} + \frac{\partial^2}{\partial y^2} + \frac{\partial^2}{\partial z^2}.$$

∇V is called the **gradient,** (if V is a scalar field). As an operator on the vector field $\mathbf{V}$, $\nabla \cdot \mathbf{V}$ is called the **divergence,** and $\nabla \times \mathbf{V}$ is called the **curl.** Del is also called **nabla.** See also Figure C-13.

delay cap: A cap which detonates a fixed time after an electrical current is applied.

delayed fission neutron (log): DFN.

delay filter: See *linear-phase filter.*

delay line: A device capable of retarding a signal by a fixed time interval. (a) A **mercury delay line** uses transducers to convert electrical signals into ultrasonic patterns which are then carried across a gap containing liquid mercury to another transducer which reconverts them into electrical signals. (b) **Electrical delay line** may use capacitive and inductive elements; coaxial cables and transmission lines involve delays by the transit time through the lines. (c) Recirculating magnetic tape or disc delays signals by the amount of time needed for the tape to pass from write to read heads.

delay time: 1. In refraction work, the additional time required to traverse a raypath over the time which would be required to traverse the horizontal component at the highest velocity encountered on the raypath. Compare *intercept time.* See Sheriff and Geldart, v. 1 (1982, p. 218–223). The concept implies that the refractor is nearly horizontal under both shotpoint and detectors. Delay time is often assigned separately to the source and geophone ends of a raypath. See Figure D-2. **2.** Delay produced by a filter; see *filter correction.* **3.** Time lag introduced by a delay cap. **4.** In induced-polarization work, the time interval between the "off" instant of the charging current and the instant a measuring voltmeter oscillograph is turned "on". Delay times up to 500 or 1000 ms may be necessary to allow dissipation of transient voltages which are not directly related to the polarization decay voltage.

delimiter: A special character in a string used to denote units (e.g., blanks delimit words, periods delimit sentences, commas delimit parameters, etc.).

Delrac: See *Decca*.

delta function: An *impulse* (q.v.).

delta *t*: Moveout or stepout, usually written Δt. **1.** The time difference between the arrival times at different geophone groups. See *dip moveout* and *dip calculation*. Delta *t* ordinarily does not imply normal moveout unless specifically stated. **2.** Interval transit time, as used with the sonic log.

demodulation: The process of retrieving an original signal from a modulated signal.

Demoivre's theorem: The relationship,

$$e^{ir\theta} = (\cos\,\theta + i\,\sin\,\theta)^r = \cos\,r\theta + i\,\sin\,r\theta,$$

where $i = (-1)^{1/2}$. Named for Abraham Demoivre (1667–1754), English mathematician.

DeMorgan's theorems: See *Boolean algebra*. Named for Augustus DeMorgan (1806–1871), English mathematician.

demultiplex: To separate the individual component channels which have been multiplexed. See *multiplexed format*.

Densilog: A *density log* (q.v.). Densilog is a Dresser Atlas tradename.

densimeter: An instrument for measuring intensity of electromagnetic (usually light) radiation, as in determining albedo from remote sensing images.

density: 1. Mass per unit volume. Commonly measured in $g/cm^3 = 10^3 kg/m^3$, often without the units being expressed explicitly. Bulk rock densities vary mainly because of porosity and are generally in the range 1.9–2.8 g/cm^3. **2.** Frequency of occurrence. **3.** The equivalent position of a color on a gray scale.

density contrast: The difference in density between two formations or rock units. Lateral density contrasts are responsible for lateral changes in the earth's gravity.

density log: A well log which records formation density. The logging tool consists of a gamma-ray source (e.g., Cs^{137}) and a detector so shielded that it records backscattered gamma rays from the formation. This secondary radiation depends on the density of electrons, which is roughly proportional to the bulk density. The compensated density-logging tool (FDC) includes a second detector which responds more to the mud cake and small borehole irregularities; its response is used to correct the readings of the main detector. See Figure D-3. Sometimes called **gamma-gamma log.** Compare *nuclear cement log* and *photon log*.

density profile: A line of gravity readings taken over a topographic feature having appreciable relief which is not associated with density variations or structure, the object being to determine the best density factor for elevation corrections. The most appropriate density is the one that minimizes the correlation of gravity values

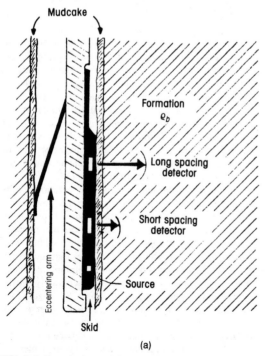

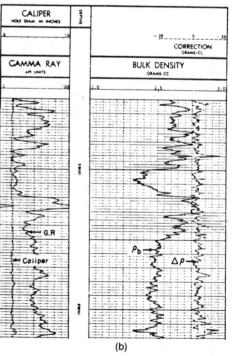

FIG. D-3. **Density log.** (a) Schematic of compensated density logging sonde. (b) Log showing the density ρ_b and the correction for mudcake, etc., $\Delta\rho$. (Courtesy Sclumberger.)

with elevation. Method devised by Lewis L. Nettleton. See **triplets** and Telford et al. (1976, p. 28-30).

dep: *Departure* (q.v.).

departure: The east or west component of a line expressed in linear units; the difference of the longitudes of the ends of the line measured at a given latitude. For a line directed toward the northeast or southeast quadrant, the departure is positive or **easting;** it is negative or **westing** for a line directed toward the southwest or northwest quadrant. Abbreviated **dep.**

departure curve: A graph which allows one to correct for measuring conditions or situations which differ from "standard." Such curves, for example, might correct well logs for differences in temperature, hole diameter, mud type, adjacent beds, invasion, etc.

depositional energy: See *energy.*

depositional remanent magnetism: DRM. See *remanent magnetism.*

depositional sequence: A stratigraphic unit composed of relatively conformable strata bounded at top and bottom by unconformities or correlative conformities.

depth controller: A device with movable vanes which fastens to a marine streamer to maintain it at a predetermined depth; see Figure D-4.

depth map: A seismic structure map that shows the vertical distance from a datum to a stratigraphic horizon in feet or meters.

depth migration: Seismic migration which allows for velocity changes in the horizontal direction. See Sheriff and Geldart, v. 2 (1983, p. 67-69).

depth of compensation: The assumed depth at which the pressure due to overlying crustal elements is constant and below which lateral density variations disappear. Involved in isostatic correction. Sometimes taken as the top of the asthenosphere. See also *isostasy.*

depth of invasion: See *invaded zone.*

depth of investigation: 1. The depth below the surface to which an exploration system can effectively explore. Depends on array design, spacing, property contrast,

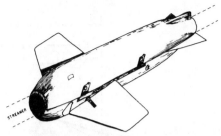

FIG. D-4. **Depth controller** is clamped around the streamer. When the hydrostatic pressure is less than the controller setting, the vane tilts so as to cause the controller to sink as it is pulled through the water. When the pressure is greater than the setting, the vane tilts the other way, causing the controller to rise. (Courtesy Conoco.)

body geometry, and signal-to-noise ratio. The maximum depth at which interfaces or the sources of anomalies are resolvable considering the signal-to-noise ratio and other measurement considerations. Also see *skin depth.* **2.** The radius about a logging sonde within which material contributes significantly to the readings from the sonde.

depth of penetration: 1. *Depth of investigation* (q.v.). **2.** *Skin depth* (q.v.).

depth phases: Waves from earthquakes which begin by traveling upward, such as *p*P, *p*S, *s*P, *s*S; **ghosts.**

depth point: 1. In reflection seismic work, the position midway between shotpoint and geophone (the **midpoint**), under which the point of reflection is located if the reflector is horizontal and if velocity layering is also horizontal. **2.** Sometimes used for the point of reflection in the subsurface. In such usage, the depth point may be different for every event, depending on the reflector's

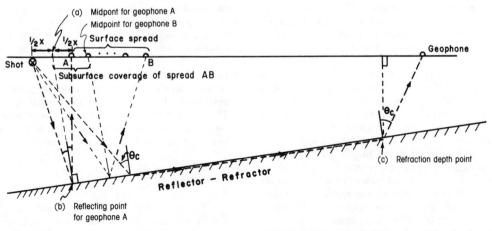

FIG. D-5. **(a)** Midpoint, **(b)** reflecting point, and **(c)** refraction **depth point**. Where the reflector dips, the reflecting point is not under the midpoint and the subsurface coverage on the reflector is not exactly the difference between midpoints.

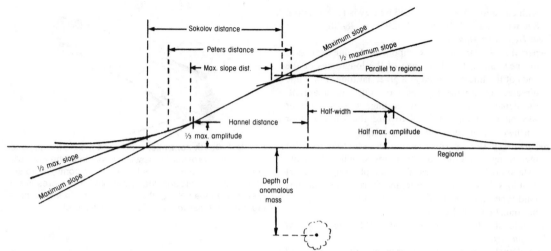

FIG. D-6. **Depth rules** showing where various measurements are made on an anomaly curve. **Sokolov distance** = horizontal distance between intersections of maximum-slope-line with regional and with line parallel to regional through the maximum; **Peters distance** = horizontal distance between points having half the maximum slope; **maximum-slope distance** = horizontal distance over which the curve is approximately a straight line having the maximum slope; **Hannel distance** = horizontal distance between points having 1/3 and maximum amplitudes; **half-width distance** = horizontal distance between points having 1/2 and maximum amplitudes.

dip and depth and the velocity distribution. **3.** In refraction work, the point for which the depth to a horizon has been calculated, usually the point where the head wave energy leaves the refractor to travel to the geophone. See Figure D-5.

depth probe: A group of profiles for which the measuring system dimensions are successively increased, designed to obtain information on the layering pattern in an area. Also called **sounding, expander,** and **depth profile. 1.** A refraction depth probe determines approximate depths and velocities of refraction markers. Also called **refraction test. 2.** An electrical-surveying technique of exploring vertically down into the earth by employing an orderly horizontal expansion of the interelectrode interval. Data from a depth probe is interpreted to give the depth to a resistivity contrast or anomalous IP material if horizontal layering exists. Also called **vertical electrical sounding (VES).** (See *two-dimensional plot.*)

depth rule: A rule relating the depth of a body to some measurement of a feature of anomaly shape. Most depth rules apply to specific source shapes such as dikes, point poles (or point masses), etc. See Figure D-6. Such rules are used in both gravity and magnetic interpretation; they include (a) **half-width rules** for the gravity effect of point masses (depth = 1.3 half-width) and horizontal line masses (depth = half-width; see Figure H-1); (b) the **straight-slope-distance rule** for the magnetic effect of dike or vertical-prism bodies (depth = the horizontal distance over which the curve approximates a straight line with the maximum slope), sometimes called the **Steenland-Vacquier rule;** (c) the **Peters' rule** for the magnetic effect of dikes (depth = ⅝ of the

horizontal distance between points where the slope is half the maximum slope); (d) the **Tiburg rule** for magnetic poles (depth = ⅔ of the horizontal distance at half the maximum amplitude); (e) the **Hannel rule** for magnetic poles (depth = half of the horizontal distance at a third the maximum amplitude); (f) the **Sokolov rule** for magnetic anomalies (depth = Sokolov distance); and other such rules.

depth section: A seismic cross-section or record section where the vertical scale is linear with depth. Usually (but not necessarily) the data have been migrated so that their horizontal and vertical locations represent as nearly as possible the true positions of features.

depth sounder: *Fathometer* (q.v.).

depth sounding: *Sounding* (q.v.).

dereverberation: *Deconvolution* (q.v.). to attenuate seismic energy which bounces in a surface water layer or other near-surface layer. Also called **deringing.** See also *Backus filter.*

deringing: See *dereverberation.*

derivative map: A map of one of the derivatives of a potential field, usually the second vertical derivative. The objective of a derivative map is to emphasize short wavelength (high frequency) anomalies. Second derivative maps are based on Laplace's equation:

$$\frac{\partial^2 \phi}{\partial z^2} = -\left[\frac{\partial^2 \phi}{\partial x^2} + \frac{\partial^2 \phi}{\partial y^2}\right].$$

The horizontal derivatives, $\partial^2 \phi / \partial x^2$ and $\partial^2 \phi / \partial y^2$, are usually estimated by finite difference methods from values measured at gridded points on a map, often using a residualizing template based on a polar representation of the Laplacian or by two-dimensional convolution with such a template.

Descartes' law: *Snell's law* (q.v.). Named for René Descartes (1596-1650), French philosopher and scientist.

design gate: The aperture or window which contains the data from which parameters are to be determined.

destructive interference: See *interference*.

det: 1. Detonator; an explosive *cap* (q.v.). **2.** *Determinant* (q.v.).

detail log: A borehole log plotted at a scale larger than 1 inch per 100 ft; specifically, an electric log at a scale of 5 inches per 100 ft.

detail survey: A survey run after a prospect has been located, the objective of which is to define details of the prospect.

detectable limit: The minimum thickness for a bed to give a seismic reflection; approximately 1/30 the dominant wavelength. Compare *resolvable limit*.

detection logging: See *exploration logging*.

detector: 1. A device which senses or measures a phenomenon. **2.** A *geophone* (q.v.).

determinant: A scalar function of a square matrix (**A**):

$$\det \mathbf{A} = \sum_{i=1}^{N} a_{ik}A_{ik},$$

where A_{ik} is the cofactor of the element a_{ik}. The cofactor is $(-1)^{i+k}$ times the matrix found by deleting the ith row and the jth column.

deterministic: From a certain set of causes, a unique situation will develop. As opposed to **probabilistic,** which leads only to the probability that certain situations will follow.

deterministic deconvolution: Deconvolution where the particulars of the filter whose effects are to be removed is known.

detonating cord: An explosive cord. A detonation at one end starts an explosion wave traveling down the cord, detonating other explosives which may be attached to the cord.

detonator: *Cap* (q.v.).

detrital remanent magnetism: See *remanent magnetism*.

development well: A well drilled within an area believed to be productive of oil, gas, or other economic resource previously discovered by an exploratory well. Compare *wildcat well*.

deviation: 1. In drilling, departure of a borehole from vertical. See *drift, directional survey,* and *rectify*. **2.** Angle with the vertical.

dextral: Rotation to the right or clockwise. A dextral strike-slip fault is also called right lateral. Opposite is **sinistral.** See Figure F-2.

DFN: Delayed fission neutron (log).

DFS: Digital (seismic) field system. Texas Instruments tradename.

DHD, DHI: Direct hydrocarbon detector or indicator; see *hydrocarbon indicators*.

diagnostic check: A routine designed to locate malfunctions.

dialog mode: See *interactive*.

diamagnetic: Having net negative magnetic susceptibility and a permeability less than that of free space (less than

unity in the cgs system). The motion of an electron about a nucleus produces a miniature circular current whose magnetic-moment vector precesses around an applied external field. This additional periodic motion produces a magnetic moment opposite in direction to the applied field. Diamagnetic effects in the earth's field rarely exceed 1 gamma. The strongest diamagnetic anomalies are probably due to salt domes. Compare *paramagnetic* and *ferromagnetic*.

diamond array: A type of geophone or source point array in which the elements are laid out on a grid of lines at about 45 degrees to the seismic line, the pattern having the general shape of a diamond. See Figure A-14.

diapir: A flow structure whose mobile core has pierced overlying rocks. Salt and shale are the most common sedimentary rocks involved in diapirs. Intrusive rocks can also form diapir-like features but 'diapiric'' is usually restricted to plastic flow.

die-away: See *pulsed neutron-capture log*.

dielectric constant: A measure of the capacity of a material to store charge when an electric field is applied. It is the dimensionless ratio of the **capacitivity** (or **permittivity,** the ratio of the electrical displacement to the electric field strength) of the material to that of free space.

dielectric log: A class of high-frequency electric logging sondes which operate at a single frequency in the MHz to low GHz range to measure formation effects of phase-shift, amplitude, and attenuation on a transmitted electromagnetic wave. Phase shift, related to dielectric permittivity, is treated as propagation time t_p, a quantity virtually independent of salinity for water but significantly lower for water than for oil, gas, or rock materials. If lithology is known, t_p (corrected for attenuation) may allow porosity determination. The dielectric log offers a means to calculate residual hydrocarbon saturation in the shallow flushed zone.

dielectric loss: The energy loss per cycle in a dielectric material due to conduction and slow polarization currents or other dissipative effects.

dielectric polarization: The response of a dielectric material to an electric field, producing an induced dipole-moment per unit volume. In an insulating dielectric material, no net electric charge need be transferred by the exciting field. By some definitions, induced polarization is a lossy type of dielectric polarization with a long time constant.

dielectric susceptibility: See *electric susceptibility*.

differential: 1. A difference between quantities. Thus, a differential voltmeter measures the difference between voltages. **2.** A differential input on a voltmeter helps reject noise that originates from the ground. See *common-mode rejection*.

differential compaction: Uneven settling of sediments as a result of loss of porosity. Differences in the irreversible volume change rocks suffer when put under pressure, as by the weight of sediments deposited on top of them. Reefs, for example, are often less compactible than surrounding shales; the greater compaction of the shales thus results in drape structures over the reef, the amount of the vertical expression of the drape features becoming gradually smaller with height above the reef.

differential curvature: For a gravitational equipotential surface, the difference between the curvature of the surface in the direction in which it curves the most and the curvature at right angles to this direction, multiplied by the gravitational constant. Measured by the torsion balance.

differential normal moveout: 1. The difference between the normal moveouts of adjacent channels. **2.** Sometimes refers to *residual normal moveout* (q.v.), the normal moveout which remains after an incorrectly assumed amount is removed. **3.** Also may refer to the difference between the normal moveout for primary events and that for multiples.

differential Omega: See *Omega*.

differential pressure: The difference between the pressure on a rock because of the weight of the overburden and

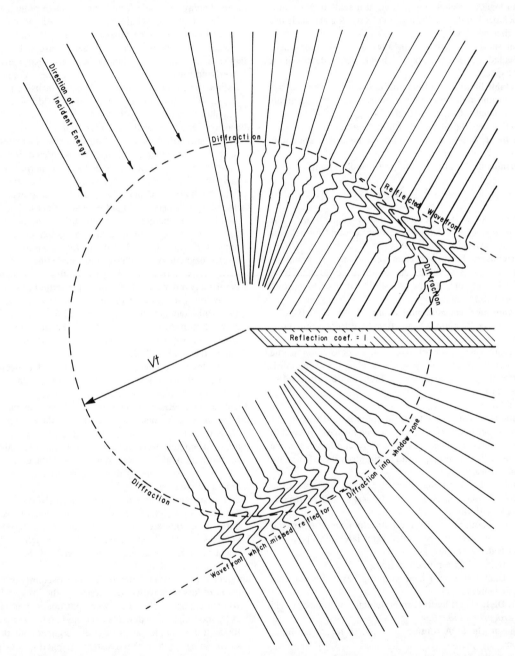

FIG. D-7. **Diffraction** of a plane wave from a semiinfinite barrier. Schematic diagram showing amplitude at time *t* after the onset of the plane wave struck the point of the barrier. Shows the reflected wavefront, the wavefront which missed the reflector, and diffractions from the reflector termination. (Courtesy Chevron Oil Co.)

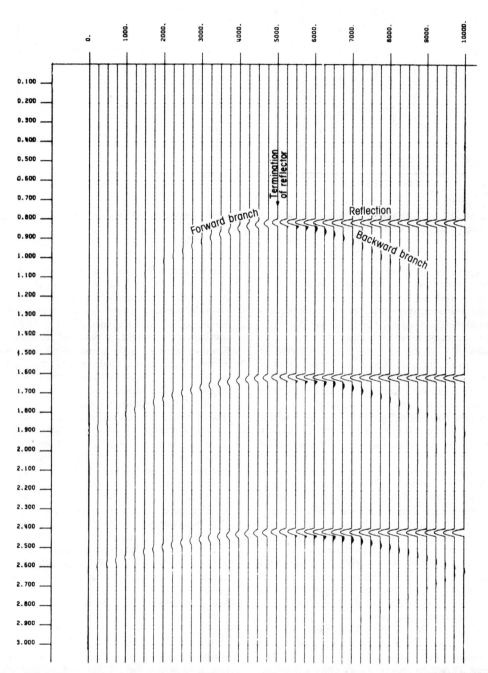

FIG. D-8. **Diffractions** generated by termination of three flat reflectors. The diffraction curvature becomes smaller with depth. The diffraction branch under the reflection (the **backward branch**) is the inverse of the branch which extends beyond the reflection (the **forward branch**). The crest of the diffraction curve locates the diffracting point and the diffraction curvature depends on the depth and the velocity above the diffracting point. The reflection amplitude decreases to one-half at the point where the reflection is tangent to the diffraction curve, and diffraction-curve amplitude is antisymmetric about this point of tangency. Amplitudes and waveshape are continuous at the point of tangency. (Courtesy Chevron Oil Co.)

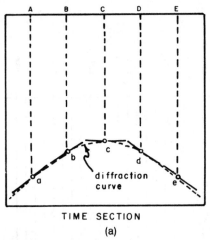

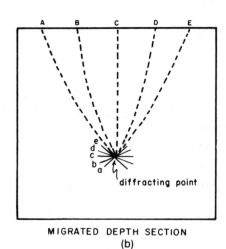

TIME SECTION   MIGRATED DEPTH SECTION

(a)   (b)

FIG. D-9. Migration of a **diffraction** by segments. (**a**) A diffraction approximated by line segments. (**b**) On a migrated section the segments form a **diffraction knot**.

the pressure of the fluid in the rock's pore space. Differential pressure is an important factor in determining seismic velocity.

differential weathering correction: The difference between the weathering corrections at two locations; for example, at two nearby geophone groups.

differentiation: 1. A mathematical operation giving the rate of change (slope) of a function with respect to some variable. The equivalent operation for discrete series is convolution with the operator $[-1, +1]$. **2.** Separation according to some criteria, such as particle size.

diffracted reflection: The diffraction resulting from reflected energy striking a diffracting point; its curvature is appropriate to the diffracting point depth, not to the arrival time.

diffraction: 1. Penetration of wave energy into areas forbidden by geometrical optics, e.g., the bending of wave energy around obstacles without obeying Snell's law, explained by Huygens' principle. The phenomenon by which energy is transmitted laterally along a wave crest. When a portion of a wave train is interrupted by a barrier, diffraction allows waves to propagate into the region of the barrier's geometric shadow. See Figure D-7. **2.** An event observed on seismic data produced by diffracted energy: see Figure D-8. Such events result at the termination of reflectors (as at faults) and are characterized on seismic records and sections by a distinctive curved alignment. A simple diffraction lies along a diffraction curve (which depends on the velocity distribution above the diffracting point). **Phantom diffractions** involve energy which reaches the diffracting point by a longer route than through the direct one (as with a diffracted reflection); they have more curvature than appropriate to their arrival time. Diffractions generated by a line source which is not at right angles to the line appear to have less curvature, becoming flatter as the line generating the diffraction becomes parallel to the line of observation. Diffracted energy shows greater curvature than a reflection (ex-

cept for reverse branches where there are buried foci). A reflection can be thought of as the interference result of diffractions from points lying on the reflector. When correctly migrated, a simple diffraction collapses at the location of the diffracting point (Figure D-9). See Sheriff and Geldart, v. 1 (1982, p. 59-64 and 102-105).

diffraction curve: A **curve of maximum convexity,** the relation between arrival time and observer position for energy which has traveled from a diffracting point. See Figure D-10. One should actually speak of a diffraction "surface" to emphasize the three-dimensional aspect. The curvature of reflected energy cannot exceed this curvature (except for reverse branches and certain situations such as diffracted reflections). Diffraction curves are specific for a particular velocity function, like the wavefront chart to which they are related and from which they can be constructed. Diffraction curves are used in identifying simple point diffractions and locating the diffracting points (see Figure M-7) and in determining velocity from diffraction curvature. Errors in interpreting diffractions can result if the diffracting point lies to the side of the seismic line, if the diffraction event results from a line diffractor which is not normal to the seismic line, or if the diffraction is not simple. See Hagedoorn (1954).

diffraction function: The function

$$(\sin \pi x)/\pi x = \text{sinc } \pi x.$$

diffraction knot: Where *diffraction* (q.v.) energy is approximated by straight line segments and migrated properly, the migrated segments cluster in an asterisk-like "knot." See Figure D-9.

diffraction overlays: A set of diffraction curves constructed for a specific velocity function. Transparent overlays are used to identify diffraction events on seismic sections.

diffraction stack: A weighted stack of all the elements along a diffraction curve, which yields a migrated section (to the extent that the data were two-dimensional); the **Kirchhoff method of migration** of reflection seismic data.

diffuse layer: The outer, more mobile ions of an electrolyte-solid interface which together with the fixed layer constitutes a **double layer**. Also called **diffuse zone, diffuse double layer,** or **outer Helmholtz double layer.**

diffusion: 1. The motion of ions or molecules in a solution resulting from the presence of a concentration gradient. 2. A method of heat conduction resulting from the motion of molecules.

diffusion impedance: See *Warburg impedance.*

diffusion rate law: See *Fick's law.*

Dighem: A helicopter-towed electromagnetic prospecting system using a suspended bird containing a vertical transmitting coil and three orthogonal receiving coils. Tradename of Dighem Ltd.

digital: Representation of quantities in discrete (quantized) units. A digital system is one in which the information is contained and manipulated as a series of discrete numbers, as opposed to an *analog* (q.v.) system, in which the information is represented by a continuous flow of the quantity constituting the signal.

digital clipping: Loss of the most significant bit of a number, such as produced by overflow. Produces a different result from ordinary clipping and generates spurious high frequencies. See Figure C-6.

digital computer: See *computer.*

digital filter box: *Convolver.* (q.v.).

digital recording: Any method of recording data in digital form, such as a series of magnetized or nonmagnetized spots coded to represent numbers.

digital-to-analog: Conversion of a digital (usually binary) number into a corresponding voltage. Often written **D/A.**

digitize: To quantize. 1. To sample a continuous voltage at discrete regular time intervals, quantize the measurements, and record the values as a sequence of numbers in bit combinations on magnetic tape. 2. To sample a function regularly. Equivalent to multiplying the function by a *comb.* 3. To convert coordinates and other parameters to a form that can be read by a digital computer, as with an *X-Y reader* (q.v.).

digitizer: Equipment for sampling curves, seismic traces, or other data recorded in analog form.

dihedral angle: The angle between two intersecting planes.

dike: A tabular body; igneous rock in the form of a slab of finite thickness which is extensive in other directions, which cuts across adjacent rock. Vertical and dipping dike models are commonly used in potential field calculations. See Figure M-9.

dilatancy: The phenomenon of volume increase due to crack opening when a rock is under triaxial loading.

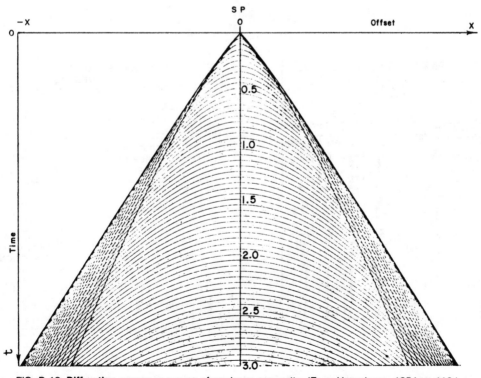

FIG. D-10. **Diffraction curves** or curves of maximum convexity. (From Hagedoorn, 1954, p. 116.)

dilatation: 1. Change in volume per unit of volume. **2.** A *rarefaction* (q.v.).

dilatational wave: *P-wave* (q.v.).

dimensional analysis: Equating units in a physical relationship so that the dimensions as well as the number values balance.

dimensionless induction number: See *induction number*.

dimensionless units: Ratios which do not depend on the units in which quantities are measured. For example, distance is often measured in terms of wavelengths, frequency is often expressed as a ratio to natural frequency, etc. Often the same as **normalized units;** see *normalize*.

dimple: A shallow velocity anomaly (such as might result from local permafrost variation) which depresses or raises all seismic data seen through it and distorts the normal moveout of deeper events by velocity focusing.

dim spot: A local decrease of the amplitude of a seismic event. Where a significant acoustic impedance contrast occurs in the absence of hydrocarbons, the presence of hydrocarbons may lessen the contrast and hence the amplitude of a reflection. Antonym: **bright spot.**

Dinoseis: A seismic energy source in which a plate is driven against the ground by a confined explosion of gas. An Arco Oil and Gas tradename.

diode transistor logic (DTL): A family of semiconductor logic formed by diode **gates** which are diode-coupled to the base of the output transistor. DTL logic is characterized by medium speed, low power dissipation, high drive capability, and low cost.

dip: 1. The angle which a plane surface makes with the horizontal. **2.** The angle which bedding makes with the horizontal. **3.** The angle which a reflector or refractor makes with the horizontal. **4. Apparent dip** is the angle between horizontal and the component of dip in the plane of a section. **5.** Electromagnetic *pitch* (q.v.).

dip calculation: Calculation of the dip or dip component of a reflecting or refracting interface from observations of the variation of arrival time of seismic events as the observing point is moved. May involve resolving *cross-spread* (q.v.) data. The dip angle for a reflection measured at the surface is usually less than the angle at the reflector because of raypath curvature.

dip filter: See *velocity filter*.

dip line: A seismic line which is perpendicular to the strike of reflecting interfaces of interest.

dip log: A *dipmeter* (q.v.) log. Diplog is a Dresser Atlas tradename.

dipmeter: 1. A well log from which the magnitude and azimuth of formation dip can be determined; see Figure D-11a. The resistivity dipmeter includes: (a) three or more microresistivity readings made using sensors distributed in azimuth about the logging sonde, (b) a reading of the azimuth of one of these, (c) a reading of the hole deviation or drift angle, (d) its bearing, and (e) one or two caliper measurements. The microresistivity curves are correlated to determine the differences in depth of bedding markers on different sides of the hole and dip calculations are based on such correlations. **2.** Earlier types of dipmeter used three SP curves, three wall scratchers, etc. **3.** A log showing the formation dips calculated from the above, such as a tadpole plot or stick plot; see Figure D-11b.

dip migration: See *migration*.

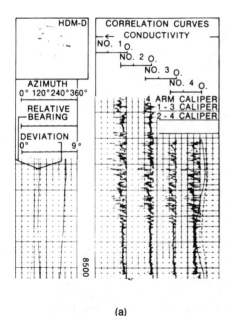

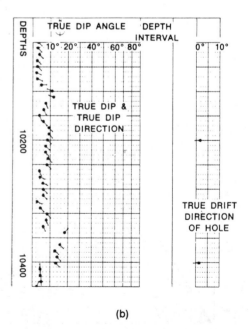

(a) (b)

FIG. D-11. **Dipmeter log. (a)** Log of field data. **(b)** Calculated dipmeter log or **tadpole plot**. (Courtesy Schlumberger.)

dip moveout: A change in the arrival time of a reflection because of dip of the reflector. The quantity $\Delta t_d/\Delta x$ in Figure R-7.

dip needle: A magnetic needle free to rotate about a horizontal axis.

dipole: 1. A pair of equal charges or poles of opposite signs, that ideally are infinitesimally close together. 2. In resistivity and IP surveying, a pair of nearby current electrodes which approximates a dipole field from a distance, or a voltage-detecting electrode pair. Where the electrode separation is large, it is sometimes called a **bipole**. 3. In electromagnetic surveying, an electric- or magnetic-field transmitting or receiving antenna which is small enough to be represented mathematically as a dipole. The near field (electric and magnetic) from a magnetic and electric dipole (respectively) varies as the inverse cube of the distance. 4. A *doublet* (q.v.) or two-stick wavelet.

dipole array: *Dipole-dipole array* (q.v.).

dipole-dipole array: In-line electrode array used in induced-polarization, electrical, and electromagnetic surveying, where both current and potential-measuring electrodes are closely spaced. See Figure A-13.

dipole field: The major part of the *magnetic field of the earth* (q.v.).

dipole moment per unit volume: A measure of the intensity of polarization of a material. Units are ampere-meters per cubic meter.

dipole strength: See *magnetic dipole.*

dip resolution: 1. Calculation of true dip from *cross-spread* (q.v.) data. 2. Recognition and separation of events having nearly the same dip.

dip shooting: A seismic field method wherein the primary concern is determining the dip of reflectors. May involve either isolated or continuous profiles or cross-spreads.

dip spectrum: 1. A graph showing the frequency with which different dips occur. 2. Sometimes used for a sonograph which displays coherent energy as a function of apparent velocity (or dip) and arrival time.

dip sweeping: See *sonogram.*

dip vector: An arrow on a map pointing in the direction of dip. Sometimes the arrow length indicates (a) the dip magnitude, sometimes (b) the contour spacing, and sometimes (c) the portion of the line over which such a dip direction is evident. Numbers written alongside the arrow may indicate the amount of dip or the rate of dip, and sometimes a grade is also indicated. Used as a method of posting data on a map prior to contouring and also as a grading method to show the distribution and reliability of the data.

Dirac comb: A *comb* (q.v.); a series of equally spaced delta functions.

Dirac function: A **delta function** or *impulse* (q.v.). Named for Paul Adrien Maurice Dirac (1902-), British physicist.

direct coupling: see *coupling.*

direct current apparent resistivity: See *apparent resistivity.*

direct detection: A measurement which may indicate the presence or absence of hydrocarbons. Sometimes considered synonomous with bright spot. Direct detection is effective under some circumstances, but no universal method has been found. Also called **direct hydrocarbon detection** and **indication (DHD** or **DHI).** See *hydrocarbon indicator.*

direct interpretation: 1. Solution of the *direct problem* (q.v.) or **forward solution.** 2. Direct mathematical solution of a potential field problem without use of precomputed curves or models.

directional charge: An explosive charge or charge array in which the explosion front travels at approximately the seismic velocity, so that energy traveling in the desired direction (usually vertically) adds up constructively as opposed to that traveling in other directions. In refraction shooting, a horizontal directional charge detonating at the refractor velocity is sometimes used to concentrate the energy traveling along the refraction path. Sometimes involves the use of broomstick charges, delay caps, impulse blasters, or sausage powder. The charge dimensions have to be at least a significant fraction of a wavelength to achieve appreciable directivity.

directional scanning: A playback process that mixes adjacent traces along various dip alignments, either directly or after inverting one of the traces. Such processes are called direct or reversed mixing. They may attenuate or enhance events with specific dips.

directional survey: 1. Measurement of **drift**, which is the azimuth and inclination of a borehole from the vertical. Often made from dipmeter survey data. Sometimes involves a continuous log and sometimes measurements made only at discrete levels. 2. An IP or resistivity-survey method starting from a position such as a drill hole to find the trend direction of an anomalous subsurface body.

directivity graph: 1. A plot (often in polar coordinates) of the relative intensity versus direction of an outgoing seismic wave such as that resulting from a directional charge or from a source pattern; see Figure D-12. The directivity results from the interference of the waves from the various components of the pattern. 2. A plot of the relative response of a geophone pattern or of directivity resulting from mixing; see Figure D-12. The response is sometimes plotted in polar form, as in Figure D-12a. Directivity graphs are often specified in various units (apparent velocity for a certain frequency, apparent wavelength, frequency arriving from a certain direction, apparent velocity, etc.). The vertical scale is often logarithmic (i.e., given in dB) rather than linear. The effect on wavelets can be very different from the effect on a sinusoidal wave train so that use of steady-state patterns can be misleading. A plot in *f-k* space is a form of directivity graph showing also the frequency dependency. See also Figure C-4.

direct memory access: 1. A method of transferring blocks of data directly between an external device and system memory without the need for CPU intervention. This method increases the data transfer rate and hence system efficiency. See *cycle stealing.* Abbreviated

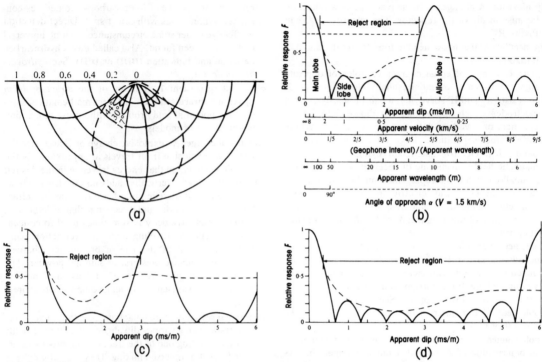

(a)

(b)

(c)

(d)

FIG. D-12. **Directivity graphs**. (**a**) Polar plot showing the relative amplitude of a radiated wave (or the relative sensitivity to waves approaching from different directions). (**b**) Directivity of 5 in-line geophones spaced 10 m apart. (**c**) Response of a tapered array of 5 geophones spaced 10 m apart and weighted 1, 2, 3, 2, 1; such weighting could be achieved with 9 geophones distributed as the weighting. (**d**) Response of 9 geophones spaced 5.5 m apart. The solid curves are for harmonic (steady-state) waves, the dashed curves for a transient with a bell-shaped spectrum peaked at 30 Hz and a width of 30 Hz. (From Sheriff and Geldart, v. 1, 1982, p. 141-142.)

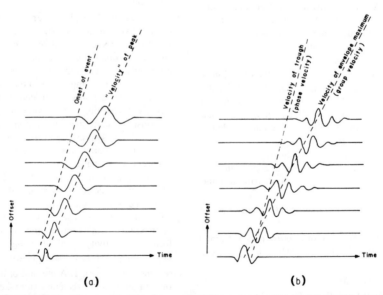

(a)

(b)

FIG. D-13. **Dispersion**. (**a**) Change in wavelet shape because of dispersion in a Voigt solid. Amplitudes have been normalized. (**b**) Change in wavelet shape because of energy shifting to later cycles. The axes of time and offset could be interchanged on either graph. (From Balch and Smolka, 1970.)

DMA. 2. Memory that can be accessed directly without going through all the memory which precedes it.

direct modeling: Calculating the effect of a *model* (q.v.).

direct problem: Computation of the effects of a certain model; as opposed to the **inverse problem** of determining the model from observation of effects. See Figure M-8. Also called **forward problem**.

direct recording: Magnetic tape recording in which the magnetization intensity is proportional to the signal strength (or signal strength plus a bias).

direct wave: A wave which travels directly by the shortest path. Other waves traveling by longer routes may arrive earlier because they travel at higher velocity.

Dirichlet conditions: The necessary and sufficient conditions for a Fourier series: In any region, (a) $f(x)$ is continuous except for a finite number of finite discontinuities, and (b) has only a finite number of maxima and minima. Named for Peter Guster Dirichlet (1805-1859), German mathematician.

dirty: Shaly; containing appreciable amount of shale dispersed in the interstices. Such shale lowers the permeability and effective porosity and affects the readings of many types of logs. A composite electrolyte system of clay and sand can be responsible for membrane polarization effects. Clay particles in the sand act as selective ion sieves, and surface conduction along the clay minerals causes low resistivity.

discette: A *floppy disk* (q.v.) also spelled **diskette**.

disc hydrophone: A piezoelectric hydrophone similar to a bender. Two piezoelectric discs are supported around their circumference so that pressure tends to bend them in, causing stresses which generate a voltage across the disc thickness.

discrete Fourier transform: A *Fourier transform* (q.v.) calculated for a wavelet over a finite interval so that values are given only for the fundamental frequency (the reciprocal of the interval) and its harmonics.

disc storage: A method of bulk storage of programs and data. The medium is a rotating circular plate coated with a magnetic material, such as iron oxide. Data are written (stored) and read (retrieved) by fixed or movable read/write heads positioned over data tracks on the surface of the disk. Addressable portions can be selected for read or write operations. Also spelled **disk.**

disharmonic folding: Folding in which there is an abrupt change in fold profile across a décollement surface.

disjunction: The logical operation, "either A or B" (written A ∪ B), performed by an OR gate. Also called **union.** See Figure G-1.

dispersion: 1. Variation of velocity with frequency. Dispersion distorts the shape of a wave train; peaks and troughs advance toward (or recede from) the beginning of the wave as it travels. Leads to the concept of **group velocity** U distinct from **phase velocity** V. Where λ = wavelength, ν = frequency,

$$V = \nu/(1/\lambda),$$

$$U = d\nu/d(1/\lambda).$$

The dispersion of seismic body waves is very small under most circumstances, but surface waves may

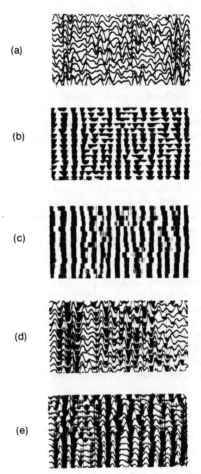

(a)

(b)

(c)

(d)

(e)

FIG. D-14. **Display modes.** (**a**) Wiggle trace. (**b**) Variable area. (**c**) Variable density. (**d**) Wiggle trace superimposed on variable area. (**e**) Wiggle trace superimposed on variable density.

show appreciable dispersion in the presence of near-surface velocity layering. See Figure D-13. The dispersion of electromagnetic body waves is large in most earth materials. With **normal dispersion**, velocity decreases with frequency; with **inverse dispersion**, velocity increases with frequency. **2.** A statistical term for the amount of deviation of a value from the norm. See *statistical measures*.

dispersion curve: A plot of wave velocity as a function of frequency. See *normal modes*.

dispersion relation: See *Hilbert transform*.

dispersive filter: A phase-shifting filter which does not affect the amplitude spectrum.

displacement: 1. The distance a particle is removed from its equilibrium position, as in the ground motion associated with a seismic wave. **2.** Relative movement of the two sides of a fault. **3.** The amount by which refraction data are displaced horizontally from the geophone positions to indicate where the head-wave energy presumably left the refractor. Sometimes called offset or transplacement. See Figure O-1.

displacement current: A current which is proportional to the time rate of change of electric flux density. In most earth materials, displacement currents are negligible compared with conduction currents for the range of frequencies used in electromagnetic methods.

display: 1. A graphic hard-copy representation of data, especially of seismic data. See Figure D-14. A graph of amplitude as a function of time for each geophone group output gives a **wiggle trace, squiggle,** or conventional display. **Variable area** display has the area under the wiggle trace shaded to make coherent events more evident; it often involves a bias and trace clipping. A **variable density** display represents amplitude values by the intensity of shades of gray. **Superimposed modes** or combined modes involve the use of both wiggle trace and variable area or variable density simultaneously; they retain many of the good features of each type display. Data are also represented by color encoding and in other ways. **2.** A soft-copy representation of data on a cathode-ray tube or similar nonpermanent device.

disseminated sulfide mineralization: Sulfide minerals scattered as specks and veinlets through rock and constituting not over 20 percent of the total volume.

distal: Referring to the portion of a sedimentary unit remote from the sediment source.

distance meter: A device for measuring line-of-sight dis-

tances, generally by transmitting a light pulse to a prism reflector at a station and timing the arrival of the reflected light.

distortion: An undesired change in waveform, as opposed to desired changes in waveshape like those from modulation. (a) **Amplitude distortion** is due to undesired amplitude-versus-frequency characteristics. (b) **Harmonic distortion** is nonlinear distortion characterized by the generation of harmonics of an input frequency. The percent harmonic distortion is a measure of fidelity; if E_f is the rms voltage of the fundamental and E_n is the rms voltage of the nth harmonic, the **percent harmonic distortion** is

$$100 \, (\Sigma E_n^2)^{1/2}/E_f.$$

(c) **Intermodular distortion** is nonlinear distortion characterized by the appearance in the output of frequencies equal to the sum and difference of integral multiples of the component frequencies present in the input. (d) **Nonlinear distortion** is caused by a deviation from a linear relationship between input and output. (e) **Phase distortion** results when phase shift is not linear with frequency over the band-pass or where it is linear but where the zero-frequency intercept is not a multiple of π. Where the intercept is a multiple of $n\pi$, the wave shape is not changed if n is even and is inverted if n is odd, although the entire waveform is delayed. *See linear-phase filter.*

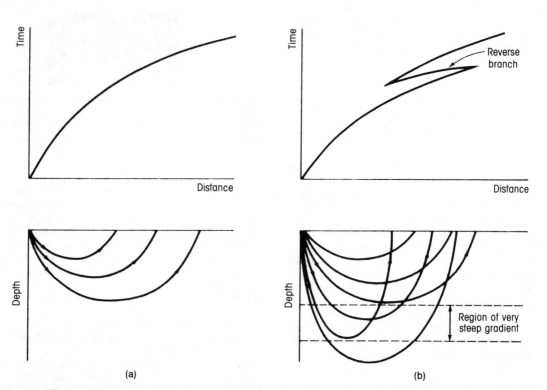

(a)   (b)

FIG. D-15. **Diving waves**. (a) Raypaths are curved because of vertical velocity gradient. The inverse slope of the time-distance curve indicates the velocity at the bottom of the travel path. (b) A large velocity gradient can produce a reverse branch to the arrival time versus distance curves. Velocity changes may be such that the time-distance curves are not continuous.

distortional wave: *S-wave* (q.v.).

distortion point: The input amplitude for which the distortion-generated third harmonic becomes a certain percentage, generally 3 percent but sometimes 1 percent. For smaller inputs the system is nearly linear.

distortion tail: A correlation ghost. Harmonic distortion with Vibroseis surface sources produces spurious correlations, especially for the second harmonic (which may be very large). Such a distortion tail follows the correlation for a downsweep and appears as a forerunner for an upsweep.

distributed: Referring to electric circuits, the smearing out of resistive, capacitive, or inductive circuit elements such as with a transmission line. Opposite of **lumped**.

distributed processing: Use of multiple, loosely coupled processing systems to accomplish a task.

distribution function: A relationship which describes the probability that a quantity will have a value less than a particular value. It is thus the cumulative integral of the probability density $P[x]$ of the random variable x (which is not necessarily a proper function):

$$F(x) = \int_0^x P[x]\ dx.$$

diurnals: Phenomena with a periodicity of about one day. Average daily changes in radio-wave propagation because of diurnal changes in the ionosphere affect radio-positioning systems. Published tables allow one to correct for major sky-wave variations with systems like Omega, but unpredictable local sky-wave variations and sun-spot effects remain.

diurnal variation: Daily fluctuations, for example, those of the geomagnetic field related principally to the tidal motion of the ionosphere (involving amplitude and phase variations with season and latitude by as much as 100 gammas).

divergence: 1. The decrease in amplitude of a wavefront because of geometrical spreading. With body waves the energy spreads out as the spherical wavefront expands, causing the energy density to vary inversely as the square of the distance (spherical divergence). With surface waves the energy density varies inversely as the distance (cylindrical divergence). Tube waves do not suffer energy loss because of divergence. Note that energy density decreases for other reasons also. 2. The divergence of a vector field is expressed in Figure C-13 for rectangular, cylindrical, and spherical coordinates.

divergence theorem: The flux ϕ through a surface (or the integral of the vector flux density g over a closed surface) equals the divergence of the flux density integrated over the volume contained by the surface. Most commonly called **Gauss theorem**.

divergent reflections: A reflection configuration (see Figure R-8) indicating differential subsidence.

diversity stack: A stack in which amplitudes which exceed some threshold are excluded. Amplitudes less than this threshold may not be affected. Used with vertical stacking to prevent occasional large bursts of noise (such as traffic noise) from dominating the stacked record, and also used with common-midpoint stacking to discriminate against ground roll and similar high-amplitude wave trains. 2. A stack in which components are weighted inversely as their mean power over certain intervals.

diving waves: Refraction in a strong velocity-gradient zone may reverse the downward component of seismic ray travel and bend the rays back to the surface. Such refraction arrivals have apparent velocity appropriate to that at their greatest depth of penetration even though they have no appreciable path through a distinctive refractor. See Figure D-15. If there are no velocity reversals, diving waves may be used to derive the velocity distribution by means of the Wiechert-Herglotz integral. See Meissner (1966) and **Blondeau method.**

divining: Use of a forked stick which allegedly bends toward water, petroleum, or other sought-for accumulations; used meaning "unscientific."

Dix formula: For reflections from a sequence of flat, parallel layers and small offsets, the interval velocity in the nth layer V_n is given by

$$V_n = [(\overline{V}_n^2 t_n - \overline{V}_{n-1}^2 t_{n-1})/(t_n - t_{n-1})]^{1/2},$$

where $\overline{V}_{n-1}$ and $\overline{V}_n$ are the stacking velocities from the datum to reflectors above and below the layer and t_{n-1} and t_n are reflection arrival times. This formula is often misused to calculate interval velocities in situations which do not satisfy Dix's assumptions. See Dix (1955).

D-layer: 1. The innermost layer of the ionosphere. It occurs at heights of 50 to 90 km during daylight hours only, reflects ELF, VLF, and LF waves, absorbs MF waves, and partially absorbs HF waves. Some consider the D-layer as starting at 70–80 km and merging with the E-layer. See Figure A-16. 2. A layer within the earth; see Figure E-1.

DMA: *Direct memory access* (q.v.).

DNMO: *Differential normal moveout* (q.v.).

doghouse: The hut (or cab) which contains seismic recording instruments in the field.

dog-leg: An abrupt angular change in direction, as in a survey traverse or in a borehole.

domain: 1. The set of elements to which a mathematical or logical variable is limited; the set on which a function is defined; the set of values which an independent variable may take. 2. The class of terms that have a given relation to something is called the "domain of that relation". Thus, when we speak of a seismic trace in the **time domain**, we mean that time is the independent variable. In the **frequency domain**, frequency is the independent variable. In the **f,k domain** or **f,k space**, frequency and wavenumber are the independent variables; etc. 3. A region of magnetic polarization in a single direction (magnetic moments parallel) which behaves as a unit during change in magnetization. 4. The areal extent of a given lithology or environment.

dome: A structure in which the beds dip in all directions away from a central area; e.g., a salt dome.

dominant frequency: The predominant frequency determined by measuring the time between successive peaks or troughs and taking the reciprocal. See Figure W-2.

dominant wavelength: The wavelength associated with the *dominant frequency* (q.v.).

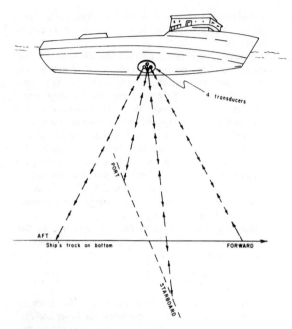

FIG. D-16. **Doppler sonar**. Four transducers send sonar beams fore and aft and to each side, and receive the reflections of these beams from the seafloor. The use of four beams allows the effects of the ship's roll, pitch, and yaw to be reduced. The ship's velocity is computed from the Doppler-effect frequency shift of the reflected beams. (Courtesy Marquardt.)

donor: See *n-type semiconductor*.

doodlebugger: A geophysicist engaged in field work. Originally used derisively to mean unscientific.

Doppler count: 1. A count of the number of cycles occurring within a certain time. The Transit satellite emits a constant frequency but the apparent frequency varies because of the velocity of the satellite with respect to the observing station (Doppler effect) (see Figure S-1). From the counts for several successive time intervals, the instant of closest approach and the distance of closest approach can be calculated. By knowing where the satellite was located at this instant, the position of the observing station then can be determined. **2.** Doppler counting is also used in connection with Doppler radar and Doppler sonar.

Doppler effect: Apparent change in frequency of a wave caused by motion of the source with respect to the receiver and/or vice versa. Frequencies are increased if the source and receiver are approaching, decreased if they are moving apart. For sound and water waves, the observed frequency ν_0 is given in terms of the source frequency ν_s, the velocity of propagation V, the velocity of the source V_s, and the velocity of the observer V_0 (components of velocity toward each other):

$$\nu_0 = \nu_s(V + V_0)/(V - V_s).$$

The relationship for light and radio waves is slightly different because of relativistic effects and the constancy of the velocity of light irrespective of motion of source and observer:

$$\nu_0 = \nu_s \left[(V + V_0 - V_s)/(V - V_0 + V_s)\right]^{1/2}.$$

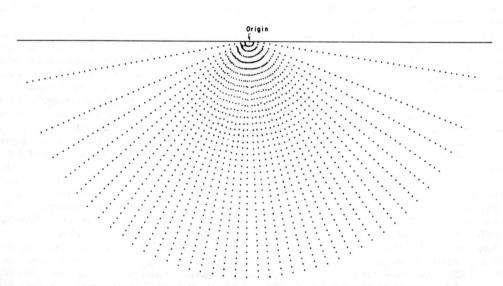

FIG. D-17. **Dot chart** (two-dimensional). The gravity effect at the origin of a mass anomaly of given cross-section is $k\rho n$, where k is a scale constant, ρ is the density contrast, n is the number of dots lying in the anomalous mass when such is superimposed on the dot chart.

See *Doppler navigation.* Named for Christian Johann Doppler (1803-1853), Austrian physicist.

Doppler navigation: Positioning in which frequency shift because of the Doppler effect is involved. Usually refers to Doppler radar (for aircraft) or Doppler sonar (for ships), occasionally to satellite navigation (see *Doppler count*).

Doppler radar: A radio-navigation system used by aircraft. A radio beam transmitted from the aircraft and reflected back by various landscape features is received at the aircraft. Because the aircraft transmitter is in motion, the signal undergoes a frequency shift (the *Doppler effect*) which is measured and converted to aircraft velocity. A computer on the aircraft determines the plane's position by integrating the velocity over time. Actually, two transmitter-receivers beam to the ground ahead of and behind the aircraft to determine the forward speed and another two beam to the right and left of the aircraft to determine the crosstrack speed. The arrangement is similar to that used with Doppler sonar (Figure D-16). The Doppler data combined with a compass heading give the direction of the aircraft. Velocity can be measured to about 0.5 percent, somewhat less over water because the surface of the water may itself be in motion.

Doppler shift: 1. Change in observed frequency because of the *Doppler effect* (q.v.). **2.** The shift toward lower frequencies which results from stretching long-offset seismic traces in the removal of normal moveout.

Doppler sonar: A sonic location system used by ships, based on the Doppler effect; see Figure D-16. Velocity measurements together with gyro-compass direction may be integrated to give position. Signals are reflected from the seafloor or from the water mass (volume reverberation or water track mode, often when the water is deeper than 600-1000 ft, resulting in poorer accuracy) depending on system power, frequency, signal characteristics, depth.

DOS: Disc *operating system* (q.v.).

dot chart: A chart used to compute the theoretical gravity (or other potential) effect of a mass distribution; see Figure D-17. A dot chart is superimposed on a scale cross-section of the mass; the number of dots which fall within the mass outline multiplied by the anomalous density is proportional to the gravity effect at the origin of the chart. The chart is then moved to a different position on the outline and the number of dots counted again to give the effect at another point, and so on for every point for which the gravity value is to be determined. Most charts assume that the mass distribution extends to infinity perpendicular to the plane of the chart and end corrections have to be applied to remove this restriction. Also called **graticule.**

dot product: The dot product of the vectors

$$\mathbf{X} = [x_1, x_2, x_3, \cdots, x_n]$$

and

$$\mathbf{Y} = [y_1, y_2, y_3, \cdots, y_n]$$

is

$$\mathbf{X} \cdot \mathbf{Y} = [x_1 y_1, x_2 y_2, x_3 y_3, \cdots, x_n y_n].$$

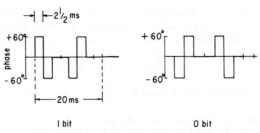

FIG. D-18. **Doublet phase-modulation** code. Used for the message from navigation satellites.

The **dot-product reverse** is

$$[x_1 y_n, x_2 y_{n-1}, x_3 y_{n-2}, \cdots, x_n y_1].$$

Compare *cross product.*

double Bouguer correction: 1. The *Bouguer correction* (q.v.) to sea level for measurements made on the ocean floor involves a correction to remove the upward attraction of the sea water above the meter and another correction to replace the sea water with the replacement density. **2.** Corrections for measurements made in mines or in boreholes are usually the result of making measurements both above and below a layer.

double coverage: *Reverse control* (q.v.).

double layer: The layers of molecular ions and charged dipoles at a solid/solution interface. It is electrically analogous to a capacitor in that there is charge separation between the solid (electrode) and the charge center of the oriented ions or dipoles. Next to an electrode there may be an adsorbed fixed layer of ions called the **inner Helmholtz double layer.** A diffuse layer (**outer Helmholtz double layer**) in the electrolyte contains an excess of ions that is usually of the same charge as the electrode but opposite to that of the fixed layer. The thickness of the double layer is less than 100 Ångstroms (10^{-8}m).

double-layer capacitance: Capacitance due to the presence of the double layer of charge at the interface between an electrode and electrolyte. In effect it is in parallel with the **Warburg impedance.**

double-layer weathering: Situation where corrections must be made for two distinctive near-surface low-velocity layers.

double precision: The retention of twice as many digits (bits) to specify a quantity as the computer normally uses.

double-run: A resurvey of a traverse to tie back to the same reference point, to reinforce the certainty that errors have not been made. Used where tie to an independent reference point is not feasible.

doublet: 1. A two-stick wavelet [*a,b*] whose *z*-transform is $a + bz$. If $|a| > |b|$, [*a,b*] is **minimum phase**; if $|a| < |b|$, **maximum phase.** Sometimes called a **dipole. 2.** See *doublet modulation.* **3.** The first derivative of the Dirac delta function.

doublet filter: A digital filter containing only two nonzero values.

doublet modulation: A phase-modulation scheme. Used to

encode the message broadcast in Transit satellite navigation. A phase shift of +60 degrees for 2.5 ms (Figure D-18) is followed by a phase shift of −60 degrees so that there is no net phase shift. A bit consists of two such doublets in opposite sense following each other by 5 ms.

Douglas sea state: A scale of sea wave heights. See Figure B-1.

downdip: The direction of the gradient, e.g., the direction of surveying in which reflectors or refractors dip toward the geophones.

downhole: Measurements made in a borehole.

downhole ground: A long electrode often attached to the logging cable some distance (perhaps 100 ft) above an electrical logging sonde or hung just below the casing, used as the reference electrode instead of a reference electrode at the surface. Used in the case of bad SP interference from electrical surface facilities or telluric currents.

downhole method: 1. Engineering seismology measurement of *P*- and *S*-wave velocities utilizing a source at the surface and a clamped triaxial geophone in a borehole. *S*-wave energy is often enhanced by use of directional sources. **2.** Induced-polarization method that explores the region near a drill hole using a single potential or current electrode in the drill hole and other electrodes on the ground surface. Compare *in-hole IP method*.

downlap: Downdip termination of strata against an unconformity at the base of a depositional unit. See Figure R-8.

downstairs: In the denominator of a fraction.

downsweep: Vibroseis sweep in which the frequency decreases with time.

downward continuation: The process of determining the value of a potential (e.g., gravitational) field at a lower elevation from values measured at a higher elevation, based on the field continuity. A potential field is not

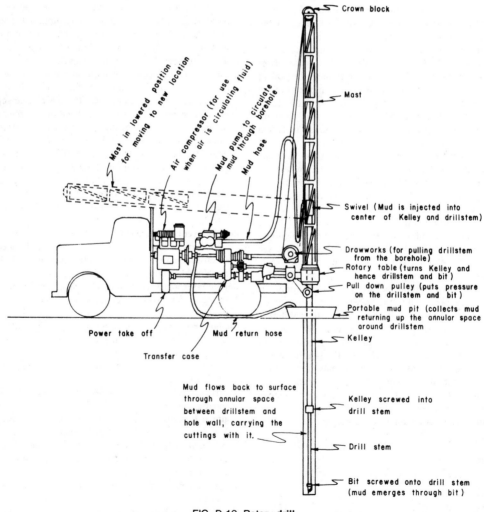

FIG. D-19. Rotary **drill**.

continuous across the boundaries of anomalous masses. As the depth from which an anomaly originates is approached, its potential field expression becomes sharper and tends to outline the mass better until the depth of the mass is reached; beyond this point the field computed by continuation becomes erratic. Noise in the data often precludes successful application. See *continuation*.

dowser: One engaged in *divining* (q.v.).

drag bit: A type of drill *bit* (q.v.). See Figure B-3.

drape: Sag in bedding around a feature such as a reef, usually as a consequence of differential compaction, sometimes because of initial dip.

drape flown: Refers to an airborne geophysical survey flown at a constant distance above mean surface elevation rather than at a constant elevation above sea level. "Mean" may be determined over different lengths.

drift: 1. A gradual and unintentional change in the reference value with respect to which measurements are made. If drift is slow and fairly steady in time, the difference produced by drift can be determined by subsequently rereading the value of the quantity being measured and prorating the difference over other readings made in between. Gravity-meter drift may be caused by gradual heating up of the meter as the day progresses, "creep" in the spring, elastic aging, hysteresis, lunar tide, etc. Drift is different from **tare**, which is a sharp, sudden change in reference value. **2.** A layer of glacial deposits. Glacial drift may vary with position and hence require a variable correction on seismic records, the effect being similar to that of a weathering layer. This often requires a double-layer weathering correction (part for the entire drift layer and part for the lower-velocity layer of the top part of the drift). **3.** The attitude of a borehole. The **drift angle** or **hole deviation** is the angle between the borehole axis and the vertical; the **drift azimuth** is the angle between a vertical plane through the borehole and north. **4.** A shoran measurement of location with respect to one fixed point. Measurement with respect to a second fixed point is called **rate** or **range**. Both drift and rate values are necessary to establish a fix.

drill: A device for boring holes. Usually means a mechanically driven rotary drill (see Figure D-19), often truck mounted but at times portable. A drill usually includes a means for rotating the drill pipe and a pump for circulation of a fluid (air or water or mud) down through the pipe to wash the cuttings away from the bit and carry them up to the surface in the annular space between the wall of the hole and the drill pipe. Water jets, augers, spudder or cable tools, and air-blast equipment are also used under certain conditions.

drill bit: See *bit*.

drill collar: A heavy thick-walled length of drill stem which is placed immediately about the bit to add weight to the bit to help it cut into formations faster. See Figure D-20.

driller's log: A record of the formations drilled through.

drilling: Electric drilling = *sounding* (q.v.).

drilling break: A change in the penetration speed of the drill bit caused by a change in formation, especially an

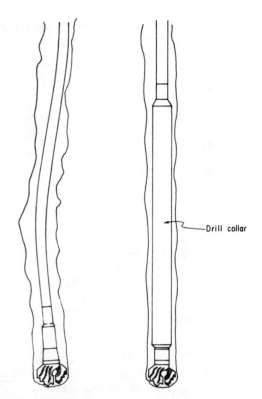

FIG. D-20. **Drill collar usage**. A heavy drill collar puts weight on the bit without bending the drill pipe. Increasing the weight on the drill bit by pushing down on the drill stem causes it to bend, resulting in a crooked hole.

increase indicative of penetration into a porous zone.

drilling-time log: A record of the time to drill a unit thickness of formation.

drill pipe: The pipe which is rotated by a drill, to which a bit is attached and through which the drilling fluid circulates.

drill-pipe log: A well log which is obtained from a logging instrument which has a self-contained recording mechanism. The log consists of an SP and short and long normals. The tool is lowered through the drill pipe and the flexible electrode assembly is pumped out through a port in the bit. The log is recorded by a tape recorder within the tool during the process of coming out of the hole. The tape is played back to obtain the log. Welex tradename.

DRM: Depositional (or detrital) remanant magnetization.

drop: 1. To drop a weight on the ground in order to generate a seismic wave; see *thumper*. **2.** To eliminate, as in *dropout* (q.v.).

drop along: Weight-drop recording for common-midpoint stack.

dropout: A loss of information upon reading or writing on magnetic tape. Usually caused by defects in the magnetic tape or dirt on the tape surface or recording head.

dropped coverage: Portion of a seismic line not shot,

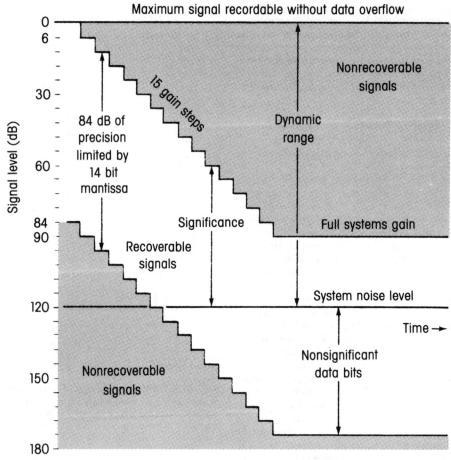

FIG. D-21. Magnitude of signals for a binary-gain recording system illustrating **dynamic range**. The shaded area represents nonrecoverable signals.

usually for permit or access reasons or because of danger of doing damage.

drop-point: The surface location where a weight drop occurs, analogous to shotpoint. See *thumper*.

Dropter: A weight dropped from a helicopter, used as a seismic source. CGG tradename.

dry hole: 1. A well judged to be incapable of producing oil, gas, or geothermal fluids in economic quantity. Operations on such a hole are terminated by its being "plugged and abandoned" (abbreviated **P&A**) instead of being "completed". **2.** Any unprofitable exploration venture.

dry steam: 1. Steam which lies above the vapor curve for water, that is, has an enthalpy greater than that for equilibrium with water at the existing pressure. **2.** Where the total mass is all steam. See *steam quality*.

DSS: *Deep seismic sounding* (q.v.).

DTL: *Diode transistor logic* (q.v.).

dual: Parallel recording from the same inputs. Parallel or dual recording is often done using different amplifier settings (different gains, filters, etc.). Refraction is often dually recorded with one recording set at high gain to yield sharp first arrivals and a parallel recording at lower gain to allow the picking of secondary arrivals.

dual induction log: An *induction log* (q.v.) consisting of two induction curves with different depths of investigation. Usually run with a resistivity device which has a shallow depth of investigation, such as a shallow Laterolog or 16 inch normal.

dual polarity section: A section in which peaks and troughs are given equal weight. With variable area, troughs are sometimes colored red and peaks blue, or some other color scheme may be used.

dummy load: 1. A load connected during calibration and maintenance. **2.** A ground-matching resistance used with pulsed-square-wave transmitters to balance the output load from the transmitter during the "power off" portion of the duty cycle.

dump: 1. To write out for examination the contents of a magnetic tape or other data storage. **2.** Unintentional shutdown of a computer, as by loss of power.

duplex: Simultaneous and independent transfer of data in both directions. Compare *half-duplex* and *simplex*.

duplication check: See *check*.

duty cycle: 1. The proportion of time a switch is "on". **2.** The percent of time in which current is delivered during a complete cycle of a transmitter (such as an IP transmitter).

DWT: A deep well thermometer, a sonde for recording temperature logs.

dyadic: A second-order tensor.

dynamic corrections: *Normal-moveout corrections* (q.v.), which depend on record time.

dynamic correlation: A velocity-analysis operation which involves crosscorrelating traces for different offsets, summing the crosscorrelations for similar pairs of traces over a number of nearby subsurface points, displaying the crosscorrelations for successive differences of offset squared, picking alignments on this display, and computing the residual normal moveout and stacking velocity for such alignments.

dynamic memory: A type of semiconductor memory in which the presence or absence of a capacitive charge represents the state of a binary storage element. The charge must be periodically refreshed.

dynamic range: 1. The ratio of the maximum reading to the minimum reading (often noise level) which can be recorded by and read from an instrument or system without change of scale. **2.** The ability of a system to record very large and very small amplitude signals and subsequently recover them. The smallest recoverable signal is often taken to be the noise level of the system, and dynamic range as the ratio of the largest signal which can be recorded with no more than a fixed amount of distortion (often 1 to 3 percent) to the rms noise; see Figure D-21. However, sometimes signals can be extracted even though they are buried in the noise. The definition sometimes considers the entire signal extraction process rather than the recording equipment only. **3.** For direct recording on magnetic tape, the noise level is for unrecorded tape; bandwidth should be specified because selected narrow bandwidths may give improved dynamic range. **4. Instantaneous dynamic range** or **significance** is the smallest signal which will cause a measurable (significant) change in the presence of a large signal. Word length imposes a limitation (13 bits represents about 84 dB).

dynamic similarity: A ratio of masses involved in physical modeling. See *modeling theory*.

dynamite: A high explosive, originally one made with nitroglycerin and a cellulose material, but now usually used for any high explosive.

Dynasource: A seismic energy source which involves an air-powered piston striking a mechanical plate. Tradename of EG&G.

E

earth: Ground, the electrical reference potential.

earth holography: Recording of the wave pattern from a constant frequency source along with a reference wave. The resulting hologram can be viewed by light to allow one to "see" the structure which generated the wave pattern. See *holography* and Hoover (1972).

Earth layering: Geophysical models of the Earth often assume that it consists of concentric, homogeneous, and isotropic layers within each of which the velocity varies smoothly. The deeper layering and variation of physical properties with depth is indicated in Figure E-1. Raypaths and wavefronts for direct *P*-waves are shown in Figure E-2. Direct *S*-waves, various reflected body waves (some are shown in Figure E-2b), and surface waves will also be observed.

Earth's magnetic field: See *magnetic field of the Earth, normal magnetic field.*

earthquake: Sudden movement of the earth resulting from the abrupt release of accumulated strain, usually a result of faulting or volcanism. An earthquake is classi-

fied according to the depth of its focus or hypocenter: **shallow** <60 km, **intermediate** <300 km, **deep** >300 km (sometimes >450 km). The **seismicity** of an area is its likelihood of having earthquakes. Some of the energy released in an earthquake is radiated as seismic energy. **Intensity** is a measure of the effects of the ground motion at a specific locality; **magnitude** is a measure of the energy released in an earthquake; see *intensity scale* and *magnitude of earthquake.*

earthquake prediction: The aspect of seismology which deals with the conditions or indications that precede an earthquake, with the objective of predicting the magnitude, time, and location of an impending shock.

earthquake seismology: Study of earthquakes and their resulting seismic waves as a means of understanding the structure of the Earth. A schematic seismic record showing several of the wave arrivals from an earthquake is shown in Figure E-3. The scheme for identifying different arrivals is discussed under *wave notation* (q.v.). The time intervals between different arrivals

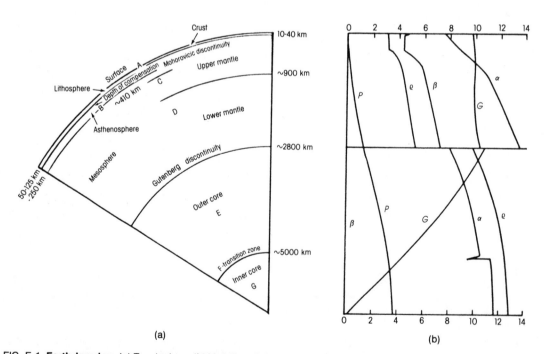

(a) (b)

FIG. E-1. **Earth-layering.** (a) Terminology. (b) Variation of physical properties with depth within the earth. P = pressure in 10^{11} Pa, ρ = density in g/cm^3, β = S-wave velocity in km/s, α = P-wave velocity in km/s, G = gravity in Gal. (After Haddon and Bullen, 1969.)

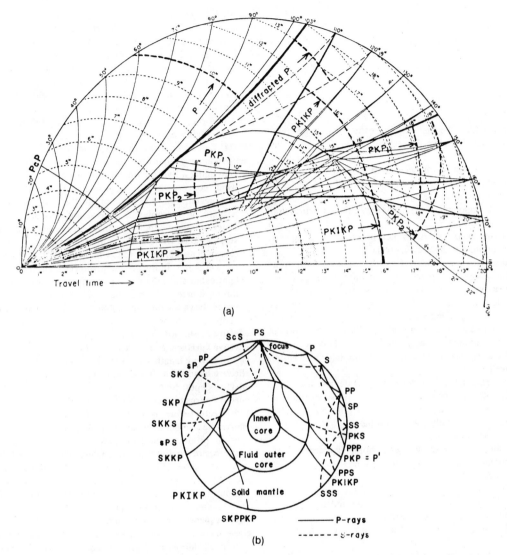

(a)

(b)

FIG. E-2. (a) **Earthquake wavefronts** and raypaths for direct *P*-waves. (From Gutenberg, 1959.) (b) Raypaths for some reflected and refracted waves, illustrating wave nomenclature. (See also *wave notation*.)

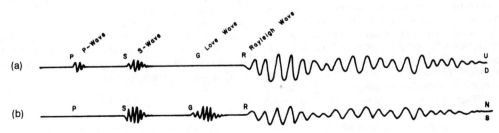

FIG. E-3. **Earthquake seismogram** schematic. (a) The vertical component. (b) One horizontal component.

FIG. E-4. **Echelon** arrangement.

depend upon the distance from the earthquake hypocenter to the observing station; see *Jeffreys-Bullen curves*.

Earth tide: Deformation because of Earth-moon-sun gravitational effects. Produces displacements of the solid Earth up to about 10 cm.

easting: Component of a survey leg in the east direction. See *departure*.

eccentering arm: A device which presses a sonde against the borehole wall. See Figure D-3.

eccentric anomaly: A way of expressing a position on an elliptical orbit as a parametric equation; see Figure E-9.

eccentricity: 1. The ratio of the focus-to-center distance to the length of the semimajor axis for an ellipse; see Figure E-9. **2.** The condition occurring when the shot is not located at (or perpendicularly offset to) the center of what is intended to be a split spread. This occurs on marine records when a separate shooting boat is employed.

Eccles-Jordan trigger: A *flip-flop* (q.v.) device.

ecf: Elevation correction factor. See *elevation correction*.

echelon: A staggered arrangement of parallel features; see Figure E-4. Usually called **en echelon**.

echo check: See *check*.

echogram: An echo sounder or fathometer record.

echo-ranging sonar: *Fathometer* (q.v.).

echo sounder: *Fathometer* (q.v.).

ecliptic: The apparent annual path of the sun among the stars. The plane of the Earth's orbit and the approximate plane of the solar system.

ecliptic coordinates: Celestial latitude and celestial longitude.

economic basement: See *basement*.

eddy current: A circulating electrical current induced in a conductive body by a time-varying magnetic field. **Lenz's law** states that the direction of eddy current flow is such as to produce a secondary magnetic field which opposes the primary field. The secondary field has a quadrature component which depends upon the ratio of the resistance to the reactance of the eddy-current path. In electromagnetic prospecting, eddy currents should be distinguished from **galvanic currents** (which are of electrochemical origin) and also from naturally occurring currents or those of natural electrochemical origin.

edit: To prepare data, text, etc., for processing. **1.** Specifically, to prepare a digital tape containing geophysical information. Editing often involves rearranging data (**reformatting**), testing data validity, deleting unwanted data (**killing**), selection of data, insertion of data (such as headers, process parameters, or instructions), breaking the data up into blocks (**gapping**), etc. Editing a digital seismic tape may involve some calculations

(such as removing binary-gain effects and substituting a correction for spherical divergence). Editing may or may not include static-shift and normal-moveout corrections. **2.** The removal of data which are judged not to be members of the set to be analyzed (because a measurement does not fit with other data). For example, the removal from navigation-satellite data of Doppler counts for which the satellite was low in the sky.

editor: A program which permits a user to create new files or to modify existing files.

EDM: Electronic distance measurement.

eel: An array of hydrophones in a separate cable. A tube containing a hydrophone group fastened to a streamer which floats on the surface or to a cable which is laid on or dragged along the seafloor. In the first instance, it permits the hydrophones to be at sufficient depth in the water to improve their response and remove them from surface noise while keeping the streamer above obstacles such as reefs or wrecks which would threaten it if it were at depth. In the second instance, it allows use of a light cable and modular replacement of arrays. In the third instance, it permits the hydrophones to float in the water above the bottom so that they have good uniform coupling to the water.

effective anisotropy: See *anisotropy*.

effective aperture: See *aperture width*.

effective array length: See *array (seismic)*.

effective bandwidth: See *bandwidth*.

effective depth: *Skin depth* (q.v.).

effective porosity: See *porosity*.

E-field ratio telluric method: See *telluric profiling*.

effort: The number of separate elements combined together to produce the final product. For example, the number of separate raypaths which are combined, that is, the number of energy sources (holes, thumps, pops, etc.) per shotpoint times the number of geophones per group times the number of records stacked or mixed. Occasionally called **multiplicity**.

eigenfrequency: See *eigenfunction*.

eigenfunction: One of a set of functions which satisfies both a differential equation and a set of boundary conditions. For example, a stretched string or an organ pipe might vibrate in a number of modes, each with a characteristic frequency (**eigenfrequency**, **eigenvalue**, or **characteristic root**). The superposition of these eigenfunctions is the general solution. The eigenfunctions corresponding to different eigenvalues are orthogonal (or independent). "Eigen" is German for "characteristic."

eigenstate: The condition of a system represented by one *eigenfunction* (q.v.).

eigenvalue: See *eigenfunction*.

eikonal equation: A form of the wave equation for harmonic waves in which the local velocity V is compared to a reference velocity V_R (analogous to comparing a velocity to the speed of light in vacuum):

$$\nabla^2 \phi = (V/V_R)^2 = n^2,$$

where n is an index of refraction and ϕ is the wave function. Valid only where the variation of properties is small within a wavelength; sometimes called the "high-frequency condition".

elastic: The ability to return to original shape after removal of a distorting stress. The return of shape is complete and essentially instantaneous rather than gradual.

elastic constants: Elasticity deals with deformations that vanish entirely upon removal of the stresses which cause them. The passage of a low-amplitude seismic wave is an example. The general elasticity tensor relating stress and strain in anisotropic media possesses 21 independent constants. In **transversely isotropic media** (where properties are the same measured in two orthogonal directions but different in the third), these reduce to five independent constants. **Isotropic media** (where properties are the same measured in any direction) have only two independent elastic constants. For small deformations, Hooke's law holds and strain is proportional to stress. The stress-strain properties of isotropic materials which obey Hooke's law are specified by **elastic moduli**. These include the following:

a. Bulk modulus k, the stress-strain ratio under simple hydrostatic pressure:

$$k = \Delta P/(\Delta V/V),$$

where ΔP = pressure change, V = volume, and ΔV = change in volume. $\Delta V/V$ is called the **dilatation**.

b. Shear modulus, rigidity modulus, or **Lamé's constant** μ, the stress-strain ratio for simple shear:

$$\mu = (\Delta F/A)/(\Delta L/L),$$

where ΔF = shearing (tangential) force, A = cross-

	E	σ	k	μ	λ
(E, σ)			$\dfrac{E}{3(1 - 2\sigma)}$	$\dfrac{E}{2(1 + \sigma)}$	$\dfrac{E\sigma}{(1 + \sigma)(1 - 2\sigma)}$
(E, k)		$\dfrac{3k - E}{6k}$		$\dfrac{3kE}{9k - E}$	$3k\left(\dfrac{3k - E}{9k - E}\right)$
(E, μ)		$\dfrac{E - 2\mu}{2\mu}$	$\dfrac{\mu E}{3(3\mu - E)}$		$\mu\left(\dfrac{E - 2\mu}{3\mu - E}\right)$
(σ, k)	$3k(1 - 2\sigma)$			$\dfrac{3k}{2}\left(\dfrac{1 - 2\sigma}{1 + \sigma}\right)$	$3k\left(\dfrac{\sigma}{1 + \sigma}\right)$
(σ, μ)	$2\mu(1 + \sigma)$		$\dfrac{2\mu(1 + \sigma)}{3(1 - 2\sigma)}$		$\mu\left(\dfrac{2\sigma}{1 - 2\sigma}\right)$
(σ, λ)	$\lambda\dfrac{(1 + \sigma)(1 - 2\sigma)}{\sigma}$		$\lambda\left(\dfrac{1 + \sigma}{3\sigma}\right)$	$\lambda\left(\dfrac{1 - 2\sigma}{2\sigma}\right)$	
(k, μ)	$\dfrac{9k\mu}{3k + \mu}$	$\dfrac{3k - 2\mu}{2(3k + \mu)}$			$k - 2\mu/3$
(k, λ)	$9k\left(\dfrac{k - \lambda}{3k - \lambda}\right)$	$\dfrac{\lambda}{3k - \lambda}$		$\tfrac{3}{2}(k - \lambda)$	
(μ, λ)	$\mu\left(\dfrac{3\lambda + 2\mu}{\lambda + \mu}\right)$	$\dfrac{\lambda}{2(\lambda + \mu)}$	$\lambda + \tfrac{2}{3}\mu$		

FIG. E-5. **Elastic constants** for isotropic media expressed in terms of each other.

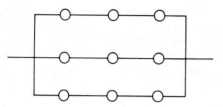

FIG. E-6. **Electrically square** arrangement of nine simliar geophones.

sectional area, L = distance between shear planes, and ΔL = shear displacement.

 c. Young's modulus or **stretch modulus** E, the stress-strain ratio when a rod is pulled or compressed:

$$E = (\Delta F/A)/(\Delta L/L),$$

where $\Delta F/A$ = stress (force per unit area), L = original length, and ΔL = change in length. $1/E$ is sometimes called **compliance**.

 d. Lamé's λ constant:

$$\lambda = k - 2\mu/3.$$

 e. Poisson's ratio σ, the ratio of transverse strain to longitudinal strain. When a rod of length L is pulled it is elongated by ΔL and its width W is contracted by ΔW:

$$\sigma = (\Delta W/W)/(\Delta L/L).$$

Poisson's ratio varies from 0 to 1/2. Poisson's ratio has the value 1/2 for fluids and 1/4 for a **Poisson solid**, a solid for which $\lambda = \mu$.

 f. Velocities of *P*- and *S*-waves, V_p and V_s: Expressed in terms of the moduli and the density ρ:

$$V_p = [(\lambda + 2\mu)/\rho]^{1/2},$$
$$V_s = (\mu/\rho)^{1/2},$$

and

$$V_s/V_p = [(0.5 - \sigma)/(1 - \sigma)]^{1/2}.$$

For isotropic media the various elastic moduli can be expressed in terms of each other; see Figure E-5.

elastic deformation: A nonpermanent deformation; the body returns to its original shape when the stress is released. Often limited to deformations in which stress and strain are linearly related in accordance with **Hooke's law**.

elastic impedance: *Acoustic impedance* (q.v.).

elastic limit: The greatest stress that can be applied without causing permanent deformation.

elastic moduli: *Elastic constants* (q.v.).

elastic wave: A seismic wave, including both *P*- and *S*-waves.

E-layer: 1. The **Heaviside layer**, a layer of very thin air about 110 km high in the ionosphere, which is ionized by sunlight and which reflects HF radio waves back to the earth during the daytime and MF waves at night. See Figure A-16 and also **F-layer** and **D-layer. 2.** A layer within the Earth; see Figure E-1.

Elcord: An explosive delay unit, used to match explosion velocity to formation velocity, the objective being to send more of the energy vertically downward. Dupont tradename.

electrically square: Consisting of a number of similar elements connected in a parallel-series arrangement so that the apparent impedance is the same as that of a single element; see Figure E-6.

electric basement: See *basement*.

electric dipole: Two equal charges q of opposite sign separated by the distance δx, giving a **dipole strength** of $q\delta x$.

electric drilling: *Electric sounding* (q.v.).

electric field: A spatial vector quantity equal to the electric potential gradient, produced by charged bodies or a time-varying magnetic field. Unit is volts per meter. The electric field **E** induced in a loop equals the negative time derivative of the magnetic flux Φ cutting the loop ($d\mathbf{l}$ is a length element of the loop):

$$\oint \mathbf{E} \cdot d\mathbf{l} = -\partial\Phi/\partial t.$$

It is also expressed in terms of the change in the magnetic induction **B** with time t.

$$\nabla \times \mathbf{E} = \partial\mathbf{B}/\partial t.$$

electric log: 1. A generic term including all electrical borehole logs (SP, normal, lateral, laterologs, induction, microresistivity logs). The two most common electrode configurations are shown in Figure E-7. **2.** Records of surface resistivity surveying; compare *electric survey*. **3.** Electrolog, a borehole log which usually consists of SP and two or more resistivity logs, such as short and long normal and long lateral logs. Electrolog is a Dresser Atlas tradename.

electric profiling: An IP, resistivity, or electromagnetic method utilizing fixed spacing of electrodes or antennas in which the measuring system is moved progressively along profile lines to detect resistivity changes along the profiles. See *moving source method*.

electric sounding: An IP, resistivity, or electromagnetic method in which electrode or antenna spacing is increased to obtain information from successively greater depths at a given surface location. Electromagnetic sounding can also be done with a fixed spacing by varying the frequency (**frequency-domain sounding**) or by exciting the earth with an electric current step (ideally) and measuring the change of the magnetic field with **time-domain sounding**. Electric sounding is intended to detect changes in the resistivity of the earth with depth at one location (assuming horizontal layering).

electric survey: 1. Measurements at or near the earth's surface of natural or induced electric fields, the objective being to map mineral concentrations or for geologic or basement mapping. See *electric profiling, electric sounding, electromagnetic method, induced-polarization method, magnetotelluric method, resistivity method, self-potential method, telluric current method.* **2.** *Electric log* (q.v.) run in a borehole.

electric susceptibility: The ratio of intensity of electric polarization to electric field, a measure of the polarization property of a dielectric. Also called **dielectric susceptibility**. The primary term of the *metal factor* (q.v.).

electric trenching: *Electric profiling* (q.v.).

electrochemical SP: The component of the SP (self-potential) comprising the sum of the liquid-junction potential and the shale potential, both of which are determined by the ratio of the activity of the formation water to that of the mud filtrate. The **liquid-junction potential** is produced at the contact between the invasion filtrate and the formation water as a result of differences in ion diffusion rates from a more concentrated to a more dilute solution (**concentration cell**). Negatively charged chloride ions have greater mobility than positive sodium ions and an excess negative charge tends to cross the boundary, resulting in a potential difference. The **shale membrane potential** or **Mounce potential** results because shale acts as a cationic membrane permitting sodium cations to flow through it but not chloride anions. The liquid-junction and shale potentials are additive. See also *SSP* and *electrokinetic potential*.

electrode: 1. A piece of metallic material that is used as an electric contact with a nonmetal. Can refer to a grounding contact, to metallic minerals in a rock, or to electric contacts in laboratory equipment. **2.** *Porous pot* (q.v.).

electrode array: See *array (electrical)*.

electrode equilibrium potential: The reversible (no energy loss) equilibrium potential across the interface between an electrode and an electrolyte, measurable when no current passes through the interface. Measured as the voltage difference between a reference electrode and the electrode in question. It is primarily due to the free energy of the electron-transfer process.

electrode impedance: 1. In electrochemistry, the total impedance across the interface between an electrode and an electrolyte. The equivalent-circuit model includes solution resistance, capacitances in the fixed and diffuse layers, and Warburg impedance. **2.** In electrical-circuit theory, the self-impedance of a single electrode or the mutual-impedance between electrodes.

electrode polarization: 1. In electrochemistry, an electrode is polarized if its potential deviates from the reversible or equilibrium value. **2.** Polarization also can be induced due to passage of current through an interface or to a change in ion concentration at an electrode surface. The amount of extra polarization is the **overvoltage** or **induced polarization** of the electrode.

electrode potential: See *electrode equilibrium potential*.

electrode resistance: The electric resistance between an electrode and the immediate surroundings; sometimes called **contact resistance**, **self resistance**, **grounding resistance**, or **mutual resistance**, depending upon the situation.

electrodialysis: Migration of charge through a membrane in an electric field.

electrodynamic geophone: Moving-coil geophone; see *geophone*.

electrofacies: The set of well-log responses which characterize a lithologic unit and permit that stratigraphic interval to be correlated with or distinguished from others.

electrokinetic potential: A voltage which results from flow of a fluid containing ions; *streaming potential* (q.v.).

Electrolog: *Electric log* (q.v.).

electrolyte: 1. A material in which the flow of electric current is accompanied by the movement of matter in

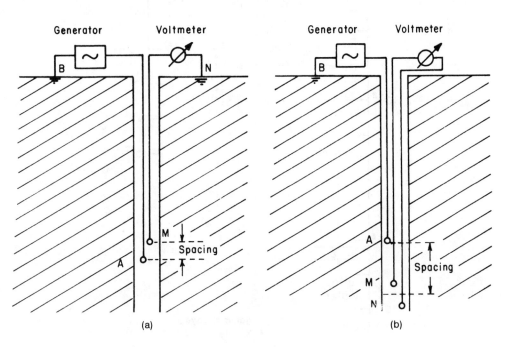

FIG. E-7. **Electric-log configurations.** (**a**) Normal log. (**b**) Lateral log.

the form of ions. **2.** Any substance that dissociates into ions.

electrolytic tank: A container holding a conductive solution in which electric model experiments can be carried out. See *analog modeling*.

electromagnetic coupling: See *coupling*.

electromagnetic method: A method in which the magnetic and/or electric fields associated with subsurface currents are measured. In general, electromagnetic methods are those in which the electric and magnetic fields

in the earth satisfy the diffusion equation (which ignores displacement currents) but not Laplace's equation (which ignores induction effects) nor the wave equation (which includes displacement currents). One normally excludes methods which use microwave or higher frequencies (and which consequently have little effective penetration) and methods which use dc or very low frequencies in which induction effects are not important (resistivity and IP methods). **Natural field** methods (such as Afmag and magnetotellurics) employ natural

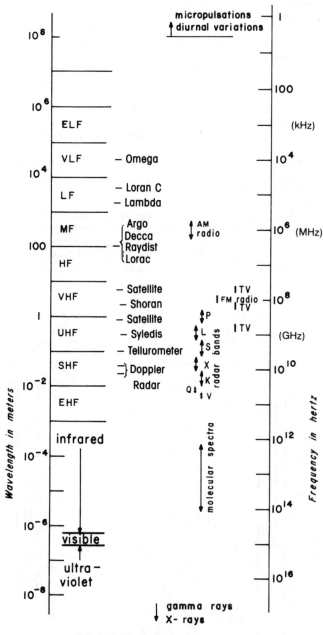

FIG. E-8. **Electromagnetic** spectrum.

energy as the source; **controlled-source methods (CSEM)** (such as loop-loop techniques) require a man-made source. **EM method.**

electromagnetic spectrum: The range of frequencies or wavelengths of electromagnetic radiation. See Figure E-8.

electron density: Number of electrons per unit volume, including both free and orbital electrons which are part of atoms. See *density log.*

electronic: Relating to devices, circuits, or systems in which conduction is primarily by electrons moving through a vacuum, gas, semiconductor, or conductor.

electronic conductor: A material such as a metal that conducts electricity by virtue of electron mobility. See *conductor.*

electron-positron pair formation: Creation of an electron and positron which may result from radiation or collisions more energetic than 1 MeV.

electron-transfer reaction: An electrode surface phenomenon involving an oxidation-reduction reaction, generating a Faradaic current.

electro-osmosis: 1. The phenomenon whereby an electric field moves a fluid through a membrane. **2.** Charge separation in an electrolyte by osmotic action.

electrostatic plotter: A rastor-oriented plotter in which the image is produced by implanting an electrostatic charge on the medium (paper or film) which is then passed through a toner.

electrostrictive: The property of a material that causes it to change dimensions when subject to an electric field. See *piezoelectric.*

electrotape: An electronic surveying instrument similar to the tellurometer. Tradename.

elevation angle: In transit satellite navigation, the angle measured from horizontal at the receiver location to a satellite. Satellite passes are often designated by the elevation angle at closest approach.

elevation correction: 1. The correction applied to reflection or refraction arrival times to reduce observations to a common reference datum. **2.** In gravity, the sum of the free-air and Bouguer corrections. The elevation correction is obtained by multiplying the difference between station and reference elevation by the **elevation correction factor (ecf):**

$$\text{ecf} = (0.0941 - 0.01276\ \rho)\ \text{mGal/ft}$$

$$= (0.3086 - 0.04185\ \rho)\ \text{mGal/m},$$

where ρ = density in g/cm^3.

elevation correction factor: See *elevation correction.* Abbreviated ecf.

elevation datum: See *datum.*

ELF: Extremely low frequency; electromagnetic energy between 30 and 300 Hz. Natural energy in this band originates mostly from lightning strikes. See *sferics* and Figure E-8.

ellipse: A conic section; see Figures C-12 and E-9.

ellipse area: A quantity measured in telluric surveys which can be interpreted in terms of variations in earth resistivity.

ellipse of polarization: See *polarization ellipse.*

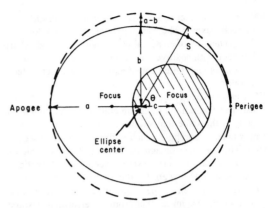

FIG. E-9. **Ellipse** terminology. An ellipse is the locus of points for which the sum of the distances from the two foci is constant. A satellite follows an elliptical path about a body at one focus. If a = semimajor axis, b = semiminor axis,

eccentricity = $\varepsilon = c/a = (2f - f^2)^{1/2} = [1 - (1/E)^2]^{1/2}$;
ellipticity = $E = a/b = 1/(1 - f) = (1 - \varepsilon^2)^{-1/2}$;
flattening = $f = (a - b)/a = 1 - 1/E$;
and $\ominus$ = the **eccentric anomaly**
when the satellite is at S. The polar equation of an ellipse with one focus at the origin is $\varrho = \varepsilon h/(1 - \varepsilon \cos \phi)$, where h = distance from the focus to a directrix line.

ellipsoid: A solid figure for which every plane cross-section is an ellipse. An **ellipsoid of revolution** can be generated by rotating an ellipse about one of its axes. An ellipsoid of revolution is usually used to approximate the *geoid* (q.v.). Some standard ellipsoids are listed in Figure G-3. See *Geodetic Reference System.*

ellipsoid of anisotropy: An equipotential surface about a point current source in an anisotropic, homogeneous medium. Such a surface is an ellipsoid of revolution, whose flattening is the **coefficient of anisotropy;** see *anisotropy.*

ellipticity: 1. The ratio of the major to minor axes of an ellipse (Figure E-9). **2.** Specifically, in electromagnetic surveying, the ratio for the polarization ellipse which can be determined by measuring the amplitudes and phases of two orthogonal components of the magnetic field.

E-log: *Electric log* (q.v.).

elongated charge: A long column of explosives used in an effort to achieve directivity. Types used include sausage powder, Elcord delay units, acoustic delay units, and broomstick units. An elongated charge is designed to direct the input pulse downward by matching the effective detonation velocity to the formation velocity, but their physical dimensions compared to a seismic wavelength are often too small to achieve significant directivity.

Eltran: An early electric transient exploration method similar to time-domain EM methods. Eltran used a large-spaced dipole-dipole electrode array for petroleum exploration.

EM: Electromagnetic; see *electromagnetic method.*

embedded wavelet: The wavelet shape which would result from reflection of an actual wave train by a single sharp interface with positive reflection coefficient. See *convolutional model.* Often called **equivalent wavelet** or **basic wavelet.**

EM coupling: See *coupling.*

emissivity: The ratio of radiation emitted to that of a black body at the same temperature.

empty hole: A borehole which is filled with air or gas.

emu: The cgs-electromagnetic system of units, which is similar to the cgs-esu system except that "practical" electrical units (except for the abampere) are used.

emulate: To have the same input/output characteristics as a different system so that from a performance standpoint it appears identical to the different system (except as regarding speed).

emulator: A program or hardware device which duplicates the instruction set of one computer on a different computer. It is used in program development for the emulated computer when that computer is not available.

end-of-file gap: An elongated gap (usually 3.75 inches long) on magnetic tape to indicate the end of a file of records.

end-of-file mark: Machine-readable mark on digital magnetic tape indicating the end of a complete group of data, such as the end of a seismic record; **EOF.**

end-on: Seismic field arrangement wherein the shot is at (or near) the end of the geophone spread. See Figure S-17. Sometimes called **end-line.**

en echelon: *Echelon* (q.v.).

energy: 1. The capacity to do work. 2. The sum of the squares of the amplitudes of the elements in a wavelet (times a proportionality constant which is often omitted). Thus if the wavelet b_t is

$$b_t = [b_0, b_1, b_2, \cdots, b_n],$$

the energy is

$$k(b_0{}^2 + b_1{}^2 + b_2{}^2 + \cdots + b_n).$$

3. **Depositional energy** refers to the kinetic energy (due to waves or currents) present in the environment in which sediments were deposited. High energy has the capability of producing good particle size sorting.

energy reflectivity: See *reflectivity.*

engineering geophysics: Use of geophysical methods to get information for civil engineering. The aim is usually to describe not only the geometry of the subsurface but its nature (for example, its elastic characteristics as determined by measurements of seismic velocities and densities). Seismic reflection and refraction, side-scan sonar, gravity, magnetic, electric, and sampling methods are employed. Shallow refraction is commonly used to give the depth to bedrock. In water-covered areas, high-powered fathometers, sparkers, gas guns, and other seismic reflection methods employing high frequencies (up to 5 kHz) are used to obtain reflections from shallow interfaces so that bedrock and the nature of fill material can be diagnosed. Such methods are also used to locate large pipelines on or buried in the sea bottom by the prominent diffractions which they generate.

Usually restricted to shallower than 1000 ft penetration.

enhanced oil recovery: **EOR** or **tertiary recovery;** techniques for producing oil left behind in the ground after *secondary recovery* (q.v.). Includes steam injection, steam flooding, miscible fluid injection, surfactant flooding, alkaline chemical injection, polymer flooding.

enhancement: Improvement of data by filtering or noise-rejection processes.

enthalpy: The internal energy of a thermodynamic system plus the product of the system's pressure and volume. The heat transferred during an isobaric process equals the change in enthalpy.

entropy: 1. A thermodynamic quantity which measures the energy which is not available. Higher entropy represents increased disorder. Entropy never decreases in a reaction, according to the **second law of thermodynamics 2.** A measure of the uncertainty in a message. If $P(m_i)$ is the probability that the message m_i has been transmitted, then the entropy H where there are i possible messages, is given by

$$H = -\Sigma P(m_i) \log_2 P(m_i).$$

The entropy of a situation with no uncertainty is zero. Entropy is a measure of the average information content of a message.

envelope: The low-frequency curves encompassing or bounding deflections of higher frequency. The curves are usually drawn by smoothly connecting adjacent peaks and adjacent troughs. See Figure C-10.

envelope amplitude: See *complex-trace analysis.*

EOR: *Enhanced oil recovery* (q.v.).

Eötvös effect: The vertical component of a Coriolis acceleration observed when measuring gravity while in motion. The meter's velocity over the surface adds vectorialy to the velocity due to the earth's rotation, varying the centrifugal acceleration and hence the apparent gravitational attraction. The Eötvös correction in mGal for a meter whose speed is V knots at an azimuth angle α and latitude ϕ is

$$E = 7.503 \ V \cos \phi \sin \alpha + 0.004154 \ V^2.$$

The Eötvös uncertainty dE in terms of direction uncertainty $d\alpha$ and speed uncertainty dV is

$$dE = (7.503 \ V \cos \phi \cos \alpha) \ d\alpha$$
$$+ (7.503 \cos \phi \sin \alpha + 0.008308 \ V)dV.$$

See Glicken (1962). Named for Baron Roland von Eötvös (1848-1919), Hungarian physicist.

Eötvös unit: A unit of gravitational gradient or curvature; 10^{-6}mGal/cm.

epeirogenic: See *tectonic types.*

ephemeral data: The part of the data broadcast by a navigation satellite which varies with each broadcast. The ephemeral data include the time since the hour or half hour and corrections to the satellite location; see Figure K-1.

ephemeris: A table showing the position of a body (such as a navigation satellite) at various times.

epicenter: The location on the earth's surface below which the first motion in an earthquake occurs. Compare *hypocenter.*

Epilog: A specific synergetic log. Tradename of Dresser Industries.

epithermal neutron: See *neutron log*.

equalization: 1. Trace equalization involves adjusting the gain of different channels so that their amplitudes are comparable, possibly so that their rms amplitudes over some analysis window are made equal. **2. Cross-equalization** involves matching the frequency spectrum of different channels to each other or to a predetermined curve, possibly including adjustments because of phase differences. **3.** Filtering to correct for frequency discrimination in recording or playback such as the linear-with-frequency response inherent in magnetic-tape pickup from direct analog recordings.

equalizing: See *deconvolution*.

equal-ripple filter: See *Chebychev array*.

equatorial array: A configuration of electrodes used in resistivity surveying. See Figure A-13.

equilibrium conditions: 1. A condition of balance at a state of minimum energy where energy is neither produced nor consumed. **2.** A condition predicted by the law of mass action where the velocities of forward and reverse reactions of a reversible process are equal.

equinoctial: The **celestial equator**; the intersection of the plane of the earth's equator and the celestial sphere.

equipotential method: Mapping the potential field produced by stationary (often remote) current electrodes. The mapping is done by moving a potential electrode over the area. Also called **equipotential survey** and **applied-potential method.** In the **equipotential-line method,** the line of points on the surface of the ground which constitutes the locus of a given voltage difference from another electrode is mapped with a "probe" electrode.

equipotential surface: The continuous surface which is everywhere perpendicular to lines of force. No work is done against the field when moving along such a surface. Mean sea level is an equipotential surface with respect to gravity. An electrical equipotential surface is everywhere perpendicular to current flow.

equipotential survey: See *equipotential method*.

equiripple response: A directivity pattern in which the minor lobes are of equal height, as results from a Chebychev array. See Figure C-4.

equivalence: See *map projection*.

equivalent circuit: An electrical circuit which has the same input-output relationship as another circuit.

equivalent electrical response: Combinations of layer resistivities and thicknesses which would produce practically indistinguishable electrical sounding responses.

equivalent velocity: 1. *Stacking velocity* (q.v.), the constant velocity which gives nearly the same normal moveout as observed in a velocity analysis. **2.** *Apparent velocity* (q.v.).

equivalent wavelet: See *embedded wavelet*.

equivalent width: The width of a boxcar with the same peak amplitude which contains the same energy.

equivoluminar wave: *S-wave* (q.v.).

ergodic: Having the same statistical properties throughout the ensemble. An ergodic system will eventually return arbitrarily close to any prior state. The statistical properties measured over sufficiently long intervals anywhere throughout the ensemble will be the same.

error: A deviation from the correct value. See *round-off error* and *truncation error*; *random* and *systematic*; *accuracy* and *probable error*.

error checking: Techniques for detecting errors which occur during the processing and transfer of data. See *check, cycle-redundancy check*, and *parity check*.

error control: A system for detecting errors and sometimes for correcting them also. See *check*.

error-detecting checks: See *check*.

error function: The error function $\mathrm{erf}(t)$ is

$$\mathrm{erf}(t) = (2/\pi^{1/2}) \int_0^t e^{-y^2}\, dy.$$

For data which have a Gaussian distribution, the probability that an error lies between $\pm a$ is $\mathrm{erf}(ha)$, where $h = $ **precision index**.

error message: A message from a computer about an incompatibility in program instructions or erroneous conditions in the data.

ERTS: See *Landsat image*.

ES: *Electric survey* (q.v.).

Euler-Cauchy method: A finite-difference method of solving differential equations. See Sheriff and Geldart, v. 2 (1983, p. 157).

Euler's identity: The relationships

$$e^{i\theta} = \cos\theta + i\sin\theta = \mathrm{cis}\,\theta,$$

$$\cos\theta = (e^{i\theta} + e^{-i\theta})/2 = \cosh(i\theta), \text{ and}$$

$$\sin\theta = (e^{i\theta} - e^{-i\theta})/2i = -i\sinh(i\theta).$$

Named for Leonhard Euler (1707-1783), Swiss mathematician.

eustatic cycle: The time interval during which a worldwide rise and fall of sea level takes place.

eutectic mixture: A mixture of two materials which has a lower melting point than either of the materials by themselves.

eV: Electron volt.

evanescent waves: Waves which fade away rapidly with distance from a boundary. See Sheriff and Geldart, v. 1 (1982, p. 42, 59)

even function: A function which has the same value when the sign of the variable is changed; i.e., $f(x) = f(-x)$. The Fourier transform of an even function is the **cosine transform**; its frequency-domain representation is zero phase. Any function can be represented as an even part and an odd part. Aeromagnetic anomalies are sometimes separated into even and odd parts for interpretation. Antonym: **odd function**.

event: A lineup on a number of traces which indicates the arrival of new seismic energy, denoted by a systematic phase or amplitude change on a seismic record; an **arrival**. May indicate a reflection, refraction, diffraction, or other type of wavefront. The distinguishing features of various types of events are discussed in Sheriff and Geldart, v. 1 (1982, p. 101-112).

Evison wave: Channel wave of *SH*-type particle motion in a low-velocity layer between two higher velocity half-spaces. Compare *Krey wave*. See Evison (1955).

exaggeration: Use of a different vertical than horizontal scale. See Figure V-5.

EXCEPT gate: A circuit with multiple inputs which functions when signal is present on one input and absent on other inputs. Also called **exclusive OR**. Differs from **OR gate**, which does not have the restriction of absent signal on the other inputs. See *gate* and Figure G-1.

excess-three code: See *binary-coded decimal*.

excess time: The vertical traveltime through the weathering minus the time it would have taken if travel had been at the subweathering velocity.

exchange current: A term in electrochemistry for the reversible electric current at an electrode that is in equilibrium with an electrolyte.

excitation-at-the-mass method: *Mise-à-la-masse method* (q.v.).

exclusive filter: A filter with a very sharp, narrow passband.

exclusive OR: *EXCEPT gate* (q.v.).

executive: An *operating system* (q.v.) or supervisor.

expand: 1. To break down into elements, as to expand in terms of frequency components. 2. To increase the gain, as with a gain control. 3. To extend a spread to longer offsets.

expander: 1. *Expanding spread* (q.v.); specifically, a **depth probe**. 2. An IP and resistivity-surveying technique in which the electrode-separation interval is successively expanded so as to achieve greater depth of exploration. Also called **sounding** or **probing**. Data from an expander is usually interpreted to give the depth to horizontal layers with contrasting physical properties, if such exist.

expanding spread: 1. A spread moved to greater offsets for successive shots from the same location, to give the equivalent of recording many geophone groups from a single shot. Used in refraction work, in noise analysis, and in velocity analysis. 2. *Depth probe* (q.v.).

expanding-spread vertical-loop technique: Electromagnetic survey method using a fixed transmitter and movable receiver, often along lines at about 45 degrees to the anticipated strike of the conductor sought.

expectation: 1. An operator which denotes the mean, or weighted mean where values are not equally probable. Usually denoted by braces, $E\{\ldots\}$. 2. In seismic processing usage it indicates the sum of elements for finite wavelets (b_t) or the mean of an infinite time series (u_t):

$$E\{b_t\} = b_0 + b_1 + b_2 + \cdots + b_n, \text{ or}$$

$$E\{u_t\} = \lim_{n \to \infty} (u_{-n} + \cdots + u_0$$

$$+ \cdots u_n)/(2n + 1).$$

expected value: *Expectation* (q.v.).

exploding reflector: A direct modeling scheme in which the model surfaces are assumed to explode at time zero with explosive strengths proportional to their reflectivity; velocities are all cut in half so that the one-way traveltime to the surface equals the two-way traveltime for coincident source-receiver pairs at the surface. A form of wave theory modeling.

exploding wire: A marine seismic energy source in which an electric arc discharge is initiated by placing between the electrodes a wire which vaporizes during the discharge. Called **WASSP**.

exploration: The search for commercial deposits of useful minerals, including hydrocarbons, geothermal resources, etc.

exploration geophysics: The application of geophysics to *exploration* (q.v.) and also to engineering and archaeology. Synonym: **applied geophysics.**

exploration logging: The use of widely spaced in-hole electrodes, perhaps combined with surface electrodes, to expand the exploration effectiveness of a drill hole. Also called **detection logging**.

exploratory well: A well drilled some distance from a previously demonstrated accumulation, or drilled to a stratigraphic objective not previously known to be productive; a **wildcat**. See Figure W-7.

exponentiate: 1. To introduce time-dependent gain in processing, that is, to multiply input values by e^{kt} where k may be either a positive or negative real number. The value of k is often determined arbitrarily or empirically. Often refers to additional gain adjustment after removal of gain effects during recording and correction for spherical divergence. 2. To magnify variations in input data so that the output is proportional to a constant raised to an exponent of the input value.

exponential decay: Decrease of amplitude proportional to $e^{-\alpha x}$ or $e^{-\beta x}$ where α and β are decay constants, $x =$ distance, and $t =$ time.

exponential ramp: A taper used at the edge of a window. To multiply values for $t > t_1$ by $e^{k(t - t_1)}$ where k is a negative real number.

extended-range shoran: An extremely sensitive shoran system which depends on energy refracted around the Earth's curvature and hence is not line-of-sight limited, as ordinary shoran is. Called **XR shoran** and has a range of the order of 250 km.

extended resolution: Efforts to increase the upper frequency response of a band-pass system in order to improve resolution. See Sheriff and Geldart, v. 1 (1982, p. 150-151).

extended spread: In-line offset spread; see *in-line offset*.

extrinsic conduction: The low-temperature ionic or electronic conduction in solid electrically conducting materials, due to weakly bonded impurities or defects. Also called **structure-sensitive conductivity**. See *intrinsic conduction*.

eyeball: To make an educated guess after casual visual inspection, without actually measuring or calculating.

F

facies: 1. The aspect of a rock reflecting conditions of its origin and distinguishing it from adjacent units. **2.** See *seismic facies*.

factor analysis: A method of finding a mathematical relationship among values. Assume that a set of observations $[x_1, \cdots, x_k]$, can be accounted for by a set of linear relationships involving the factors $[z_1, \cdots, z_p]$, where $p < k$:

$$x_i = a_{i1}z_1 + a_{i2}z_2 + \cdots + a_{ip}z_p + E_i,$$

where E_i is an error term (also called the **specific factor**).

Factor analysis is a method of finding the parameters a_{ij}. Also called **multivariant analysis.**

fade: *Mute* (q.v.).

failed arm: That arm of a *triple junction* (q.v.) which ceased to be a spreading center; **aulacogen.**

fairing: 1. Adjusting to produce a regular surface of correct form. Specifically, smoothing to remove errors of measurement or noise in the data, so as to reveal the significant portion. **2.** Irregularities which induce turbulent flow. Drag on a cable pulled transversely through the water is less for turbulent than for nonturbulent

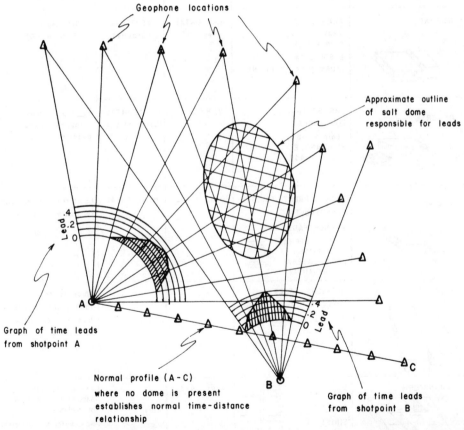

FIG. F-1. **Fan shooting**. Detectors are located roughly on the arc of a circle (centered at the shotpoint) in different directions. Expected arrival times are determined from a normal traveltime curve. (versus distance) established from a calibration or normal profile shot where no local high-velocity body is present. An early arrival (**lead**) with respect to the normal indicates that part of the travel path is at an abnormally high velocity, signifying the present of a local body such as a dome of high-velocity salt. This early method was used to locate salt domes within the thick low-velocity Gulf Coast sediments. (From Nettleton, 1940.)

flow. Irregularities (such as short strings fastened to the cable) create small vortices which reduce fluid friction. A pulled cable without fairing alternately tends to shed large vortices and then have spurts of motion, causing vibration and inducing noise.

fairway: 1. The region within which effort is to be concentrated, such as (in velocity analysis) a band of possible velocities within which one searches for velocity picks from normal-moveout measurements. **2.** A trend of hydrocarbon accumulations.

FAL: *Formation-analysis log* (q.v.).

false color: The use of colors to represent different frequency bands (or other measurable characteristics) where the colors are not those naturally characteristic. For example, infrared differences displayed as parts of the visible spectrum.

faltung: Folding, *convolution* (q.v.). Faltung is German for "folding."

Famous: French-American Mid-Ocean Undersea Survey, a 1974 investigation of the mid-Atlantic ridge using submersibles.

fan filter: *Velocity filter* (q.v.).

fan shooting: Refraction technique to search an area for local high-velocity bodies. See Figure F-1.

fantom: *Phantom* (q.v.).

Faradaic: Pertaining to an electrochemical electron-transfer reaction as in the oxidation-reduction process in an electrolytic cell.

Faradaic path: Passage of current at an electrode by the conversion of atom to ion or vice versa; i.e., as the result of an electrochemical reaction. See *Warburg impedance*. A non-Faradaic path involves the ionic layers which are adjacent to the electrode acting as a condenser.

Faraday's law of induction: The voltage E (in volts) induced in a wire of length l (in meters) cutting a

FAULT TYPE	RELATED TERMS	STRESS DIRECTION		CHARACTERISTICS
		MINIMUM	MAXIMUM	
NORMAL 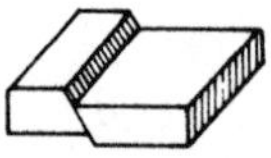	TENSION FAULT GRAVITY FAULT SLIP FAULT LISTRIC FAULT (CURVED FAULT PLANE)	HORIZONTAL (Tension)	VERTICAL (Gravity)	Dip usually 75° to 40°
REVERSE 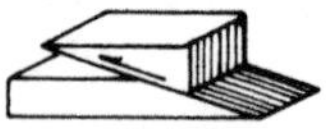	THRUST FAULT LOW ANGLE (dip < 45°) HIGH ANGLE (dip > 45°)	VERTICAL	HORIZONTAL (Compression)	Fault plane may disappear along bedding
STRIKE - SLIP 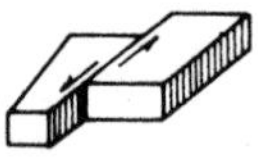	TRANSCURRENT FAULT TEAR FAULT WRENCH FAULT RIGHT LATERAL (Dextral) LEFT LATERAL (Sinistral)	HORIZONTAL	HORIZONTAL	Fault trace often 30° to maximum stress
ROTATIONAL 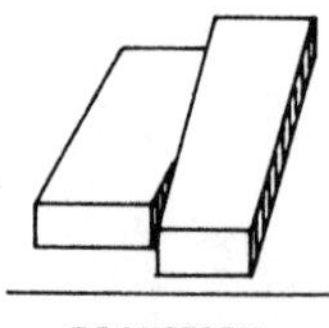	SCISSORS FAULT HINGE FAULT			Throw varies along fault strike; may vary from normal throw to reverse.
TRANSFORM 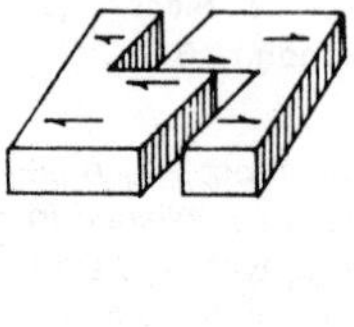	DEXTRAL SINISTRAL	HORIZONTAL		Associated with separation or collision of plates New material fills rifts between separating plates or one plate rides up on another if plates collide.

FIG. F-2. **Fault types.**

magnetic field of strength B (in gauss) at the velocity V (in m/s) is:

$$E = BVl = -d\phi/dt,$$

where $d\phi/dt$ = rate of cutting of lines of magnetic flux in maxwell/s and the minus is inserted because of Lenz's law. Named for Michael Faraday (1791-1867), English physicist.

far-field: Field remote from the source. Spherical waves involve terms which decrease inversely as the distance traveled and also terms which decrease inversely as the distance squared (see Sheriff and Geldart, v. 1 (1982, p. 47-48); hence the relative importance of effects differ with distance. "Far-field" implies that distances are sufficiently great that the terms which depend on the inverse of the distance squared are unimportant. The far-field represents radiated energy. If the distance from the source is R and the wavelength is λ, then far-field implies that $R \gg \lambda$. Compare *near-field*.

fast Fourier transform: An algorithm (such as the Cooley-Tukey method) which accomplishes the discrete Fourier transform in a high-speed digital computer more rapidly than direct evaluation. Most involve iterative methods and take advantage of mathematical symmetry and redundancy. See Sheriff and Geldart, v. 2 (1983, p. 179-180). **FFT.**

fathom: 6 ft or 1.8288 m.

fathometer: A device for measuring water depth by timing sonic reflections from the water bottom; an **echo sounder.**

fatigue: See *hole fatigue.*

fault: 1. A displacement of rocks along a shear surface. See Figures F-2 and F-3. The surface along which the displacement occurs is called the **fault plane** (often a curved surface and not "plane" in the geometric sense). The **dip** of the fault plane is the angle which it makes with the horizontal; the angle with the vertical is called the **hade.** The **trace** of a fault is the line which the fault plane makes with a surface (often the surface of the ground or a bed). Faults are classified as normal, reverse or strike-slip, depending on the relative motion along the fault plane. A fourth type of fault associated with plate movement is a *transform fault* (q.v.). A **hinge** or **scissors fault** is produced by rotation of the blocks across the fault about an axis perpendicular to the fault plane so that throw varies systematically along the fault trace. Primary faults may produce secondary stresses which produce secondary faults (which may be of different type). Thus thrusting may produce tensions which cause secondary normal faults. Faulting and folding commonly are responses to the same stresses; see Figures F-13 to F-15. Faulting during sediment deposition (**growth faulting**) often affects the stratigraphy. Thus beds may abruptly thicken and become more sandy across a normal growth fault. Faults evidence themselves in seismic data principally by: (a) abrupt termination of events, (b) diffractions, (c) changes in dip, flattening or steepening, (d) distortions of dips seen through the fault, (e) cut-out of data beneath the fault or in the shadow-zone of the fault, (f) changes in the

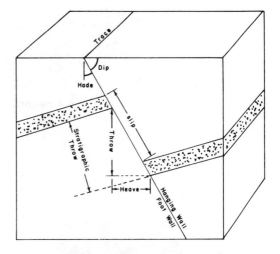

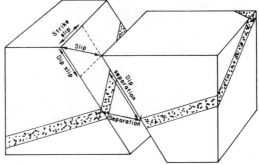

FIG. F-3. **Fault** nomenclature.

pattern of events across the fault, and (g) occasionally a reflection from the fault plane; see Sheriff and Geldart, v.2 (1983, p. 97-105). **2.** Evidence in gravity or magnetic data of the edge of a roughly horizontal slab with density or susceptibility different from that of horizontally adjacent material.

fault-plane solutions: Resolution of the direction of slippage along a fault determined from variations in the direction of first motion at stations distributed in different directions from the epicenter.

Faust's equations.: Empirical relationships between seismic velocity V in ft/s, geologic age T, true formation resistivity ρ in ohm feet, and depth of burial z in feet:

$$V = K(zT)^{1/6},$$
$$V = 2 \times 10^3(\rho z)^{1/6}.$$

z is sometimes taken to be the maximum depth to which the formation has ever been buried, not necessarily its present depth. See Faust (1951, 1953). The reciprocal of the second equation above is sometimes written with three constants (k,a,b) to be evaluated empirically from sonic-log transit time Δt data:

$$1/V = \Delta t = k\rho^{-a}z^{-b}.$$

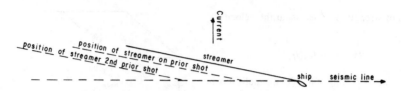

FIG. F-4. **Feathering** of a marine streamer.

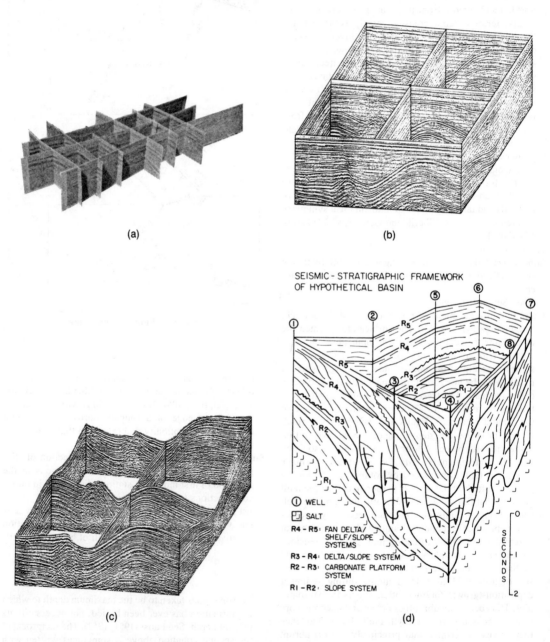

(a)

(b)

SEISMIC - STRATIGRAPHIC FRAMEWORK
OF HYPOTHETICAL BASIN

① WELL

SALT

R4 - R5: FAN DELTA/
SHELF/SLOPE
SYSTEMS

R3 - R4: DELTA/SLOPE SYSTEM

R2 - R3: CARBONATE PLATFORM
SYSTEM

R1 - R2: SLOPE SYSTEM

(c)

(d)

FIG. F-5. (a) **Fence diagram** physically composited from grid of seismic record sections. (Courtesy Chevron Oil Co.)
(b) Computer-drawn isometric fence diagram of six seismic lines. (Courtesy Seiscom-Delta-United.) (c) Same as (b)
except with data above some picked horizon removed. (d) Fence diagram to show spatial relationship of data from a series
of eight wells. (From Brown and Fisher, 1977, p. 222.)

FDC: Compensated formation-density log; see *density log*. Schlumberger tradename.

FDL: Formation-density log; see *density log*.

feathering: En echelon arrangement of successive spreads, as produced in marine shooting when a cross-current causes the cable to drift at an angle to the seismic line. See Figures F-4. and T-3a.

feather edge (of pinchout): The line of disappearance of a wedge of material. Evidence of the wedge's presence in seismic data disappears when the wedge gets too thin to be detected but before the wedge itself disappears.

feather pattern: A weighted or tapered pattern of geophones within a group (or of shotholes, weight-drop points, etc.), such that contributions of the elements of the pattern decrease with distance from its center.

feedback: The use of part of the output of a system as a partial input. **Negative feedback**, where part of the output is fed out-of-phase into the input, is used to attenuate variations such as for self-correcting or control purposes. Used in AGC systems. **Positive feedback**, where a portion of the output is fed in-phase into the input, is used to produce oscillations.

feedback filter: *Recursive filter* (q.v.).

Fejer kernel window: A window shaped according to

$$\text{sinc}^2(\pi t/T) = (T/\pi t)^2 \sin^2(\pi t/T).$$

fence diagram: Network of simplified sections, usually displayed in vertical isometric projection, to illustrate variations in the third dimension. A fence-diagram is shown in Figure F-5.

fence effect: An IP, resistivity, or electromagnetic anomaly produced by the presence of a nearby grounded conductor, such as a metal fence. The nature of fences is usually indicated on IP maps; for example, "wooden post fence", "steel post fence", etc.

Fermat path: *Raypath* (q.v.).

Fermat's principle: The seismic raypath between two points is that for which the first-order variation of traveltime with respect to all neighboring paths is zero. It is sometimes phrased as that path for which the traveltime is a minimum (or, in certain cases, a stationary value or a maximum) compared with all neighboring paths. If the intervening media have different speeds, the path will not be straight, but will be such that the over-all traveltime is minimized (usually). The resulting raypath is also called the **least-time path** or **brachistochrone**. Snell's law follows from Fermat's principle. Named for Pierre Fermat (1601-1665), French mathematician.

ferrimagnetism: Property of some spinel-structured ferrites which show both ferromagnetic and antiferromagnetic properties because ionic interactions favor both parallel and antiparallel alignment of group (domain) magnetic moments. Ferrimagnetic substances include distinct ferromagnetic sublattices which couple antiferromagnetically so that the observed magnetism is the difference.

ferromagnetic: Having positive and relatively large susceptibility and generally large hysteresis and remanence. In ferromagnetic materials the atoms interact and atomic magnetic moments couple so that groups of atoms (**domains**) behave collectively and orient in a parallel configuration. As the temperature of such materials rises to the Curie point, the thermal energy of the atoms becomes sufficient to overcome the coupling energy and the material behaves paramagnetically. See also *diamagnetic* and *paramagnetic*.

FET: *Field-effect transistor* (q.v.).

fetch: The action of obtaining an instruction from a stored program and decoding the instruction. Also refers to that portion of a computer's instruction cycle where this action is performed.

FFI: *Free fluid index* (q.v.). See also *nuclear-magnetism log*.

FFT: *Fast Fourier transform* (q.v.).

Fick's law: The **diffusion-rate law** of electrochemistry: the time rate of change of diffusion flux is proportional to the concentration gradient and the activity of the electrode.

fiducials: 1. Points accepted as fixed bases of reference. 2. Marks which indicate points of simultaneity; e.g., a mark on a magnetic-intensity record showing which point corresponds to a point on an altimeter record and to a point on the map or navigation records which were made at the same time. 3. Time marks on a seismic record.

fiducial time: Arrival time on a seismic record with respect to a datum plane.

field: 1. That space in which an effect, e.g., gravity or magnetism, is measurable. Fields are characterized by continuity, i.e., there is one and only one value associated with every location within the space. 2. The outdoors, where geophysical surveys are made. 3. A large tract or area containing valuable minerals, such as a coal field or an oil field. 4. One or more columns on a punched card or other storage device where related arrangements of characters or digits represent a quantity, amount, name, identity, etc.

field balance: See *Schmidt field balance*.

field continuation: See *continuation*.

field-effect transistor: FET. A transistor whose internal operation is unipolar in nature; widely used in integrated circuits due to small size, low power dissipation, ease of manufacture, and low cost. A semiconducting device which uses as input a transverse electric field to vary its conductance and thus control its output current. Ideally it is a voltage-controlled current source.

field filter: A band-pass filter (usually) used in recording seismic or other data.

field impedance: see *impedance*.

field intensity: Force per unit. For a magnetic field, force per unit magnetic pole (or current per unit length); for a gravitational field, force per unit mass (or acceleration); for an electric field, force per unit charge (or voltage gradient). Also called **field strength**. Sometimes expressed as **flux density**, lines of force per unit area.

field tape: A magnetic tape containing geophysical data recorded in the field. As opposed to a processed tape on which the data have been modified by computer processing.

field timing: The timing signals recorded in the field,

High-pass (low-cut) filter Low-pass (high-cut) filter Band-pass filter	Attenuates high and/or low frequencies
Notch filter	Attenuates narrow band of frequencies
Hi-line balancing	Adjusts resistive/reactive impedance
Spiking deconvolution Whitening	Builds up all frequency components within specified band-pass to same amplitude
Predictive deconvolution	Removes repetitive aspects after time lag
Optimum filtering Wiener filtering	Filters to give results as close as possible to some desired output subject to constraints
Wavelet processing	Determines or changes embedded wavelet
Maximum entropy filtering	Produces result as unpredictable as possible
Minimum entropy filtering	Maximizes spikey character of output
Homomorphic deconvolution	Lifters in the cepstral domain
Stacking	Attenuates out-of-register components
Velocity filtering	(Multichannel filter) Attenuates certain apparent velocities (or certain apparent dips)
Coherency filtering	(Multichannel filter) Attenuates where certain coherence tests are not satisfied
Automatic picking	(Multichannel filter) Eliminates data which fail certain coherency and amplitude tests
Spatial filtering	Performed by discrete sampling in space
Werner filtering	Inverts magnetic data

FIG. F-6. Types of **filtering**.

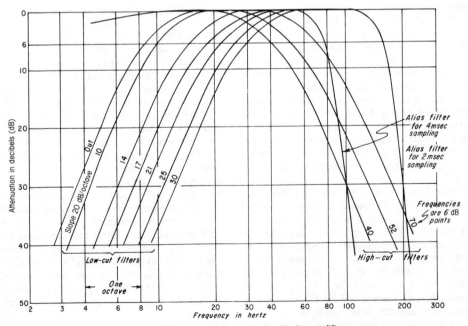

FIG. F-7. **Filter curves** for typical seismic amplifier.

usually displayed on a field monitor record as timing lines, on an analog field tape as a 100 Hz timing signal, and on digital tape as timing words or words counted from the time break. Commonly originates from a crystal-controlled oscillator or from a tuning fork. See Figure R-5.

figure of the Earth: *Geoid* (q.v.).

file: **1.** A collection of related records or program instructions treated as a unit. **2.** A set of records on a recording medium (such as magnetic tape) that are delimited by end-of-file marks.

file protect: An interlock which prevents destroying information already stored by writing new information on top of it.

filter: **1.** A part of a system that discriminates against some of the information entering it. The discrimination is usually on the basis of frequency, although other bases such as wavelength, moveout, coherence, or amplitude may be used. See Figure F-6. **2.** **Linear filtering** is called *convolution* (q.v.). A linear filter may be characterized by its impulse response or more usually by its frequency-domain transfer characteristics (amplitude and phase response) as a function of frequency. Filter characteristics are designated by specifying the frequencies at which the amplitude is down by 3 dB (70 percent or half power) and by the slope of the cutoff. Thus "14/18-56/36" specifies a low-cut filter down 3 dB at 14 Hz with an 18 dB/octave slope and a high-cut filter down 3 dB at 56 Hz with a 36 dB/octave slope. Typical seismic filter curves are shown in Figure F-7. **3.** Some passive electrical filters are shown in Figure F-8. Simple RC or RL filters have slopes of 6 dB/octave, but they may be cascaded to achieve higher slopes. Thus a double-section simple filter may have a slope of 12 dB/octave. A K-type filter may have a slope of 18 dB/octave; these may also be cascaded. See also *high-cut filter* (= low-pass filter) and *low-cut filter* (= high-pass filter). **4.** *Alias filters* (q.v.) are very sharp high-cut filters designed to prevent aliasing. **5.** **Band-pass filters** are often specified by listing successively their low-cut and high-cut component filters. The order of the specification is sometimes reversed. **6.** **Notch filters** reject sharply a very narrow band of frequencies. The M-derived filter (Figure F-8) is such a type; it is usually used in conjunction with another filter, e.g., MK. **7.** **Digital filters** provide a means of filtering data numerically in the time domain by summing weighted samples at successive time increments, i.e., by *convolution* (q.v.). Digital filtering permits one to filter easily in accord with arbitrarily chosen characteristics which might prove difficult or impossible to achieve with physical circuit components. **8.** Specific types of filters used in electric circuits include the **Butterworth filter** (see Sheriff and Geldart, v.2 1983, p. 188-189), a low-pass filter design with flat response, and the **Chebychev filter (Tchebyscheff filter)**, a band-pass filter with a steep rolloff, characterized by a uniform ripple in the passband. **9.** Filtering can be accomplished by optical methods as well as by electrical and digital methods; see *laserscan*. **10.** See also *inverse filter* and Sheriff and Geldart, v.2 (1983, p. 33, 40, 42). **11.** Stacking is a

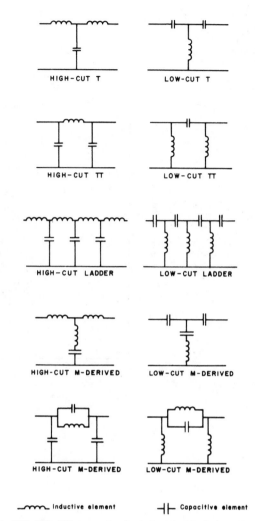

HIGH-CUT T   LOW-CUT T

HIGH-CUT TT   LOW-CUT TT

HIGH-CUT LADDER   LOW-CUT LADDER

HIGH-CUT M-DERIVED   LOW-CUT M-DERIVED

HIGH-CUT M-DERIVED   LOW-CUT M-DERIVED

⌢⌢⌢ Inductive element   ⊣⊢ Capacitive element

FIG. F-8. **Filters** made of passive electrical elements.

filtering process; see Sheriff and Geldart, v. 1 (1982, p 150-151). **12.** To remove solids from a suspension by passage through a sieve.

filter cake: *Mud cake* (q.v.).

filter correction: Correction of record times to compensate for time delays associated with the use of different filters.

filtering: The attenuation of certain components of a signal based on some measurable property. Usually implies that the measurable property is frequency. May be done by analog methods (often electrically) or numerically. See *filter* and Figure F-6.

filter panel: A display showing data filtered by various narrow band-pass filters to see the effects of different passbands on various events. Also called **frequency slices.**

filtrate: Fluid which has passed through a filter. Specifically, the drilling fluid which has passed through the *mud cake* (q.v.).

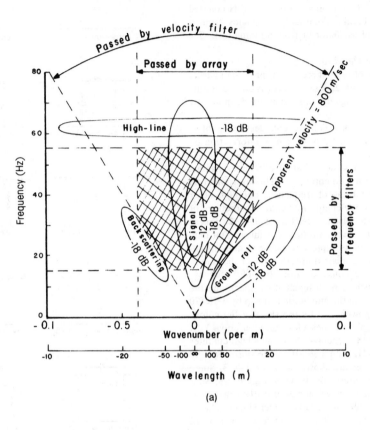

(a)

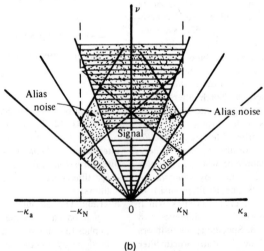

(b)

FIG. F-9 **f, k plot**. (**a**) The region passed by array filter, frequency filter, and velocity filter is cross-hatched. Radial lines through the origin represent constant apparent velocity ($V = \nu/x$). (**b**) Data beyond the Nyquist wavenumber x_N (a consequence of discrete spatial sampling) folds back (**aliases**) and may get mixed up with the signal.

fingers: Probes on the sonde of a *caliper log* (q.v.).

finite-difference method: 1. A scheme to solve differential equations by substituting difference quotients for derivatives and then using these equations to approximate a solution. **2.** Method of approximating a derivative by taking the difference of the function at two discrete points. See Sheriff and Geldart, v.2 (1983, p. 155-158).

finite-element method: A numerical scheme for approximating a solution to differential equations by using small enough domains or elements so that the solution involves nearly linear relationships.

firing: Generating a seismic wave. Originally meant "by means of an explosion", but now includes any means.

firing rate: The rate at which a transducer, sparker, or other energy source is discharged.

firmware: A computer program that is implemented in a type of hardware, such as read-only memory.

first arrival: *First break* (q.v.).

first break: The first recorded signal attributable to seismic-wave travel from a known source. First breaks on reflection records are used for information about the weathering. An initial compression usually shows as a downkick. First-break times are used in static correlation and in head wave interpretation.

first-break intercept-time method: A method of making static corrections based on first breaks; see Figure S-22.

first-order triangulation: See *triangulation*.

first point of Aries: See *Aries*.

fish: A sensor which is towed in the water, such as used with side-scan sonar.

fishtail bit: A drilling bit used to penetrate soft formations. Also called "drag bit". See Figure B-3.

fish trap: Conservation agent attached to a seismic crew (usually a marsh or water crew).

fix: A determination of location, as by the intersection of two *lines-of-position* (q.v.). A fix is made without reference to a former position. Fixes are determined by terrestrial, electronic, or astronomic means. See Figure L-5.

fixed data: Data which are constant for a collection of data and thus need not be input to a computer more than once.

fixed layer: A compact layer of ions and molecules held in place on an electrode or solid by chemical or electrostatic adsorption forces. Also called the **bound layer** or **inner Helmholtz double-layer.**

fixed-layer capacitance: Capacitance due to the presence of fixed-layer ions.

fixed point: 1. A method of data representation in which the location of the radix point (decimal point) is fixed. Compare *floating point*. **2.** Sometimes implies integer operation.

fixed-source method: A profiling method in which the source of energy is stationary and the receiver is moved about to explore the area. Electromagnetic sounding techniques are often fixed-source methods. Compare *moving-source method*.

***f–k* analysis:** Frequency-wavenumber analysis, an examination of how data sort into distinguishable sets in the frequency-wavenumber domain. Used to examine the

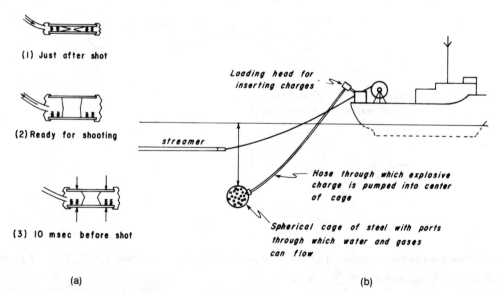

(1) Just after shot

(2) Ready for shooting

(3) 10 msec before shot

(a)

Loading head for inserting charges

streamer

Hose through which explosive charge is pumped into center of cage

Spherical cage of steel with ports through which water and gases can flow

(b)

FIG. F-10. **(a) Flexichoc;** (1) Two plates are separated by compressed air until they lock into position; (2) the air between them is pumped out, creating a vacuum; (3) on signal, the plates are unlocked and water pressure forces them together; the inrush of water into the consequent empty space creates a shock wave in the water. **(b) Flexotir;** small charges (about 2 oz) are propelled through a rubber hose by water under pressure into a steel cage where they are detonated; holes in the cage allow water repelled by the explosion to flow out and in, dissipating the bubble energy. (Courtesy Compagnie Générale de Géophysique.)

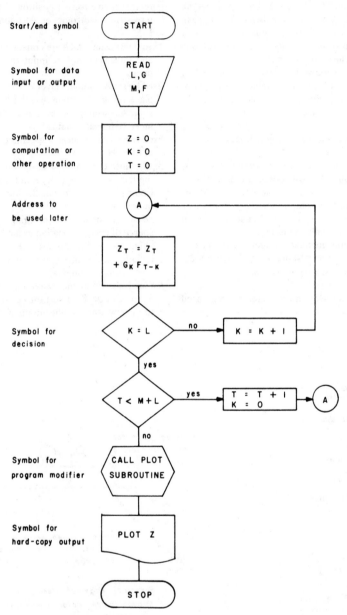

FIG. F-11. **Flow chart** for the convolution of $g(t)$ with $f(t)$. Input: $G = [G_0, G_1, ..., G_L]$; filter: $F = [F_0, F_1; ...., F_M]$; output: $Z = [Z_0, Z_1, ..., Z_{L+M}]$ where $Z_i = \sum\limits_{k=0}^{L+M} G_k F_{i\text{-}k}$.

direction and apparent velocity of seismic waves. See Figure F-9.

_f–k_ filter: *Velocity filter* (q.v.).

_f–k_ migration: *Migration* (q.v.) in the frequency-wavenumber domain. See Sheriff and Geldart, v.2 (1983, p. 62-63).

_f-k_ space: A domain in which the independent variables are frequency (*f*) and wavenumber (*k*), the result of a two-dimensional Fourier transform of a seismic record or seismic section. Seismic data analysis sometimes involves *f–k* plots in which energy density within a given time interval is plotted and contoured on a frequency-versus-wavenumber basis. Sometimes used in velocity-filter design. See Figure F-9.

flag: 1. A bit attached to a computer word to indicate the boundary of a field. **2.** An indicator to tell a later part of a computer program about a condition which occurred earlier in the program.

flagging: Strips of plastic, cloth, or paper used to mark instrument or shot locations.

flat spot: A horizontal seismic reflection not conformable with other reflections and attributed to an interface between two fluids such as gas and water or gas and oil. See *hydrocarbon indicator*.

flattened section: A datumized section. **1.** A seismic record section in which a particular record event has been made flat (or sometimes merely smooth) by introducing arbitrary time shifts. Sometimes used as a method of determining static corrections. If flattened on an event which represents a bed which was deposited horizontally, the flattened section resembles a paleosection showing the attitude of deeper events at the time of deposition of the flattened bed (assuming that data migration and nonprimary reflection events do not cause problems). Useful in studies of variations in the time interval between events. **2.** A *paleosection* (q.v.).

flattening: The ratio of the difference between the major and minor axes of an ellipse to the major axis; see Figure E-9. The flattening of the Earth is about 1/295.25; see also Figure G-3.

flat-topping: Reducing values which exceed some chosen value to the chosen value; clipping. May be caused by loss of sensitivity due to saturation of some part of the measuring system. Compare *digital clipping* and see Figure C-6.

F-layer: One of the layers of ionized air in the ionosphere which reflects back radio waves up to about 50 MHz. In the daytime the *F*-layer subdivides into two layers, the lower of which (F_1) is usually 175 to 250 km high and is found only during daylight. The higher F_2 layer at 250 to 400 km is present both night and day and is the principal reflector of HF radio waves at night.

Flexichoc: An implosive energy source for marine shooting. See Figure F-10a. CGG tradename.

Flexotir: A seismic method for marine shooting whereby charges are detonated in a steel cage containing holes through which the water repelled by the explosion flows out and in, thus dissipating some of the energy in the bubble of waste gases and attenuating the bubble effect. This permits charges to be fired much deeper than otherwise (at about 40 ft) which greatly increases the shot efficiency. See Figure F-10b. Also called **cage shooting**. IFP tradename.

flexural wave: A normal mode in a thin plate with motion antisymmetric about the median surface of the plate.

flip-flop: A bistable oscillator; a device with exactly two stable states. Used to store one bit of information.

floating: Not electrically connected to ground or to the system reference voltage.

floating charge: A seismic charge which is not as deep as intended and thought to be. Characterized by early uphole time and delayed reflection times.

floating datum: A variable reference surface used in areas of variable topography. The elevation of the floating datum is varied to minimize the illusion of subsurface structure which would be caused by surface relief if a constant-elevation datum were used. The topography may produce velocity variations because of the increased loading or the topography may merely reflect structure which also affects velocity distribution. A **one-third floating datum** is sometimes used in which the reference reflects one-third of the surface elevation above a flat plane. Tilted datums are also used sometimes. The floating datum is merely one way of introducing static corrections so that the velocity distribution beneath the datum is simplified, usually by assuming no horizontal variation. More elegant methods of dealing with lateral velocity variations are now used, thus lessening use of floating datums.

floating point: A number expressed by the significant figures times a base raised to a power. Thus 139000 might be written as 1.390×10^5 to indicate four significant figures. Writing numbers in floating point format prevents the loss of significant figures in case the number becomes too small or too large for a fixed register. Computers usually use bases that are a power of 2 rather than the base 10.

floppy disk: A disk in a disk storage memory which is flexible rather than rigid. A **diskette**.

flower structure: A type of geologic structure which results from strike-slip movement associated with convergence.

flow chart: A diagram showing the operations involved in a process. Often used to show the steps of a computer program or steps in processing and their sequence. See Figures F-11 and P-9.

flowmeter: A device which measures the flow of fluid. In borehole studies the flowmeter may be lowered through the flow stream or set in one spot with a packer. See *spinner survey*.

fluid sampler: See *formation tester*.

fluid-travel log: A recording of the flow of fluid past a logging tool. A radioactive slug is injected into the fluid stream through ports in the logging tool and the radioactive counting rate is measured as a function of time at two detector positions. Used to locate leaks.

fluid wave: *Acoustic wave* (q.v.).

flushed zone: See *invaded zone* and Figure I-5.

flute: *Mute* (q.v.).

flutter: Noise caused by variations in speed of magnetic tape; wow.

flux: 1. A representation of magnetic, electrical, etc., lines

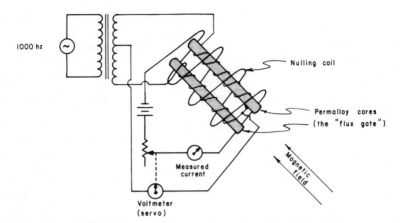

FIG. F-12. **Fluxgate magnetometer**. Two permalloy cores approach saturation in the weak magnetic field of the Earth. A 1000-Hz cyclic field superimposed by a coil around the core completes saturation. The place in the energizing cycle at which saturation is reached is a measure of the strength of the Earth's field. A secondary coil detects the changes in flux. Two parallel cores are used with windings in opposite directions and the difference is measured. A current through an additional winding nullifies most of the background magnetic field so that the magnetometer is sensitive to small changes in the Earth's field. The current through the nulling coil is a measure of the magnetic field strength.

of force where the flux density is proportional to field intensity. **2.** A flow of ideas.

flux density: See *field intensity*.

fluxgate magnetometer: An instrument capable of detecting changes in the magnetic field of the order of 0.2 gamma. See Figure F-12. The magnetometer measures the magnetic field component along the axis of its core and must be oriented with the field if the total intensity is to be measured. With the Gulf magnetometer this is accomplished by using three mutually perpendicular fluxgate instruments and servomechanisms which vary the orientation so as to minimize the magnetic field in two of these, thus maximizing the field in the third. Compare *proton magnetometer*, *optically pumped magnetometer*, and *Squid* (cryogenic) *magnetometer*.

flyer: A number of geophones permanently connected at intervals along a short cable, used for one geophone group. A typical flyer might contain, for example, 6 geophones connected in series at 20 ft intervals, with clips at the end for connecting it to the main cable which carries the signal to the recording equipment.

FM: *Frequency modulation* (q.v.).

focus: 1. The location of the first rupture in an earthquake; the **hypocenter**. **2.** See *buried focus*.

focused log: A laterolog or guard log. Sometimes refers to FoRxo, microlaterolog, or proximity log.

fold: 1. The multiplicity of common-midpoint data. Where the midpoint is the same for 12 offset distances, e.g., the stack is referred to as "12-fold". **2.** See also *folding*.

fold-back: 1. A cable which is doubled back on itself such

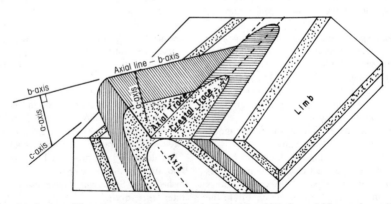

FIG. F-13. **Folding** terminology. The **plunge** of the anticline is the angle which the axial line makes with horizontal. The **b-axis** is the direction of the axial line, the **a-axis** is in the plane containing the axial line and the axis of the fold, and the **c-axis** is perpendicular to this plane.

that two or more geophone stations are located at the same position on the ground. Used for making comparisons (for example, of different types of geophone arrays) on the same shot. **2.** If too many cables are connected together and too many geophone groups are connected, groups at two different locations may feed the same channel, producing confused results. This is sometimes called fold-back.

folding: 1. Frequency folding such as results from inadequate sampling, producing *alias* (q.v.) problems. **2.** *Convolution* (q.v.). **3.** Bending of geological strata. **Primary folding** is response to deep seated forces with a strong horizontal component. **Gravitational folding** results from downward sliding and flow, which are secondary results of uplift and tilting. Local folding can be caused by compaction or by the upwelling of salt or igneous rocks. The upraised part of a fold forms an **anticline**, the downwarped portion a **syncline**. Folding terminology is shown in Figure F-13. Materials respond differently to the same stresses (and stress durations) and cleavage, flow and faulting are usually associated with folding; see Figure F-14. **Competent** beds tend to retain their thickness in folding; they govern the folding wavelength, which is of the order of 25 times the thickness of the most competent member. **Incompetent** beds flow in response to folding stresses, but the distinction with competent beds is gradational. Two folding styles are illustrated in Figure F-15. Folding is **disharmonic** where the folding of one bed is not geometrically related to the folding of nearby beds, incompetent beds intervening between them.

folding frequency: The Nyquist frequency; see *alias*.

fold test: Variation in the direction of remanent magnetism in a rock which has been folded; it can indicate whether the magnetization predated the folding or not.

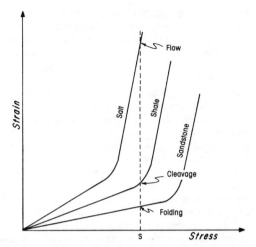

FIG. F-14. **Folding**, cleavage, or flow may result from stress. The stress duration is involved as well as the stress magnitude. (From deSitter, 1964.)

foreshock: An earthquake which precedes a larger earthquake.

foreign: 1. Not part of a system. **2.** Not part of one's own company.

foresight: A surveying measurement with the objective of determining the position and elevation of the stadia rod with respect to the survey instrument.

format: The arrangement of data (as on a magnetic tape). Involves the placement of bits of different significance, number of bits per byte and bytes per word, parity,

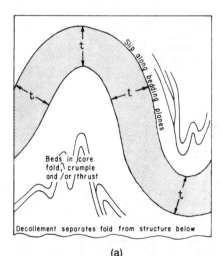

(a)

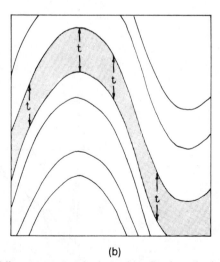

(b)

FIG. F-15. **Folding** styles. (**a**) In **concentric** or **flexural-slip folding**, competent beds tend to slip along the bedding planes and retain constant thickness perpendicular to the bedding, whereas incompetent beds fold and fault and separate the fold from deeper structure. (**b**) In **similar folding**, beds tend to retain their vertical thickness, thus thinning on the sides of the fold. The thickness *t* tends to be constant.

multiplexing arrangement, timing and gain information, record identification, and other auxiliary data, gaps, start-stop codes, etc. Standard formats are specified by SEG, 1980. Specific computer systems may use their own format so that the first step with new data may be to reformat it.

formation: A lithological unit with some distinction which permits identification. It is not necessarily a time unit. May be composed of several "members" and be part of a larger "group."

formation-analysis log: A computed log of apparent fluid resistivity and apparent porosity based on induction log and either sonic- or density-log data. Abbreviated **FAL**; also called **R_{wa}-analysis log**.

formation-density log: See *density log*. Abbreviated **FDL**.

formation evaluation: The analysis and interpretation of well-log data, drill-stem tests, cores, drill cuttings, etc., in terms of the nature of the formations and their fluid content. The objectives of formation evaluation are to ascertain if commercially productive hydrocarbons are present and, if so, the best means for their recovery; other objectives are to derive lithology and other information on formation characteristics for use in further exploration. See Figure W-8.

formation factor: The ratio of the resistivity of a formation to the resistivity of the water with which it is saturated. See *Archie's formula*.

formation tester: A tool run on a wireline to obtain samples of formation fluid. Hydrostatic, flow, and shut-in pressures are recorded.

form factor: 1. Geometric factor, the geometric multiplying factor which depends on the type of electrode array and interval being used. 2. The type curve for a profile across an idealized body, e.g., across a sphere.

Fortran: FORmula TRANslation; a high-level language designed to simplify programming for digital computers. Designed for solving algebraic problems and for scientific procedural programming.

forward bias: See *bias*.

forward solution: Solution of a *direct problem* (q.v.), such as predicting the electric potential for a given distribution of resistivity current sources. See Figure M-8.

FoRxo: See *microlaterolog*. Welex tradename.

fossil remanence: See *remanent magnetism*.

Foucault current: See *induction log*. Named for Jean Bernard Leon Foucault (1819-1868), French physicist.

foundation coefficient: A coefficient expressing how many times stronger is the effect of an earthquake in a given rock than it would have been for undisturbed crystalline rock.

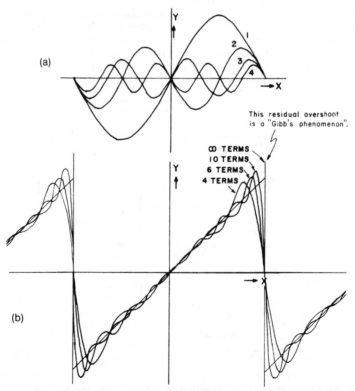

FIG. F-16. (a) **Fourier analysis** involves finding frequency components for a waveform. For a sawtooth waveform, the first four components are: sin x; $-(1/2)$ sin $2x$; $(1/3)$ sin $3x$; $-(1/4)$ sin $4x$. (b) **Fourier synthesis** involves superimposing the components to reconstitute the waveform.

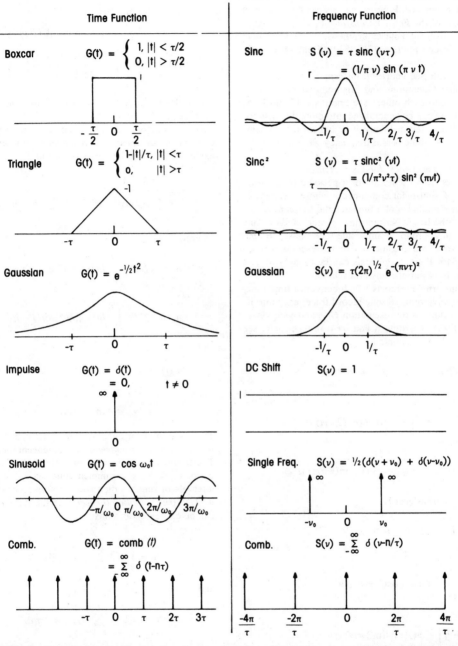

Time Function	Frequency Function

Boxcar $\quad G(t) = \begin{cases} 1, & |t| < \tau/2 \\ 0, & |t| > \tau/2 \end{cases}$

Sinc $\quad S(\nu) = \tau \, \text{sinc}\, (\nu\tau)$
$$= (1/\pi\,\nu)\, \sin\,(\pi\,\nu\,t)$$

Triangle $\quad G(t) = \begin{cases} 1-|t|/\tau, & |t| < \tau \\ 0, & |t| > \tau \end{cases}$

Sinc² $\quad S(\nu) = \tau \, \text{sinc}^2\, (\nu t)$
$$= (1/\pi^2\nu^2\tau)\, \sin^2\,(\pi\nu t)$$

Gaussian $\quad G(t) = e^{-\frac{1}{2}t^2}$

Gaussian $\quad S(\nu) = \tau(2\pi)^{1/2}\, e^{-(\pi\nu\tau)^2}$

Impulse $\quad G(t) = \delta(t)$
$$= 0, \qquad t \neq 0$$

DC Shift $\quad S(\nu) = 1$

Sinusoid $\quad G(t) = \cos\,\omega_0 t$

Single Freq. $\quad S(\nu) = \frac{1}{2}(\delta(\nu+\nu_0) + \delta(\nu-\nu_0))$

Comb. $\quad G(t) = \text{comb}\,(t)$
$$= \sum_{-\infty}^{\infty} \delta\,(t-n\tau)$$

Comb. $\quad S(\nu) = \sum_{-\infty}^{\infty} \delta\,(\nu-n/\tau)$

FIG. F-17. **Fourier transform** pairs. The time functions on the left are Fourier transforms of the frequency functions on the right and vice versa. Many more transform pairs could be shown. The above are all even functions and hence have zero phase. Transforms for real odd functions are imaginary, i.e., have a phase shift of $+\pi/2$. Transforms of functions which are neither odd nor even involve variation of phase with frequency. Note: $\nu = 1/t$,

Fourier analysis: The analytical representation of a waveform as a weighted sum of sinusoidal functions. Determining the amplitude and phase of cosine (or sine) waves of different frequencies into which a waveform can be decomposed. Fourier analysis can be thought of as a subset of the *Fourier transform* (q.v.). See Figure F-16. Opposite of **Fourier synthesis.** Named for Jean Baptiste Joseph Fourier (1768-1830), French mathematician.

Fourier integral: See **Fourier transform**.

Fourier pairs: Operations and functions which Fourier-transform into each other. See Figures F-17 and F-18. Fourier pairs can be generalized into more dimensions, such as illustrated in Figure F-19 (where the domains might be offset-time versus frequency-wavenumber (*f–k* space).

Fourier plane: *Frequency domain* (q.v.).

Fourier series: Representation of a periodic function by the sum of sinusoidal components whose frequencies are integral multiples of a fundamental frequency.

Fourier synthesis: Superimposing cosine and/or sine waves with appropriate amplitude (and phase) to construct a waveform (or time-domain representation). See Figure F-16. Fourier synthesis can be thought of as a subset of the Fourier-transform relations.

Fourier transform: Formulas which convert a time function $g(t)$ (waveform, seismic record trace, etc.) into its frequency-domain representation $G(\nu)$ and vice versa. $G(\nu)$ and $g(t)$ constitute a **Fourier-transform pair**; see Figure F-17. An example is

$$g(t) \leftrightarrow G(\nu) = \int_{-\infty}^{\infty} g(t)\, e^{-j2\pi\nu t}\, dt$$

$$= \int_{-\infty}^{\infty} g(t) \cos(2\pi\nu t)\, dt$$

$$+ j\int_{-\infty}^{\infty} g(t) \sin(2\pi\nu t)\, dt.$$

The **inverse transform** is

$$g(t) = \int_{-\infty}^{\infty} G(\nu)\, e^{-j2\pi\nu t}\, d\nu$$

$$= \int_{-\infty}^{\infty} G(\nu) \cos(2\pi\nu t)\, d\nu$$

$$- j\int_{-\infty}^{\infty} G(\nu) \sin(2\pi\nu t)\, d\nu.$$

Finding $G(\nu)$ from $g(t)$ is usually called **Fourier analysis** and finding $g(t)$ from $G(\nu)$ is called **Fourier synthesis.** $G(\nu)$ is the **complex spectrum**, the real part being the **cosine transform** and the imaginary part the **sine transform** whenever $g(t)$ is real. Another expression for $G(\nu)$ is

$$G(\nu) = |A(\nu)|e^{j\gamma(\nu)},$$

where the functions $A(\nu)$ and $\gamma(\nu)$ are real and are, respectively, the **amplitude spectrum** and the **phase spectrum** of $g(t)$.

$$A(\nu) = \{[\text{Real part of } G(\nu)]^2$$
$$+ [\text{Imaginary part of } G(\nu)]^2\}^{1/2};$$

$$\gamma(\nu) = \tan^{-1}\left[\frac{\text{Imaginary part}}{\text{Real part}}\right];$$

γ is in the 1st or 2nd quadrant if the imaginary part is positive, in the 1st or 4th quadrant if the real part is positive. A record trace $h(t)$ which extends only from 0 to T may be assumed to be repeated indefinitely and so expanded in a Fourier series of period T:

$$h(t) = \sum_{0}^{\infty} a_n \cos(2\pi nt/T) + \sum_{0}^{\infty} b_n \sin(2\pi nt/T)$$

where

$$a_n = (2/T)\int_{0}^{T} h(t) \cos(2\pi nt/T)\, dt,$$

$$b_n = (2/T)\int_{0}^{T} h(t) \sin(2\pi nt/T)\, dt,$$

and

$$h(t) \leftrightarrow H_n = |A_n|e^{i\gamma_n}.$$
$$A_n = (a_n^2 + b_n^2)^{1/2},$$
$$\gamma_n = \tan^{-1}(b_n/a_n).$$

The same rules for quadrants apply to γ_n as expressed for $\gamma(\nu)$. a_0 is the **zero-frequency component** (or **dc shift**). The frequency spectrum is discrete if the function is periodic. If h_t is a sampled time series sampled at intervals of time t_2, then we can stop summing when $n > 2T/t_2$ (see *sampling theorem*). In this case, a_n and b_n can be expressed as sums:

$$a_n = (2/T)\sum_{t=0}^{T} H_t \cos(2\pi nt/T),$$

and

$$b_n = (2/T)\sum_{t=1}^{T} H_t \sin(2\pi nt/T).$$

See also *phase response* and *fast Fourier transform.* Operations in one domain have equivalent operations in the transform domain (Figure F-18). Computations can sometimes be carried out more economically in one domain than the other and Fourier transforms provide a means of accomplishing this. The Fourier transform relations can be generalized for more than one dimension (see Figure F-19). For example,

$$G(\kappa, w) = \int_{-\infty}^{\infty} \int_{-\infty}^{\infty} g(x, t) \, e^{-j(\kappa x + \omega t)} \, dx \, dt,$$

$$g(x, t) = \int_{-\infty}^{\infty} \int_{-\infty}^{\infty} G(\kappa, \omega) \, e^{j(\kappa x + \omega t)} \, d\kappa \, d\omega.$$

Fourier transforms are discussed in Sheriff and Geldart, v. 2 (1983, p. 26-27 and 160-167).

four-way dip: Dip calculated from a cross-spread (especially where both in-line and cross-spreads are splits). See Figure C-17.

fracture log: A well log of the cumulative amplitude of the wave arrivals from a sonic logging tool during a certain gate time. A fracture zone attenuates the acoustic energy.

Fraunhofer diffraction: The special case of *Fresnel diffraction* (q.v.) at a distance approaching infinity (or by using a lens to make rays parallel). Named for Joseph von Fraunhofer (1787-1826), German physicist.

free-air anomaly: Gravity data which have been corrected for latitude and elevation (free-air correction) but not for the density of the rock between datum and the plane of measurement (Bouguer correction). Measures the attraction because of the mass of the subadjacent earth.

free-air correction: 1. A correction for the elevation of a gravity measurement required because the measurement was made at a different distance from the center of the earth than the datum. The first term of the free-air correction is 0.09406 mGal/ft or 0.3086 mGal/m. **2.** In Turam, normalizing a ratio of successive measurements by dividing by the ratio of the calculated free-space vertical magnetic field. Compare *normal correction*.

free-field: Not relating to a preassigned or fixed field or format.

free-fluid index: The percent of the bulk volume occupied

Time operation	Frequency operation	Significance
Linear addition $ag(t) - bh(t)$	Linear addition $aG(\nu) - bH(\nu)$	Linearity and superposition apply in both domains. The spectrum of a linear sum of functions is the linear sum of their spectra (if spectra are complex, rules of addition of complex quantities apply). Any function may be regarded as a sum of component parts and the spectrum is the sum of the component spectra.
Scale change $g(kt)$	Inverse scale change $\dfrac{1}{k} G\left(\dfrac{\nu}{k}\right)$	Time-bandwidth invariance. Compressing a time function expands its spectrum in frequency and reduces it in amplitude by the same factor. The amplitude reduces because less energy is spread over a greater bandwidth. The case where $k = -1$ reverses the function in time, interchanging positive and negative frequencies, and for real time functions, reversing the phase.
Convolution $g(t) * h(t)$	Multiplication $G(\nu)H(\nu)$	The spectrum of the convolution of two time functions is the product of their spectra. If the spectra are complex, multiplication is equivalent to multiplying amplitude spectra and adding phase spectra.
Multiplication $g(t)h(t)$	Convolution $G(\nu) * H(\nu)$	The spectrum of the product of two time functions is the convolution of their spectra.
Delay $g(t - t_0)$	Linear added phase $e^{-j2\pi\nu_0 t} G(\nu)$	Delaying a function by a time t_0 multiplies its spectrum by $e^{-j2\pi\nu_0 t}$, thus adding a linear phase $\theta = -2\pi\nu t_0$ to the original phase. Conversely, a linear phase filter produces a delay of $-d\theta/d\nu = 2\pi t_0$.
Complex modulation $e^{j2\pi\nu_0 t} g(t)$	Shift of spectrum $G(\nu-\nu_0)$	Multiplying a time function by $e^{j2\pi\nu_0 t}$ shifts its spectrum to center about ν_0 rather than zero frequency.

FIG. F-18. Equivalence of **Fourier transform** operations. Doing the "time operation" is equivalent to doing the "frequency operation" on the transform of the data. Note: $g(t) \leftrightarrow G(\nu)$, $h(t) \leftrightarrow H(\nu)$.

TIME DOMAIN

FREQUENCY DOMAIN

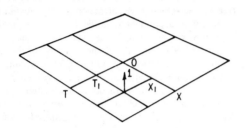 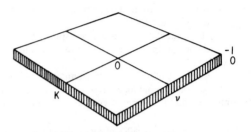

An impulse, $\delta(X-X_1, T-T_1)$, transforms into a dc shift plus a phase shift.

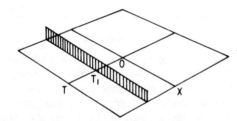 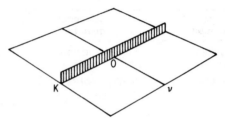

A one-dimensional impulse, $\delta(T-T_1)$, transforms into a one-dimensional impulse, $\delta(\nu)$, plus a phase shift.

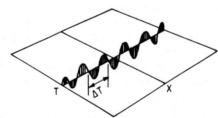 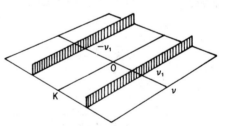

A sinusoid, $A\cos(2\Pi T/\Delta T + \emptyset)\,\delta(X)$ transforms into two one-dimensional impulses, $\delta(\nu_1-\nu_1)$ and $\delta(\nu_1+\nu_1)$, where $\nu_1 = 1/\Delta T$.

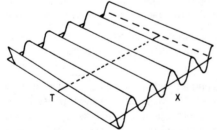 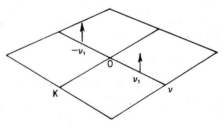

A sinusoidal surface transforms into two impulses, $\delta(F-F_1, K)$ and $\delta(F+F_1, K)$

FIG. F-19. **Fourier transform pairs** in two dimensions. (From Lindseth, 1970.)

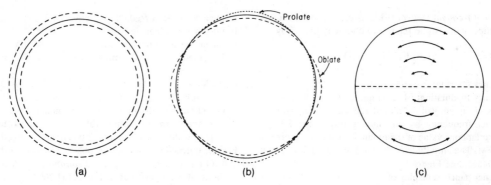

FIG. F-20. **Free oscillation of the earth**. The simplest modes are (**a**) radial, (**b**) spherical, and (**c**) toroidal oscillation. (From Stacey, 1969.)

by fluids which are free to flow as measured by the *nuclear-magnetism log* (q.v.). Abbreviated **FFI**. Gas gives a low FFI.

free nutation of the earth: See *Chandler wobble*.

free oscillation of the earth: A simple change-of-shape oscillation of the whole earth. The period is 53 minutes in the lowest mode. See Figure F-20.

free-space field: The field about an antenna in the absence of nearby conductors. See *primary field*.

frequency: 1. Symbol, ν. The repetition rate of a periodic waveform, measured in "per second" or hertz. The reciprocal of period. Compare *spatial frequency*. **2.** **Angular frequency** ω, measured in radians/s, is to frequency ν as 2π:

$$\omega = 2\pi\nu.$$

3. The **dominant frequency** of wavelets refers to an approximate repetition (the reciprocal of the peak-to-peak time interval) even though the entire wavelet does not repeat; see Figure W-2. **4.** The **frequency content** of a waveform refers to the amplitudes of the sinusoidal components into which the waveform can be decomposed by Fourier analysis, even where there is nothing repetitive about the waveform itself. For mathematical symmetry, the Fourier integral is usually written for frequencies from $-\infty$ to $+\infty$. Negative frequencies can be thought of as the repetition rate where one counts backward in time.

frequency domain: A representation in which frequency is the independent variable; the Fourier transform variable when transforming from time.

frequency-domain method: A method of potential-field analysis in which parameters of interest are estimated from characteristics of amplitude and phase spectra. In magnetic induced-polarization measurements, frequency-domain parameters measured include relative phase shift and percent frequency effect (PFE). Involves the variation of apparent resistivity with frequency.

frequency-domain migration: *Migration* (q.v.) in which the wave equation $\phi(x, z, t)$ is transformed into the frequency-wavenumber domain,

$$\phi(x, 0, t) \rightarrow \Phi(\kappa_x, \kappa_z, \omega),$$

various operations are performed, and then it is transformed back,

$$\Phi(\kappa_x, \kappa_z, \omega) \rightarrow \phi(x, z, 0),$$

which is the sought-for solution. See Sheriff and Geldart, v.2 (1983, p. 62-63).

frequency effect: The difference between resistivity (or voltage) measurements (ρ_1 and ρ_2) made at two frequencies f_1 and f_2, divided by the resistivity (or voltage) at one of the frequencies. Abbreviated **FE**.

$$FE = (\rho_2 - \rho_1)/\rho_1.$$

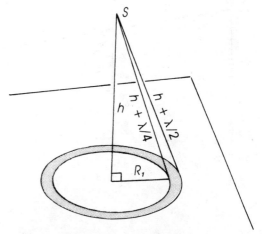

FIG. F-21. **Fresnel zones**. For coincident source and receiver at S, the first Fresnel zone radius is R_1 (perpendicular to h). The second Fresnel zone is·the shaded annular ring. Higher order zones (not shown) are also annular rings. The dominant wavelength is λ.

Percent frequency effect (**PFE**) is often used. Frequency effect can also be related to chargeability *M* by

$$M = FE/(1 + FE).$$

frequency filtering: *Filtering* (q.v.).

frequency modulation: Modulation in which the instantaneous frequency of the modulated wave differs from the carrier frequency by an amount proportional to the instantaneous value of the modulating wave. Abbreviated **FM.** The amplitude of the modulated wave is usually constant. See Figure M-10.

frequency panel: A display of a set of sections filtered with adjacent narrow band-pass filters.

frequency response: The characteristics of a system as a function of frequency. See *Fourier transform.*

frequency slices: A *frequency panel* (q.v.).

frequency sounding: See *geometric sounding.*

frequency spectrum: The characteristics of a waveform described as a function of frequency. See *Fourier transform.*

fresh: Very low in dissolved salts. Sometimes used comparatively with respect to normal sea water (which is 35 000 parts of dissolved salts per million), sometimes used comparing formation water with mud and filtrate. "Fresh water" has less than 2 000 ppm dissolved salts.

Fresnel diffraction: Diffraction observed close to the diffracting object. Compare *Fraunhofer diffraction.* Named for Augustin Jean Fresnel (1788-1827), French physicist.

Fresnel zone: 1. The portion of a reflector from which reflected energy can reach a detector within one-half

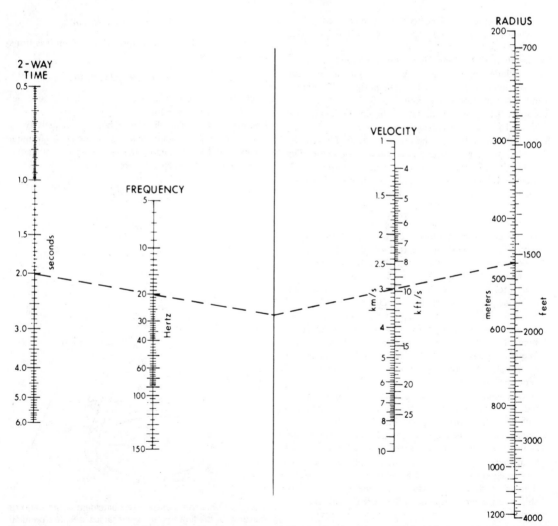

FIG. F-22. Nomogram for determining **Fresnel-zone** radius. A straight line connecting two-way traveltime and frequency intersects the central line at the same point as a straight line connecting velocity and the radius of the zone. For example, a 20 Hz reflection at 2.0 s and a velocity of 3.0 km/s has a Fresnel-zone radius of 470 m.

wavelength of the first reflected energy. Reflected energy from such a zone interferes constructively. The central **first Fresnel zone** is circular for a horizontal reflector; successive Fresnel zones are annular rings of successively larger radii; see Figure F-21. **2.** The first Fresnel zone only, the portion of a plane reflector mainly effective in generating a reflection. For a harmonic wave, the effects of successive zones generally cancel each other. A nomogram for determining the radius of the first Fresnel zone is shown in Figure F-22. See Sheriff and Geldart, v.1 (1982, p. 119-122). **3.** A **Fresnel lens** has alternate zones blacked out so that only constructive zones pass light.

front end: The portion of a seismic line ahead of the source point (in the direction of travel along the line).

frontier area: A relatively unexplored area (at least by the organization classifying it as frontier).

front-loaded: *Minimum-phase* (q.v.).

frost breaks: The effect of repetitive shots at random times following a shot, produced by ice fracturing when shooting in permafrost. Also called **ice noise.** See Sheriff and Geldart, v.2 (1983, p. 14).

FT: Field tape; a magnetic-tape record of geophysical data made directly "in the field" as opposed to a processed tape which is made from information which originates on another tape.

full duplex: A communication medium and protocol that support transmission in both directions simultaneously.

full-waveform log: A display of the acoustic wave train in wiggle form, as opposed to the similar sort of display in variable density form. Used in the 3-D log, microseismogram, or VDL log. See also *sonic log, fracture log,* and *cement-bond log.* Similar to *signature log* and *character log.*

full-wave rectifier: An electronic device which uses both positive and negative polarities of alternating current to produce direct current.

fundamental: The lowest frequency of a periodic function.

fundamental strength: The stress a material is able to withstand over a long time under a given set of conditions (temperature, pressure, solution, etc.) without deforming continuously.

G

G: Designation for **Love waves**, also designated L_Q. See also *wave notation* and Figure E-3.

GAG: *Geophysical Analysis Group* (q.v.).

GAGC: Ganged automatic gain control; see *gain control*.

gain: An increase (or change) in signal amplitude (or power) from one point in a circuit or system to another, often from system input to output.

gain control: 1. Control for varying the amplification or attenuation of an amplifier, used to compensate for variations in input signal strength. Gain control is often automatic, using a feedback loop whereby the output level controls the gain so as to keep the output level within certain limits. Used in many types of amplifers. At maximum gain the amplifier has maximum sensitivity, whereas at lower gains it will accept larger inputs without distortion. **2.** Gain controls used in seismic amplifiers include (a) **automatic gain control** in which the gain of each channel is controlled automatically and independently of other channels; (b) **ganged automatic gain control** in which the gain of all channels is the same although automatically determined, the basis being one single channel or the average energy level of a number of the channels; (c) **preset** or **programmed gain control** in which the gain as a function of record time is determined arbitrarily prior to the shot; (d) **binary gain control** in which gain is allowed to vary only by factors of two, but the time at which the gain changes is determined automatically; the times at which the changes occur are recorded so that the gain effects can be removed in processing (see Figures B-2 and D-21); (e) **quaternary gain control** in which gain is allowed to vary only by factors of four; and (f) **instantaneous-floating-point control** in which the gain is determined for each sample based on the amplitude of that channel without prejudice by earlier samples or the amplitude of other channels.

gain trace: A trace on a seismic record which indicates the gain (amplification) used on one or more channels. See Figure R-5.

Gal: A unit of acceleration or of gravitational force per mass, used in gravity measurements. One Gal = 1 cm/s^2 = 10^{-2} m/s^2 = 10^{-2} newton/kg. The Earth's nominal gravity is 980 Gal. Named for Galileo Galilei (1564-1642), Italian physicist.

galvanic contact: An actual electrical contact with the ground, as opposed to inducing electric current flow by induction. Called a **conduction contact** or an **ohmic contact** if linear and rectification is not involved. Named for Luigi Galvani (1737-1798), Italian anatomist.

galvanometer: A device to measure small currents. A coil suspended in a constant magnetic field rotates through an angle proportional to the electrical current flowing through the coil. A part of many seismic cameras. Often abbreviated **galvo**.

gamma: A unit of magnetic field. A gamma = 10^{-5} gauss = 10^{-9} tesla = 1 nanotesla.

gamma configuration: See *array (electrical)*.

gamma-gamma log: See *density log*.

gamma-ray log: A well log which records natural radioactivity. Also a generic term for any logging system based on gamma irradiation and measurement of resulting reactions. **1.** In sediments the log mainly reflects shale content because minerals containing radioactive isotopes (the most common of which is potassium) tend to concentrate in clays and shales. Volcanic ash, granite wash, and some salt deposits also give significant gamma-ray readings. The log often functions as a substitute for the SP for correlation purposes in cased holes, in conductive muds in open holes, and for thick carbonate intervals. See Figures D-3, L-3, and N-2. **2.** Used in exploration for radioactive minerals such as the uranium minerals. **3.** See *natural gamma-ray spectroscopy log, induced gamma-ray spectroscopy log*.

gamma-ray surveying: Measurement of naturally occurring gamma rays in the search for radioactive materials. Portable gamma-ray spectrometers incorporate energy-discrimination ability which permits distinguishing between radio materials. Thus one might measure all gamma rays, only those above 1.3 MeV (for potassium + uranium + thorium), only those above 1.6 MeV (for uranium + thorium), or only those above 2.5 MeV (for thorium).

ganged gain control: *Gain control* (q.v.) in which the gain of several channels varies in the same way.

gap: 1. Shotpoint gap, a larger interval between geophone groups on either side of the shotpoint than between other groups. **2. Interrecord gap,** an interval of space on a digital magnetic tape during which no information is recorded. Serves to indicate the beginning of a new record or a new block of data.

gapped deconvolution: Deconvolution with a *gapped operator* (q.v.).

gapped operator: A deconvolution operator which has groups of nonzero filter elements appreciably separated from each other by zeros. Used to attenuate multiples which involve water-layer peg-legs in deep water. See Sheriff and Geldart, v. 2 (1983, p. 43).

Gardner method: A refraction interpretation method which involves separating intercept time into constituent delay times associated with the shot and geophone ends of the trajectory. See Gardner (1939).

Gardner's rule: The empirical relationship that density is proportional to the 1/4 power of *P*-wave velocity. See Gardner et al. (1974).

gas chimney: A region of low-concentration gas escaping and migrating upward from a hydrocarbon accumulation. Generally shows as a region of severely deteriorated data quality, often associated with low velocity and velocity sags (push down) underneath the chimney.

gas exploder: A seismic energy source in which a mixture of propane or butane with oxygen or air is exploded under the water. Also called **gas gun**. Compare *Dinoseis*.

gas gun: *Gas exploder* (q.v.).

gas hydrate: Hydrated methane in a solid state. Such crystalline, ice-like clathrate compounds can exist under certain conditions of low temperature and high pressure, as in areas of deep water. See Sheriff and Geldart, v. 2 (1983, p. 15-16).

gas seep: A place where gas bubbles are entering the water column, which can sometimes be seen on profiler data. Often results in a wipeout of coherent reflection energy from underneath the seep region.

Gassmann equation: Seismic *P*-wave velocity through a minimum-volume packing of uniform spheres is proportional to

$$[E^2 z/(1 - \sigma^2)^2]^{1/6},$$

where E = Young's modulus, z = pressure (or depth where proportional to pressure), σ = Poisson's ratio. See Gassmann (1951).

GASSP: Gas source seismic profiler; a *gas exploder* (q.v.) system in which gas is exploded in a long rubber tube which expands because of the explosion. Shell Development tradename.

gas well: See *GOR*.

gate: 1. The interval of record time over which a function (such as an autocorrelation or crosscorrelation) is evaluated. Also called **window**. A gate where the boundaries are abrupt is called a *boxcar* (q.v.). **2.** A circuit with several inputs and one output which is used in digital logic (computer design). **Truth tables** for inputs A and B to certain types of gates (assuming 0 indicates "no signal" or "false" and 1 indicates "signal" or "true") are:

A	B	$\overline{A}$	AND	OR	EXCEPT	NOR	NAND
0	0	1	0	0	0	1	1
1	0	0	0	1	1	0	1
0	1	1	0	1	1	0	1
1	1	0	1	1	0	0	0

A superscript line as in $\overline{A}$ indicates "inversion" or "not A", which is also indicated in circuit diagrams by a small circle. See Figure G-1 and *Boolean algebra*.

gather: A side-by-side display of seismic traces which have some acquisition coordinate in common. A **common-midpoint** or **CDP gather** displays data for the same midpoint, usually after correction for normal moveout and statics. A **common-offset** or **common-range gather** displays data for the same offset for a sequence of nearby midpoints.

gauge pressure: Pressure above one atmosphere.

gauss: The cgs-emu unit of magnetic induction (or flux density) **B.** It is a measure of the number of magnetic lines of force per unit area. 1 gauss = 1 maxwell/cm^2 = 10^5 gammas = 10^{-4} tesla = 10^{-4} weber/m^2. "Gauss" is also used in the cgs system as a unit for magnetization, or dipole moment per unit volume. In cgs units, a magnetizing force (in oersteds) gives rise to a flux density or field (in gauss), with the values being equal in magnitude if in free space. Named for Karl Friedrich Gauss (1777-1855), German mathematician.

Gauss error function: See *error function*.

Gaussian distribution: A **normal** or **bell-shaped distribution:** A set of values so distributed about a mean value

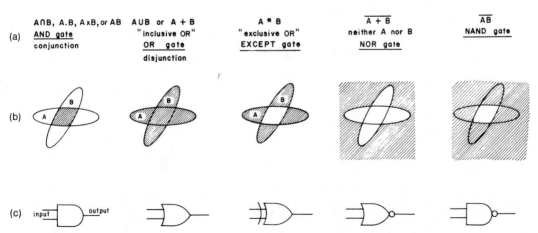

FIG. G-1. **Gate types.** (**a**) Word and symbol representations. (**b**) **Venn-diagram** representations. If the condition where A is true is denoted by one closed outline and the condition for B by another, then the condition for certain situations is illustrated by cross-hatching. (**c**) Circuit-diagram representations. Gates may have more than two inputs.

m that the probability $\varepsilon(\Delta a)$ of a value lying within a small interval Δa centered at the point a is

$$\varepsilon(\Delta a) = (e^{-1/2(a-m)^2 \sigma^2}) \Delta a / (2\pi)^{1/2} \sigma,$$

where σ is the standard error. $\varepsilon(\Delta a)$ is called the **error function**.

Gaussian window: A window whose shape is Gaussian. One of its properties is that its Fourier transform is also Gaussian:

$$e^{-at^2} \leftrightarrow (\pi/a)^{1/2} e^{-\omega^2/4a}.$$

The double-headed arrow indicates "transforms to".

Gauss's theorem: The total flux ϕ through any closed surface is equal to $4\pi k$ times the source strength m enclosed by the surface:

$$\phi = 4\pi k m = \oint_s \mathbf{g} \cdot \mathbf{ds} = \oint_s \nabla U \cdot \mathbf{ds}$$

$$= \int_v \nabla \cdot g \, dv = 4\pi k \int_v \rho dV.$$

(The 4π is often deleted in the mks system.) k is a constant which depends on the units of measure. This can also be expressed in terms of the flux density or field strength $\mathbf{g}$, the source density ρ, and the potential U. ϕ may be electrical flux if m is electrical charge; in the mks system, ϕ is in webers if m is in coulombs and $k \approx 9 \times 10^9$. Or ϕ may be gravitational flux if m is mass,

in which case $k = -\gamma$, where γ is the gravitational constant. Or ϕ may be magnetic flux if m is magnetic poles. Also called **Gauss' law**. The equality between the surface and volume integrals involving g is also called the *divergence theorem* (q.v.).

GCR: *Group-coded recording* (q.v.).

GCT: Greenwich civil (mean) time, usually written *GMT* (q.v.).

Geertsma formula: An equation for predicting the *P*-wave velocity in a porous fluid-filled rock. See Geertsma (1981).

gelatin: An explosive, often of the dynamite type.

generalized reciprocal method: A seismic refraction method for delineating undulating refractors from in-line reversed spreads. See Palmer (1980).

generation of computers: See *computer generation*.

geocentric latitude: The angle between a line through the center of the earth and a plane through the equator. Compare *geodetic latitude* and Figure G-2.

geochronology: See *geophysics*.

geocosmogony: See *geophysics*.

geodesic: The shortest distance between two points subject to some constraint, such as being on the surface of the Earth.

geodesy: The study of the Earth's form and gravitational field. Involves the location of points on the Earth with respect to reference systems.

geodetic latitude: Ordinary or **geographic latitude**, the angle between a tangent to the ellipsoid which approxi-

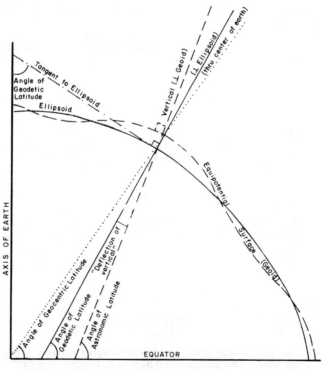

FIG. G-2. **Geocentric, geodetic,** and **astronomic latitude**.

mates the Earth's shape and the Earth's axis; see Figure G-2. Geodetic latitude differs from "geocentric latitude" ψ because of the ellipticity of the earth. If the geodetic latitude is ϕ, then $\phi - \psi = 11.7 \sin 2\phi$ (approximately) in minutes of arc. The maximum difference (at 45 degrees) is about 21.5 km. Geodetic latitude also differs from "astronomic latitude" in places where mass distribution causes a vertical line to point other than perpendicularly to the ellipsoid. Sometimes a location is projected parallel to the Earth's axis onto a sphere whose radius is the ellipsoid's major axis and the angle between a radius to this point and the axis is the **reduced latitude**.

Geodetic Reference System: The Earth spheroid adopted for gravity data reduction purposes. Based on the flattening value of 1/295.25, established by satellite measurements. Abbreviated **GRS67**. See *International Gravity Formula* and Woollard (1979).

geodetic system: Reference for latitude and longitude.

Location systems in different parts of the world are based on different reference positions and ellipsoid assumptions. The major systems are listed in Figure G-3.

Geodynamics Project: An international program of research (1971-1980) on the dynamics of the earth, with emphasis on deep-seated origins of geologic phenomena, especially movements of the lithosphere.

Geoflex: A seismic land source in which explosive detonating cord is buried by a plowlike device. Imperial Chemical Industries tradename. Compare *Aquaseis*.

Geograph: *Thumper* (q.v.) or the weight-drop method. Tradename of Mandrel Industries.

geographic latitude: *Geodetic latitude* (q.v.).

geoid: The sea-level equipotential surface to which the direction of gravity is everywhere perpendicular. An oblate ellipsoid of revolution (the **ellipsoid** or **spheroid**) which approximates the geoid is the reference for geodetic latitude determinations.

Datum	Ellipsoid	Equatorial radius, m	$\dfrac{1}{\text{Flattening}}$	Comments
North American "Meades Ranch"	Clarke 1866	6 378 206	295	Used throughout Western Hemisphere
European "Pelmert Tower, Potsdam"	International 1910 (Potsdam)	6 378 388	297	Used in most of Eurasia and Africa
Russian "Pulkovo"	Bessel 1841	6 377 397	299.2	
	Krassouski 1938	6 378 245	298.3	
Tokyo	Bessel 1841	6 377 397	299.2	
Indian "Kalimpur"	Everest 1830	6 377 276	300.8	
	International 1910	6 378 388	297	
	Clarke 1880	6 378 249	293.5	Used in Africa
	Australian	6 378 160	298.25	
	Airy	6 377 563	271	
	Fischer	6 378 155	298.3	
	Malayan	6 377 304	300	
	Heiskanen 1926	6 378 397	297.0	
	Hough (U.S. Army 1956)	6 378 260	297.0	
	APL (Applied Physics Lab)	6 378 144	298.23	Navigation satellite
North American 1983	IUGG 1979	6 378 137	298.257	To be used in North America after 1985
World Geodetic System 1972	WGS 1972	6 378 137	298.26	Datum used for satellite positioning

FIG. G-3. Major **geodetic** systems.

geoidal height: The height of the geoid above the reference ellipsoid measured perpendicular to the reference ellipsoid.

geologic age: See Appendix L.

geologic basement: See *basement*.

geomagnetic pole: The pole of best-fitting dipolar magnetic field to the Earth's magnetic field. See *magnetic pole*.

geomagnetic reversal: Change in the polarity of the Earth's magnetic field from its present polarity. This has occurred several times in the Earth's history.

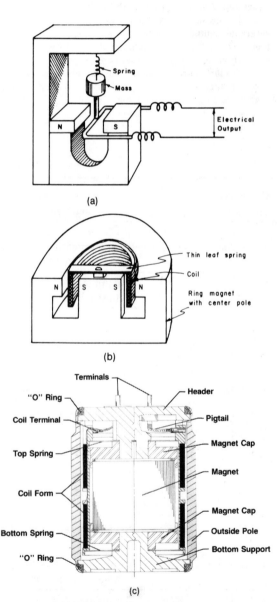

(a)

(b)

(c)

FIG. G-4. **Geophone**. (**a**) Schematic of essential elements: a wire (or coil) with inertial mass which remains steady as the magnet moves. (**b**) Half of a moving-coil geophone. (**c**) Cutaway of digital-grade geophone. (Courtesy Geo Space.)

geomagnetic-variation method: Studies of deep conductivity variations in the Earth (particularly lateral variations) by simultaneously observing variations in the three components of the magnetic field at different stations, as the field varies naturally.

geometric factor: 1. The geometry-dependent weighting factor for determining how the conductivities of each medium in the vicinity of a sonde affects apparent-conductivity measurements; especially used with the induction log. **2.** A numerical factor used to multiply the voltage-to-current ratio from measurements between electrodes to give apparent resistivity. Geometric factor is dependent on the type of electrode array and spacing used. Also called **geometric constant, array factor,** or **form factor**. See Figure A-12.

geomagnetic pole: See *magnetic field of the earth*.

geometric shadow: The area (volume) outlined by drawing straight lines paralleling the direction of wave approach through the extremities of a structure. It differs from the actual shadow because of diffraction and refraction.

geometric similarity: Said of two systems in which corresponding angles are equal and lengths proportional. Involves the ratio of length involved in physical modeling. See *modeling theory*.

geometric sounding: A resistivity or electromagnetic depth sounding in which the geometry is varied while the frequency is held constant, as opposed to **frequency sounding** in which the geometry is held constant and the frequency is varied.

geometric spreading: See *spherical divergence*.

geophone: The instrument used to transform seismic energy into electrical voltage; a **seismometer, seis, detector, receiver, bug, jug,** or **pickup**. Most land geophones are of the moving coil type; see Figure G-4. A coil is suspended by springs in a magnetic field (the magnet may be integral with the case of the instrument). A seismic wave moves the case and the magnet but the coil remains relatively stationary because of its inertia. The relative movement of magnetic field with respect to the coil generates a voltage across the coil, the voltage being proportional to the relative velocity of the coil with respect to the magnet (when above the natural frequency of the geophone). Below the natural frequency, the output (for input of constant velocity of magnet motion) is proportional to frequency and hence to the acceleration involved in the seismic wave passage. See Evenden et al. (1971). Compare *hydrophone* and *streamer*.

geophone array: The use of areal, linear, or (occasionally) vertical patterns with more than one geophone per channel. Used to discriminate against events with certain apparent wavelengths. See Figure A-14 and *array (seismic)*.

geophone cable: Insulated cable to which geophone groups are connected.

geophone distance: Usually *group interval* (q.v.), sometimes *geophone interval* or *geophone offset* (q.v.).

geophone distortion: Waveshape changes produced by nonlinear response of a geophone. Very small with modern geophones.

geophone interval: 1. The distance between adjacent geophones within a group. **2.** Sometimes used for **group interval**, the separation between the centers of adjacent geophone groups.

geophone offset: The distance from the source point to a geophone or to the center of a geophone group.

geophone pattern: See *array (seismic)*.

geophone planter: A device or a person that positions geophones for receiving seismic signals, especially used for planting phones several feet deep in marsh.

geophone station: The location of a geophone or (more commonly) of the center of a geophone array.

Geophysical Analysis Group: GAG; a research project at Massachusetts Institute of Technology during 1952-57 which applied communication theory to seismic analysis. See Flinn et al. (1967).

geophysical exploration: Making and interpreting measurements of physical properties of the Earth to determine subsurface conditions, usually with an economic objective, e.g., discovery of fuel or mineral deposits. Properties measured include seismic traveltime and waveshape changes, electric potential differences, magnetic and gravitational field strength, temperature, etc. Synonyms: **applied geophysics, geophysical prospecting**.

geophysical survey: A program of *geophysical exploration* (q.v.). See *electric survey, geothermal prospecting, gravity survey, magnetic survey, reflection survey, refraction survey, remote sensing,* and *well log*.

geophysicist: One who studies the physical properties of the Earth or applies physical measurements to geologic problems; a specialist in geophysics.

geophysics: 1. The study of the Earth by quantitative physical methods, especially by seismic reflection and refraction, gravity, magnetic, electrical, electromagnetic, and radioactivity methods. **2.** The application of physical principles to studies of the Earth. Includes the branches of (a) **seismology** (earthquakes and elastic waves); (b) **geothermometry** (heating of the Earth, heat flow, and volcanology and hot springs); (c) **hydrology** (ground and surface water, sometimes including glaciology); (d) **physical oceanography**; (e) **meteorology**; (f) **gravity** and **geodesy** (the Earth's gravitational field and the size and form of the Earth); (g) **atmospheric electricity** and **terrestrial magnetism** (including ionosphere, Van Allen belts, telluric currents, etc.); (h) **tectonophysics** (geological processes in the Earth); and (i) **exploration** and **engineering geophysics**. **Geochronology** (the dating of Earth history) and **geocosmogony** (the origin of the Earth) are sometimes added to the foregoing list. **3.** Often refers to solid-earth geophysics only, thus excluding (c), (d), (e), and portions of other subjects from the above list. **4. Exploration geophysics** is the use of seismic, gravity, magnetic, electrical, electromagnetic, etc., methods in the search for oil, gas, minerals, water, etc., with the objective of economic exploitation.

Geosat: An Earth satellite designed to acquire geologic information. Usually refers to the Geosat Committee which tries to persuade NASA and other bodies to incorporate features in other satellites which will provide useful geologic information.

geostatic pressure: Lithostatic pressure, the pressure due to the weight of the overlying rock, which generally is different from the pressure of fluids in the rock's pore space. See *abnormally high pressure*.

geosyncline: A subsiding area of extensive sediment and/or volcanic accumulation; a more-or-less continually sinking area, usually roughly linear.

geothermal gradient: The rate of change of temperature with depth in the Earth. It averages about 30°C/km at shallow depths.

geothermal heat flow: The heat flow from the Earth's interior per unit area per unit time. The product of thermal conductivity and thermal gradient. See *HFU*.

geothermal prospecting: Prospecting for high-temperature water and/or steam close to the surface, that can be utilized profitably for electric power generation and/or direct heat utilization. Measurements of variations in Earth temperature which are not attributable to variations in solar heating. Geothermal methods also may be used to locate geologic features which affect heat flow (salt domes, dikes, faults, etc.) or groundwater variations. Diurnal temperature variations penetrate to about 1 m (and annual temperature variations to 20 m). See Poley and van Steveninck (1970).

geothermometry: See *geophysics*.

Geovision: The seismic system developed by Frank Rieber in the late 1930s which included photographic recording and playback with summing as a function of apparent dip.

ghost: 1. Energy which travels upward from an energy release and then is reflected downward, such as occurs at the base of the weathering or at the surface. Ghost energy usually joins with the down-traveling wave train to change the effective waveshape. Sometimes called **secondary reflection** (which is also applied to other multiples). **2.** Energy reflected from the water surface before being picked up by a submerged streamer. **3.** A *correlation ghost* (q.v.) which results from harmonic distortion when using Vibroseis. **4.** A reflection of light from the front side of a mirror as well as from the silvered back side, thus producing a double image.

Gibbs' phenomenon: When a waveform which includes a discontinuity (or whose derivatives are discontinuous) is Fourier synthesized, the fit is poor near the discontinuity. As the number of frequency components included in the synthesis increases, the region of poor fit becomes narrower, but some overshoot at discontinuities continues. The poor fit sometimes is called **Gibbs' ears.** Named for Josiah Willard Gibbs (1839-1903), American mathematician and physicist. See Figure F-16 and Sheriff and Geldart, v.2 (1983, p. 166-167).

giga: A prefix meaning 10^9.

GIGO: "Garbage in, garbage out". A "principle" in data processing which emphasizes that meaningful data must be input if a meaningful result is to be obtained.

gimbals: A device for supporting an instrument so that it will remain essentially horizontal even when the support tips.

Gish-Rooney method: 1. An electrical-surveying method in which the polarity of both current flow and potential-measuring electrodes is reversed frequently to cancel the effects of electrode polarization. **2.** A resistivity-interpretation method. Mainly obsolete.

glitch: An unexpected and usually random event that may alter data or functions performed on data. A transient spike.

global: Common to the entire system or entire data set.

Global Positioning System: The **GPS Navstar** positioning system planned by the U.S. Defense Dept. to provide location determination by observation of 18 satellites (3 each in 6 different orbital planes) at 20 000 km altitude, and will employ two frequencies in the 10-20 GHz band. The system is to be continuously available worldwide and to provide three-dimensional accuracy of 16 m. It is to become fully operational in 1988.

G-log: *Seismic log* (q.v.). GSI tradename.

GMT: Greenwich mean (civil) time, the international reference time, the time at the observatory at Greenwich, England.

gnomonic projection: See *map projection*.

goodness of fit: See *chi square*.

GOR: Gas-to-oil ratio; the ratio of the gas to the oil which a well produces. A well is usually classed as an **oil well** if its GOR $< 15\,000$ ft^3/bbl, as a **gas well** if GOR $> 150\,000$ ft^3/bbl, or as a **condensate well** if intermediate.

Goupillaud medium: A medium of parallel layers of such thickness that the traveltime perpendicularly through each is equal. Used in computing synthetic seismograms. See Goupillaud (1961).

GPS: *Global Positioning System* (q.v.).

grab sampler: See *corer*.

graben: A down-dropped block bounded by normal faults, often relatively long and narrow. Compare *horst*.

gradient: 1. The first derivative or rate of change of one variable with respect to another variable, often with respect to distance. For example, the change in gravity, temperature, magnetic susceptibility, or electrical potential with respect to horizontal or vertical distance. Sometimes measured with a *gradiometer* (q.v.). **2.** The operation which finds the gradient from a potential function:

$$\text{Gradient of } U = \text{grad } U = \nabla U$$

$$= \mathbf{i}\,\frac{\partial U}{\partial x} + \mathbf{j}\,\frac{\partial U}{\partial y} + \mathbf{k}\,\frac{\partial U}{\partial z}.$$

3. The component of the gradient in an arbitrary direction, as the "vertical gradient" of gravity.

gradient array: See *array (electrical)*.

grading: Indicating the relative reliability of data or of an interpretation, an important aspect of interpretation. Usually a subjective process. Sometimes grading systems employ letters: vg = very good, g = good, f = fair, p = poor, vp = very poor, and ? = questionable. Coherency-measuring criteria are sometimes used to make grading quantitative. Contours are often graded by varying the type of line: solid contour = reliable; dashed contour = less reliable or interpolated between data; and dotted contour = speculative.

gradiomanometer: A device for determining the density of the wellbore fluid by measuring the vertical pressure at two points.

gradiometer: A device for measuring the gradient of a potential field. **1.** An arrangement of two magnetometers, one above the other, so that the difference in their readings is proportional to the vertical gradient of the magnetic field. **2.** A three-arm torsion balance which is sensitive to gravity gradients but not to curvature.

grain: 1. Alignment of features which define a direction. **2.** A small particle.

grammar: Rules for a computer language.

graphic log: *Sample log* (q.v.).

graticule: 1. A template for graphically integrating gravity or similar data. See also *dot chart* and *zone chart*. **2.** A grid network such as lines representing parallels and meridians on a plotting sheet.

gravimeter: An instrument for measuring variations in gravitational attraction; a **gravity meter**. Most gravimeters are of the unstable or astatic type. The gravitational force on a mass in the meter is balanced by a spring arrangement. A third force is provided which acts when the system is not in equilibrium; it intensifies the effect

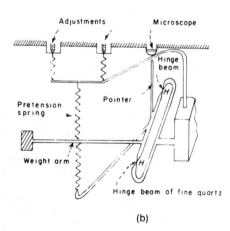

Adjusting screw to null instrument by changing support of main spring

Light beam for indicating when beam is horizontal

Main spring

Beam θ_1

Weight

Weight when balanced

Hinge

Mirror

(a)

Adjustments   Microscope

Hinge beam

Pointer

Pretension spring

H

Weight arm

H

Hinge beam of fine quartz

(b)

FIG. G-5. **Gravimeters.** (a) LaCoste and Romberg schematic. (b) Worden schematic.

of changes of gravity and increases the sensitivity of the system. Usually a **zero-length spring** is used; it has a stress-strain curve which passes through zero length when projected back to zero strain. See Figure G-5.

gravitational constant: The proportionality constant γ in **Newton's law of universal gravitation:** the gravitational force F between two point masses m_1 and m_2 can be related to the distance r between them:

$$F = \gamma m_1 m_2 / r^2.$$

γ has the value 6.670×10^{-11} newton m^2/kg^2. The gravitational acceleration g is the force per unit mass (the force on the mass m_1 which is in the gravimeter),

$$g = F/m_1 = \gamma m_2 / r^2,$$

which therefore is a measure of the effect of the other mass m_2 (e.g., the mass of the Earth).

gravitational potential: The negative of the work required to move a unit mass from infinity to a given point against gravitational forces. In the field of a point mass m, a distance r away, this is $\gamma m / r$ where γ is the gravitational constant. Also called **Newtonian potential.**

gravity: The force of attraction between bodies because of their masses. Usually measured as the acceleration of gravity. See *gravitational constant.*

gravity anomaly: 1. The difference between the gravity which is observed and that expected from a model. **2.** *Bouguer anomaly* (q.v.). **3.** *Free-air anomaly* (q.v.).

gravity corer: See *corer.*

gravity meter: *Gravimeter* (q.v.).

gravity reduction: Applying Bouguer, free-air, isostatic, latitude, or terrain corrections to gravity measurements.

gravity standard: The International Gravity Standardization Network 1971 (IGSN71) is now standard. See *International gravity formula* and Woollard (1979).

gravity survey: Measurements of the gravitational field at various locations over an area of interest. The objective in exploration work is to associate variations with differences in the distribution of densities and hence of rock types. Occasionally the whole gravitational field is measured (as with a pendulum) or derivatives of the gravitational field (as with a torsion balance), but usually the difference between the gravity field at two points is measured (as with a *gravimeter*, q.v.). Gravity data usually are displayed as Bouguer or free-air anomaly maps.

gravity unit: A unit of gravitational acceleration, equal to 0.1 mGal or 10^{-6} m/s^2.

gray code: A binary number code in which successive numbers differ by only one bit; see Figure N-4. The gray code is used in error minimization because the number of bit changes is the same for a step change regardless of the magnitude of the quantity.

grazing incidence: A raypath tangent to an interface.

Green's equivalent layer: See *surface density.*

Green's theorem: A form of the divergence theorem relating a volume integral to surface integrals. If F and G are two scalar functions, then

$$\iiint_V (F\nabla^2 G - G\nabla^2 F)\, dv$$

$$= \iint_S (F\nabla G - G\nabla F) \cdot ds,$$

where dv is a volume element and ds is a surface element. Named for George Green (1793-1841), English mathematician.

Greenwich: 1. Longitude measured with respect to the prime meridian which passes through the Royal Astronomic Observatory at Greenwich, England. **2.** The time at the Greenwich observatory; **GMT**, sometimes **GCT.**

Greenwich hour angle: See *hour angle.*

Gregory-Newton formula: A relationship used to interpolate between sample values. See Sheriff and Geldart, v.2 (1983, p. 156).

grey level: In variable-area and variable-density record sections, the overall "greyness" of the section is an important psychological factor affecting the section's interpretability.

grid effect: 1. Systematic error introduced in interpolating values at grid points where observations have not been made. Such errors tend to create false anomalies along certain orientations. **2.** False anomaly trends which may be created when grid residual maps are made with templates which are not azimuthally symmetrical.

grid residual: A method of emphasizing anomalies of a certain size in a potential-field map. A grid (usually square or trigonal) is drawn on a contour map and values are determined at the grid intersections by interpolation. The residual at one of the grid intersections is the value at that point less the average at other intersections a fixed distance away. Averages at several distances may be used and weighted to approximate second-derivative or other functions. The process of making grid residuals is also called **map convolution** (a two-dimensional convolution) because it represents map data convolved with a residualizing operator (or template). See also *residualize.*

grid smoothing: A method of smoothing sharp irregularities in potential-field measurements which arise from very shallow disturbances. A grid is drawn on a contour map and the average of values a fixed small distance away is taken as the smoothed value at the grid intersection.

GRM: *Generalized reciprocal method* (q.v.).

ground: A point in an electrical circuit used as a common reference point, often the conducting chassis on which the electrical circuit is physically mounted. It is frequently, but not necessarily, connected to the earth by a low-resistance conductor.

ground coupling: The mechanical connection of a geophone to the earth. Most often a spike on the geophone base is pressed into the earth. The geophone **plant.**

ground loop: The feature of an electrical circuit in which the circuit is connected to the common conductor (ground) at two or more points, thus forming closed loop circuits in which the common conductor is a part. Current flow in these loops may result in the "ground"

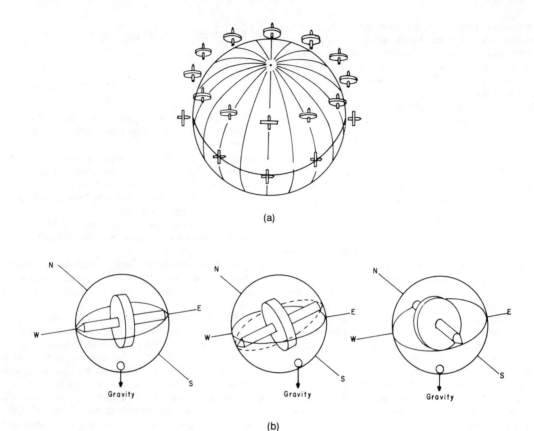

(a)

(b)

FIG. G-6. **Gyrocompass**. (**a**) A gyroscope tends to maintain its orientation in space, making it appear to change direction as seen from the rotating earth. (**b**) A weight on the vertical circle makes a gyroscope into a gyrocompass. Tilt of the gyroscope axis because of rotation of the earth tends to raise the weight, thus exerting a torque which makes the gyroscope precess unless the axis is aligned north-south. This "meridian-seeking" ability is the essence of a gyrocompass.

being at different potential levels at different points, an undesirable feature.

ground mix: The use of an array or pattern of sources or geophones distributed over a sizable surface area. The objective of arrays usually is to have vertically traveling reflection energy add up in-phase while horizontally traveling energy and random noise partially cancel, in effect the entire array acting as one large source or geophone. Small time shifts between the array elements will result in frequency filtering action. The term is sometimes reserved for situations where adjacent geophone or source patterns actually overlap.

ground noise survey: *Noise survey* (q.v.).

ground roll: Surface-wave energy which travels along or near the surface of the ground. It is usually characterized by relatively low velocity, low frequency, and high amplitude. Ground roll tends to mask desired reflection signals. Source and geophone patterns, frequency filtering, *f–k* filtering and stacking are used to discriminate against it. Rayleigh-type waves are usually the main

source and ground roll is sometimes called **pseudo-Rayleigh wave**. See Figure R-2.

ground truth: 1. Data obtained on the ground concerning the significance of albedo anomalies observed in remote sensing, to help interpretation. **2.** Data from a ground monitor used to show that extraneous events did not occur during the acquisition of airborne data, as use of a ground magnetometer to ascertain the absence of disturbing magnetic storms.

ground unrest: Background or ambient noise, such as produced by wind, microseisms, etc.

group: The various geophones which collectively feed a single channel. The number of geophones may vary from one to several hundred. A large group is sometimes called a **patch**. See *array (seismic)*.

group-coded recording: NRZI recording technique that employs special data encoding, check characters and data grouping to enhance reliability. Used to record 6250 bpi 9-track tape.

group delay: The time delay associated with a geophone

group, the delay (or advance) being produced by the group's elevation, near-surface conditions, and/or the choice of reference datum.

group interval: The horizontal distance between the centers of adjacent geophone groups.

group velocity: The velocity with which the energy in a wave train travels. In dispersive media where velocity varies with frequency, the wave train changes shape as it progresses so that individual wave crests travel at a different velocity (the **phase velocity** V) than does the envelope of the wave train. The velocity of the envelope is the group velocity U:

$$U = V - \lambda(dV/d\lambda) = V + \nu(dV/d\nu),$$

where λ = wavelength and ν = frequency. *Normal mode propagation* (q.v.) also results in dispersion and thus different values for group and phase velocities. See also Figure D-13 and compare *phase velocity*.

growth fault: See *fault*.

GRS67: See *International Gravity Formula*.

GTO: Gate turn-off switch; see *controlled rectifier*.

g.u.: Gravity unit; 0.1 milligal.

guard electrodes: Extra electrodes whose function is to focus the current flow. Also called **bucking electrodes**. See *laterolog*.

guard log: A log made with *guard electrodes* (q.v.). A **laterolog** or **focused log**.

guest: A mineral introduced into and usually replacing another mineral.

guided wave: 1. A *channel wave* (q.v.). 2. An interface wave or *surface wave* (q.v.).

Gulf magnetometer: A type of *fluxgate magnetometer* (q.v.).

gun: 1. An *air gun* (q.v.), a seismic energy source from which a bubble of highly compressed air is released. 2. A *gas gun* (q.v.), a seismic energy source in which an explosive gas mixture is detonated. 3. A *water gun* (q.v.), a seismic source in which a volume of water is suddenly projected into the water. 4. A device for obtaining sidewall cores in a borehole. 5. A device used to perforate or open holes in casing so that fluid can flow into the borehole.

Gutenberg-Weichert discontinuity: The boundary between the Earth's mantle and core. See Figure E-1.

guyot: A flat-topped *seamount* (q.v.). Named for Arnold Henry Guyot (1807-1884), Swiss-American geologist.

G-wave: A long-period (40 to 300 s) *Love wave* (q.v.), usually restricted to an oceanic path. Named for Gutenberg. Velocity is often nearly constant at 4.4 km/s so the wave appears nearly impulsive.

gyrocompass: A gimbal-mounted gyroscope incorporating unbalanced masses which make the axis of rotation precess about true north. See Figure G-6. If a torque tries to change the plane of rotation of the gyroscope, the gyroscope axis will rotate about an axis which is perpendicular to both the gyroscope's axis of spin and the torque; this is called **precession**.

H

H: See *H-type section.*

hachure: A short line or mark along a contour or fault trace which points in the down-dip direction or toward smaller values.

hade: The complement of dip; see Figure F-3.

Hagedoorn method: *Plus-minus method* (q.v.) of refraction interpretation. See Hagedoorn (1959).

Hales method: A graphical refraction interpretation method, particularly useful where the refractor changes depth markedly, such as where there is considerable relief or over large faults, but with constant velocity above the refractor. See Hales (1958) or Sheriff and Geldart, v.1 (1982, p. 255-257).

half adder: A circuit with two inputs (A and B) and two outputs, sum and carry (S and C). Its truth table is:

A	B	C	S
0	0	0	0
1	0	0	1
0	1	0	1
1	1	1	0

C is an AND gate and S is an EXCEPT gate. See *gate.*

half adjust: Rounding in which the value of a particular digit determines whether a one shall be added to the next-higher significant digit.

half-duplex: A system in which transmission can occur in only one direction at any time. Transmissions in opposite directions alternate. Compare *duplex.*

half-plane: A plane which exists everywhere to one side of a line but not on the other side.

half-maximum distance: *Half-width* (q.v.).

half-power point: The frequency value on an amplitude response curve for which the amplitude reaches $1/\sqrt{2}$ or 70.7 percent.

half-space: A mathematical model bounded only by one plane surface, i.e., the model is so large in other dimensions that only the one boundary affects the results. Properties within the model are usually assumed to be homogeneous and isotropic, although other models are also used.

half-width: (a) For a gravity or magnetic anomaly, the horizontal distance across the peak (or trough) at the half-maximum (or half-minimum) amplitude; see Figure H-1a. Used in estimating the depth to the center of mass (for gravity data). In magnetic interpretation, the half peak width is approximately the depth to the center of a spherical body. (b) For a semiinfinite horizontal slab (''fault'' anomaly), half the width between points where the anomaly is one-quarter and three-quarters amplitude; see Figure H-1b. See also *depth rule.*

Hall effect: A transverse potential develops across a semiconductor or strip of metal which carries a current when located in a strong magnetic field.

halo: See *halo effect.*

halo effect: 1. Many residual and second-derivative methods produce a ring or halo of opposite sign around an anomaly, reflecting the opposite field curvature around the periphery of the anomaly. Halos do not represent separate anomalous masses. Halos can be reduced or eliminated by biasing. **2.** A ring anomaly, claimed to be characteristic of certain electromagnetic or geochemical effects of structures or hydrocarbon accumulations.

hammer: A hammer striking a steel plate is used as a seismic source for shallow refraction measurements. The hammer incorporates a switch which starts a timer when the hammer strikes; the timer is stopped when energy received by a geophone reaches some threshold value.

Hammer chart: A template for making gravity terrain corrections. See Hammer (1940).

Hamming: Smoothing with weights 0.23, 0.54, 0.23. See *Hamming function.*

Hamming function: A function used to shape the cutoff of a window or gate in data processing in order to avoid undesirable truncation effects of sharp truncation. The Hamming function (different from the Hanning function) is

$$0.54 + 0.46 \cos \phi, \ -\pi < \phi < \pi$$
$$0, \ \phi < -\pi \text{ or } \phi > \pi,$$

where ϕ increases linearly from $-\pi$ to π across the window. For the kth element out of n elements within the window, $\phi = 2\pi k/(n + 1) - \pi$. See also Figure W-11. Named after R. W. Hamming. Compare *Hanning function.*

handshake: The sequence of signals required for communication between system functions. The I/O bus protocol for a system defines its handshaking requirements. For example, asynchronous I/O systems require a response (reply) to each signal to complete an I/O operation.

hands-off tuning: Automatic tuning of a side-scan sonar. Tradename of Klein Associates.

Hankel transform: The Hankel transform of order m of the real function $f(t)$ is

$$F(s) = \int_0^\infty f(t) \ t \ J_m(st) \ dt,$$

where J_m is the m-order Bessel function. Also called a **Bessel transform.**

Hannell rule: See *depth rule.*

Hanning: Smoothing with weights 0.25, 0.50, 0.25.

Hanning function: A function used to shape the cutoff of a window in data processing to avoid undesirable effects

116

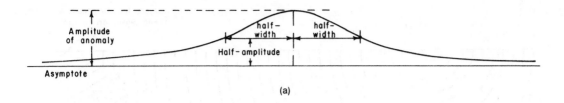

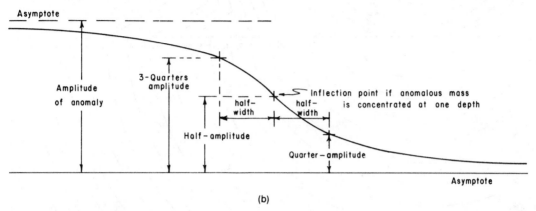

FIG. H-1. **Half-width**. (a) Gravity anomaly due to a point or line element (sphere or horizontal cylinder). Depth to center of sphere = 1.305 x half-width; depth to center of cylinder = half-width. (b) Gravity anomaly due to a thin semiinfinite slab (step or fault). Depth to center of anomalous mass = half-width.

of sharp truncation. The Hanning function is

$$1/2 + (1/2) \cos \phi, -\pi < \phi < \pi,$$
$$0, \phi < -\pi \text{ or } \phi > \pi$$

See also Figure W-11. Named after J. von Hann. Compare *Hamming function*.

hard copy: A printed (or otherwise displayed on paper) copy of data in human-readable form, such as a paper copy of an image on a computer terminal screen.

hardware: Equipment, especially computing-machine equipment.

hard-wired logic: A group of logic circuits permanently interconnected to perform a specific function.

harmonic: 1. A frequency which is a simple multiple of a fundamental frequency. The third harmonic, for example, has a frequency three times that of the fundamental. **2.** Two frequencies are **harmonically related** if they are each harmonics of a common fundamental. **3.** Any component of a Fourier series except the fundamental.

harmonic analysis: Decomposing a periodic (or aperiodic) waveform into constituent cosine waves, i.e., into a Fourier series (or its Fourier transform).

harmonic distortion: *Distortion* (q.v.) characterized by the generation of harmonics of input frequencies.

harmonic function: A function that repeats after successive equal intervals of the arguments. A solution of Laplace's equation.

hash total: Summation check; see *check*.

Haskell matrix: See *Thomson-Haskell method*.

haversine: Haversine $\alpha = (1 - \cos \alpha)/2$.

Hayford modification: See *isostasy*.

HDDR: High-density data recording, a system where digital data are recorded sequentially on a track at 8000 bpi; a half-inch tape contains 14 tracks. Gus Manufacturing tradename.

HDT: High-resolution dipmeter; a *dipmeter* (q.v.) which records four microresistivity curves and has an additional electrode on one pad which yields another curve at displaced depth. The displaced-depth curve is used to correct for variations in logging sonde speed.

head: A *magnetic head* (q.v.).

head-check pulse: An impulse applied simultaneously to all channels of an analog magnetic recorder so that the alignment of magnetic heads can be checked. See Figure R-5.

header: The identification information and tabulation of parameters which precede data, as on magnetic tape.

head wave: A wave characterized by entering and leaving a high-velocity medium at the critical angle. See Figures C-15, H-2, and T-6. Also called a **refraction, Mintrop, von Schmidt, conical wave**.

heat conductivity: *Thermal conductivity* (q.v.).

heat flow unit: *HFU* (q.v.).

Heaviside function: A *step function* (q.v.).

Heaviside layer: *E-layer* (q.v.). Named for Oliver Heaviside (1850-1925), English physicist.

heighting: Determining the difference in elevation be-

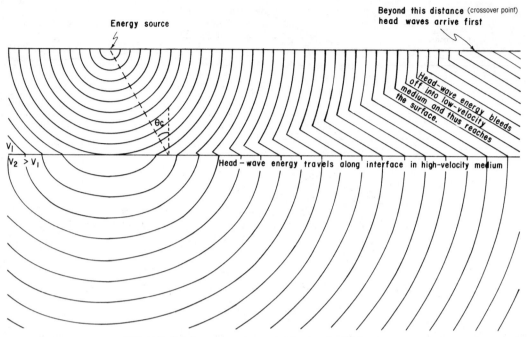

FIG. H-2. **Head waves** and wavefronts showing first energy.

tween two stations, often determined trigonometrically by measuring the distance and the angle between the line of sight and horizontal.

height of instrument: The elevation of a survey-instrument telescope with respect to the ground level. Abbreviated **HI**. Used with theodolites, levels, alidades, etc.

Heiskannen modification: See *isostasy*.

FIG. H-3. **Herringbone effect**. Flight lines *e* and *i* are displaced southward from their correct positions, *n* is displaced eastward, thus producing fictitious anomalies. The effect on the contours is called "herringbone."

Helmholtz coil: A pair of parallel, equal diameter coaxial coils separated by a distance equal to their radius, which provides a nearly constant magnetic field over a large volume between the coils. It permits an accurate calculation of the magnetic field between the coils and is used in calibration of magnetometers and nulling an ambient magnetic field for magnetic measurements. Named for Hermann Ludwig von Helmholtz (1821-1894), German scientist.

Helmholtz double layer: See *double layer*.

Helmholtz equation: The space-dependent form of the wave equation for a wave harmonic in time:

$$(\nabla^2 + \kappa^2)\,\psi = 0,$$

where $\kappa = \omega/V$, ω = angular frequency, and V = velocity.

Helmholtz separation method: A method of separating scalar and vector potentials into other scalar and vector functions which facilitates mathematical solution of the wave equation. See Sheriff and Geldart, v.1 (1982, p. 39).

HEM: 1. Helicopter electromagnetics. **2.** *Horizontal-loop electromagnetic method* (q.v.).

Hermitian matrix: A matrix which equals the transpose of its conjugate:

$$(\mathbf{A}^*)^T = \mathbf{A}.$$

It has real eigenvalues and can be diagonalized by similarity transformations. Named for Charles Hermite (1822-1901), French mathematician.

herringbone: A pattern of systematic deviation of contours on a contour map produced when one line of data is systematically mislocated or has systematic bias. See Figure H-3.

hertz: A unit of frequency. Abbreviated **Hz.** Hertz is the same as cycles per second = cps. Named after Heinrich Rudolph Hertz (1857-1894), German physicist who discovered electromagnetic waves.

Hertz equation: A relation for the radius of contact of elastic spheres when under pressure. See Love (1944).

heterogeneity: Lack of spatial uniformity. Opposite of **homogeneity**.

heuristic: 1. A method or scheme used for teaching. 2. Pertaining to learning, especially a "trial and error" method.

hexadecimal: A radix 16 number system. See Figure N-4.

hexagonal packing: The most compact arrangement for a packing of uniform spheres. See Sheriff and Geldart, v.2 (1983, p. 5-6).

HFU: Heat flow unit: 10^{-2} cal/m^2 s = 41.86 mW/m^2. The mean heat flow of the earth is 1.2–1.5 HFU. Heat flow ranges from about 0.9 in shield areas to over 2 HFU in Cenozoic volcanic areas; mid-ocean ridge values reach 8 HFU.

HI: *Height of instrument* (q.v.).

hiatus: An interval of time not represented by strata. May be depositional (because strata were never deposited) or erosional (because they were removed subsequent to deposition).

hidden layer: A layer which cannot be detected by refraction methods. See Figure H-4. **1.** A layer of lower velocity lying beneath a layer of higher velocity. **2.** A layer which is too thin or has insufficient velocity contrast to give a distinct arrival (sometimes insufficient to give a first arrival). See *blind zone*.

Hi-Fix: See *Decca*. Decca Survey tradename.

high-cut filter: A *filter* (q.v.) that transmits frequencies below a given cutoff frequency and substantially attenuates frequencies above the cutoff. The same as **low-pass filter**.

high level language: A language to simplify programming. A single source statement usually translates into many computer instructions. Compare *assembler language*.

high-line: Voltages induced in seismic cables or instruments by nearby electric transmission lines. Characterized by the frequency of power transmission (usually 60 or 50 Hz, or $16\frac{2}{3}$ Hz for some electric railways) or its harmonics. Coupling may be capacitive, by electromagnetic induction (especially if the transmission lines are unbalanced), or by leakage currents from ground-return systems. The problem is most severe when the seismic cables and geophones are not well insulated.

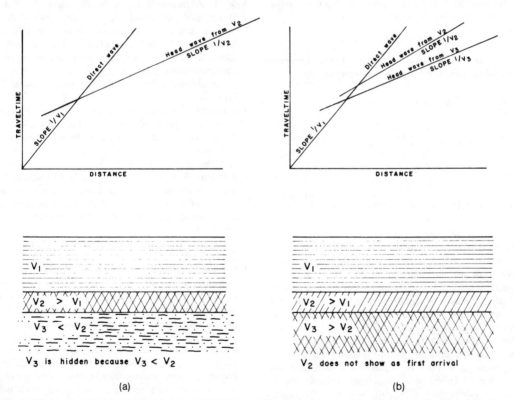

FIG. H-4. **Hidden layers**. (a) A layer whose velocity is lower than that of an overlying layer (velocity inversion) does not produce a head wave. (b) The head wave from a thin layer without sufficient velocity contrast may not produce a first arrival.

high-line eliminator: A part of seismic recording equipment used to attenuate high-line interference. May consist of bridges which balance the voltage across the input transformers with respect to a center tap at ground potential, the assumption being that voltages with respect to ground are noise. Such bridges usually have two adjustments, for the resistive and reactive component of the high-line-induced voltages. High-line interference is also sometimes reduced with a notch filter which removes a narrow band of frequencies around the high-line frequency. See also *humbucking*.

high-pass filter: A filter that passes without significant attenuation frequencies above some cutoff frequency while attenuating lower frequencies. The same as **low-cut filter**.

high-resolution: Seismic frequencies above the normal exploration range, recorded with the objective of improving resolution, especially of shallow events. Abbreviated **HR**. Usually implies frequencies from 75–150 Hz, sometimes to 500 Hz or higher.

high-resolution thermometer: A small-diameter fast-response thermometer for logging open or cased boreholes with a temperature resolution of 0.5°F

high-speed layer: A layer in which the speed of wave propagation is greater than in the layer above it and which therefore can carry head-wave (refraction) energy.

Hilbert transform: Given an $h(t)$ which is nonsingular at $t = 0$ and which is a causal response so that $h(t) = 0$ for $t \geq 0$, then its Fourier transform,

$$H(\omega) = R(\omega) + jX(\omega)$$

(where ω = angular frequency) has the special property known as the Hilbert transform, expressed by

$$X(\omega) = -(1/\pi)R(\omega)*(1/\omega)$$

$$= -(1/\pi)\ \mathcal{P}\int_{-\infty}^{\infty} \frac{R(y)\ dy}{\omega - y},$$

and

$$R(\omega) = (1/\pi)X(\omega)*(1/\omega) = (1/\pi)\ \mathcal{P}\int_{-\infty}^{\infty} \frac{X(y)\ dy}{\omega - y},$$

where $\mathcal{P}$ denotes the Cauchy principal value at discontinuities. If $H(\omega)$ vanishes for $\omega < 0$, its Fourier transform $h(t) + jx(t)$ has $h(t)$ and $x(t)$ forming a **Hilbert transform pair**; $h(t)$ and $x(t)$ have the same amplitude spectrum but differ in phase by 90 degrees. [$h(t) + jx(t)$] is called the **analytic signal** belonging to $h(t)$, and $x(t)$ is the **quadrature signal** corresponding to $h(t)$. Often used in *complex trace analysis* (q.v.).

Hilbert-transform technique: A technique for determining the phase of a minimum-phase function from its power spectrum. It is used in computing a deconvolution operator. Given the power spectrum $P(\nu)$ and that the wavelet is minimum phase, the wavelet's frequency-domain representation $W(\nu)$ is

$$W(\nu) = A(\nu)\ e^{j\gamma(\nu)} = |P(\nu)|^{1/2}e^{j\gamma(\nu)}.$$

The amplitude $A(\nu)$ is the square root of the power spectrum. Taking the logarithm of both sides,

$$\ln W(\nu) = (1/2)\ln P(\nu) + j\gamma(\nu),$$

splits the function into real and imaginary parts. To be minimum phase, the function must be analytic in the lower half-plane. Then the Hilbert transform can be used to find the phase $\gamma(\nu)$ from $\ln P(\nu)/2$:

$$\gamma(\nu) = (1/2)\ln P(\nu) * 1/(\pi\nu).$$

Since the amplitude and phase are known, the Fourier transform can be computed and the time-domain expression for $W(t)$ determined. See Sheriff and Geldart, v.2 (1983, p. 171-172).

hi-line: See *high-line*.

hinge: The part of a fold where the curvature is greatest.

hinge fault: A fault where the blocks across the fault have rotated about an axis perpendicular to the fault plane. Thus the throw varies along the fault strike. Also called **scissors fault**. See Figure F-2.

hiran: High-precision *shoran* (q.v.).

histogram: A multiple bar diagram showing the relative population of a sequence of regularly arranged classes.

hodograph: 1. The figure described by the terminus of a moving vector. 2. A plot of the motion of a point as a function of time. See Figure R-2. 3. A time-distance curve.

hole: 1. A mobile vacancy in the electronic valence structure of a semiconductor (an atom with less than its normal number of electrons). *P*-type semiconductors have an excess of holes. An electron from a neighboring atom can fill the hole so that in effect the hole moves to the neighboring atom. The apparent movement of holes in an electric field is equivalent to a current. 2. A borehole.

hole blow: 1. Ejection of water, mud, and sometimes rocks from the shothole as a result of the shot explosion. 2. Noise on a seismic record caused by such ejection; see *hole noise*.

hole deviation: See *drift*.

hole fatigue: A delay between the detonation of a shot and the initiation of the seismic impulse from it; a consequence of changes in the shot environment (usually formation of a cavity) produced by an earlier shot in the same hole.

hole logging: Drill-hole IP or resistivity surveying. See *hole probe*.

hole noise: Noise from the shot, caused by *hole blow* (q.v.) or rumbling around in the borehole of the gases resulting from an explosion. Hole noise may last for several seconds. It is often excessively strong on geophone groups near the shothole and attenuates rapidly with distance.

hole plug: A device used to plug a shothole after shooting. Usually the plug is pushed far enough into the hole to prevent its being dislodged and earth is shoveled over it level with the surrounding ground. Also called **bridge plug**. Hole plugs are also used to close shotholes temporarily between the drilling/loading operation and the detonation of the charge.

hole probe: A drill-hole IP or resistivity survey in which closely spaced in-hole electrodes are used to determine the electrical properties of rock near the drill hole. Also called **electric logging, IP** or **resistivity logging,** and **hole logging.**

Hollerith code: The code established by Hollerith to designate numbers and letters on punched cards.

hologram: A recording of the amplitude and phase distribution of a wave disturbance. See Figure H-5.

holography: Recording of the intensity of the interference pattern which results from the scattering of coherent radiation and a reference beam. **Optical holography** usually involves photographing the interference from laser light (in order to achieve a coherent source); subsequent illumination of the photographic plate allows one to "see" the photographed object in three dimensions. See Figure H-5 and also *earth holography*.

homocline: A region of broadly uniform dip.

homogeneous: The same throughout; uniformity of a physical property throughout a material.

homomorphic deconvolution: Removal of the effects of an earlier filter in the cepstral domain; see *cepstrum*.

homomorphism: A relationship between two algebraic systems of the same type that preserves the algebraic operation. A correspondence between elements of two sets D (the domain) and R (the range) such that each element of D determines a unique element of R and each element of R is the correspondent of at least one element of D.

Hooke's law: Stress is proportional to strain. Valid for many materials for small strains. See *elastic constants*. Named for Robert Hooke (1635-1703), English physicist.

hop: Travel of a radio wave to the ionosphere and back to Earth.

horizon: The surface separating two different rock layers. Where such a surface (even though not itself identified) is associated with a reflection which can be carried over a large area, a map based on the reflection event may be called a **horizon map,** sometimes contrasted to a phantom map (see *phantom*).

horizon-slice map: Display of the data from a three-dimensional set of data which lie on the same reflecting horizon, thus showing areal variations in amplitude. Compare *time-slice map* (q.v.); see also Figures T-4 and T-8.

horizontal cylinder: A model used in calculating potential-field effects; a horizontal cylinder so long that the ends of the cylinder do not produce any effects. The model is equivalent to a horizontal line whose mass is ρA per unit of length, when ρ is the density contrast and A is the cross-sectional area of the cylinder.

horizontal-dipole sounding: Electromagnetic sounding configuration using either a transmitting coil with its axis horizontal or a horizontal grounded wire.

horizontal-loop method: An electromagnetic survey method wherein the planes of the transmitter and receiver coils are horizontal and the coils are carried a fixed distance apart. In-phase and quadrature components are measured by the receiver. Also called **slingram.**

horizontal microcode: Low-level code in which each in-

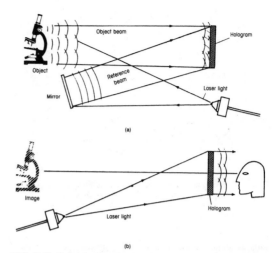

FIG. H-5. **Holography. (a)** Image is formed by interference between laser light reflected from the object and reference beam. **(b)** Image is reconstructed by shining laser light onto interference pattern recorded on film.

struction word may initiate multiple primitive operations.

horizontal mixing: 1. Horizontal stacking or *common-midpoint stacking* (q.v.). **2.** Occasionally, *ground mixing* (q.v.).

horizontal profiling: See *profiling*.

horizontal section: A *time-slice map* (q.v.).

horizontal stacking: *Common-midpoint stacking* (q.v.).

horst: A crustal block raised up with respect to neighboring blocks by normal faulting. A horst is usually long compared to its width. Compare *graben*.

hot shot: 1. To carry out a short urgent program, often without moving the field camp or crew headquarters. **2.** A daily living allowance paid crew members on such a program. **3.** Slang for an expert in a particular field.

hot spot: A localized high heat-flow region with a deep cause, often associated with volcanism and other geothermal activity. Hawaii and Yellowstone are presumably such hot spots. The **hot-spot hypothesis** uses hot spots as a fixed reference frame for determining plate motion with respect to them.

hot-wire analyzer: A device used to detect hydrocarbon gases returned to the surface by the drilling mud. Basically a Wheatstone bridge, two arms of which are kept at a high temperature. Hydrocarbon gases become oxidized as they pass over one arm, which increases its temperature, changes its resistance, and unbalances the bridge. The hot-wire analyzer response is usually plotted in well log format and called a *mud log* (q.v.).

hour angle: Angular distance of a body west of the projection of a meridian onto the celestial sphere. **Local hour angle** is the angle between a body and the projection of the observer's meridian onto the celestial sphere. **Greenwich hour angle** is with respect to the projection of the Greenwich meridian. **Sidereal hour angle** is angular distance west (unlike the others which are measured east) of the vernal equinox; it is thus the supplement of right ascension.

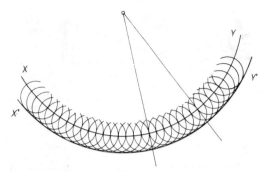

FIG. H-6. **Huygens' principle**. XY = wavefront at t, $X'Y'$ = wavefront at $t + \Delta t$. Radii of small circles = $V\Delta t$.

housekeeping: Administrative, accounting, or overhead operations for a computer.

HR: *High resolution* (q.v.).

H-type section: A three-layer resistivity model in which the middle layer is more conductive than the layers above or below it. See Figure T-5.

hue: The spectral value of a color. The other two color parameters are *saturation* and *density* (q.v.).

hum: Electrical interference or noise occurring at the power-line frequency or its harmonics. See *high-line*.

Humble formula: A special form of *Archie's formula* (q.v.). Named for Humble Oil Co.

humbucking: An arrangement to reduce electromagnetic pickup, especially from power lines. Humbucking geophones involve two coils wound in opposite directions and so connected that electromagnetic-pickup voltages have opposite polarity while seismically induced voltages are in-phase.

hundred-percent section: A seismic record section which provides continuous coverage but does not utilize data redundancy.

hunting: Following a desired course in an oscillatory manner, successively correcting course in opposite directions. In "hunting a course", one veers slightly to the right of the desired course, then slightly to the left, etc., and so is never very far off position. A characteristic of an underdamped servo-system.

Huygens' principle: The concept that every point on an advancing wavefront can be regarded as the source of a secondary wave and that a later wavefront is the envelope tangent to all the secondary waves. See Figure H-6. Named for Christian Huygens (1629-1695), Dutch mathematician.

H-wave: *Hydrodynamic wave* (q.v.).

hybrid migration: Seismic migration which involves flipping back and forth between domains (time, frequency, wavenumber domains), so as to take advantage of the strong points of each while minimizing effects of their limitations.

hybrid scale: A scale used with laterologs which is linear with resistivity for low resistivities and linear with conductivity (the reciprocal of resistivity) for high resistivities. See Figure L-3.

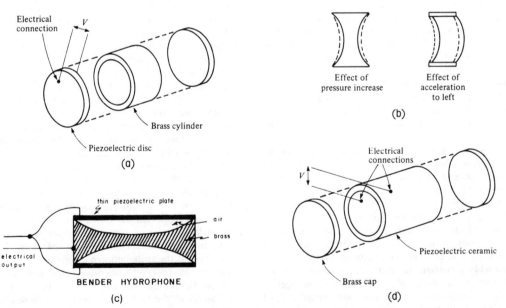

FIG. H-7. Piezoelectric **hydrophones**. (**a**) A disc generates a voltage across opposite faces when bent. (**b**) Acceleration-cancelling feature of the disc hydrophone. (**c**) Bender hydrophone. (**d**) Cylindrical hydrophone. (From Sheriff and Geldart, V. 1, 1982.)

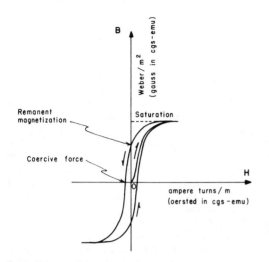

FIG. H-8. **Hysteresis loop**. As an applied field H is changed, the magnetization B lags behind.

hybrid spread: A geophone spread with unequally spaced groups.

hydrate reflection: A reflection from the base of a zone containing methyl hydrate, found just below the sea floor in deep water. See Sheriff and Geldart, v.2 (1983, p. 15).

hydrocarbon indicator: A measurement which indicates the presence or absence of a hydrocarbon accumulation. Indicators include local amplitude increase (**bright spot**) or decrease (**dim spot**), phase change, frequency change (especially a local lowering of frequency), a horizontal event because of a gas-water, gas-oil or oil-water contact (**flat spot**), lower velocity than laterally equivalent sediments, a decrease in amplitude below the accumulation (**shadow zone**), an apparent sag below because of increased time in transiting the accumulation (**velocity sag**). See Sheriff (1980, chap. 9) and Sheriff and Geldart, v.2 (1983, p. 141-143).

hydrocarbon saturation: Fraction of the pore volume filled with hydrocarbons.

Hydrodist: A short-range (~40 km) radio-positioning system operating in the 3 GHz range. Tellurometer tradename.

hydrodynamic wave: 1. H-wave; a seismic surface wave similar to a Rayleigh wave except that it moves in the opposite sense (that is, forward at its "up" position). Also called a **Sezawa M_2 wave. 2.** Waves on the surface of a fluid.

hydrogen index: The ratio of hydrogen per unit volume compared with that in fresh water. Neutron log response depends mainly on the hydrogen index.

hydrology: See *geophysics*.

hydromagnetics: See *magnetohydrodynamics*.

hydrophone: A detector which is sensitive to variations in pressure, as opposed to a geophone which is sensitive to particle motion. Used when the detector can be placed below a few feet of water, as in marine or marsh work or as a well seismometer. Some hydrophones operate because of *magnetostriction* (q.v.) but most are *piezoelectric* (q.v.). Piezoelectric hydrophones include benders, disc hydrophones, and cylindrical hydrophones (see Figure H-7). The sensing element is usually a piezoelectric ceramic material such as barium titanate, lead zirconate, or lead metaniobate. Piezoelectric hydrophones are high-impedance devices and signals may be passed through preamplifiers or impedance-matching transformers before transmission through the streamer to the recording instruments. Compare **geophone**.

Hydrosein: A marine seismic source utilizing the implosion which results as two plates are driven suddenly apart, creating a void between them into which water rushes. Western Geophysical Co. tradename.

hydrostatic head: The height of a column of water extending to the surface.

hydrostatic pressure: See *abnormally high pressure*.

Hydrotrac: A medium-frequency radio-positioning system.

hyperbolic line of position: A *line of position* (q.v.) determined by measuring the difference in distance to two fixed points.

hyperbolic functions: Functions which satisfy the relations:

$$\sin jx = -j \sinh x; \quad \sinh jx = j \sin x;$$
$$\cos jx = \cosh x; \quad \cosh jx = \cos x;$$
$$\cosh^{-1}x = \ln[x + (x^2 - 1)^{1/2}];$$
$$\sinh^{-1}x = \ln[x + (x^2 + 1)^{1/2}].$$

hyperbolic search: A search for coherency among traces in a common-midpoint gather along a hyperbolic trajectory such as normal moveout should produce. The objective is a measure of the best value of normal moveout. Used in many velocity-analysis methods.

hypocenter: An earthquake **focus;** the point at which the first motion in an earthquake originates. The projection on the surface of the earth is the **epicenter.**

hysteresis: 1. A phenomenon exhibited by a system or material in which response depends nonlinearly on past responses. A property which has been changed will not return to the original state after the cause of the change has been removed. **2.** Especially the effect where the magnetization produced by an applied field lags behind the field; see Figure H-8. This involves energy loss. When the applied field returns to zero, the residual magnetism which is retained is called **remanent magnetism.** The magnetic field intensity required to reduce the remanent magnetization to zero is the **coercive force.**

Hz: Hertz, a unit of frequency; cycles/second.

I

i: 1. Symbol to indicate $(-1)^{1/2}$; j is also used. **2.** A unit vector in the x-direction, as in the operator del.

IC: Integrated circuit; a solid-state device containing more than one circuit element. Synonym: **chip**.

ice-bridge effect: In cold areas an ice plug may form in the top of a shothole, confining the gases from the explosion and producing secondary shocks similar to bubble pulses.

ice-noise: 1. Seismic noise resulting from expansion and contraction caused by solar heating. Also noise generated by differential movement of ice floes. **2.** The effect of repetitive shots at random times following the shot, produced by ice fracturing when shooting in permafrost. See Sheriff and Geldart, v.2 (1983, p. 14).

ideal body: A simple model such as a point mass, line mass, cylinder, sphere, vertical step, etc. See Figure M-9.

ideal polarized electrode: A metal-to-electrolyte contact at which no charge crosses the interface. As charge accumulates, the electrode interface behaves like a capacitor without leakage. No chemical reaction takes place and there is no exchange current or faradaic process. This condition is approximated when high-overvoltage, nonreactive metals are at equilibrium with an electrolyte.

IES: Induction electrical survey: A borehole log which usually includes SP, 16-inch normal, and deep-investigation induction logs. Also abbreviated **IEL**.

iff: If and only if.

IFP: *Instantaneous floating point* (q.v.).

IGRF: International Geomagnetic Reference Field; a long-wavelength regional magnetic field determined by an international committee about every five years (e.g., 1965, 1975, 1980); expected secular changes are included.

IGSN71: International Gravity Standardization Net 1971; see Woollard (1979).

IGY: *International Geophysical Year* (q.v.).

II(t): Unit *boxcar* (q.v.).

III(t): *Comb* (q.v.) or shah.

ILd: Deep investigation *induction log* (q.v.).

ILm: Medium investigation *induction log* (q.v.).

image: 1. A method used in tracing raypaths through a constant velocity medium with plane boundaries. An image of a source is located as far below a reflector as the source is above it; raypaths from the image then approach geophones as would reflected energy. See Figure I-1. **2.** A method used in electrical modeling whereby boundaries are replaced by source images so that the potential and current distribution in the zone of interest is unchanged. The objective is to facilitate computation of the potential distribution.

image point: The apparent location of an *image* (q.v.); see Figure I-1.

imaginary: The part of a complex number that involves the factor $(-1)^{1/2}$; the out-of-phase component. In impedance, the reactive component.

imaging: 1. *Migration* (q.v.). **2.** See *remote sensing*.

imbedded wavelet: *Embedded wavelet* (q.v.).

impact blaster: A blaster which fires when it senses sudden motion as from the passage of a seismic wave.

impactor: A device which strikes the ground to compact the earth, used also as a source in the Mini-Sosie method; see *sosie*.

impedance: 1. The apparent resistance to the flow of alternating current, analogous to resistance in a dc circuit. Impedance is (in general) complex, of magnitude Z with a phase angle γ. These can be expressed in terms of resistance R (in ohms), inductive reactance $X_L = 2\pi\nu L$, and capacitive reactance $X_C = 1/2\pi\nu C$:

$$Z = [R^2 + (X_L - X_C)^2]^{1/2},$$

$$\gamma = \tan^{-1}[(X_L - X_C)/R].$$

Z is in ohms when frequency ν is in hertz, L is inductance in henrys, and C is capacitance in farads. **2.** Ratio of the pressure to the volume displacement at a given surface in a sound-transmitting medium; the product of density and velocity or **acoustic impedance. 3.** In magnetotelluric exploration, if $\mathbf{E}_x$ is the electric-field component in an arbitrary horizontal direction and $\mathbf{H}_y$ is the magnetic-field associated with it (in the orthogonal horizontal direction), the **wave impedance** (or **field impedance** or **surface impedance**) Z_x is:

$$Z_s = \mathbf{E}_x/\mathbf{H}_y.$$

The surface impedance is a complex number because of phase differences between $\mathbf{E}_X$ and H_y. Measurements of surface impedance versus frequency can be interpreted in terms of the electrical properties of the subsurface.

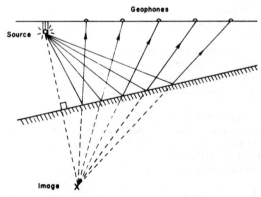

FIG. I-1. Reflected rays reaching the geophones from the source appear the same as direct rays from the **image point**. The use of straight raypaths implies constant velocity.

The surface impedance is a tensor if the conductivity structure is nonlayered. The reciprocal of impedance is called **admittance**.

impedance matching: Making the impedances of two connecting circuits be complex conjugates of each other. Impedance matching gives maximum transfer of power and other benefits.

imploder: A marine seismic energy source which creates a void into which the surrounding water rushes. See *implosion*.

implosion: Collapse into a region of very low pressure; **cavitation**. The creation of such a region under water (as with the Hydrosein or Flexichoc) causes water to rush in with great force and the collision of the in-rushing water on itself generates a seismic shock wave. The outrush of water propelled by a bubble of high-pressure gas (as from an underwater explosion or from an air gun) leaves behind a region at the very low vapor pressure of water, into which the water subsequently collapses, resulting in the *bubble effect* (q.v.).

impulse: 1. The limit of a pulse of unit area as its width approaches zero and its height approaches infinity. Also called **Dirac function** and **delta function** and symbolized by $\delta(t)$. It has a value at only one instant and unit energy content:

$$\delta(t - a) = 0, \text{ if } t \neq a,$$

and

$$\int_{-\infty}^{\infty} \delta(t) \, dt = 1.$$

The impulse is sometimes called the **unit impulse** because its energy is unity. The essential characteristic of an impulse (which is sometimes used as a definition) is expressed by

$$\int_{-\infty}^{\infty} f(t) \, \delta(t - a) \, dt = f(a).$$

An impulse contains all frequencies in equal proportions at zero phase. In digital form, an impulse δ_t is:

$$\delta_{t-a} = \begin{cases} 1 \text{ if } t = a, \\ 0 \text{ if } t \neq a. \end{cases}$$

See *impulse response* and compare *Kronecker delta*. **2.** A pulse which is of sufficiently short time-duration that its waveshape is of no consequence.

impulse blaster: A device which fires an electrical blasting cap when it senses a shock wave. Used to fire a second charge when the shock from the initial explosion reaches it so that the downgoing waves add in-phase.

impulse response: The response of a system to input of an *impulse* (q.v.). Also called **memory function**. The impulse response characterizes a linear system and contains the same information as the frequency-domain **transfer function**, which is the Fourier transform of the impulse response. The output of a linear system is given by the input to the system convolved with the system's impulse response.

impulsive source: A source which produces a very sharp wave of very short duration and which simulates the generation of an *impulse* (q.v.). An explosion is an example of such a source.

incident angle: The angle which a raypath makes with a perpendicular to an interface, which is the same as that which a wavefront makes with the interface in isotropic media.

inclination: 1. The angle between a line's direction and the horizontal; e.g., **magnetic inclination**, the angle at which magnetic lines of force dip. **2.** The dip of a plane (bed, fault, or other tabular body) measured from the horizontal. The attitude of the plane may be characterized by the direction of a line normal to it and the inclination of the plane by the angle between its normal and vertical.

inclinometer: 1. A device for measuring hole inclination and azimuth. See *directional survey*. **2.** A device for measuring the pitch and roll of a ship. Usually either pendulous or gyroscopic. **3.** A surveying instrument which measures the angle between the horizontal and the line of sight.

inclusive OR gate: See *OR gate*.

incoherent light: Light which is composed of many frequency components, random in phase.

incompressibility modulus: Modulus of volume elasticity, **k**; see *elastic constants*.

incompetent: See *folding*.

independent: Not expressible in terms of each other. Two quantities are statistically independent if they possess a joint distribution such that knowledge of one does not give information about the distribution of the other. Equations are independent if their *Jacobian* (q.v.) does not vanish.

independent variable: In the equation $y = f(x)$, x, the argument of f, is the independent variable.

index: A symbol or number to identify one element out of a set, such as an element in a matrix.

index factor: A constant which, when multiplied by certain measurements made on potential-field anomalies, gives an estimate as to the depth of the anomalous mass (sometimes the maximum depth at which the anomalous mass could be located). Used in magnetic and gravity interpretation. See *depth rule*.

index of refraction: A ratio of two phase velocities. In electromagnetics (including light), the ratio of velocity in a medium to that in a vacuum. In acoustics (including seismics), the ratio to that in water.

index word: In computing, the location where the address portion of an instruction can be modified so that a number of operations can be performed repeatedly.

indirect address: In computing, an address that refers to a storage location that contains another address.

induced-current dipole moment per unit volume: A vector parameter describing induced-polarization properties **P** as a function of chargeability M and current density **J**:

$$\mathbf{P} = -M\mathbf{J}.$$

induced gamma-ray spectroscopy log: Bombardment by high-energy neutrons causes elements to emit gamma

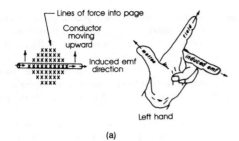

(a)

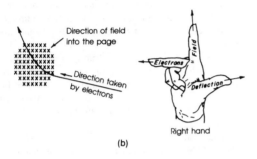

(b)

FIG. I-2. (a) **Induction** and (b) **motor rules**. Another convention uses the second finger to indicate the direction of positive current flow (which is opposite to the direction of electron movement) in which case the left hand is used for motors and the right hand for generators.

rays of characteristic energy. Borehole logging of the gamma energy spectrum resulting from a 14 MeV pulsed neutron source allows identification and analysis as ratios of formation fluid and rock elements. A log for carbon/oxygen distinguishes hydrocarbon from water and hence determines saturation independent of salinity. Silicon/calcium discriminates sandstones from carbonates. Measurement may include spectra of both inelastic (fast-neutron) and capture-gamma (thermalneutron) reactions. See *neutron activation log*.

induced polarization: 1. Usually abbreviated **IP**. An exploration method involving measurement of the slow decay of voltage in the ground following the cessation of an excitation current pulse (**time-domain method**) or lowfrequency (below 100 Hz) variations of earth impedance (**frequency-domain method**). Also known as the **overvoltage method**. Refers particularly to electrode polarization (overvoltage) and *membrane polarization* (q.v.) of the earth. Also called **induced potential, overvoltage**, or **interfacial polarization**. Various electrode configurations are used; see Figures A-12 and A-13. **2.** The production of a double layer of charge at mineral interfaces or of changes in such double layers as a result of applied electric or magnetic fields. Compare *SP* (spontaneous potential).

inductance: The capability of an electric circuit to induce an electromotive force (emf) within the circuit. Measured in henrys.

induction: 1. The process by which a magnetizable body becomes magnetized by merely placing it in a magnetic field. **2.** The process by which a body becomes electrified by merely placing it in an electric field. **3.** The process by which electric currents are initiated in a conductor by merely placing it in an electromagnetic field. According to Faraday's law of induction, a voltage **E** is generated by varying the magnetic flux:

$$\mathbf{E} = -d\phi/dt = -L \ dI/dt,$$

where **E** = voltage, $d\phi/dt$ = time rate of change of magnetic flux in webers/sec, L = inductance in henrys, and dI/dt = time rate of change of current in ampere/s. The vector directions are shown in Figure I-2.

induction electrical survey: *IES* (q.v.).

induction log: An electrical conductivity/resistivity well log based on electromagnetic-induction principles. See Figure I-3. A high-frequency alternating current of constant intensity induces current flow in the formation (**Foucault current**). This current (also called **ground loop**) causes an alternating magnetic field which produces a current in a receiving coil. The receiving-coil current is nearly proportional to the conductivity of the formation. Induction sondes may have several transmitting and receiving coils to produce a highly focused log. An induction log can be recorded whether the borehole fluid is conductive or nonconductive, as in oil-base muds or gas. A dual induction log measures different depths of penetration; **ILd** indicates deep and **ILm** medium penetration.

induction method: An electromagnetic method in which eddy currents are induced in the earth by a time-varying magnetic field. The term is usually applied only to those electromagnetic methods for which the receiver is in the near-field or induction zone of the transmitter.

induction number: A dimensionless parameter which determines the response of an electromagnetic system. It is equal to 2π times a characteristic length of the system divided by the wavelength. As examples, the response of a two-loop system on a homogeneous earth is given by $(\sigma\mu\omega)^{1/2}R$ and on a thin layer by $(\sigma\mu\omega tR)^{1/2}$, where σ is the conductivity, μ the magnetic permeability, ω the angular frequency, t thickness of the sheet, and R the source-detector loop separation. A full-scale electromagnetic system and a scale-model analog have the same response if their induction numbers are equal.

inductive coupling: See *coupling*.

inductivity: Magnetic permeability; the three-dimensional inductance of a material. Free space (and nonmagnetic rock) has an inductivity of 1.257×10^{-6} henrys per meter.

inertial navigation: A dead-reckoning method of determining position in which accelerations are measured with very sensitive accelerometers mounted on a gyroscopically stabilized platform and doubly integrated to give position. Inertial systems tend to accumulate error and hence need to be updated frequently. Usually used in conjunction with other positioning systems rather than on a stand-alone basis.

infinite dike: See *tabular body*.

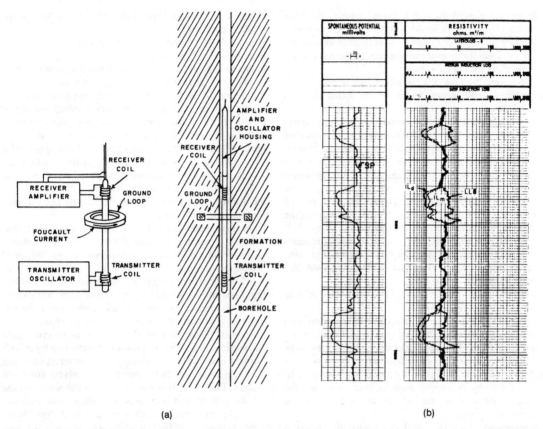

(a)　　　　　　　　　　　　　　　　(b)

FIG. I-3. **Induction log**. (a) Schematic two-coil induction logging sonde. (b) Dual-induction log. IL_d is the deep-induction log. IL_m the medium-induction log. The separation of the resistivity curves shows variation of resistivity with distance from the borehole, probably because of invasion. Laterolog-8 (LL8) and SP logs were run at the same time. (Courtesy Schlumberger.)

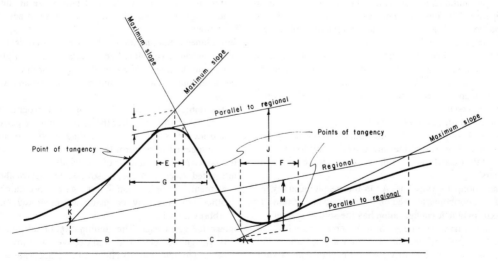

FIG. I-4. **Inflection-tangent-intersection method**. The letters indicate the various parameters which are measured. (From Naudy, 1970.)

infinite electrode: A (usually) fixed, remote electrode which is very far from the field surveying electrodes that are successively moved. Ideally it is located far enough from the measurement electrodes to have negligible effect on the measurements.

infinitely long: So long that end effects are trivial.

infinitesimal strain theory: Strains so small that Hooke's law holds.

inflection-tangent-intersection method: A magnetic interpretation method which involves measuring a number of shape features on a profile across a magnetic anomaly and making the best fit of these measurements to theoretical values for a vertical dike. Both distance measurements and amplitude measurements may be made as shown in Figure I-4. These values are plotted on a logarithmic scale which slides along a graph of the theoretical values until the best match is achieved. A good fit permits one to determine depth, width, magnetic inclination, and magnetization of an equivalent vertical dike. Also called **ITI method** and **Naudi method**. See Naudy (1970).

information retrieval: Searching large quantities of data for wanted information.

infrared: The part of the electromagnetic spectrum with wavelengths between those of visible light and 10^{-3}m; see figure E-8. **Thermal infrared** implies wavelengths longer than those of normal heat radiation.

in-hole IP method: Technique for measuring near-hole IP and resistivity properties using at least one potential and/or one current electrode in the hole. Compare *downhole method*.

inhomogeneity: Lack of spatial uniformity of a physical property. Also called **heterogeneity**.

initial blanking: *Mute* (q.v.).

initial condition: A constraint that describes a system at time zero.

initialize: To set the initial value. Quantities being altered in iterative processing (as might happen in a Fortran "DO loop") may require a starting value assigned them by an "initializing" statement before the loop (as in Figure F-11). Some systems force initial values of zero and do not require initializing statements unless this starting condition is not acceptable.

initial suppression: Attenuation at the beginning of a seismic recording. Used to subdue the amplitude of the noise prior to the first breaks or to prevent overload from high-amplitude first-break energy. Also called **presuppression**.

initial transient: See *secondary voltage*.

injection: Insertion of orbit information into a navigation satellite. Transit satellites are observed by fixed tracking stations and the orbit data are updated every 12-16 hours.

in-line: Along a survey line. An **in-line geophone array** has all the geophones located on the seismic line. An **in-line electromagnetic configuration** has the source and receiver on the traverse line. (Not to be confused with "on-line" versus "off-line" in computer applications.)

in-line offset: A *spread* (q.v.) which is shot from a shotpoint which is separated (**offset**) from the nearest active point on the spread by an appreciable distance (more than a few hundred feet) along the line of the spread. See Figure S-17.

in-line telluric method: See *telluric profiling*.

inner product: *Dot product* (q.v.).

in-phase: 1. The condition in which two waves of the same frequency have the same phase. For two waves of different frequencies, a temporary in-phase condition occurs when crests (or troughs) occur simultaneously in both. **2.** Electrical signal with the same phase angle as that of the exciting signal or comparison signal.

input: 1. The current, voltage, or driving force applied to a circuit, system, or device. **2.** The terminals where the input to a system is applied. **3.** Data to be processed. **4.** To transfer data into a computer.

input filter: An electrical filter placed between the geophone and the amplifier.

input impedance: The impedance across the input terminals of an electrical circuit. Where the input impedance of a measuring device is much higher (say, by 100 times) than the impedance being measured, the measurement will not be altered greatly by the presence of the device.

Input system: A time-domain electromagnetic survey system in which measurements are made during the off-periods between source pulses. Barringer tradename.

in-seam methods: Use of channel waves to investigate changes (such as interruptions by faults) in a layer with lower velocity than underlying and overlying beds, such as coal measures. Both source and receivers are in the layer. Both reflection (receivers near the source in the same gallery) and transmission (sources and receivers in different galleries) techniques are used. See Sheriff and Geldart, v.1 (1982, p. 152-154).

in situ: Material in its original location. Used in connection with measurements of properties of material which do not involve moving the material (and risk altering it thereby). For example, in-situ velocity measurements would be made on rock in place in contrast to measurements made on a sample of the rock in a laboratory.

in-situ modeling: The interpretation of gravity or magnetic data in which the model field points are at the same locations (including elevations) as those of actual field stations.

instantaneous floating point: IFP; see *gain control*.

instantaneous frequency: See *complex-trace analysis*.

instantaneous phase: See *complex-trace analysis*.

instantaneous velocity section: See *velocity* and *seismic log*.

instruction: The basic part of a computer program which specifies to the computer the operation to be performed and identifies and locates the data, device, or mechanism needed to perform the operation.

insulator: A nonconductor of electricity.

integrated circuit: A single solid-state electrical-circuit element. Complete amplifiers, gates, oscillators and other modules may be made as integrated circuits. Abbreviated IC.

integrated geophysics: The combination of seismic, gravity, magnetic, electrical, radioactive, well-log, and/or other geologic data to effect a more accurate or complete interpretation than any one data set could provide.

integrated navigation system: A combination of positioning systems in a synergetic manner. Specifically, the combination of satellite navigation with Doppler sonar and gyrocompass (and other subsystems), or of satellite navigation with radio navigation. The Doppler sonar and gyrocompass (or radio navigation) provide location information between satellite fixes and also provide the velocity information required for an accurate satellite fix. The satellite fixes are used to give the reference locations and remove cumulative error effects in the Doppler sonar and gyrocompass (or uncertainties in the radio navigation).

integration: *Mixing* (q.v.).

integration of chargeability: Measurement of the area under an IP decay curve by integrating the decay voltage with time. When normalized by dividing by the primary voltage, this measurement is one definition of chargeability. The areas under several successive decay curves can be averaged to improve the precision of the measurement.

intelligent terminal: A computer terminal which includes a minicomputer or microcomputer so that it is capable of simple computation operations on its own, but is also an input/output station for a large computer.

intensity: 1. The rate of flow of seismic energy through a unit area perpendicular to the direction of wave travel. For a seismic wave, intensity is proportional to the square of the amplitude of displacement or velocity. 2. A measurement of the effects of an earthquake at a particular place; a measurement of the amount of shaking. The intensity depends not only upon the strength of the earthquake (**earthquake magnitude**) but also upon the distance from the epicenter, the local geology at the point, the nature of the surface materials, the construction of buildings. See also *intensity scale*. 3. Often refers to electric or magnetic field strength; see *electric field* or *magnetic field*.

intensity of magnetization: Magnetic moment per unit volume (occasionally, per unit mass). Includes induced and remanent components.

intensity scale: A standard of measurement of earthquake *intensity* (q.v.). Among systems used are the Mercalli scale, the modified Mercalli scale (see Figure M-5), and the Rossi-Forel scale. Compare *Richter scale*, a measurement of magnitude rather than of intensity.

interactive: A process in which a human is involved on-line. A human at a terminal inputs instructions and the computer responds with a display of the results of executing the instructions. "Interactive" implies that the computer is on-line and that the operator waits at the terminal to get the response, so that he can modify the instructions if the response is not satisfactory; that is, the computer and operator are in a **dialogue mode.** Occasionally the use of off-line terminals is also called interactive.

interbed multiple: *Pegleg multiple* (q.v.).

intercept distortion: Change in waveshape produced by a system with linear phase response where the intercept value of the phase-response curve at zero frequency is other than $2n\pi$ where n is any integer. Results from frequency components being shifted with respect to each other.

intercept method: A method of computing near-surface corrections from the intercept time at zero distance on a time-distance plot of first breaks.

intercept time: The time obtained by extrapolating the refraction alignment on a refraction time-distance $(t-x)$ plot back to zero offset. See Figure C-15.

interface: 1. The common surface separating two different media in contact. 2. The contact or connecting element between two computing machines or components by means of which information is passed between the two. **Interface devices** are used to reformat data (such as to convert serial data bits to parallel bits or from analog-to-digital form or vice versa), to hold data until needed (**buffer** function), and sometimes to do simple operations like summing (counting). 3. To make the output of one device acceptable as input of another device.

interface wave: *Surface wave* (q.v.).

interfacial polarization: A dielectric property due to conductivity contrasts in a material. See *induced polarization*.

interference: 1. The superposition of two or more waveforms. Interference is **constructive** where the waveforms are in phase (so peaks add to peaks), **destructive** where 180 degrees out of phase (so peaks cancel troughs). 2. Signals from another source (e.g., atmospheric static) which tend to obscure a desired signal.

interior angle: See *angles (surveying)* and Figure A-9.

interlacing: 1. Displaying data from deep and shallow shots side by side to aid in identifying reflections and ghost effects. 2. In refresh graphics, the use of alternate scan lines on successive refresh frames.

interlock: A device (such as a switch) which prevents operation if some condition is not satisfied.

interlocking: 1. Involving energy which has traversed the same raypath, usually in opposite directions. Two seismic records are interlocking if some geophone group of the one record occupies the source location of the interlocking record, and vice versa. Reflection events on interlocking traces should have the same arrival time. 2. Interdependent controlling of several units whose functions have to be coordinated for proper (or safe) operation.

intermediate storage: Use of magnetic tape, discs, or drums for temporary storage of seismic data during digital or analog processing. Abbreviated **ISD.**

intermodular distortion: See *distortion*.

International Active-Sun Years: A program of studying solar-terrestrial phenomena during sunspot maxima.

International Geomagnetic Reference Field: *IGRF* (q.v.).

International Geophysical Year: A program of geophysical observation from July 1, 1957 to Dec. 31, 1958, near sunspot maximum; **IGY.** The program included large-scale investigations in Antartica.

International Gravity Formula: The Geodetic Reference System 1967 (GRS67) gravity at the latitude ϕ is

$$978\ 031.846\ (1\ +\ 0.005\ 278\ 895\ \sin^2\phi$$
$$+\ 0.000\ 023\ 462\ \sin^4\phi\ \text{mGal.}$$

See *latitude correction* and Woollard (1979).

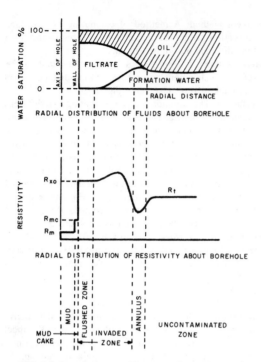

FIG. I-5. **Invaded zone**. (Courtesy Schlumberger.)

International Gravity Standarization Net: See Woollard (1979).

International Years of the Quiet Sun: A program during 1964-1965 of studying solar-terrestrial phenomena during sunspot minima.

interpolation: Determining values at locations where they have not been measured or specified from nearby values. See also *bilinear interpolation* and *spline*.

interpreter: 1. One who determines the geological significance of geophysical data. **2.** A machine which reads coded information (such as punched cards) and prints out the translation. **3.** A computing-machine routine which translates and executes each source-language statement before translating and executing the next one, as with programs written in Basic. Compare *compiler*.

interrogate: To make a simple inquiry to the system, from which a quick, short answer is expected.

inter-record gap: 1. The space between records on a magnetic tape. Also called the **interblock gap. 2.** The distance tape will travel from complete stop to operational speed.

interrupt: A signal which advises that some operation needs to be done. Interrupts allow peripheral on-line equipment to advise a computer's control that it has data to be read or requires instructions or data. Compare *flag*.

intersection: *Conjunction* (q.v.). See also Figure G-1.

interstitial water: Water in the interstices or pore spaces in a formation.

interval density: The density of an interval of rock integrated from gamma-gamma log data or determined by a borehole gravity meter; **apparent density**.

interval time: The time difference between two reflection events. Interval time may vary because of changes in velocity from one material to another, changes in thickness such as produced by differential compaction, changes in original section thickness (with stratigraphic implications), or changes subsequent to deposition because of erosion, solution, flow, or structural deformation. See also *interval transit time*.

interval transit time: The traveltime of a wave over a unit distance, hence the reciprocal of wave velocity. Measured by a sonic log, usually in microseconds per foot.

interval velocity: The velocity of an interval in the subsurface measured by determining the traveltime over a depth interval along some raypath. **1.** In sonic log determinations, the interval may be 1 to 3 ft; in well shooting it may be 1000 ft or more. Usually refers to *P*-wave velocity. **2.** The average velocity of the interval in the subsurface between two reflections. Often used for velocity calculated by the *Dix formula* (q.v.) from velocities measured from normal moveout, which implies horizontal constant-velocity layers.

intrabasement anomaly: A local anomaly caused by magnetic polarization variation wholly within the basement complex. Generally the depth to the bottom of such a body is more than twice the depth to its top.

intrinsic conduction: Conduction due to major components of the material, as opposed to conduction due to impurities or imperfections. At high temperatures intrinsic conduction dominates other conduction modes.

intrinsic dispersion: The variation with frequency of seismic velocity in an inelastic material because of the inelasticity, as distinguished from the geometric dispersion associated with the physical configuration of the material. Intrinsic dispersion accompanies attenuation.

intrinsic IP: The true induced polarization of a specific material or geologic unit.

intrinsic thermal-neutron decay-time: The true decay-time of the formation as opposed to the measured decay-time which is subject to hole and diffusion effects. See *neutron-lifetime log*.

invaded zone: The portion about a well bore into which drilling fluid has penetrated, displacing some of the formation fluids. Invasion takes place in porous, permeable zones because the pressure of the mud is greater than that of the formation fluids. See Figure I-5. As the mud penetrates into the formation, portions of the mud **(mud cake)** build on the formation wall, limiting further flow of mud fluid **(filtrate)** into the formation. Directly behind the mud cake is a **flushed zone** from which almost all of the formation water and most of the hydrocarbons have been displaced by filtrate. The invasion process alters the distribution of resistivities and other properties and consequently the value which logs read. The **depth of invasion** is the equivalent depth in an idealized model rather than the maximum depth reached by filtrate. In oil-bearing intervals, the filtrate may push a bank of formation water ahead of it to produce a relatively low-resistivity **annulus** which is especially important with deep-investigation induction logs.

inverse dispersion: Dispersion in which velocity decreases with frequency, so the frequency of a wave train increases with time. Antonym: *normal dispersion.*

inverse filter: A filter with characteristics complementary to another filter so that when used in series with the other filter no frequency-selective filtering occurs (except for overall time delay) over some band-pass. See *deconvolution.*

inverse Fourier transform: See *Fourier transform.*

inverse magnetostriction: See *magnetostriction.*

inverse matrix: The inverse of a square matrix **A** is the matrix which yields the identity matrix when multiplied by **A**, that is, $\mathbf{A}^{-1}\mathbf{A} = \mathbf{I}$. The inverse matrix can be found by dividing the adjoint (or adjunct) of **A** by the determinant of **A**. A nonsquare matrix **B** can be squared up by multiplying by its transpose, that is,

$$[\mathbf{B}^T\mathbf{B}]^{-1}\mathbf{B}^T\mathbf{B} = \mathbf{I},$$

and the inverse can then be found by the same procedure as indicated for **A**.

inverse modeling: Determining a model which could have given rise to observed effects; solution of the *inverse problem* (q.v.). Usually inverse modeling is not unique. Also called **inversion**.

inverse problem: The determination of a distribution of parameters whose calculated response matches observations within given tolerance. In contrast to the **direct**, **forward**, or **normal problem**, which involves calculating what would have been observed from a given model. See Figure M-8.

inverse-square law: 1. The magnitude of a potential field surrounding a unit element varies inversely as the square of the distance from the element:

$$\text{Field} = km/r^2.$$

Applies to a gravitational field where the element is mass and k is the *gravitational constant* (q.v.); to a magnetic field and where the element m is a magnetic pole and k is 10^{-7} webers/ampere meter, where pole strength m is in ampere meters; and to an electrostatic field where the element m is an electrical charge (m in coulombs) and k is 9×10^9 newtons/coulomb. For the electrostatic case, it is also called **Coulomb's law. 2.** The energy density of a seismic body wave from a point source in a homogeneous isotropic medium varies inversely as the square of the distance from a point source. Also called **spherical divergence. 3.** The intensity of electromagnetic energy (light or radio) varies inversely as the square of the distance from a source.

inversion: 1. The process of solving the *inverse problem* (q.v.). **2.** Calculating acoustic impedance (or velocity) from a seismic trace, taken as representing the earth's reflectivity; the result is a *seismic log* (q.v.). **3.** Finding the reciprocal. **4.** The "inverse of A" is "not $A = \overline{A}$"; see *gate*.

I/O: Input/output; refers to the input or output functions of a computing machine.

IOM: Input/output module.

ion exchange: The property of some minerals (particularly clays) which enable them to absorb certain anions and cations and retain them in a state whereby they can be exchanged for other anions and cations in solution. Ion exchange is a diffusion process and its rate depends on ion mobility.

ionic conductor: See *conductor.*

ion mobility: Ease of movement of ions in an electric field, measured by the ratio of ion velocity to electric-field strength.

ionosphere: That part of the Earth's atmosphere which includes several layers of ionized gas at heights of 80 to 400 km, which bends certain radio waves back toward the Earth. It is divided into several layers according to the types and concentration of ions. See Figure A-16, *D-layer*, *E-layer*, *F-layer*, and *refraction correction.*

IP: *Induced polarization* (q.v.).

ips: Inches per second.

IP susceptibility: A term used as a measure of induced polarization, implying an analogy with other types of polarization such as induced magnetic effects.

IRM: Isothermal remanent magnetization.

irreducible water saturation: The fraction of the pore volume occupied by water in a reservoir at maximum hydrocarbon saturation. It represents water which has not been displaced by hydrocarbons because it has been trapped by adhering to rock surfaces, trapped in small pore spaces and narrow interstices, etc. Irreducible water saturation is an equilibrium situation. It differs from residual water saturation, the value measured by core analysis, because of filtrate invasion and the gas expansion that occurs when the core is removed from the bottom of the hole to the surface.

irrotational wave: *P-wave* (q.v.).

IR survey: A remote-sensing method using either infrared radiation reflected from the surface after being beamed downward from an aircraft (*thermal infrared*, q.v.), or occasionally natural infrared radiation. Infrared surveys are usually flown at night to reduce natural radiation and solar effects.

ISD: Intermediate storage device; see *intermediate storage.*

iso-: A prefix meaning "equal," used in conjunction with other words to denote contour lines through points on maps at which the measured value is the same. Examples include the following: (a) **isoanomaly:** A line on a map connecting points of equal anomaly, used especially for maps showing magnetic or gravity anomalies; also spelled "isanomaly". (b) **isobar:** Line joining places of equal barometric pressure. (c) **isobath:** Line of equal water depth. (d) **isochore:** Line of equal interval between two beds or two seismic events; differs from isopach in that it may express variations in many units and the effects of unconformities, whereas isopach implies variations within a single unit; isochore is now falling into disuse. (e) **isochron:** Line of equal reflection time, equal time difference, or equal delay-time. (f) **isocline:** Line of equal magnetic inclination. (g) **isogal:** Line of equal gravity anomaly. (h) **isogam:** Line of equal magnetic intensity. (i) **isogon:** Line of equal magnetic

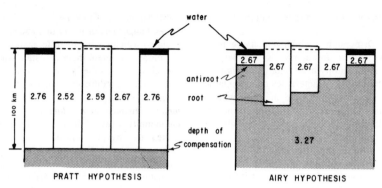

FIG. I-6. **Isostasy** concepts.

declination. (j) **isohyet:** Line of equal amounts of rainfall. (k) **iso-J:** Line of equal values of the Jacobian or relative ellipse area used in telluric surveying. (l) **isopach:** see separate entry. (m) **isopleth:** Line of equal geochemical measurement (such as element or isotope ratio). (n) **isopor:** Line of equal secular change, such as equal annual change of isogonic or isoclinic lines. (o) **isorad:** Line of equal radioactivity or lines of equal gamma radiation. (p) **isoseismal:** Lines between regions of different intensity for a given earthquake; such a line might indicate the boundary between zones IV and V on the Mercalli scale, for example. (q) **isotherm:** Line of equal temperature. (r) **isotime:** see separate entry.

isochronous surface: *Time surface* (q.v.).

isopach: 1. A contour which denotes points of equal thickness of a rock type, formation, group of formations, etc. 2. A contour which denotes points of equal difference in seismic arrival times for two reflection events, thus indicating constant thickness if the velocity is constant. **Isotime** is a more proper term to use but isopach is in common usage. 3. A contour denoting equal vertical distance, thus not necessarily corrected for the dip of the bedding.

isostasy: The gravitational balance of large portions of the Earth's crust as though they were floating on a denser underlying layer (the asthenosphere). Isostasy accounts for major topography. (a) The **Pratt hypothesis** assumes density variations so that areas of less-dense crust rise topographically above areas of more-dense (see Figure I-6). (b) The **Hayford modification** requires that the pressure be balanced at the "depth of compensation". (c) The **Airy hypothesis** varies the thickness of crustal blocks of constant density so that the thicker parts ride higher; thus mountainous areas are compensated by deep **roots** and deep ocean basins by **antiroots.** Ocean basins thus have antiroots at 6–8 km with the roots of mountains extending to 50–60 km. (d) The **Heiskanen modification** permits density to vary but compensates 2/3 of the topography with roots. (e) The **Vening Meinesz hypothesis** allows some of the balance to be accommodated laterally by the surrounding region rather than in the vertical direction only. The **radius of regionality** specifies the size of the region over which compensation is distributed; it is of the order of 200 km.

isostatic correction: A correction to gravity data to compensate for lateral density or thickness variations between large blocks of the Earth's crust. The correction assumes an isostatic model; it is made from elevation data and water-depth data using zone charts. See Heiskanen and Vening Meinesz (1958, p. 159-170).

isostatic rebound: Isostatic adjustment after removing or imposing a stress. From sea-level changes attributed to the melting of Pleistocene continental glaciers, the **isostatic-rebound relaxation time** is of the order of 4000 years, corresponding to a viscosity of 4×10^{22} poise.

isothermal remanent magnetism: IRM. See *remanent magnetism.*

isotime: Isochron. 1. The time interval between two reflections. Isotime changes sometimes indicate stratigraphic changes, reef buildups, variations in salt thickness, etc. The isotime may vary because of variation in the velocity or the thickness or both. 2. Contours of reflection time, time intervals, etc.

isotropic: Having the same physical properties regardless of the direction in which they are measured. Compare *anisotropy.*

isovelocity surfaces: Surfaces of constant seismic velocity. In the absence of structural uplift and lateral variations of lithology, isovelocity surfaces are apt to be nearly horizontal planes. However, where structural uplift occurs, isovelocity surfaces tend to follow structure but with less relief than the structure. Changes in velocity bend seismic rays and hence alter apparent structure.

iterative 1. A procedure which repeats until some condition is satisfied; see *loop.* **2.** Processing by successive approximations, each based on the preceding, in such a way as to converge onto the desired solution.

iterative modeling: An interpretation technique for solving the *inverse problem* (q.v.) by successive approximations; generally performed on a computer.

ITI method: *Inflection-tangent-intersection method* (q.v.).

I-wave: A *P*-wave in the earth's inner core. Compare *K-wave.*

J

J: *Jacobian* (q.v.).

j: 1. Symbol to indicate $(-1)^{1/2}$; i is also used. **2.** Multiplication by j rotates a complex vector by 90 degrees. **3.** A unit vector in the y-direction.

jack: The receptacle into which an electrical plug fits.

Jacobian: For a set of transform equations $y_i = y_i(x_1, x_2, \cdots, x_n)$, the Jacobian matrix **J** is:

$$\mathbf{J} = \|\mathbf{J}_{ik}\| = \|\partial y_i / \partial x_k.\|$$

Vanishing of the determinant of the Jacobian shows that relations are not independent. For the 2×2 matrix,

$$\left\| \begin{matrix} a & b \\ c & d \end{matrix} \right\|,$$

independence is shown by $ad - bc \neq 0$. Named for Karl Gustav Jacob Jacobi (1804-1851), German mathematician.

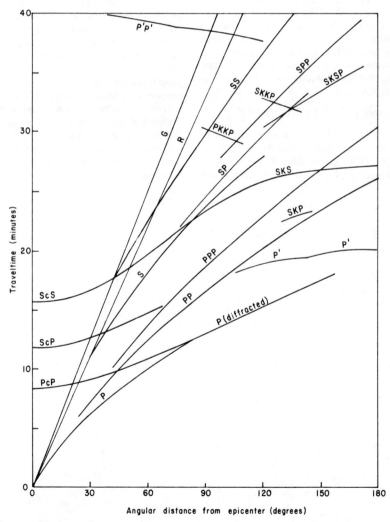

FIG. J-1. **Jeffreys-Bullen traveltime curves** for some waves from a shallow earthquake. See Figure E-2 and *wave notation* regarding labeling of the curves. (From Seismological Tables, 1940.)

Jacob's staff: A pointed stick about 5 ft long used to support a surveyor's compass.

Janus configuration: An arrangement of transducers aimed in opposite directions, as the forward-aft (or port-starboard) pair of Doppler-sonar transducers (see Figure D-16). Named for Janus, the Roman god of doorways who had two faces so he could watch in opposite directions.

JCL: *Job-control language* (q.v.).

Jeffreys-Bullen curves: Relationships between the arrival time of various modes of seismic waves and the distance (angular distance) from the epicenter of the earthquake which generated them. See Figure J-1.

jet: To drill a shothole by pumping water (or mud) down the drill stem or drill casing, where the water flow rather than the grinding action of a bit is the main force for removal of the material.

job-control language: A method of telling a computer what jobs to do, in what order to do them, what are the requirements of a job (e.g., what tapes to use), and what to do with the results. Abbreviated JCL.

Johnson noise: Noise resulting from the random thermal energy of conduction electrons. Also called **thermal noise.** For an electrical element, the mean square voltage due to Johnson noise is proportional to the absolute temperature K, the bandwidth B (in hertz), and the resistance R (in ohms):

$$\text{rms noise voltage} = 7.4 \times 10^{-12}(KBR)^{1/2}.$$

JOIDES: Joint Oceanographic Institutions for Deep Earth Sampling, a program to obtain cores of the sediments in the deep oceans. Holes drilled from the ship *Glomar Challenger* did much to prove plate tectonics and hence has had tremendous impact on geology and geophysics.

joint inversion: *Mutual inversion* (q.v.).

joint probability: The probability of simultaneous occurrence of values of two or more quantities.

Josephson junction: A thin insulator separating two superconducting materials, through which electron pairs can tunnel. See Josephson (1965, p. 419).

joystick: An interactive control for moving a cursor on a video display so that something can be done to the matrix element corresponding to the cursor's location.

jug: *Geophone* (q.v.). Derives from an early oil-damped geophone (oil jug).

jug hustler: One who lays out and picks up the seismic spread and geophones.

jug line: 1. Cable connecting geophones to instruments. 2. The *spread* (q.v.).

jug planter: *Geophone planter* (q.v.).

Julian day: 1. The number of a day within a calendar year, referred to Greenwich. 2. The day number since noon, January 1, 4713 BC, the beginning of the Julian period.

jump: An instruction which can cause the computer to fetch the next instruction from a location other than the next sequential location. Synonym: **branch.**

jump a leg: To miscorrelate one or more cycles. See also *leg*.

jump correlation: See *correlation*.

jumper: 1. A relatively short electrical connection. 2. A connection different from the ordinary. 3. A temporary electrical connection used to bypass part of a circuit. 4. A seismic cable without takeouts for geophone connections, used to connect with the main spread cable when it is otherwise difficult to reach because of access problems.

K

κ: **1.** 2π times the wavenumber; the **wavenumber** is the number of waves per unit distance, the reciprocal of wavelength:

$$\text{wavenumber} = 1/\lambda = \kappa/2\pi = \nu/V,$$

where λ = wavelength, ν = frequency, and V = velocity. Thus k is to wavenumber in the spatial sense as angular frequency ω is to frequency ν in the time sense. **2.** Some authors use $\kappa = 1/\lambda$ instead of the above. **3.** Seismic usage often implies **apparent wavenumber**, λ_a being the apparent wavelength and V_a the apparent velocity. If this definition is used, κ varies with the angle between the raypath and the line of measure (the line of the spread, usually).

k: **1.** *Resistivity-contrast factor* (q.v.). **2.** A unit vector in the z-direction. **3.** See *k-type section*.

Kalman filter: A recursive filtering scheme applicable to nonsteady-state linear systems. A system is often described by a set of differential equations involving orthogonal state variables. The errors in each measurement and in the model which the set of equations describes are assumed to be independent and Gaussian. The filter estimates the error based on prior measurements and incorporates each new measurement in future error estimates. Kalman filtering is used in real-time reduction of integrated satellite-navigation data and in some seismic-filtering schemes. See Bayless and Brigham (1970).

kappa meter: An instrument for measuring magnetic susceptibility. Tradename of ABEM Stockholm.

kataseism: Earth movement toward the focus of an earthquake. Antonym: **anaseism.**

K.B.: *Kelly bushing* (q.v.).

K-band: Radar band (10.0 – 36.0 GHz) which is sensitive to vegetation; used in remote sensing and Doppler-radar.

kelly: The unit at the top of the drill pipe which transmits the rotary motion of the rotary table to the drill pipe. The kelly is usually a square, grooved, or hexagonal member supported at the upper end by the swivel and secured into and supporting the drill pipe at the lower end.

kelly bushing: The journal box insert in the rotary table of a drilling rig through which the *kelly* (q.v.) passes. Its upper surface is commonly the reference datum for well logs and other measurements in a well bore. Abbreviated **K.B.**

kelvin: The SI temperature unit referenced to absolute zero (0 K) with intervals equal to the degree celsius (formerly called centigrade); 273.15 K = 0° C. Note "degree" or the degree symbol ° is not used with K and that kelvin is not capitalized (see Appendix A). Named for Baron William Thomson Kelvin (1824-1907), English physicist.

Kelvin material: A material in which the stress depends both on the strain and the rate-of-change of strain.

Kepler coordinates: The quantities which describe the elliptical orbit of a satellite with respect to the celestial sphere. Navigation satellites broadcast their Kepler coordinates and ephemeral information which allow the calculation of their location at any given instant (see Figure K-1). Named for Johann Kepler (1571-1630), German astronomer.

Kepler's laws: **First law:** Every planet follows an elliptical path with the sun at one focus. **Second law:** A line from the center of the sun to the center of a planet sweeps the same area in the same time; hence the planet travels fastest when closest to the sun. **Third law:** The square of the period of revolution of a planet is proportional to the cube of its mean distance from the sun.

kernel function: **1.** For any linear transform,

$$F(\xi) = \int K(x, \xi) f(x) \, dx,$$

$K(x,\xi)$ is the "kernel." **2.** A mathematical function of resistivity and depth which can be calculated from apparent resistivity data, from which one tries to derive the resistivity stratification. Koefoed (1965, p. 568-591) derives the kernel function for Schlumberger-configuration data and Paul (1968, p. 159-162) for Wenner-configuration data. The derivation of resistivity stratification from the kernel is shown by Pekeris (1940) and Vozoff (1956). The electric potential V at the surface of a horizontally layered earth due to a dc point-source which is also located at the surface was expressed by Stefanesco in 1930 as a Hankel integral,

$$V = C/r + 2C \int_{-\infty}^{\infty} K(\lambda) \, J_0 \, (\lambda r) \, d\lambda,$$

where r is the distance from the point source to the observation point, $J_0(\lambda r)$ is a Bessel function, λ is a phantom variable of integration, C is a constant, and $K(\lambda)$ is the kernel function. Also called **Stefanesco function.**

key bed: A reflection with sufficient distinguishing characteristics to make it easily identifiable for use in correlations.

key punch: An electromechanical device for encoding data on punched cards, as for input to a computer.

keyseat: Where a borehole changes direction so that the drilling pipe rubs the side of the borehole, the effect is to wear a groove in the borehole wall. The drill pipe may fit in this groove but the groove may be smaller than the couplings between units of drill pipe, making it difficult to pull the drill pipe.

k-factor: See Figure S-22.

kick: **1.** Onset of a transient wave, also called *break* (q.v.). **2.** Sudden drop in drilling fluid pressure caused by fluid

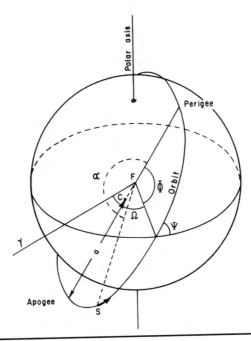

	C = Center of the eliptical orbit.	of perigee and (2) period of satellite.
	F = Focus = center of Earth for navigation satellites.	To locate rotating Earth:
	S = Satellite.	Right ascension of Greenwich at time of perigee.

To define plane of orbit:
- Ω = Right ascension of the ascending node of the orbit.
- Ψ = Inclination of the orbital plane; approximately 90 degrees for polar orbit. (Cos Ψ and sin Ψ are broadcast for Transit satellites.)

To define shape of orbit:
- Φ = Argument of perigee.
- a = Semimajor axis of orbit.
- e = Eccentricity = $\overline{CF}/a$.

To locate satellite in orbit:
- α = Angular position of satellite in orbit; calculated from (1) Greenwich time

To locate rotating Earth:
Right ascension of Greenwich at time of perigee.

To allow for precession:
- dΩ/dt = Rate of change of right ascension.
- dΦ/dt = Precessional rate of perigee.

Also broadcast are:
- Time of last injection.
- Satellite frequency.

Ephemeral information:
- Greenwich time.
- Correction to the eccentricity anomaly.
- Correction to the semimajor axis.
- Distance out of the orbital plane.

FIG. K-1. **Kepler coordinates** and satellite ephemeral information. ϒ is the first line of Aries or the vernal equinox.

penetrating a porous formation. **3.** Sudden change in drilling fluid pressure caused by gas entering the drilling fluid from a porous formation.

kill: To set a trace (or portion of a trace) equal to zero.

K-index: A measure of the average intensity of magnetic disturbances in time, such as magnetic storms, but excluding diurnal and lunar time variations.

kinematic similarity: A ratio of times in physical modeling. See *modeling theory.*

Kirchhoff diffraction equation: A form of *Kirchhoff's equation* (q.v.) expressing the wave amplitude at a point *P* which is several wavelengths from a diffracting aperture in the plane *B*. Referring to Figure K-2, if α and β are the angles between the normal to *B* and the rays from *Q* and *P* to *B*, the lengths of the rays being *r* and *s*, then the wave amplitude ϕ at *P* can be expressed as an integral over the aperture area *s*:

$$\phi = -(1/2\lambda) \iint_s (1/rs)(\cos \alpha - \cos \beta)\, e^{j(r+s)/\lambda}\, ds.$$

Kirchhoff's equation: An integral form of the wave equation expressing the wave function at the point P, ψ_p, in source-free space in terms of the values of ψ and its derivative on a surrounding surface S at the preceding time $(t - r/V)$:

$$\psi_p = (1/4\pi) \iint \{[\psi]\, \partial(1/r)/\partial n$$
$$- (1/Vr)(\partial r/\partial n)\,[\partial\psi/\partial t] - (1/r)\,[\partial\psi/\partial n]\}\, ds.$$

The terms in brackets are evaluated at time $(t - r/V)$, which is called **retarded time**. r is the distance from P to points on the surface S, and n is a unit vector normal to S. The Kirchhoff integral equation used in migration can be written

$$\psi(x, z, t) = (z/\pi)\int [1/r^3 - (2/Vr^2) \cdot$$
$$\cdot\, (\partial/\partial t)\psi(x', 0, t + \tau)]dx',$$

where x' is position at $z = 0$, τ is the two-way time $2r/V$, and r is the distance from x' to x. For r much longer than a wavelength this simplifies to the Rayleigh-Sommerfeld approximation,

$$\psi(x, T, t) = -(2T/\pi V^2)\int (1/T^2) \cdot$$
$$\cdot\, (\partial/\partial t)\psi (x', 0, t + T)\, dx',$$

where $T = 2z/V = $ vertical traveltime. This expresses migration by integration along a diffraction curve.

Kirchhoff's laws: First law: The vector sum of all currents into any junction point is zero. **Second law:** The vector sum of all voltages around a closed loop is zero. Kirchhoff's laws apply to three-dimensional materials such as the earth as well as to conventional electrical circuits. Named for Gustav Robert Kirchhoff (1824-1887), German physicist.

Kirchhoff's method of migration: *Migration* (q.v.) by integrating along diffraction curves, in effect integrating with the *Kirchhoff equation* (q.v.). See Sheriff and Geldart, v.2 (1983, p. 60-62).

κ-κ domain: A wave field in which the independent variables are wavenumber in the x-direction and wavenumber in the z-direction (depth). Similar to the f-k domain except with the frequency variable replaced by $2\pi\kappa V$ where $V = $ velocity.

Klauder wavelet: The autocorrelation of a Vibroseis sweep. The *Vibroseis* process of injecting a sweep of frequencies into the ground and then correlating with

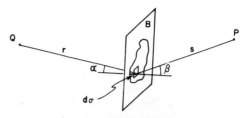

FIG. K-2. Symbols used in **Kirchhoff diffraction equation.**

the sweep pattern to yield a seismic record is equivalent to convolving the earth's reflectivity with the autocorrelation of the Vibroseis sweep, so that the Klauder wavelet is in effect the seismic waveform for correlated Vibroseis records.

knot: A nautical mile per hour = 1.1508 statute mile per hour = 1.852 km/hour = 101.27 ft/minute = 0.5148 m/s.

Knott's equations: Equations governing the partition of energy between reflected and refracted plane P- and S-waves at a plane interface, as a function of the velocities, elastic constants, and densities in the two media and of the incident angle. See Sheriff and Geldart, v.1 (1982, p. 65-66) and *Zoeppritz's equations.*

Koenigsberger ratio: The ratio of the remanent magnetization to the product of susceptibility and the earth's magnetic field strength; symbol: Q. See Hood (1964).

Krey wave: Channel wave involving P-SV particle motion in a low-velocity layer between two higher velocity half-spaces. Named for Theodore C. Krey, German geophysicist.

Kronecker delta: A unit tensor, whose diagonal elements are unity and whose other elements are zero. Usually written as δ_{ij}. Compare *impulse.*

K-type section: A *three-layer resistivity model* (q.v.) in which the middle layer is more resistive than the layers above or below it. See Figure T-5.

kurtosis: A measure of the sharpness of a population distribution. If x_{10}, x_{25}, x_{75}, and x_{90} are the values for which 10, 25, 75, and 90 percent of the population is smaller, respectively, then

$$\text{kurtosis} = (x_{75} - x_{25})/2(x_{90} - x_{10}).$$

k-vane: A depressor paravane used to pull down a sidescan sonar or other underwater towed device. Also called **k-wing.** Tradename of Klein Associates.

K-wave: A P-wave in the earth's outer core. Compare *I-wave.*

L

L: 1. A surface wave. (*L* stands for "long" waves.) L_Q denotes a Love (Querwellen) wave, L_R a Rayleigh wave, but Q and R are preferred to L_Q and L_R. See *wave notation*. **2.** The area above a decay curve from 0.45 to 1.75 s (induced-polarization usage):

$$\int_{0.45}^{1.75} (V_{0.45} - V(t))\, dt.$$

ladder network: A step-type electrical network composed of *H*, *L*, *T*, or pi sections connected one after the other. Often used so that one can pick off different voltages at different levels with the same impedance. Ladder filters are shown in Figure F-8.

lag: 1. A difference in the time of two events. **2.** A delay in the arrival time of seismic events. Lagging refraction or reflection arrivals may indicate subsurface structure or delay caused by weathering variations, phase shifts in filtering, shothole fatigue, etc. Negative of **lead**. **3.** The phase angle by which the current is behind the emf in an inductive circuit. **4.** The time delay between the breaking of the bridgewire in a detonating cap and the resulting explosion. **5.** To be behind.

lagged product: The product of two values corresponding to different times.

Lagrange interpolation formula: A method of calculating a polynomial for interpolating between a set of values which are not necessarily equally spaced.

$$y_1 = y(x_1),\ y_2 = y(x_2),\ \cdots,\ y_n = y(x_n);$$

$$y(x) = \frac{(x - x_2) \cdots (x - x_n)}{(x_1 - x_2) \cdots (x_1 - x_n)}\, y_1$$

$$+\ \frac{(x - x_1)(x - x_3) \cdots (x - x_n)}{(x_2 - x_1)(x_2 - x_3) \cdots (x_2 - x_n)}\, y_2 + \cdots$$

$$+\ \frac{(x - x_1)(x - x_2) \cdots (x - x_{n-1})}{(x_n - x_1)(x_n - x_2) \cdots (x_n - x_{n-1})}\, y_n;$$

that is, in the factors multiplying y_k the factor $(x - x_k)$ is omitted. Named for Joseph Louis Lagrange (1736-1813), French mathematician.

lambda: A medium-range positioning system in which the mobile station is the master and which involves the transmission of two frequencies (100 to 200 kHz) to remove lane ambiguity. Two small fixed stations use reception from the mobile master to control phase-locked transmitters. The use of phase-locked oscillators permits continuous transmission despite short losses of signal such as might result from sky-wave interference. A development of range-range *Decca* (q.v.).

Lambert conformal conic projection: A conformal secant (or tangent) conic map in which the earth's features are projected radially from the earth's center onto a cone which intersects the earth along two (one) standard parallels; see Figure M-3. Parallels are thus the arcs of circles and meridians are straight lines and angles are preserved but the scale varies except along parallels. Lambert projections are used as the standard map references in some states. **Lambert coordinates** on such a map are rectangular grid coordinates with respect to an arbitrary reference point; they are not oriented precisely north-south (except along the reference meridian) nor east-west. (The **Lambert equal-area** map is an azimuthal projection.) Named for Johann Heinrich Lambert (1728-1777), German physicist.

Lambert coordinates: See *Lambert conformal conic projection*.

Lamb's problem: An investigation of the effects of seismic disturbances initiated by a point source on the surface of a semiinfinite, perfectly elastic medium. Named for Sir Horace Lamb (1849-1934), English mathematician.

Lamb wave: A type of guided wave in a thin layer (thickness $< \lambda$).

Lamé constants: See *elastic constants*. Named for Gabriel Lamé (1795-1870), French mathematician.

lampitude: The cepstrum-domain equivalent of amplitude. See *cepstrum*. A permutation of the letters in "amplitude."

Lancing: Emplacing small explosive charges ($\sim$100 g) about 2 m deep with a 1 inch diameter spear. Prakla-Seismos tradename.

Landsat image: A "photograph" synthesized from measurements made by a Landsat satellite, often shown in false color. A full scene is 185×185 km represented by 3240 pixels east-west and 2340 scans north-south. **Landsat bands** for multispectral scanners are 4: 05 to 0.6 μm (green); 5: 0.6 to 0.7 μm (red); 6: 0.7 to 0.8 μm (infrared); 7: 0.8 to 1.1 μm (infrared).

lane: The unit of measuring position with standing-wave radio-positioning systems. In phase-comparison (CW) systems, the lane is the distance represented by one cycle of the standing-wave interference pattern resulting from two radiated waves. Its apparent wavelength is not constant but depends on the position within the network. Phase-comparison systems permit location within a lane but do not necessarily determine in which lane. The lane ambiguity has to be resolved either by counting the number of lanes which have been crossed since a known location or by some additional measurement, such as measurements at different frequencies. Lane shape depends on the configuration of the system. Circular, hyperbolic, and occasionally elliptical systems are used, as are combinations of these. See Figure L-5.

langley: A measure of heat flow, a calorie/cm^2. Named for Samuel Pierpont Langley (1834-1906), American physicist.

language: A method of communicating instructions to a computer. Involves words and associated rules (**gram-**

mar). Computer languages include Algol (algorithmic language), Cobol (common business oriented language), Fortran (formula translation) in several versions, Basic, PL/1, etc.

Laplace's equation: A differential equation which describes certain behavior at points in free space. The **Laplacian** $\nabla^2 U$ of a potential function U vanishes in space which contains neither sources nor sinks. (∇ is the operator "del".) In rectangular coordinates,

$$\nabla^2 U = \frac{\partial^2 U}{\partial x^2} + \frac{\partial^2 U}{\partial y^2} + \frac{\partial^2 U}{\partial z^2} = 0.$$

Compare *Poisson's equation*. Named for Pierre Simon Laplace (1749-1827), French mathematician.

Laplace transform: The linear transforms

$$F(s) = \int_{-\infty}^{\infty} f(t)\, e^{-st}\, dt,$$

and

$$f(t) = (1/2\pi i) \int_{c-i}^{c+i} F(s)\, e^{st}\, ds.$$

s is a complex number and t is a real one. When the limits of integration are $\pm\infty$, the transform is **two-sided**. The two-sided Laplace transform becomes identical with the Fourier transform when s is purely imaginary. More often the **one-sided transform** is used, especially in the study of transient waveforms. In this case the integral is

$$F(s) = \lim_{h\to 0} \int_{h}^{\infty} f(t)\, e^{-st}\, dt,$$

and

$$f(t) = \frac{1}{2\pi i} \int_{c-i}^{c+i} F(s)\, e^{st}\, ds.$$

The one-sided transform is often written with limits 0 to ∞, the limit being implied. Laplace transforms may not exist for all values of s, and hence many Laplace transforms are limited to **strips of convergence**, the ranges of values for the real part of s for which the above integrals are finite. The Laplace transform domain is often called the **s-plane**. See Sheriff and Geldart, v.2 (1983, p. 172-174).

Laplacian: See *Laplace's equation*.

large aperture seismic array: LASA. A geophone array in Montana set up to detect nuclear explosions and distinguish them from earthquakes. LASA consists of 21 subarrays of 25 detectors each, the subarrays being of the order of 7 km in diameter and the LASA itself is about 200 km in diameter. See Figure L-1. Other large-aperture arrays include NORSAR in southern Norway and UKAEA (United Kingdom Atomic Energy Authority) arrays in Southern Scotland; Yellowknife in Canada, Australia, and India. Also called **phase array stations**.

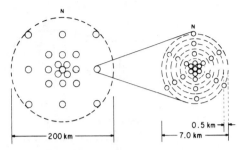

FIG. L-1. **LASA, large-aperture seismic array**. Each subarray is composed of sensors as shown to the right. (From White, 1969.)

large scale integration: High-density integrated circuits used for complex logic functions. Circuits can range up to several thousand transistors on a silicon chip only 0.1 inch square. Abbreviated **LSI**.

Larmor frequency: The frequency with which gyromagnetic moments precess in a magnetic field. Atoms and nuclei possess magnetic moments because of their spin and precess like small gyroscopes about the direction of an externally applied steady magnetic field (such as the Earth's field). Radio-frequency energy at right angles to the steady field will be absorbed because of resonance when the RF-frequency equals the precession frequency. This principle is involved in *proton-resonance* and *optically pumped magnetometers* and in the *nuclear-magnetism log* (q.v.). Named for Sir Joseph Larmor (1857-1942), English mathematician.

LASA: *Large-aperture seismic array* (q.v.).

laser: Light amplification by the stimulated emission of radiation. A laser beam consists of highly coherent light waves.

laserscan: An optical process whereby a seismic record section undergoes two Fourier transformations to make another record section. A lens accomplishes the Fourier transform; see Figure L-2. If a grating is present in the object plane at the focal point of the lens, parallel rays will emerge and constructive interference will occur only at l_0, l_1, l_2, etc., separated by a distance Z which depends on the grating spacing. In optics, one usually thinks of white light separating into its frequency components as a result of passing through a uniform grating. In the LaserScan, monochromatic light passed through a nonuniform "grating" (e.g., a seismic record section) separates into the spacing components of the record section (like f,k space). One lens thus accomplishes a Fourier transform of the record section and a second lens synthesizes the components back into a record section. By blocking certain components in the transform plane their effect can be eliminated in the image plane. A laserscan can accomplish frequency or velocity filtering. LaserScan is a tradename of Conductron Corp. See Dobrin et al. (1965).

laser surveying: Use of a transit which transmits a beam of laser light to a mirror (or corner reflector) on a rod and times the round-trip traveltime to determine the distance.

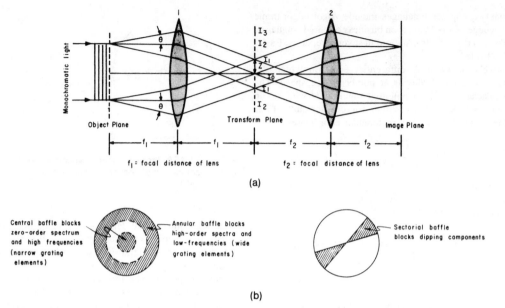

FIG. L-2. **Laserscan** principle. (**a**) Each lens in effect Fourier-transforms the data. (**b**) Baffles acomplish filtering when inserted in transform plane.

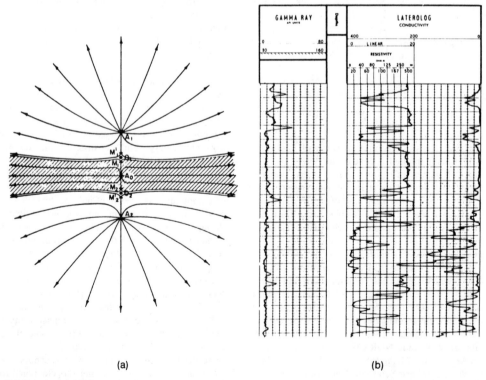

FIG. L-3. **Laterolog**. (**a**) Schematic of Laterolog-7 showing current flow lines. (**b**) Laterolog plotted on hybrid scale. (Courtesy Schlumberger.)

	Mean thickness	P-wave velocity, km/s
——sea level——		
Water	4.5 km	1.5
——sea floor——		
Layer 1 (sediments)	0.4	1.6–2.5
——often rough interface ——		
Layer 2 (basement basalt)	1.5	4.0–6.0
——gradational boundary ——		
Layer 3 (oceanic gabbro)	5.0	6.4–7.0
——Moho——		
Upper mantle		7.4–8.6

Within layer 1 in North Atlantic:
Horizon A (about 300 m below sea floor) Eocene chert

A*	Tertiary volcanic clays/black clays
β	Black clays/Jurassic or Cretaceous limestones
B	Layer 1/layer 2

FIG. L-4. Oceanic **layering** terminology.

latency: A delay encountered when waiting for a specific response. Latency is caused by propagation delays and the queuing of disks or tapes when randomly addressed. For a disk, the time to move the disk arm (**seek latency**) and the time waiting for the desired data to pass the head (**rotational latency**). See also *access time*.

latent root: *Eigenvalue* (q.v.).

lateral: A type of resistivity log, now obsolete. A constant current is passed between an electrode A in the borehole and a remote electrode while the potential difference is measured across two electrodes, M and N, located on the sonde; see Figure E-7. The MN distance is small compared to the **spacing**, which is the distance between the current electrode and the midpoint between the potential electrodes, usually 16 to 22 ft, often 18 ft 8 inches (5.69 m). A **short lateral** sometimes uses a spacing of 6 to 9 ft (1.8 to 2.7 m). The potential electrodes are usually below the current electrode, but on a **reciprocal sonde** the functions are interchanged so that the potential electrodes are above the current electrode.

lateral variations: Changes in a horizontal direction.

laterlog: A resistivity log made with a sonde that is focused by use of guard or bucking electrodes which force the "surveying current" to flow nearly at right angles to the logging sonde. Also called **guard log** or **focused log**. One type uses three electrodes plus guard electrodes. In the laterolog-7 (Figure L-3), sufficient current is fed into the bucking electrodes A_1 and A_2 so that the current from electrode A flows out nearly at right angles to the sonde (cross-hatched area in Figure L-3). Sensing electrodes M_1', M_1, M_2, and M_2', are used to adjust the bucking-electrode currents until this flow condition exists. The laterolog-3 and guard log use long bucking electrodes above and below the current electrode. Laterolog is a Schlumberger tradename.

latitude: 1. The angle between a tangent to the ellipsoid and the earth's axis; **geodetic** or **geographic latitude**. See Figure G-2. **2.** Distance north (positive) or south (negative) of a reference point or of an east-west reference line. Used in the latitude-departure survey method, where the distances are called **northing** or **southing. 3.** A smoothed magnetic inclination; see *magnetic latitude*.

latitude correction: 1. A correction to gravity data because of (a) variation in centrifugal force resulting from the Earth's rotation, as the distance to the Earth's axis varies with latitude ϕ; and (b) variation of the Earth's radius because of polar flattening. The **Geodetic Reference System 1967 (GRS67)** gives

$$g = 978{,}031.846(1 + 0.005278895 \sin^2 \phi$$
$$+ 0.000023462 \sin^4 \phi) \text{ mGal.}$$

The latitude correction amounts to $1.3082 \sin 2\phi$ mGal/mile $= 0.813 \sin 2\phi$ mGal/km. **2.** A gyrocompass correction for the rotation of the horizontal north vector as a function of latitude. (The horizontal north vector is tangent to the Earth and hence the rotation is the result of Earth curvature.)

law of reflection: The angle of reflection = angle of incidence for the same wave mode in an isotropic medium. This is a special case of Snell's law. The more general form of *Snell's law* (q.v.) must be used for the wave generated by mode conversion upon reflection.

law of refraction: See *Snell's law*.

law of tangents: Electrical current lines at a boundary are bent such that

$$\rho_1 \tan \alpha_1 = \rho_2 \tan \alpha_2,$$

where ρ_1, ρ_2 are the resistivities of the two media and α_1, α_2 are the angles which the current flow makes with a normal to the interface.

layer: For terminology of layering beneath the oceans, see Figure L-4; for deeper layering, see *Earth layering* and Figure E-1.

layered earth: 1. An idealized model of the Earth consisting of a number of horizontal homogeneous layers above a homogeneous half-space. **2.** A similar idealized model but using spherical shells to deal with problems where Earth curvature is important.

layer stripping: See *stripping*.

layout chart: *Stacking chart* (q.v.).

L-band: Radar frequencies between 390 and 1550 μHz; see Figure E-8.

lead: 1. An indication of interesting structural or other geological conditions. **2.** Amount of time by which one event is ahead of another, or by which an arrival is ahead of its "normal" arrival time. Negative of **lag. 3.** The phase angle by which current is ahead of emf in a capacitive electrical circuit. **4.** An electrical conductor for connecting to electrical devices. **5.** To be ahead of.

leakage: Low electrical resistance to ground where there should be high resistance, as with a wet seismic cable.

leaking mode: 1. A seismic wave which is imperfectly trapped between reflecting strata; energy escapes across a layer boundary. **2.** Head-wave energy which leaks through a refractor. The amplitude of head waves within a refractor (for energy incident at the critical angle or greater) decreases exponentially away from the interface. If the refractor is thin, some of the energy will "leak" through the refractor and appear as seismic waves below the refractor. **3.** Propagation in imperfect wave guides. **4.** Coherent noise produced by energy bouncing at incident angles smaller than the critical angle within beds which act as waveguides for larger angles.

leap frog: A survey technique in which two units alternately take the lead; e.g., the rodman precedes the transit instrument man, then the transit instrument man precedes the rodman, etc.

least absolute deviation fit: An l_p fit with $p=1$. The l_1 fit is the least mean deviation solution of a problem and corresponds to the maximum-likelihood estimate when the errors have a Laplace (double exponential) distribution. The best l_1 estimate to a set of numbers X_i is the median.

least-squares filter: *Wiener filter* (q.v.).

least-squares fit: An analytic function which approximates a set of data such that the sum of the squares of the "distances" from the observed points to the curve is a minimum. (Usually implies deviation measurements along paths x = constant; other criteria are sometimes used.) One must determine the functional form of the fit (whether linear, quadratic, etc.) and what is to be minimized in order to define the problem. For example, different velocity functions result depending on whether seismic time-depth data or velocity-depth data are fitted, or if the data are weighted or differently distributed in depth. Least-squares fitting is the same as the l_p fit with $p=2$. The l_2 fit is the least-variance solution and corresponds to the maximum-likelihood estimate when the errors have a Gaussian (normal) distribution. The best l_2 estimate to a set of numbers is the average of the

numbers. See Sheriff and Geldart, v.2 (1983, p. 152-155).

least-time path: The path between two points which takes the least time to traverse, subject to certain constraints. The path which a seismic ray takes according to *Fermat's principle* (q.v.) (although Fermat's principle may dictate a path other than the least-time path). Also called **minimum-time path** or **brachistochrone**. The raypath will generally be curved or bent because of velocity variations.

LED: Light-emitting diode.

Lee partitioning method: A variation of the *Wenner electrode array* (q.v.). An additional electrode is placed midway between the potential electrodes and the potential measured is between it and each of the other potential electrodes.

left-hand rule: A rule which gives the direction of an induced emf in a conductor which cuts a magnetic field. See Figure I-2.

left-lateral fault: See Figure F-2.

leg: A cycle of more-or-less periodic motion. When following a seismic event from trace to trace or from record to record, one usually concentrates on a particular trough or peak of the energy. If an erroneous correlation is made into an adjacent trough or peak, one has **jumped a leg**.

Legendre theorem: 1. The potential due to a disk at points away from its axis can be found in terms of the potential on the axis. The solution on the axis can be determined in closed mathematical form, while the solution away from the axis can be found in terms of a Legendre series. **2.** For a spherical triangle which is small compared with the spherical radius, the sides bear the same length relationship as a plane triangle whose corresponding angles are smaller by a third of the *spherical excess* (q.v.). This theorem is used in correcting plane surveying for earth curvature. Named for Adrien Marie Legendre (1752-1833), French mathematician.

leg function: A curve composed of a series of segments.

leggy: The character of a wave train which includes several cycles with significant amplitude. Legginess is produced by too narrow a filter band-pass.

Leibnitz rule: A formula for the nth derivative of the product of two functions f and g:

$$d^n(fg)/dx^n = \sum [n:K] \, (d^{n-K}f/dx^{n-K})(d^K g/dx^K),$$

where $[n:K] = n!/[(n-K)!K!]$. Named for Gottfried Wilhelm Leibnitz (1646-1716), German mathematician.

Lenz's law: An induced emf is in such a direction as to generate a magnetic field which opposes the change which induced it. Named for Heinrich Friedrich Emil Lenz (1804-1865), Russian physicist.

level: 1. A survey device used to ascertain which point on a survey rod is at the same elevation as the instrument. **2.** Amplitude, as in "a potentiometer controls the voltage level." **3.** A track on digital magnetic tape, as in "7-level recording" where seven magnetic heads spaced across the tape width record seven bits of information at one time (i.e., in one byte).

Levinson algorithm: An algorithm used to solve the normal equations for a Wiener optimum filter; see Sheriff and Geldart, v.1 (1982, p. 182-183).

L$_g$-wave: A short-period guided surface wave which travels in the continental crust. The "g" refers to granitic layer.

library: A collection of programs and data which a computer system has available to use.

liftering: The cepstrum-domain equivalent of filtering. A permutation of the letters in "filtering."

line: 1. A linear array of observation points, such as a seismic line. **2.** Equipment which is directly connected to and controlled by a central controller is **on-line**, that not so controlled is **off-line. 3.** A channel. **4.** In a power spectrum, the contribution of a single frequency. Physically, the contribution of a very narrow frequency band.

lineagenic: See *tectonic types*.

linear: Having a straight line relationship; $x = a + by$, where a and b are constants. See also *linear system*.

linear circuit: A *linear system* (q.v.).

linear device: An electronic device containing linear circuits.

linear filter: See *convolution*.

linearity: That property of a filter, amplifier, or intrinsic property which indicates that the output increases proportionally to the input. The range of linearity is usually limited to certain amplitudes and frequencies. For example, (a) the proportional relationship between induced polarization and current density, in which chargeability is constant, or (b) the symmetrical identity between voltage versus time of IP charge and decay curves, or (c) the proportionality between decay-curve amplitude and polarizing voltage.

linearly independent: The property of not being expressible as a linear combination of other elements. Thus a set of equations (or solutions) is independent if none of them can be expressed as a linear combination of the others. A set of linear functions.

$$f_i(x_1, x_2, \cdots x_n) = \Sigma a_{ij}x_j,$$

is linearly independent if the determinant does not vanish (i.e., if $|a_{ij}| \neq 0$); see *Jacobian*.

linear-phase filter: A mixed-phase filter which shifts frequency components proportional to frequency, thus introducing a constant delay but no change of waveshape. The phase-shift versus frequency graph is linear over the band-pass and has an intercept which is a multiple of 2π. Also called a **delay filter**. See Figure P-1 and *phase characteristics*. Such a filter produces no phase distortion. If the intercept is an odd multiple of π, it will invert the wavelet.

linear ramp: See *ramp*.

linear sweep: Vibroseis signal where the vibrator frequency varies linearly with time, that is, the sweep amplitude is

$$A \cos 2\pi(\nu_0 t + bt^2) \text{ for } \textbf{upsweep,}$$

or

$$A \cos 2\pi(\nu_0 t - bt^2) \text{ for } \textbf{downsweep.}$$

where ν_0 is the starting frequency, and b is the rate of change of frequency $d\nu/dt$. See Figure V-7.

linear system: 1. A system whose output is linearly related to its input. If a linear system is excited by a sine wave of frequency ν_1, the output will contain only the frequency ν_1; the amplitude and phase may be changed, however. The **rule of scaling** (if A results in B, then kA results in kB, k being any constant) and the **rule of superposition** (if A results in B and C results in D, then $A + C$ results in $B + D$) apply. **2.** An electrical circuit whose impedance is independent of applied voltage (or current).

line-mile: A unit of measure for geophysical work indicating continuous coverage over one mile of seismic line or one mile of aeromagnetic data.

line of force: A curved line in a three-dimensional potential field such that a tangent anywhere along the line is in

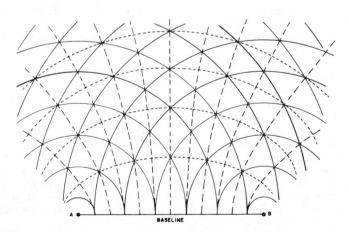

FIG. L-5. **Lines of position** (LOP). The solid lines indicate constant range from A (or B) and are circles. The dashed lines indicate constant difference in range from A and B and are hyperbolas. The short dashed lines indicate that the sum of the ranges from A and B is constant and are ellipses.

the direction of the force on a charge (pole, mass, etc.) at that location. The density of lines of force is called the **flux density** or **field strength**. Used in connection with electric fields, magnetic fields, gravitational fields, etc. See *Gauss law*.

line of position: The locus of equal values measured with a positioning system, such as a line of equal phase difference with Raydist or Omega. Lines which represent constant range from a reference location are circles (Figure L-5); lines which represent constant difference in ranges are hyperbolas; lines which represent constant sum of ranges are ellipses. The intersection of two lines of position determines a location **fix**. Positions may be determined by the intersection of two families of circles, by two families of hyperbolas, by a combination of circles and hyperbolas, etc. Abbreviated **LOP**.

line of sight: The straight-line distance from an object to an observer. High-frequency (short-wavelength) radio waves travel such raypaths so that the curvature of the earth limits the range which can be achieved. (The "effective" earth curvature differs somewhat from the actual earth curvature; see *radio earth*.) Line-of-sight range R is given in terms of the height of transmitting and receiving antennas h_t and h_r:

$$R = k(h_t^{1/2} + h_r^{1/2}).$$

The constant k is 1.22 nautical miles when h is given in feet, 4.08 km when h is given in meters. Refraction increases the effective range of radio waves depending on their wavelength; the sensitivity of the detecting system also affects the range.

line printer: A computer output device which prints a line of characters simultaneously (usually).

line source: A source of energy which can be treated mathematically as though it were condensed into an infinitely long line in three dimensions or a point in two dimensions. (a) In the seismic method, events which appear to emanate from a line, such as a diffraction from a fault. (b) In gravity and magnetic methods, line sources represent a concentration of mass or magnetized matter into a point in two dimensions (such as a horizontal cylinder treated as having the mass concentrated along the axis of the cylinder). (c) In electrical methods, a long current-carrying wire can often be regarded as a line source. Usually the current is taken to be equiphase along the wire. In the near-field zone, the magnetic field varies inversely with distance for a line source, inversely with distance squared for an electric dipole, and inversely with distance cubed for a magnetic dipole.

line spectrum: A frequency versus amplitude plot that indicates that only certain frequencies are present in the signal rather than that the spectrum is continuous.

lineup: In-phase alignment across the traces of a seismogram, showing coherent energy.

linkage: 1. The instructions which connect one program to another, providing continuity of execution between the programs. **2.** The convention for exchanging control and data between a subroutine and the module calling it.

liquid-junction potential: Liquid-boundary potential; see *electrochemical SP*.

Lissajous figure: The steady pattern on an oscilloscope when periodic waves which are harmonics of a common frequency are applied to the horizontal and vertical plates. Such patterns are used to ascertain that the frequencies are exactly the ratios of small integers and to determine their relative phase relationship.

list: 1. Overall average tilt of a ship to starboard or port (about an axis in the principal direction of motion). As opposed to **roll**, which is periodic motion about this axis. **2.** See *listing*.

listening period: The time between periodic inputs. In time-domain IP surveying, responses are measured during "listening periods" between periods of current application. In Doppler-sonar navigation, the frequency is measured during the listening periods between transmissions.

listing: A hard-copy list of a computer program or data.

listric surface: A curved fault (fracture) surface which becomes less steep as one goes deeper. Because of the curvature, rotation of the down-dropped block accompanies slippage along the fault.

lithologic log: A log showing lithology as a function of depth in a borehole. Sometimes a **strip log** based on samples, sometimes interpreted from other borehole logs.

lithosphere: The upper 100 km (approximately) of the Earth which is relatively rigid, thus including the Earth's crust and upper mantle. Characterized by relatively low attenuation of seismic waves (high Q). Underlain by the asthenosphere and the mesosphere. See Figure E-1.

lithostatic pressure: Pressure produced by the weight of overlying rock. See *abnormally high pressure*.

littoral: The depth zone between high and low water; coastal.

little slam: *Small slam* (q.v.).

live: Responsive. A seismic channel is "live" if it is responsive to input energy. As opposed to **dead**.

LLI: *Log-level indicator* (q.v.).

ln: Natural logarithm; logarithm to the base e; $\ln x = 2.3026 \log_{10} x$.

Lloyd mirror effect: Interference between a sea-surface ghost and sound following a direct path along the sea floor. The separation of interference bands depends on offset.

LNG: Liquified natural gas.

load: 1. The power which a device consumes or delivers. **2.** An impedance connected across an output. **3.** To place the explosive in a shothole. **4.** To input program and/or data into a computer.

loading pole: A pole (usually in 10 ft sections) for placing explosive charge in a shothole.

load-point marker: A marker (such as a band of metal foil affixed on a digital magnetic tape) which indicates to the computer the beginning of information on the tape. See also *end-of-reel marker*.

lobes: Passbands in a directivity graph. See Figure D-12. The main pass region is the **main lobe** and smaller pass regions are called **side lobes**. Used in connection with seismic directivity, radio antenna patterns, etc.

local gravity: Bouguer value from which the regional has been subtracted; *residual* (q.v.).

local hour angle: See *hour angle*.

local magnetic anomalies: Anomalies of restricted areal distribution caused by the magnetization of units in the uppermost parts of the Earth's crust.

lock on: Establishment of phase agreement between an oscillator in a receiver and a received radio signal. Phase-lock loops are used to maintain phase relationships when the received signal temporarily drops out.

locus: The set of all points which satisfy a given requirement. Thus a circle or sphere is the locus of points equidistant from a particular point.

log: 1. A record of measurements or observations, especially those made in a borehole. See *well log*. **2.** An instrument for measuring a vessel's speed or distance traveled or both.

logarithmic contour interval: Plotting of data on a logarithmic scale is sometimes used where properties of materials vary by many orders of magnitude. Resistivity and IP data often are contoured in intervals which are approximately logarithmic (or geometric), such as 1, 2, 5, 10, 20, 50, 100; or 1, 3, 10, 30, 100, 300.

logarithmic decrement: The natural logarithm of the ratio of the amplitudes of two successive cycles of a harmonic event. Where the amplitude decay is because of absorption, the logarithmic decrement can be related to Q and to the absorption coefficient.

logging: Measuring the physical properties of the material around a borehole. See *well log*.

logical path: The precise sequence of instructions executed by a computer. The logical path may be controlled by a series of conditional tests applied at various points.

log-level indicator: A trace which indicates the log of the gain of one amplifier channel under AGC control. Abbreviated **LLI**. See Figure R-5.

log-normal: A statistical distribution which, when plotted logarithmically, has the appearance of a normal Gaussian-distribution curve.

long count: See *satellite navigation*.

longitudinal: Along the major axis of a feature; as opposed to transverse. The longitudinal axis of a structural feature (such as an anticline) is its **b-axis**.

longitudinal conductance: Product of average conductivity and thickness of a layer. Measured in siemen.

longitudinal parity check: The bits in each column along the length of a magnetic tape are counted and a parity bit is recorded at the end of the record block. When the record is read, the bits are again counted as a check. The parity track as well as data tracks are checked in this manner. Also called **longitudinal redundancy check**. It is used in NRZI recording. **LPC**.

longitudinal wave: *P-wave* (q.v.).

long normal: A *normal* (q.v.) resistivity log made with the A and M electrodes in the sonde 64 inches (1.63 m) apart.

long-path multiple: A seismic reflection whose travel path is much longer than required for a primary reflection from the deepest interface reached. A long-path multiple tends to appear as a separate event rather than

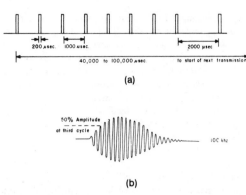

(a)

(b)

FIG. L-6. **Loran C.** (**a**) Pulse transmission pattern; slaves do not transmit the ninth pulse. Phase coding of successive pulses identifies stations and helps in sky-wave identification. (**b**) Shape of one of the Loran-C pulses.

blending into the tail of the primary. For example, the energy might be reflected by a deep reflecting interface, then at or near the surface, and again by the same or another deep interface. See Figure M-12.

long-spaced sonic log: A *sonic log* (q.v.) run with a tool having a spacing (see Figure S-12) of 8–12 ft versus 3–5 ft for the normal sonic tool. Designed to measure formation properties beyond that portion possibly altered by invasion.

long wave: *Surface wave* (q.v.).

loop: 1. Field observations which begin and end at the same point with a number of intervening observations. Obtaining data in loops is useful in correcting for drift in gravity-meter observations or diurnal variation in magnetometer surveys, and in detecting faults or other cause of misclosure in seismic work. **2.** An electrical circuit which provides feedback, as an AGC loop. **3.** A part of a computer program in which the last instruction is to repeat the preceding series of operations (with or without modification) until some particular condition is reached. **4.** Transmitting or receiving coil used in electromagnetic surveying.

LOP: *Line of position* (q.v.).

Lorac: A medium-range surveying system involving the phase comparison of beat frequencies of radio waves transmitted over different paths, similar to Raydist, Decca, and Toran. Lorac uses a fixed reference station which observes the transmissions of the center station and "red" and "green" end stations and broadcasts information about the phase differences of the CW stations on a different frequency. Seismograph Service Corp. tradename.

Loran: Long-range navigation. One of several government-maintained long-range pulse-type electronic positioning systems. Hyperbolic lines of position are determined by measuring the differences in the times of reception of synchronized pulse signals from fixed transmitters at known geographic positions. **1. Loran A** (now obsolete) involved the transmission of pulses of 1850, 1900, and 1950 kHz energy from fixed transmitters. **2. Loran B** was an experimental loran system. **3. Loran C** operates at 100 kHz and has long wavelength and long range. It allows radiofrequency cycle-matching to provide increased accuracy; see Figure L-6. The Loran C master station transmits groups of pulses at a repetition period which characterizes the network. The repetition period and transmission frequency are controlled by ultrastable cesium frequency standards. Two or more slave stations transmit pulses which are also ultrastable and synchronized with the master station. At a user's mobile station, the time difference between the master and the slave station is measured, thus defining hyperbolic lines of position which locate the mobile station. **4. Loran C** may also be operated in the **rho-rho** mode if the mobile station includes a stable atomic-frequency clock which is synchronized with the standard clocks at the transmitting stations so that the traveltime from two stations can be determined, thus giving the range to these stations. The rho-rho mode permits location by the intersection of circular lines of position. If the range to three stations can be determined, the system is called **rho-rho-rho. 5. Loran D** is similar to Loran C but uses shorter base lines (distances between master and slave stations) and has been used for military purposes only.

LOS: *Line of sight* (q.v.).

losser: A circuit element which attenuates the gain upon proper instruction; used in AGC circuits.

loss tangent: A measure of dielectric loss defined by the relationship $\tan \delta = \sigma/\epsilon\omega$, where σ is the loss tangent, ϵ is the dielectric permittivity, and ω is angular frequency.

Lotem: Long-offset transient electromagnetic sounding. See *electric sounding*.

Love wave: A surface seismic channel wave associated with a surface layer which has rigidity, characterized by horizontal motion perpendicular to the direction of propagation with no vertical motion. Designated **Q-wave, Querwellen wave,** L_Q**-wave,** *G***-wave** or *SH***-wave.** Love waves may also be thought of as channel waves in the upper layer. Love waves can travel by different modes, designated by the number of nodal planes within the layer. However, usually only the zero mode is observed. The dispersion of Love waves can be used to calculate the thickness of the surface layer. Earthquake Love waves have velocities up to 4.5 km/s, faster than Rayleigh waves. See Figure E-3 and Sheriff and Geldart, v. 1 (1982, p. 49, 51). Named for A.E.H. Love, English mathematician.

low: An area in which beds are structurally lower than in neighboring areas; a syncline or structural depression.

low-amplitude display: Seismic data displayed at low gain so that the strongest events are not overdriven and that their detail is evident.

low-cut filter: A filter that transmits frequencies above a given cutoff frequency and substantially attenuates lower frequencies. Same as **high-pass filter.**

low-frequency shadow: A region of lowered instantaneous frequency which often lies immediately under a hydrocarbon accumulation.

low-level multiplexing: Multiplexing between the preamplifier and first amplifier stage.

low-pass filter: A filter that passes frequencies below some

cutoff frequency while substantially attenuating higher frequencies. Same as **high-cut filter.**

low-velocity layer: 1. A near-surface belt of very low-velocity material often abbreviated **LVL**; also called **weathering.** The LVL is very important in seismic interpretation because it can have marked effect on the arrival times of reflections. The low-velocity zone often varies in thickness, lithology, density, velocity, and attenuation effects. The velocity of the layer is comparatively low, commonly of the order of 500 m/s. The velocity and thickness of the layer are the two parameters which have the greatest bearing on seismic interpretation. **2.** A layer of velocity lower than that of shallower refractors. See *blind zone*. **3.** Any layer bounded on both sides by layers of higher velocity. Such a layer can carry *channel waves* (q.v.). **4.** The B-layer in the upper mantle (see Figure E-1) from 60 to 250 km deep, where velocities are about 6 percent lower than in the outermost mantle. **5.** The region just inside the earth's core.

low-velocity layer correction: Weathering correction; a correction which is added to the arrival time of a reflection to give the arrival time which would have been observed if source and receiver had been located on the datum surface with no low-velocity layer present.

loxodrome: *Rhumb line* (q.v.).

LPC: Longitudinal parity check. See *check*.

l_p **fit:** The parameters which produce estimations y_i to a set of data points $\hat{y}_i$ such that the differences between the data and the estimates (the errors $e_i = y_i - \hat{y}_i$) minimizes

$$\psi = \sum_{i=1}^{n} |w_i \, e_i|^p,$$

where w_i are weighting factors. If $p = 1$, this yields the least-absolute deviation fit; for $p = 2$, least-squares; for $p = $ infinity, the **minimax** or **Chebychev results.**

LPG: Liquified petroleum gas.

L_Q **wave:** *Love wave* (q.v.).

L_R: Rayleigh wave (q.v.).

LSB: Least significant bit.

LSI: *Large-scale integration* (q.v.) or large-scale integrated circuit.

L-spread: 1. A seismic spread in which the shotpoint is offset by an appreciable distance perpendicularly to the spread line, the shotpoint being opposite one of the end geophone groups. See Figure S-17. **2.** A spread often laid out for noise studies in which about half of the length is in-line with the shot and the remainder perpendicular.

lumped circuit: An electrical network composed of specific resistance, capacitance and other elements, as opposed to distributed resistance and capacitance as in a transmission line, or as opposed to a solid-state system.

LVL: *Low-velocity layer* (q.v.) or weathering.

*L***-waves:** Long waves; seismic **surface waves** of long wavelength from earthquakes. L_Q denotes a Love Wave, L_R a Rayleigh wave.

lystric: *Listric surface* (q.v.).

M

m: Milli, a prefix indicating 10^{-3}.

M: The earthquake phase with the maximum amplitude on the seismogram. Now obsolete.

ma: Subscript used with log terms to indicate the rock matrix.

machine language: An instruction code usually created by assemblers or compilers that is directly executable by a computing machine. **Object programs** are in machine language.

Maclaurin series: A special case of a *Taylor series* (q.v.) in which expansion is about the origin. Named for Colin Maclaurin (1698-1746), Scottish mathematician.

macroscopic anisotropy: The situation where measurement of a physical property perpendicular to the bedding differs from measurement parallel to the bedding because of the inclusion of interbeds of markedly different property; see *anisotropy*.

macroscopic cross-section: See *capture cross-section*.

MAE: An analog time-domain process for reducing the effect of singing. See *Backus filter*. GSI tradename.

mafic: Rich in magnesium and iron silicates and other dark minerals; as opposed to acidic.

Magnedisc: A magnetic tape record shaped like a large disc. GSI tradename.

magnetically quiet: Having ambient magnetic variations less than tens of gammas.

magnetic anomaly: The difference between observed and theoretical or predicted magnetic values. A **residual magnetic anomaly** is the magnetic anomaly components which remain after removal of the longer wavelength regional portion.

magnetic artifacts: See *artificial magnetic anomalies*.

magnetic basement: The upper surface of extensive crystalline rocks having magnetic susceptibilities which are large compared with those of sediments. Often but not necessarily coincident with the geologic *basement* (q.v.). Generally excludes magnetic sediments, thin volcanic and other high-susceptibility rocks intruded into the sedimentary section, but thick volcanic rocks in the sedimentary section would be classed as magnetic basement where the magnetic effects of deeper bodies would not be resolvable.

magnetic cleaning: Removing the ''soft'' secondary magnetization of a sample so the ''hard'' primary magnetization can be studied.

magnetic core: A small toroidal ferrite ring used to store a bit of information in a computer's rapid-access memory. Each core is a few hundredths' of an inch in diameter. The tiny cores are strung on a grid of wires and the direction of current flow determines the direction of magnetization in the core; reversing the current direction in the wire reverses the direction of magnetization or polarity. Second generation technology now largely superseded by solid-state devices.

magnetic dip: Magnetic inclination; see *inclination*.

magnetic dipole: One ampere flowing in a circular coil of area 1 m^2. For two magnetic poles of strength p separated by an infinitesimal distance δx, the **dipole**

strength is $p\ \delta x$. **1.** In geophysical exploration, refers to a source of electromagnetic energy created by an alternating current in a single or multiturn loop carried from an aircraft or laid out on the ground. See *magnetic moment*. **2.** The magnetically polarized nature of rocks and ore bodies.

magnetic disk: A thin metal disk coated with magnetic recording material, usually with 4 to 25 disks mounted on a vertical shaft, separated so that read-write heads can enter between them and read from the top or bottom of any disk. Data are stored as magnetized spots in concentric tracks (up to 250/surface) and remain until new data are written on top of the old data. Data can be stored sequentially or randomly. The read/write heads are sometimes movable, sometimes fixed. Disk storage devices with movable heads have longer access times than those with fixed heads or drum storage devices (which also have fixed heads).

magnetic disturbance: *Magnetic storm* (q.v.).

magnetic drum: A rotating cylinder of magnetic material upon which data can be stored. An intermediate-access storage mainly made obsolete by solid-state memories configured to appear as drums. Used for program storage, program modification data, and temporary storage for high-activity random-access operations involving limited amounts of data.

magnetic equator: The line on the surface of the earth where a magnetic needle remains horizontal, that is, where magnetic lines of force are horizontal. Local field irregularities are often ignored. Also called the **aclinic line**.

magnetic field: The space through which influence on a magnet is exerted. The torque at any point in space which would tend to orient a current-carrying coil or magnet if it were located at that point. A vector quantity also called the **magnetic flux density** or the **magnetic induction** and symbolized by **B**. The unit of measure for **B** in the SI and mks systems is the tesla and in the cgs system the gauss or gamma (1 tesla = 1 weber/m^2 = 1 newton/amp-m = 10^4 gauss = 10^4 maxwell/cm^2 = 10^9 gamma). **B** is given in terms of the force $d\mathbf{F}$ produced on a small element of length $d\mathbf{l}$ which is carrying a current I:

$$d\mathbf{F} = I\ d\mathbf{l} \times \mathbf{B}.$$

B is related to the magnetizing force **H** by a constant of the medium called the permeability μ:

$$\mathbf{B} = \mu\mu_0\mathbf{H}.$$

The unit of measure for **H** in the SI or mks system is ampere turns/meter and in the cgs system, the oersted (1 amp turn/m = $4\pi 10^{-3}$ gilbert/cm). μ_0 is called the **permeability of free space**; $\mu_0 = 4\pi\ 10^{-7}$ weber/amp-m in SI units and 1 gauss/oersted in cgs units. **B** can also be expressed as

$$\mathbf{B} = \mu_0(\mathbf{H} + \mathbf{M}) \text{ in SI units,}$$

$$\mathbf{B} = \mu_0(\mathbf{H} + 4\pi\mathbf{I}) \text{ in cgs units,}$$

where **M** and **I** are called the magnetization or intensity of magnetization. A magnetic field of magnitude given by the inverse-square law surrounds a magnetic pole, and a magnetic field given by Ampere's law surrounds an electric current. **H** is sometimes called the magnetic field. See Figure M-1.

magnetic field of the earth: The earth's magnetic field is often represented by a dipole at the earth's center. The intersections of the axis of this dipole with the earth's surface are the *geomagnetic poles*. A better representation is given by a dipole about 400 km from the center. The locations where the magnetic dip is 90 degrees are called the *magnetic dip poles*. The portion of the earth's field not representable by a dipole is called the *nondipole field*. Time-varying components of the earth's field are shown in Figure M-2. See also *normal magnetic field*.

magnetic flux: The flux through a surface is the integral over the surface of the normal component of the magnetic induction; expressed in webers in mks units or in maxwells in cgs units (1 weber = 10^8 maxwells).

magnetic head: An electromagnet used for reading, recording or erasing signals on a magnetic medium.

magnetic inclination: See *inclination*.

Term	SI (International System)		cgs System	
	Symbol	Unit	Unit	Symbol
Magnetic field strength Magnetic intensity Magnetic induction Magnetic flux density	$\mathbf{B} = \mu_0(\mathbf{H} + \mathbf{M})$ $= \phi/\text{area}$	1 tesla = 1 weber/m^2 = 1 newton/amp-m	= 10^4 gauss = 10^9 gamma = 10^4 maxwell/cm^2	$\mathbf{B} = \mathbf{H} + 4\pi\mathbf{I}$
Magnetic dipole (dipole moment)	$m = pd$	1 ampere-meter^2	= 10^{10} pole cm	pd
Magnetic pole	$p = \mathbf{M} \cdot \mathbf{A}$	1 ampere-meter	= 10^8 unit poles	p
Magnetic flux Magnetic line of force	$\phi = \mathbf{B} \cdot \mathbf{A}$	1 weber = volt second	= 10^8 maxwell	ϕ
Magnetomotive force Magnetic potential	**MMF**	1 ampere turn	= 0.4π gilbert = 1.26 gilbert	**MMF**
Magnetizing force	$\mathbf{H} = \mathbf{B}/\mu\mu_0$	1 ampere turn/m	= $4\pi10^{-3}$ oersted = 0.0126 oersted = $4\pi10^{-3}$ gilbert/cm	$\mathbf{H} = \mathbf{B}/\mu'$
Magnetization Magnetic dipole moment per unit volume Magnetic polarization Magnetization intensity	$\mathbf{M} = k\mathbf{H}$ $= m/V$	ampere-m^2/m^3 = ampere/m		$\mathbf{I} = k'\mathbf{H}$
Magnetic permeability			gauss/oersted	$\mu' = |\mathbf{B}|/|\mathbf{H}|$
Relative permeability	$\mu = |\mathbf{B}|/\mu_0|\mathbf{H}|$	dimensionless	dimensionless	$\mu = |\mu'|$
Magnetic susceptibility	$k = |\mathbf{M}|/|\mathbf{H}|$ $= \mu - 1$	dimensionless k_{SI}	dimensionless = $4\pi k'$cgs	$k' = |\mathbf{I}|/|\mathbf{H}|$
Reluctance	$R = \mathbf{MMF}/\phi$	1 ampere turn/ weber	= $4\pi10^{-7}$ gilbert 1 maxwell	$R = \mathbf{MMF}/\phi$
Inductance	$L = \Delta\phi/\Delta i$	1 henry = 1 weber/amp		

An equal sign relates the magnitude of SI and cgs units.

μ_0 = permeability of free space = $4\pi10^{-7}$ weber/ampere meter
$= 12.57 \times 10^{-7}$ weber/ampere meter
$= 12.57 \times 10^{-7}$ henry/meter
= 1 gauss/oersted.

FIG. M-1. **Magnetic** quantities and units.

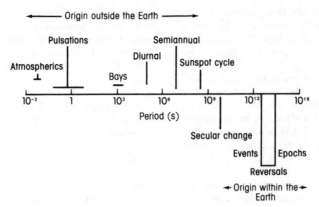

FIG. M-2. Time-varying components of the **magnetic field of the earth**.

magnetic induction: See *magnetic field*.

magnetic intensity: Magnetic-field strength; see *magnetic field*.

magnetic interpretation methods: The objective of magnetic data interpretation usually is to locate the anomalous magnetic material, its depth, dimensions, and magnetization. Most petroleum search use of aeromagnetics involves determining the depth to the top of anomalous bodies (and thereby inferring the depth to magnetic basement) and the preparation of a contour map of the magnetic basement from the results. Inverse solutions involve various auxiliary conditions in order to achieve a solution. Various shape measurements are made on magnetic profiles or maps and used in conjunction with *depth rules* (q.v.). Two-dimensional convolution operations such as calculation of a second-vertical derivative map, downward continuation, or reduction to the pole sometimes are used to help locate anomalous bodies and determine their shape. Sometimes interpretation involves comparison with the fields over known areas or comparison against the fields of model anomalies shown in a catalog of master curves. The most common magnetic interpretation models are a dipping dike and vertical prism for intrabasement bodies and a thin magnetic layer for structural features. Iterative methods involve calculating the field which a model would produce, comparing it with the observed field, and then iterating until a satisfactory degree of fit between model field and observed field is achieved.

magnetic latitude: 1. The angle of magnetic inclination determined on a smoothed regional basis rather than locally at a point. **2.** The angle having a tangent equal to half that of the magnetic dip.

magnetic meridian: The direction of the horizontal component of the earth's magnetic field; the direction of magnetic north.

magnetic moment: The strength of a *magnetic dipole* (q.v.).

magnetic permeability: The ratio of the magnetic induction **B** to the inducing field strength **H**: denoted by the symbol μ:

$$\mu = \mathbf{B}/\mu_0\mathbf{H}.$$

μ_0 is the permeability of free space = $4\pi10^{-7}$ weber/ampere meter (or henrys/meter) in SI system, and 1 gauss/oersted in the cgs system, so that the permeability μ is dimensionless. The quantity $\mu\mu_0$ is sometimes considered the permeability (especially in the cgs system).

magnetic polarization: *Magnetization* (q.v.).

magnetic pole: 1. One of the two points near opposite ends of a magnet toward which the magnetic lines of force are oriented and concentrated. If the magnet is permitted to rotate about its center, the pole which points in the direction of the earth's north magnetic pole is the *north-seeking* or **positive pole**; the other pole is the *south-seeking* or **negative pole**. **2.** The **pole strength** of a magnetized bar of cross-section A perpendicular to the magnetization M is MA. See Figure M-1. **3.** See *magnetic field of the earth*.

magnetic potential: The product of the current and the solid angle Ω subtended by a coil divided by 4π (the 4π enters because a sphere subtends the angle 4π). If there are several coils, their individual magnetic potentials (which are scalars) are added. The magnetizing force **H** is the negative gradient of the magnetic potential, a scalar representing the work done against the magnetic field to bring a unit magnetic pole to the point. A **magnetic vector potential** is a vector field whose curl gives the magnetic induction.

magnetic resonance: Interaction between the magnetic moments (electron spin and/or nuclear spin) of atoms with an external magnetic field. Magnetic resonance is basic to the operation of the *proton-resonance magnetometer* and *optically pumped magnetometer* (q.v.). See also *nuclear-magnetism log* and *Larmor frequency*.

magnetic shield: High-permeability container which isolates its interior from external magnetizing forces.

magnetic signature: The shape of a magnetic anomaly.

magnetic storm: A period of rapid, irregular, transient fluctuations of the magnetic field which are greater in magnitude, more irregular, and of higher frequency than diurnal variations. These occur most commonly during unusual sunspot activity as a result of bombardment of the earth by high-energy particles from the sun.

Magnetic storms commonly have amplitudes of 50 to 200 gammas, occasionally thousands of gammas, and their duration is often several days. Magnetic prospecting usually has to be suspended during magnetic storms.

magnetic survey: Measurements of the magnetic field or its components (such as the vertical component) at a series of different locations over an area of interest, usually with the objective of locating concentrations of magnetic materials or of determining depth to basement. Differences from the normal field are attributed to variations in the distribution of materials having different susceptibility.

magnetic susceptibility: A measure of the degree to which a substance may be magnetized; the ratio k or k' of the magnetization M or I to the magnetizing force H that is responsible for it:

$$kH = M \text{ in the SI system,}$$

$$k'H = I \text{ in the cgs system.}$$

The susceptibility is dimensionless but of different magnitude in the two systems:

$$k = 4\pi k'.$$

The susceptibility is related to the magnetic permeability μ

$$k = \mu - 1,$$

$$k' = (\mu - 1)/4\pi.$$

Susceptibility in cgs units is sometimes measured in units of 10^{-6} ("micro-cgs"). Rock susceptibility usually ranges from 0 to 0.01 cgs units (0 to 10 000 micro-cgs).

magnetic tape: A sheet or strip of plastic (such as Mylar) coated with a magnetically sensitive material on which information can be stored in the form of magnetization patterns. Half-inch tape containing 9 tracks across is usual for seismic computers, although 7-track half-inch and 21-track one-inch tapes have also been used. Analog seismic tapes occur in a variety of sizes and formats.

magnetic tape transport: A device for writing or reading magnetic tape data.

magnetization: Magnetic moment per unit volume (occasionally per unit mass), a vector quantity. Also called **magnetic polarization** or **intensity of magnetization**. Designated by symbols M or I. A measure of the effect of the medium on the magnetic field B when subject to a magnetizing force H:

$$B = \mu_0(H + M) \text{ in SI system,}$$

$$B = H + 4\pi I \text{ in cgs system,}$$

where μ_o is the permeability of free space. The proportionality between magnetization and H is the *magnetic susceptibility* (q.v.), k or k'.

magnetizing force: A measure of the influence of a magnet in the surrounding space. Symbol, H. See *magnetic field* and Figure M-1.

magnetohydrodynamics: Phenomena associated with the motion of an electrically conducting fluid (such as a liquid metal or an ionized gas) through a magnetic field. Also called **hydromagnetics.**

magnetohydrodynamic theory: The theory that coupling between the mechanical and electrodynamic forces in the fluid core is responsible for the Earth's main magnetic field.

magnetometer: An instrument for measuring magnetic-field strength. Ground magnetometers sometimes measure the vertical component of the magnetic field, sometimes a horizontal component, sometimes the total field. Most airborne magnetometers are of three types: (a) *fluxgate*, (b) *proton-resonance*, or (c) *optically pumped* (see individual entries); all measure the total-field intensity. Vector and vertical-component airborne magnetometers are also used occasionally. See also *variometer* and *Squid magnetometer.*

magnetometric induced-polarization method: An *induced-polarization* (q.v.) method which uses the survey procedures of the *magnetometric resistivity method* (q.v.).

magnetometric resistivity method: A method of electrical surveying in which the ground is energized with commutated direct current through a pair of widely spaced electrodes and the anomalous conductivity distribution is surveyed by measuring the secondary magnetic field arising from current flow. The magnetic measurement direction is perpendicular to the line between electrodes. This technique is used to explore beneath a conductive surface layer.

magnetosphere: The space pervaded by the Earth's magnetic field. Usually the space from the Earth to more than 10 Earth radii.

magnetostriction: Change in the strain of a magnetic material as a result of changes in magnetization. The dependence of magnetization (susceptibility or remanence) on applied stress is termed **inverse magnetostriction** or **piezomagnetism.** Magnetostrictive acoustical sources and hydrophones are extremely rugged. A magnetostrictive hydrophone might consist of a coil of wire wrapped around a cylinder of magnetostrictive material. A pressure wave acting radially induces hoop-stresses in the core which changes its permeability and thereby the flux linking the coil wrapped around it. The change of flux induces a voltage which is proportional to the derivative of the pressure-wave signal.

magnetotelluric method: A method in which orthogonal components of the horizontal electric and magnetic fields induced by natural primary sources are measured simultaneously as a function of frequency. Abbreviated **MT**. Apparent resistivity as a function of frequency ν is calculated

$$\rho_a = (1/\mu\nu)(E_i/H_j)^2.$$

Also, $\rho_a = 0.2Z^2/\nu$ where Z is the *Cagniard impedance* (q.v.) or *tensor impedance* (q.v.); see Vozoff (1972). Resistivity as a function of depth can be calculated for a layered earth. For a nonlayered earth, two apparent resistivity curves result from rotating the MT tensor impedance and interpretation is more involved (see *tensor magnetotelluric method*). The predominant sources of energy for magnetotelluric measurements are micropulsations having frequencies of less than 1 Hz. Sometimes magnetotelluric measurements are made at audio frequencies using energy from sferics; the method is then referred to as the **audiomagnetotelluric method (AMT)**. See *telluric current method*.

magnetotelluric noise: Unwanted voltages in the earth due to electrical discharges in thunderstorms, power lines, ionospheric currents, or magnetospheric currents.

magnitude of earthquake: A measure of the strength of an earthquake or the strain energy released by it, as determined by seismographic observations. Magnitude is a rating of an earthquake independent of the place of observation. The **Richter scale** of magnitude indicates the logarithm of the maximum amplitude observed (or which might have been observed) on an instrument of specified type 100 km from the epicenter. Empirical tables correct observations at other distances. Each step of one in magnitude means multiplying the amplitude by 10. The largest earthquakes are of magnitude about 9, and zero represented the smallest recorded earthquakes when the scale was devised. Various relations are used to give the order of magnitude of the energy released in an earthquake from the magnitude M. Roughly, the energy E in ergs is given approximately by $\log E = 10 + 2M$; Bath (1966) gives $\log_{10} E = 12.24 + 1.44$ M ergs. Distinction is sometimes made between magnitude based on body-wave versus surface-wave measurements. Microearthquakes can have negative Richter values.

main beam: *Main lobe* (q.v.).

main frame: 1. The main part of a computer system. Typically, the main frame refers to the central processor unit and main memory. **2.** A large computer, typically requiring special power installation and controlled environment.

main lobe: The portion of a directivity graph which indicates the continuous band of directions (or apparent wavelengths) in which the greatest energy is radiated (for a source) or which undergoes least attenuation (for a receiver array). See Figure D-12. Also called **main beam**.

majority vote: Determination of the most probable value of a series of measurements as the value which occurs most frequently rather than by some sort of averaging. Used in satellite navigation where the message is assumed to be that which was observed most often, each bit position being majority-voted separately.

make up: 1. A seismic shot which has been moved to a station other than its normal location, usually for safety reasons. **2.** To assemble explosive components so as to make the assembly explosive (for example, putting cap in primer or booster and attaching to the main charge).

mantissa stack: Where a series of measurements are expressed as logarithms, the sum of their decimal parts without including the whole numbers which precede the decimal place. If the series is semiperiodic but with variable amplitude, such a sum deemphasizes amplitude-variation effects.

mantle: The part of the Earth's interior between the core and the crust. The upper surface of the mantle is the Moho discontinuity characterized by a sharp increase in P-wave velocity to 8.1 ± 0.2 km/s. The density of the mantle is about $3.3 - 3.4$ g/cm^3, and the mantle is essentially nonmagnetic. The mantle includes the lower lithosphere and the asthenosphere. Below the mantle is the core, the base being the Gutenberg discontinuity. See Figure E-1.

map: 1. To transform information from one form to another. **2.** The product of such a transformation. The transformation may involve the geographical distribution of observations, or of calculations based on observations, as a Bouguer anomaly map or a seismic reflection map. The distribution may be with respect to variables other than geographic, as "to map from the time to the frequency domain." Transformations may involve a one point to one point correspondence or one to several (involving multibranched surfaces). Examples of multibranched maps might be the map of a geologic formation in the vicinity of a reverse fault where the same formation contact lies at two (or more) depths, or a seismic reflection time map in a buried-focus situation where the same reflector can be seen in several directions from the same observation point. **3.** To plot.

map convolution: A two-dimensional convolution often applied to potential field maps, whereby each point on the map grid is replaced by a weighted sum of the values at other grid points.

$$\psi_{x,y} = \sum_{\alpha} \sum_{\beta} f_{\alpha\beta} \phi_{x-\alpha, y-\beta}.$$

$\psi_{x,y}$ are the output values, $\phi_{x,y}$ the input values, and $f_{\alpha,\beta}$ is the weighting scheme, called the **template**. Simple **residual maps** are made by subtracting an average of values around the point from the value at the point. The values for different distances may be weighted and sometimes a bias is included, (i.e., $\Sigma\Sigma f_{\alpha,\beta} \neq 0$), so that the residual does not change sign very often. By weighting the points to give horizontal derivatives and using Laplace's equation, second-vertical-derivative maps may be made. Other weighting schemes can be used for field continuation, wavelength filtering, etc. See *grid residual*.

map migration: The procedure of going from an **unmigrated seismic map** (where data are plotted at midpoints) to a **migrated map** intended to indicate the correct shape and location of mapped features.

map projection: A scheme for displaying the Earth's curved surface on a plane surface. Some of the more common projections are shown in Figure M-3. Distortions of one sort or another are inevitable. **Equivalence** is the projection property wherein the product of orthogonal scale factors is maintained constant so that areas are preserved. **Conformality** is the property wherein angles are preserved. **Standard lines** are great or small circles along which the scale is a uniform constant. (a) **Tangent projections** are projections onto a surface (plane, cylinder, or cone) which is tangent to the earth; (b) **secant projections** are those made onto a surface which intersects the earth. (c) A **transverse projection** has its axis perpendicular to the Earth's axis (sometimes merely at any angle to the Earth's axis). (d) An **azimuthal projection** is onto a tangent plane; distortions increase as the distance from the point of tangency. (e) A **stereographic projection** is both azimuthal and conformal. (f) A **polyconic** has a straight central meridian and each parallel is the arc of a circle and is standard; the scale along meridians is therefore variable and the map is neither equivalent nor conformal. (g) **Transverse Mercator projections** (onto a cylinder at right angles to the earth's axis) and *Lambert conic projections* (q.v.) form many official coordinate systems; the *Universal Transverse Mercator* (q.v.) is one standard system in widespread use.

marker bed: 1. A bed or sequence of beds which yields a characteristic reflection over a more-or-less extensive area. **2.** A bed which accounts for a characteristic segment of a refraction time-distance curve and which can be followed over a reasonably extensive area. **3.** A bed with distinctive magnetic or electrical effects.

marker velocity: The velocity with which head (refracted) waves are transmitted along a marker bed.

Markovian variable: A random variable which has a distribution of values which depends only on the preceding sample. Compare *stochastic*.

marsh buggy: A self-propelled vehicle designed to operate over marsh or extemely soft ground, characterized by very low ground pressure. Some have large wheels with very wide treads, tracks, or buoyant wheels or tanks so that they will float in water.

Marthor: A hammer source for generating *S*-waves. CGG tradename.

mask: To eliminate contributions coming from certain locations.

masked layer: *Hidden layer* (q.v.).

masking effect: The effect whereby a highly conductive layer near the surface dominates resistivity measurements so as to make undetectable the effects of deeper resistivity variations.

massive sulfides: Rocks that are more than 20 percent sulfides by volume, as opposed to disseminated sulfides which involve electrical phenomena that behave like massive metallic substances.

master curve: See *type curve*.

master station: 1. A radio transmitting station of a positioning system net which controls synchronization of other transmitters in the net (**slave stations**). **2.** The control station of a network even where synchronization is not maintained.

matched filter: A filter which maximizes the output in response to a signal of particular shape. The elements of a matched filter are the elements of the signal in reverse order; hence, filtering with a matched filter is equivalent to crosscorrelating with the signal. Used where the waveform of the signal is known, as in deconvolving

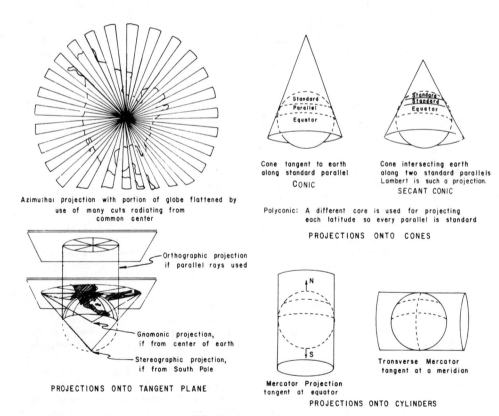

Azimuthal projection with portion of globe flattened by use of many cuts radiating from common center

Orthographic projection if parallel rays used

Gnomonic projection, if from center of earth

Stereographic projection, if from South Pole

PROJECTIONS ONTO TANGENT PLANE

Cone tangent to earth along standard parallel
CONIC

Cone intersecting earth along two standard parallels Lambert is such a projection.
SECANT CONIC

Polyconic: A different core is used for projecting each latitude so every parallel is standard

PROJECTIONS ONTO CONES

Mercator Projection tangent at equator

Transverse Mercator tangent at a meridian

PROJECTIONS ONTO CYLINDERS

FIG. M-3. **Map projections**.

Vibroseis data. Also called **crosscorrelation filter** and **correlator**. The matched filter has the same amplitude-frequency response and the negative phase-frequency response (reversed in sign) as the waveform to which it is matched. A matched filter is the most powerful filter for identifying the presence of a given waveform in the presence of additive noise. See Anstey (1964) and Treitel and Robinson (1969).

matrix: A rectangular array of numbers, called elements (which may be complex), which obeys certain rules. An $m \times n$ matrix $\mathbf{A} =$ has m rows and n columns:

$$\mathbf{A} = \|a_{ij}\| = \begin{Vmatrix} a_{11} & a_{12} \cdots a_{1n} \\ a_{21} & a_{22} \cdots a_{2n} \\ \cdots & \cdots \\ a_{m1} & a_{m2} \cdots a_{mn} \end{Vmatrix}.$$

A **null matrix O** has all elements $= 0$, and is of any size. The **identity matrix I** is square, has ones on the principal diagonal and zeros for all other elements, and is of any size. A **column vector** has only one column, a **row vector** only one row. A **square matrix** has the same number of rows as columns. The **order** of the square matrix $m \times m$ is m. A **diagonal matrix** has zeros for all elements not on the principal diagonal (a_{ii}). The **transpose** $\mathbf{A}^T$ of a matrix has the rows and columns interchanged, and hence is of $n \times m$ size:

$$\mathbf{A}^T = \begin{Vmatrix} a_{11} & a_{21} \cdots a_{m1} \\ a_{12} & a_{22} \cdots a_{m2} \\ \cdots & \cdots \\ a_{1n} & a_{2n} \cdots a_{mn} \end{Vmatrix}.$$

A **symmetric matrix** equals its transpose $\mathbf{A} = \mathbf{A}^T$. A **skew-symmetric matrix** equals the negative of its transpose $\mathbf{A} = -\mathbf{A}^T$. The **cofactor** of the element a_{ij} of a square matrix is $(-1)^{i+j}$ times the determinant which is given by deleting the ith row and the jth column. The **adjoint** or **adjugate**, adj $\mathbf{A}$, of a square matrix is the transpose of the matrix where each element is replaced by its cofactor. The **determinant**, det $\mathbf{A}$, of a square matrix is a single number and not itself a matrix.

$$\det \mathbf{A} = \sum_i a_{ik}A_{ik} = \sum_k a_{ik}A_{ik},$$

where A_{ik} is the cofactor of a_{ik}. The **inverse** of a matrix multiplied by the matrix yields the identity matrix

$$\mathbf{A}^{-1}\mathbf{A} = \mathbf{A}\mathbf{A}^{-1} = \mathbf{I}.$$

The inverse may be found by dividing the adjoint by the determinant:

$$\mathbf{A}^{-1} = (\text{adj } \mathbf{A})/(\det \mathbf{A}).$$

An **orthogonal matrix** has its inverse equal to its transpose: $\mathbf{A}^T = \mathbf{A}^{-1}$. Vectors $\mathbf{K}_i$ and $\mathbf{K}_j$ are **orthogonal** if $\mathbf{K}_i^T\mathbf{K}_j = 0$. A **Hermitian matrix** is a square matrix of complex numbers which equals the transpose of its complex conjugate: $\mathbf{H}^{*T} = \mathbf{H}$. A **Toeplitz matrix** has

identical elements on a diagonal. A **minor** is a square determinant formed by deleting rows and/or columns. The **rank** of a matrix is the order of the largest nonvanishing minor. Matrices of the same size may be added:

$$\mathbf{C} = \mathbf{A} + \mathbf{B} \text{ where } c_{ij} = a_{ij} + b_{ij}.$$

Matrices may be multiplied by a scalar:

$$\mathbf{D} = k\mathbf{A} \text{ where } d_{ij} = ka_{ij}.$$

A matrix with m columns ($\mathbf{A}$) may multiply (from the left) a matrix with m rows ($\mathbf{B}$):

$$\mathbf{E} = \mathbf{AB} \text{ where } e_{ij} = \sum_{k=1}^{m} a_{ik}b_{kj}.$$

If $\mathbf{A}$ is of size $p \times m$ and $\mathbf{B}$ is $m \times n$, $\mathbf{E}$ is of size $p \times n$. Multiplication is (in general) not commutative.

$$(\mathbf{AB})^T = \mathbf{B}^T\mathbf{A}^T,$$

and

$$(\mathbf{AB})^{-1} = \mathbf{B}^{-1}(\mathbf{A}^T)^{-1}.$$

If $\mathbf{X} = \mathbf{AY}$ is a set of linear equations, the unique solution is $\mathbf{Y} = \mathbf{A}^{-1}\mathbf{X}$, if det $\mathbf{A} \neq 0$. The **rank** of $\mathbf{A}$ is the number of such equations which are linearly independent. If the equations are inconsistent, the least-squares solution is

$$\mathbf{Y} = (\mathbf{A}^T\mathbf{A})^{-1}\mathbf{A}^T\mathbf{X}.$$

If $\mathbf{A}$ is square matrix of order n, the **eigenvalues** or **latent roots** λ are such that $\mathbf{A} \mathbf{q} = \lambda \mathbf{q}$. There are n such eigenvalues (some of which may be identical). Identical eigenvalues indicate degeneracy. The **eigenvector** $\mathbf{q}_i$ corresponding to λ_i consists of the n cofactors of a row of the matrix $\mathbf{A} - \lambda \mathbf{I}$. If $\mathbf{A}$ is symmetric, the matrix of the mutually orthogonal eigenvectors $\mathbf{q}_i$,

$$\Lambda = \mathbf{q}_i, \mathbf{q}_2, \cdots, \mathbf{q}_n,$$

has the property that $(\Lambda^T\mathbf{A} \Lambda)$ is diagonal. Matrices are used in electromagnetic work, in wave propagation studies, in multichannel processing, and elsewhere. See Sheriff and Geldart, v. 2 (1983, p. 149-151).

maximum: An anomalous area in which measurements show larger values than in neighboring areas, as a gravity maximum.

maximum convexity: The curvature on a seismic record section of a diffraction from a point in the plane of the section. This is the greatest convexity a primary coherent wave train can have except for reverse branches in buried-focus situations, diffracted reflections, and certain other events involving complex raypaths. See *diffraction curve*.

maximum-delay: *Maximum-phase* (q.v.).

maximum depth: In gravity and magnetic interpretation, the limiting depth below which the bodies causing an observed anomaly cannot lie.

maximum entropy filtering: Filtering which attempts to produce an output which is as unpredictable as possible but which has the same autocorrelation as the input.

See *entropy* and Sheriff and Geldart, v. 2 (1983, p. 191-193).

maximum entropy spectral estimate: A method to determine a finite number of Fourier-series coefficients of a nonnegative periodic function under the assumption that the function is rational in $e^{-i\omega T}$ (where T = period, ω = angular frequency) with constant numerator.

maximum-energy-sum filter: *Output-energy filter* (q.v.).

maximum-phase: A two-term wavelet $[a,b]$ is maximum-phase if $|a| < |b|$. Any wavelet is maximum phase if the two-term wavelets into which its Z-transform can be factored are each maximum phase. Same as **maximum-delay**. See *phase characteristics* and Figure P-1 and compare *minimum phase*.

Maxipulse: A marine seismic source involving detonation of small explosive charges at 7 to 15 m depth. Such a source generates a sequence of bubble pulses which must be removed subsequently by processing. Western Geophysical Co. tradename.

Maxwell's equations: A set of differential equations to which electric and magnetic fields are subject at every ordinary point:

$$\nabla \times \mathbf{E} = \frac{-\partial \mathbf{B}}{\partial t}, \nabla \cdot \mathbf{D} = \rho,$$

and

$$\nabla \times \mathbf{H} = \mathbf{J} + \frac{\partial \mathbf{D}}{\partial t}, \nabla \cdot \mathbf{B} = 0,$$

where $\mathbf{E}$ is the electric field intensity, $\mathbf{H}$ is the magnetizing force, $\mathbf{B}$ is the magnetic field strength, $\mathbf{D}$ is the electric displacement, $\mathbf{J}$ is the current density, and ρ the charge density. In the cgs system, $1/c$ (where c = velocity of light in a vacuum) precedes the time derivatives and 4π precedes the $\mathbf{J}$ and ρ. These relations can also be expressed by an equivalent system of integral equations. Developed by James Clerk Maxwell (1831-1879), English physicist.

md: Millidarcy or 1/1000 darcy.

M discontinuity: Mohorovičić discontinuity or *Moho* (q.v.).

mean: Average. The **arithmetic mean** or **arithmetic average** of n values x_i is

$$(1/n) \sum x_i.$$

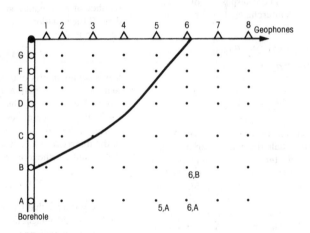

(a)

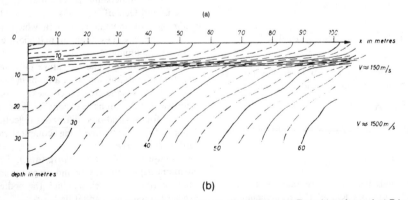

(b)

FIG. M-4. **Meissner technique** for empirical construction of wavefront chart. (a) Traveltime from shot B in the borehole into geophone 6 is posted at (6,B), etc; contouring the results gives the wavefronts which would have resulted from a shot at the top of the borehole (for uniform horizontal layering). (b) Example of a chart showing the effect of a leached salt bed. (After Meissner, 1965.)

The **geometric mean** is

$$[(1/n) \sum x_i^2]^{1/2}.$$

A **weighted mean** is

$$(\sum w_i x_i)/(\sum w_i),$$

where w_i = weights and $\sum w_i = 1$.

Meander: *Crooked-line* (q.v.). Prakla-Seismos tradename.

mean deviation: See *statistical measures*.

measurement-while-drilling log: MWD. Bottom-hole data are acquired incrementally from sensors located in the drill string near the bit in a drilling well. Measurements may include directional information (hole inclination, azimuth, tool facing), drilling parameters (bottom-hole temperature, pressure, torque, weight-on-bit, rpm), rig safety, formation evaluation and correlation data (formation resistivity and gamma-ray logging). Data are transmitted to the surface in real time by pressure pulses through the mud inside the drill pipe (timed amplitude and phase encoding). Telemetry by conductor cable integrated with the drill pipe or temporary digital recording at the sensor for later wire-line retrieval are alternative data recovery methods.

mechanical sources: Seismic sources such as vibrators and weight droppers.

mechanical seismograph: A seismic detector in which (except for use of an optical lever arm) amplification of the ground motion is accomplished by mechanical means. Extensively used in early seismic prospecting.

median: The value which half of the members of a set exceed and half are smaller. Values equal to the median may be placed in either subset to achieve this.

Meisner technique: Determining wavefronts from shots at various depths into a spread of geophones. Used to study near-surface anisotropy and S-waves. See Figure M-4 and Meissner (1965).

Meisner wave: *Head wave* (q.v.).

M-electrode: Potential drop is measured between the M- and N-electrodes in electrical-resistivity measurements. See Figure E-7.

membrane polarization: The induced-polarization effect primarily due to restrictions of ion mobility as opposed to electrode polarization. See also *normal effect*.

membrane potential: See *electrochemical*.

memory: A part of a computing machine in which data can be stored and from which it can later be retrieved. Primary or main memory is usually a solid-state type (MOS or bipolar), secondary memory is usually magnetic disk or magnetic tape.

memory function: *Impulse response* (q.v.) or memory curve of a filter or system.

menu: A list of program options available.

Mercalli scale: A descriptive scale which indicates the degree of shaking at a specific location as a result of an earthquake. The original scale was devised in 1902. An abridged version of the **modified Mercalli scale** (devised in 1931) is given in Figure M-5. Compare *magnitude*. Named for Giuseppi Mercalli (1850-1914), Italian geologist.

Mercator projection: A conformal cylindrical *map projec-*

I. Not felt
II. Felt by persons at rest.
III. Hanging objects swing; vibration like passing of light trucks.
IV. Vibration like passing of heavy trucks.
V. Felt outdoors; sleepers awakened; liquids disturbed; unstable objects displaced.
VI. Felt by all; glassware broken; books off shelves.
VII. Difficult to stand; noticed in motor cars; damage to some masonry; weak chimneys broken at roof line.
VIII. Partial collapse of masonry; twisting, fall of chimneys; frame houses moved on foundations.
IX. General panic; general damage to foundations; underground pipes broken; conspicuous cracks in ground.
X. Most structures destroyed; large landslides; water thrown on banks.
XI. Rails bent greatly; underground pipelines out of service.
XII. Damage nearly total.

FIG. M-5. Modified **Mercalli** scale (abridged) of the intensity of earthquake effects.

tion (q.v.) developed on a cylinder tangent along the equator with the expansion of the meridians equal to that of the parallels. See Figure M-3. Named for Gerhardus Mercator (1512-1594), Flemish mathematician.

mercury delay line: See *delay line*.

merge zone: A region where two sets of parameters are used, their relative weighting depending on the location within the zone.

mesh: A 2- or 3-dimensional grid used to approximate a continuous or semicontinuous surface or volume for computer modeling.

mesosphere: The lowest of three zones into which the outer part of the Earth is divided, overlain by the asthenosphere and the lithosphere. The core is below the mesosphere. See Figure E-1.

message: The desired information being sought. For example a "satellite message."

metal factor: A measure of the total change in conductivity or capacitivity of a rock, used in the interpretation of IP data; often written **MF**. Metal factor is the *percent-frequency-effect* (q.v.), normalized by dividing by the measured resistivity (ρ in ohm-feet) and multiplied by a constant:

$$MF = 2\pi 10^3 (PFE)/\rho.$$

1. Originally metal factor was defined as

$$MF = 2\pi 10^5 (\rho_{dc} = \rho_{ac})/\rho_{ac}\rho_{dc},$$

where ρ_{dc} is the low-frequency resistivity and ρ_{ac} is the high-frequency resistivity in ohm-feet. **2.** In the frequency domain the definition used is

$$MF = 2\pi10^5(\rho_{dc} - \rho_{ac})/\rho_{ac}{}^2.$$

3. In the time domain, the metal factor is

$$MF = 2000 \, M/\rho_{dc}$$

where M is chargeability in millivolt-seconds per volt and ρ_{dc} is in ohm-meters. This unit is similar to Keller's parameter, *specific capacity* (q.v.) or static capacity. **4.** A constant times the ratio (sometimes called **electric susceptibility** or **capacitivity**) between induced-current dipole moment per unit volume (P) and electric field (E):

$$MF = 2 \times 10^6 P/E.$$

The metal factor has units of conductivity. It is also called **metallic-conduction factor.**

metallic-conduction factor: *Metal factor* (q.v.)

metal-oxide semiconductor: A field-effect transistor (FET) or integrated circuit characterized by extremely high input resistance. Abbreviated **MOS.**

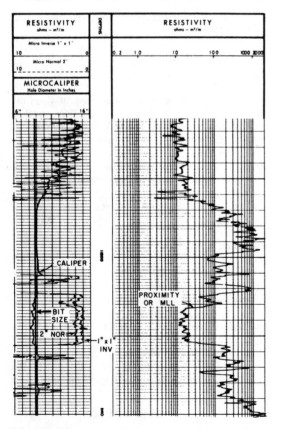

FIG. M-6. **Microlog** (micronormal and microinverse) and microcaliper log (on left) and microlaterolog (on right). (Courtesy Schlumberger.)

meteorology: *See geophysics.*

methane hydrate: See *gas hydrate.*

method of intersection: See *triangulation.*

method of least squares: See *least-squares fit.*

MeV: Million electron volts

mf: Subscript used with log terms to indicate values for the mud filtrate.

MF: *Metal factor* (q.v.).

mGal: Milligal: 10^{-3} Gal or 10^{-5} m/s^2. A unit of acceleration used with gravity measurements. Sometimes written **mG.**

mho: A unit of conductance or admittance, the reciprocal of **ohm.** Also called **siemens.**

mho per meter: A unit of conductivity; the conductivity for which a meter cube offers a resistance of one ohm between opposite faces. Reciprocal of ohm-meter.

mickey-mouse: Improvised; a short-cut method which may sacrifice rigor.

micro: A prefix meaning 10^{-6}. Symbol: μ.

microcode: A set of primitive control functions performable by a computer. Microcode is not generally accessible to the programmer. See *microprogram.*

microcomputer: A class of computer having all major central processor functions contained on a single printed-circuit board. Microcomputers are typically implemented by a small number of LSI circuits.

microcracks: A theory to explain the effect of pressure on the velocity of nonporous rocks. Minute cracks develop upon cooling because of differences between thermal expansion characteristics among the minerals of which the rock is composed. See Gardner et al. (1974).

microearthquake: A discrete earthquake event of low magnitude (Richter magnitude < 3).

microgal = 10^{-6} Gal = 10^{-8} m/s^2.

microinverse: See *microlog* (q.v.).

microlaterolog: A *microresistivity log* (q.v.) of the *laterolog* (q.v.) type. A bucking electrode and two monitor electrodes are arranged concentrically on a pad which is pressed against the formation. Similar to minifocused log, FoRxo log, or trumpet log. See Figure M-6. Microlaterolog is a Schlumberger tradename.

microlog: A type of microresistivity log using three button electrodes spaced in a line 1 inch apart and located on a pad which is pressed against the borehole wall. The lower electrode is the "A" current electrode. The potential of the upper electrode with respect to a reference electrode on the surface gives a **2-inch micronormal** and the difference between the two upper electrodes gives a $1\frac{1}{2}$-inch **microinverse** (lateral-type measurement). Because the mud cake usually has appreciably smaller resistivity than the formation, the microinverse will read less than the micronormal when mud cake is present. This difference (called **separation**) indicates a permeable formation. A caliper log is usually recorded at the same time. Similar to the contact log or Minilog. Microlog is a Schlumberger tradename. See Figure M-6.

micronormal: See *microlog.*

micro-omega: See *omega.*

microphone detector: A seismic detector utilizing contact resistance as part of the vibration-detecting element.

microphonics: Electrical noise generated by mechanical vibration.

microprocessor: A single LSI circuit which performs the functions of a **CPU**. Characteristics of a microprocessor include small size, inclusion in a single integrated circuit or set of integrated circuits, and low cost.

microprogram: 1. A hardware program which controls how a computer functions. It determines how a computer interprets an instruction in machine language. **2.** A software program constructed from the basic subcommands of a computer which the system hardware translates into machine subcommands. A microprogram provides a means of building various instruction combinations out of the subcommand structure of the computer.

micropulsations: Small amplitude fluctuations in the Earth's magnetic field, usually in the frequency range from 0.01 to 3 Hz and usually with amplitudes less than 10 gamma. Micropulsations having amplitudes up to tens of gammas result from interactions between plasma emitted from the sun (solar wind) and the Earth's field. Micropulsations are classified as continuous (p_c), irregular (p_i), pearl (p_p), etc. See also *bay* and *magnetic storm*.

microresistivity log: A well log designed to measure the resistivity of the flushed zone about a borehole, recorded with electrodes on a pad pressed against the borehole wall. See *microlog* and *microlaterolog*.

microscopic anisotropy: See *anisotropy*.

Microseismogram log: Similar to the variable-density log or *3-D log*; see Figure C-2. Microseismogram is a Welex tradename.

microseism: Feeble earth tremors due to natural causes such as wind, water, waves, etc.

microspread: A spread with very short geophone group intervals (1 to 15 ft), used in *noise analysis* (q.v.).

midpoint: The point midway between a source and a geophone.

migration: 1. Originally, rearrangement of interpreted seismic data so that reflections and diffractions are plotted at the locations of the reflectors and diffracting points rather than with respect to observation points. Now generally a computer operation on raw data to accomplish this purpose. Also called **imaging**. The measurements on which migration is based are the magnitude and direction of the apparent dip (which defines the raypath) and arrival time (which defines the distance along the raypath). The apparent dip is also called **dip moveout** and **apparent velocity**. Some migration schemes make use of the complementary properties of wavefront and diffraction curves shown in Figure M-7. Migration is usually two-dimensional because only the apparent dip component in the line direction is known. Three-dimensional migration is conceptually a simple extension of 2-D methods, but this extension to the additional dimension may involve a large multiplication in the number of computer operations required. Even where the apparent dip azimuth is known, migration is still often done two-dimensionally in two steps, first migrating in one direction and then migrating this intermediate result in the cross direction (**double 2-D migration**). Migration by computer is accomplished by integration along diffraction curves (**Kirchhoff migration**), by numerical finite-difference downward-continu-

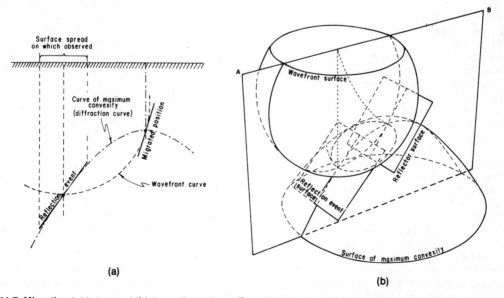

(a)   **(b)**

FIG. M-7. **Migration (a)** in two and **(b)** three dimensions. (From Hagedoorn, 1954.) A point in unmigrated space migrates to a wavefront surface, and a point in migrated space specifies a diffraction surface. Unmigrated reflections are tangent to the diffraction surface and migrated reflectors to the wavefront surface. The shape of the wavefront and diffraction surfaces depends on the velocity distribution above the reflecting point. Lateral velocity variations not only can distort the shape of these surfaces but shift the intersection of the surfaces away from the diffraction crest.

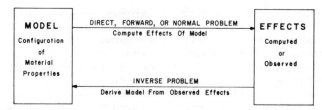

FIG. M-8. A **model** and its effects.

ation of the wave equation, and by equivalent operations in frequency-wavenumber or other domains (**frequency-domain migration**). Lateral variation of velocity affects the migration and usually ray tracing is used to determine migrated positions. See Sheriff and Geldart, v. 2 (1983, p. 59-71). **2.** Movement of ions in a solution because of an electric-field gradient; called **mobility. 3.** Movement of hydrocarbons between the locales of their generation and accumulation.

mil: 0.001 inch.

milli-: A prefix meaning 10^{-3}. Symbol m. Seismic events are often timed in milliseconds (ms).

milliard: A thousand million; 10^9.

milligal: A unit of acceleration used with gravity measurements; 10^{-5} m/s^2. Abbreviated mGal, sometimes mG.

milliradian: A unit of angle (or phase) measurement equal to 0.0573 degrees of arc. One degree equals 17.45 milliradians. Abbreviated mrad.

millisecond: 1. A thousandth of a second; abbreviated ms. **2.** A unit of chargeability, the area under the decay curve of a pulsed $(+, 0, -, 0)$ square wave. See Figure S-18 and *chargeability*.

Milne's method: A technique for the numerical solution of differential equations. See Sheriff and Geldart, v. 2 (1983, p. 157).

minicomputer: A class of computer without power or cooling requirements greater than normally available in an air-conditioned office building.

Minilog: See *microlog*. Dresser Atlas tradename.

minimax criterion: See l_p *fit*.

minimum: An anomalous area in which measurements show smaller values than in neighboring areas, as a "gravity minimum".

minimum-delay: *Minimum-phase* (q.v.).

minimum-entropy filtering: Linear filtering which maximizes the "spiky" characteristic of the output. See Wiggins (1978) and Sheriff and Geldart, v. 2 (1983, p. 193-194).

minimum-phase: 1. A two-term wavelet, **doublet** or **couplet** $[a, b]$ is minimum-phase if $|a| > |b|$; same as **minimum-delay**. Any wavelet may be represented as the convolution of couplets; the wavelet is minimum-phase if all the couplets of which it is composed are minimum-phase. For example, the z-transform of a wavelet might be $6 + z - z^2$ which can be expressed as $(3 - z)(2 + z)$, each of which is minimum-phase. Minimum-phase is sometimes expressed as the situation where all roots are outside the unit circle in the z-plane, or where there

are no zeros in the right half of the Laplace transform S-plane. See Sheriff and Geldart, v. 2 (1983. p. 180-184). A minimum-phase wavelet is sometimes called **front-loaded** because its energy is concentrated in the front end of the pulse. Maximum-phase or maximum-delay is the other extreme, and mixed-phase is intermediate. See Figure P-1 and *phase characteristics*. **2.** A multichannel matrix of vectors is minimum-phase if its determinant (which can also be expressed as the product of couplets) is minimum-phase. A multichannel response produced by impulsive inputs might be:

| | Input | |
Output	Channel 1	Channel 2
Channel 1	$(2 + z)$	(z)
Channel 2	(1)	$(6 + z)$

which has the determinant $12 + 7z + z^2 = (3 + z)(4 + z)$ that is minimum-phase; hence the multichannel response is minimum-phase. **3.** A **minimum-phase filter** is that one of the set of possible filters with identical amplitude response, which delays the energy the least; it also is called the **minimum-delay filter**. If the input to a minimum-phase filter is itself minimum-phase, then the output will also be minimum phase. Many of the filtering actions to which seismic signals are subjected are minimum-phase and much of the filtering done in digital processing is minimum-phase.

minimum-time path: *Least-time path* (q.v.).

Miniranger: An ultra-high-frequency radio-positioning system that operates in rho-rho mode used for short (line of sight) ranges. Motorola tradename.

Mini-Sosie: A shallow seismic method (see *Sosie*) employing random impacts from soil-compaction tampers ("wackers") as the energy source. Tradename of SNPA.

minor: See *matrix*.

Mintrop wave: *Head wave* (q.v.). Named for Ludger Mintrop (1880-1956), German geophysicist who developed the refraction seismic method.

minus values: See *plus-minus method*.

Mirragraph: An optical recording playback system used experimentally in the 1940-1950s. Western Electric tradename.

mirror: *Reflector* (q.v.).

misclosure: *Mis-tie* (q.v.).

mise-à-la-masse method: An electrical-exploration method in which one current electrode is positioned in a conducting mineral either in outcrop or in a borehole. The other current electrode is a great distance away and the potential electrodes are moved about with the objective of mapping the mineral deposit. Also called **excitation-at-the-mass method**.

mis-tie: **1.** The difference obtained on carrying a reflection, phantom, or some other measured quantity around a loop. **2.** The difference of values at identical points on intersecting lines or of values determined by independent methods. **3.** The difference from zero of the algebraic sum of measured differences around a loop.

mixed-delay: Mixed-phase; see *phase characteristics*.

mixed-phase: A wavelet for which one or more of the component two-element wavelets into which it can be factored are minimum-phase and one or more is maximum-phase. Same as **mixed delay**. See *phase characteristics*.

mixing: Combining the energy of different channels, generally to cancel noise. Mixing usually implies that no time shifting is involved before the data are combined. Also called **compositing**. **1.** Modern processing combines (though usually not called mixing) by various *stacking* (q.v.) operations; see Figure S-19. **2.** Mixed records may preserve two or more traces in unmixed form so that these traces are not distorted by the mixing. In **simple 50 percent mixing** the output traces contain equal contributions of energy from adjacent input channels; in **mixing toward the shot** the traces farthest from the shot are unmixed. In **taper mixing**, input channels contribute to the output in inverse proportion to their proximity. Mix does not usually

carry across the center of the record (that is, between channels on opposite sides of the source) or from one record to another.

mks: The meter-kilogram-second-ampere system of units, now replaced by SI, Système International, which is identical in many regards. See *Appendix A* and Figure M-1.

MM scale: Modified *Mercalli scale* (q.v.).

mnemonic: An easily remembered code word such as symbolic designations of instructions to a computer.

mobility: The velocity of charge carriers per unit electric field. Usually refers to movement of ions in a solution. Also called **migration**.

mode: Manner of behaving. **1.** A form of behavior, as in a "Love wave of the first mode." **2.** A method of operation, as in "Loran C in the rho-rho mode" or "Doppler-sonar in the water-bottom mode." **3.** See *statistical measures*.

mode conversion: Conversion of *P*-wave energy into *S*-wave energy or vice versa by oblique incidence on an interface. See *Zoeppritz's equations*.

model: A concept from which one can deduce effects for comparison to observations; used to develop a better understanding of observations. The "model may be conceptual, physical, or mathematical; see Figure M-8. For example, one might calculate the gravity, magnetic, or seismic effects for an assumed geologic structure and compare these with actual data. Agreement between observations and effects derived from a model does not "prove" that the model represents the actual situation; geophysical interpretation problems almost always lack

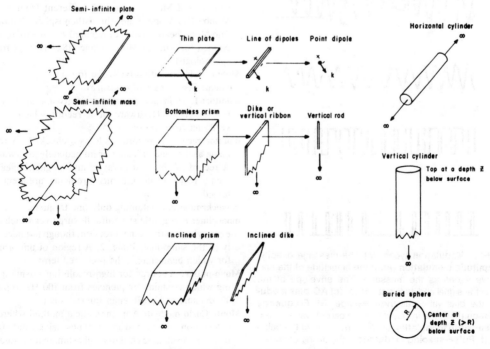

FIG. M-9. **Models** of simple types.

uniqueness. Some simple mathematical models are shown in Figure M-9. The effects of more complicated models may be collected into a catalog of master curves or type curves for use in comparison with observed effects. See *synthetic seismograms, numerical modeling, physical modeling*, and Sheriff and Geldart, v. 2 (1983, p. 117-125).

modeling theory: Significant physical properties must have certain ratios of dimensions for physical models to be realistic representations. Three ratios may be selected independently: **geometric similarity**, a ratio of length; **dynamic similarity**, a ratio of mass; **kinematic similarity**, a ratio of time. Values of these fix other model ratios. See Sheriff and Geldart, v. 2 (1983, p. 118-119).

modem: A device which converts a digitized code into an audio signal which is more suitable for transmission over voice-grade telephone lines, or for converting the audio signal back into digitized code suitable for computer operations. Contraction of modulator-demodulator. Compare *acoustic coupler*.

modified Mercalli scale: See *Mercalli scale* and Figure M-5.

modified Schmidt diagram: See *Schmidt diagram*.

modulation: 1. The process by which some characteristic of one signal is varied in accordance with another signal. Examples are shown in Figure M-10. **2.** A measure of the intensity of magnetization impressed on a direct recording magnetic tape, often expressed as a percentage of the amount which will produce certain harmonic distortion. **3.** Sometimes used to imply the number of bits used to represent a maximum voltage in digital recording. See also *doublet modulation*.

modulus: 1. The absolute magnitude of a complex number. If the complex number is $x + iy$, the modulus is $(x^2 + y^2)^{1/2}$. **2.** A number that measures a force or coefficient pertaining to a physical property, as in **bulk modulus** and **Young's modulus**; see *elastic constants*.

modulus of compression: Bulk modulus; see *elastic constants*.

modulus of elasticity: See *elastic constants*.

modulus of rigidity: Shear modulus; see *elastic constants*.

modulus of volume elasticity: See *elastic constants*.

Moho: Mohorovičić discontinuity, the seismic discontinuity which separates the Earth's crust and mantle. Situated 25–40 km below the continents, 5–8 km below the ocean floor, and 50–60 km below certain mountain ranges. Characterized by a fairly abrupt increase of *P*-wave velocity from 6.5–7.2 to 7.8–8.5 km/s, an increase of *S*-wave velocity from 3.7–3.8 to 4.8 km/s, and an increase in density from approximately 2.9 to 3.3 g/cm³. Continental Moho is probably different from oceanic Moho. The Moho is not the asthenosphere boundary. The Moho head wave is designated P_n. See Figure E-1. Named for Andrija Mohorovičić (1847-1936), Croatian seismologist.

Mohorovičić discontinuity: *Moho* (q.v.).

moment: See *statistical measures*.

monitor: 1. To inspect in order to verify that an operation is correct. **2.** Hardware or a record which permits verifying correctness.

monitor record: A record made as a check. **1.** A record made at the time of a shot or immediately afterward. **2.** A record of shipboard gravity and magnetic measurements made while the measurements are being obtained.

monochromatic: Containing only one frequency.

monocline: 1. A gentle structural flexure over which dip is everywhere in the same direction, though not necessarily of the same magnitude. **2.** A region of uniform dip, for which **homocline** is the preferred term.

Mono-pulser: A source for high-resolution seismic profiling which contains frequencies from 100 Hz to 6 kHz. Tradename of Van Reenan International.

Monte Carlo method: A mathematical method whereby a calculation is repeated many times using random values. The result gives a statistical estimate for a solution.

monument: An identifiable point on the ground to which

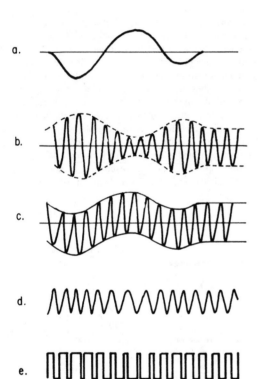

FIG. M-10. **Modulation** types. (**a**) The message directly. (**b**) **Amplitude modulation** (AM): the amplitude of the carrier wave varies as the message. The envelope of the carrier is the signal and its mirror image. (**c**) **AC-bias modulation**: the bias varies as the message. (**d**) **Frequency modulation** (FM): the frequency of the carrier wave varies. (**e**) **Pulse-width modulation** (PWM): the width of the pulse varies. (**f**) **Pulse-spacing modulation**: the timing of pulses varies.

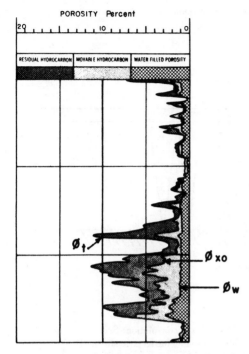

POROSITY Percent

FIG. M-11. **Movable-oil plot**. (Courtesy Schlumberger,)

surveys can be tied. Intended to be permanent. May be an inscribed tablet on concrete, a steel fence picket with identification attached, etc. A **bench mark**.

moon-position camera: A device for photographing the moon against a background of stars; used for determining geodetic location.

MOP: *Movable-oil plot* (q.v.).

MOS: *Metal-oxide semiconductor* (q.v.).

motor rule: A rule for finding the direction in which a current will be deflected in a magnetic field. See Figure I-2.

Mounce potential: See *electrochemical SP*.

movable heads: 1. Magnetic heads which can move with record time. Used with analog recording to apply normal-moveout (and sometimes static) corrections. **2.** The read-write heads of a magnetic disk that can be moved across the surface of the disk to access data at a selected radial position.

movable-oil plot: A well log calculated from other logs on which three porosity curves are plotted: "total" porosity ϕ_t, such as is derived from the sonic log; "apparent water-filled" porosity ϕ_w derived from a deep-investigation resistivity device such as the laterolog; and "apparent water-filled porosity of the flushed zone" ϕ_{x0} derived from a shallow investigation resistivity device such as the microlaterolog. The separation between the first two curves indicates the volume fraction of hydrocarbons in the noninvaded zone and the separation between the last two curves indicate the volume fraction of movable oil. The remainder represents resid-

ual hydrocarbons left in the invaded zone. Abbreviated **MOP**. See Figure M-11.

moveout: Stepout, the difference in arrival time at different geophone positions. **1.** Arrival times differ because of source-to-geophone distance differences (*normal moveout*, q.v.), because of reflector dip (**dip moveout**), and because of elevation and weathering variations (*statics*, q.v.). See also *delta-t*. **2.** Dip movement alone; see Figure R-7.

moveout filtering: *Apparent velocity filtering* (q.v.).

moveout scan: 1. Different amounts of normal moveout are successively applied to common-midpoint gathers which are then stacked; used to determine the optimum normal moveout for emphasizing certain events. See *velocity analysis*. **2.** Different amounts of dip moveout are introduced successively in making an *f-k analysis* (q.v.).

moving-coil geophone: An electrodynamic detector of seismic waves. See *geophone*.

moving source method: A profiling method in which a fixed source and receiver configuration is moved about to explore an area. Usually applied to electromagnetic methods for which the free-space coupling between transmitter and receiver is fixed (unlike a *fixed-source method*, q.v.).

ms: Millisecond or 10^{-3} second.

MSB: Most significant bit.

MSI: Medium-scale integration; a type of integrated circuit.

MS pickup: Magnetostrictive geophone or other transducer.

MSS: *Multispectral scanner* (q.v.).

MT: *Magnetotelluric* (q.v.).

MTBF: Mean time between failures.

MTTR: Mean time to recovery.

mud: An aqueous suspension used in rotary drilling. Mud is pumped down through the drill pipe and up through the annular space between it and the walls of the hole. The most common bases of drilling muds are bentonite, lime, and barite in a finely divided state. The mud helps remove cuttings, prevent caving, seal off porous zones and hold back formation fluids.

mud cake: Filter cake, the residue deposited on the borehole wall as the mud loses **filtrate** (the liquid portion of mud) into porous, permeable formations. The mud cake generally has very low permeability and hence retards further loss of fluid to the formation. See *invaded zone*.

mud cup: A container used to measure mud resistivity.

mud filtrate: Fluid which enters permeable formations from the mud, leaving a mud cake on the borehole wall. See *invaded zone*.

mud log: 1. As a rotary well is drilled, samples of the circulating mud and its suspended drill cuttings are examined for lithology and hydrocarbons by visual observation (including ultra-violet fluoroscopy) and for gas-phase constituents by a Wheatstone bridge, "hot-wire" partition gas chromatograph or hydrogen-flame ionization analyzer. A mud log is a plot of such measurements, together with a lithologic log and a drilling-time log (showing rate of penetration). Used to detect

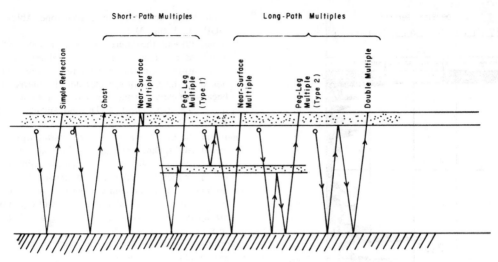

FIG. M-12. **Multiple** types.

fluids which have entered the mud from the formations. Hydrocarbons may be evidenced by fluorescence, by chromatographic analysis, gas, and other ways. The mud is also monitored for salinity and viscosity to indicate water loss or "cut." Plots of such data are usually included with a sample log and drill-time log. **2.** A log made with a microlog sonde with the arms collapsed so that the measuring pad is not pressed against the borehole wall. Measures resistivity of the mud at in-hole conditions.

mud penetrator: A piezoelectric seismic source generating in the 3.5–7.0 kHz range, used in shallow-penetration profiling.

multichannel filtering: 1. Filtering wherein the filter characteristics are partially based on the characteristics of other channels. **2.** A multichannel filter is often expressed as a matrix of output impulse responses on various channels to impulsive inputs on various channels:

	Input		
Output	Channel 1	Channel 2	Channel 3
Channel 1	f_{11}	f_{12}	f_{13}
Channel 2	f_{21}	f_{22}	f_{23}
Channel 3	f_{31}	f_{32}	f_{33}

multichannel processing: Data processing in which data from different input channels (different geophone groups, for example) are combined in some manner. Multichannel methods are involved in determining static trace shift corrections, normal-moveout and stacking velocity, stacking, apparent-velocity (dip) filtering, coherency filtering (including picking), migration, and other processes.

multidimensional convolution: See *convolution*.

multidimensional Fourier transform: See *Fourier transform*.

multidrop: A telecommunication system in which multiple devices are served by a single physical line.

multiple: Seismic energy which has been reflected more than once. While virtually all seismic energy involves some multiples, the important distinction is between long-path and short-path multiples: a **long-path multiple** arrives as a distinct event whereas a **short-path multiple** arrives so soon after the primary that it merely adds tail to the primary. Short-path multiples may obscure stratigraphic detail even where their attitude indicates the attitude of the appropriate portion of the section (and where structural aspects are not affected significantly). The attitude of long-path multiples is apt to not be representative of the portion of the section associated with their arrival time. Usually long-path multiples have traveled more in the slower (shallower) part of the section than primaries with the same arrival time, so that they ordinarily show more normal moveout and can be attenuated by common-midpoint stacking. See Figure M-12 and Sheriff and Geldart, v. 1 (1982, p. 105-110).

multiple branches: The situation where $f(\alpha)$ has more than one possible value for the same value of α. For example, more than one reflection is obtained from a reflector in the buried focus situation (see Figure B-8).

multiple coverage: Seismic arrangement whereby the same portion of the subsurface is involved in several records, as with CDP shooting. The redundancy of measurements permits various types of noise to be attenuated in processing.

multiple geophones: A number of geophones (a **group**) feeding a single channel; see *array (seismic)*. Used (a) to attenuate ground roll and other undesirable energy which approaches the spread more or less horizontally (see Figure D-12); (b) to improve the signal-to-noise ratio by increasing the sampling and thereby randomizing planting factors, noncoherent energy, etc.; and (c) to increase sensitivity.

multiple reflections: See *multiple*.

multiple regression: A mathematical procedure for finding the empirical equations which best fit a set of data in the least-squares sense. See *factor analysis*.

multiple shotholes: Two or more shotholes which are shot simultaneously. Usually used for the same purposes as *multiple geophones* (q.v.), i.e., with the holes spaced to minimize the effects of surface waves on the spread.

multiple-shot tool: A device run in a borehole to measure the direction of the borehole at several levels. See *directional survey*.

multiplex: 1. A process which permits transmitting several channels of information over a single channel without crossfeed. Usually different input channels are sampled in sequence at regular intervals and the samples are fed into a single output channel; digital seismic tapes are multiplexed in this way. Multiplexing can also be done by using different carrier frequencies for different information channels and in other ways. **2.** A stereoscopic plotting instrument used in preparing topographic maps by stereophotogrammetry.

multiplexed format: A **time-sequential format**, a data sequence in which the first sample of channel 1 is followed by the first sample of channel 2, then the first sample of channel 3, etc., until the first sample of all channels is given; then follows the second sample of channel 1, the second sample of channel 2, etc. As opposed to **trace-sequential format** in which the first sample of channel 1 is followed by the second sample of channel 1, etc., until all of channel 1 is given, followed by channel 2, etc. If an array of data is thought of as a matrix,

	Sample 1	Sample 2	Sample 3	$\cdots$	Sample n
Channel 1	a_1	a_2	a_3	$\cdots$	a_n
Channel 2	b_1	b_2	b_3	$\cdots$	b_n
Channel k	k_1	k_2	k_3	$\cdots$	k_n

then multiplexed format may be thought of as reading by columns and trace-sequential format as reading by rows. The data output of digital recording equipment is usually in multiplexed format whereas most data processing is done in trace-sequential format. Converting from the one format to the other (**demultiplexing**) is one of the first steps in data processing and usually part of the edit routine. Such a format conversion is called **matrix transposition**.

multiplicity: 1. The number of common-midpoint traces which sample essentially the same portion of reflector but with different offsets. For example, "12-fold" common-midpoint recording involves recording each subsurface point 12 times, once from each of 12 different offset distances. **2.** The number of independent raypaths which add together to provide a single output trace, also called *effort* (q.v.).

multiprocessing: A processing method in which program tasks are divided among a number of independent CPUs with the tasks being executed simultaneously.

multiprogramming: A programming technique in which two or more programs are operated on a time-sharing basis, usually under control of a monitor which determines when execution of one program stops and another begins. Also called **multitasking**.

multispectral scanner: A device which determines the amplitude in each of several frequency windows of a series of samples from successive locations. *Landsat*, (q.v.), includes such a device which looks at areas in four bands: green (band 4), red (band 5), a narrow near-infrared band (6), and a broader infrared band (7). Abbreviated **MSS**.

multivariant analysis: *Factor analysis* (q.v.).

mute: To change the relative contribution of components of a stack with record time. **1.** In the early part of the record, long offset traces may be muted or excluded from the stack because they are dominated by refraction arrivals or because their frequency content after NMO correction is appreciably lower than other traces. Where they begin to contribute, the transition may be either abrupt or gradual and may distort design for deconvolution or other processing operators. See Figure V-3. **2.** Muting may be done over certain time intervals to keep ground roll, air waves, or noise bursts out of the stack. See also *tail mute* and compare *diversity stack*. Also called **fade**.

mutual: Relations between circuits, such as the mutual inductance, capacitance or resistance (impedance) between the transmitter and receiver circuits of an IP survey system. See *coupling*.

mutual inversion: The simultaneous *inversion* (q.v.) of two independent data sets, as (for example) gravity and seismic data, to achieve a compatible model; **joint inversion**.

mutual resistance: See *electrode resistance*.

Mylar: A polyester film of high strength and dimensional stability, used as a base stock for drafting, light-sensitive film and magnetic tape. Tradename of DuPont Co.

myriameter waves: Electromagnetic waves in the $10^4 - 10^5$ m range (3 to 30 kHz, VLF).

Myriaseis: A telemetry seismic system. IFP tradename.

N

nabla: Del, symbol ∇. See *del*.

nadir: The point on the celestial sphere 180 degrees from *zenith* (q.v.).

Nafe-Drake relation: A postulated relation between density and *P*-wave velocity. See Figure N-1.

namelist: A semifree keyword format for giving parameter values to a computer. A namelist might look something like this:

&list TIME = 3, NTRACE = 6, V = 5000, &END.

NAND gate: The negative of an AND gate. A circuit with multiple inputs which functions unless signal is present at all inputs. See *gate* and Figure G-1.

nano: A prefix meaning 10^{-9}.

nanotesla: A unit of magnetic flux density. A nanotesla = 10^{-9} tesla = 1 gamma. Abbreviated nT.

nanosecond: 10^{-9}s.

natural frequency: The oscillation or vibration frequency of a system in the absence of an oscillatory disturbing force; **eigenfrequency**.

natural gamma-ray spectroscopy log: The spectrum of natural gamma-ray energy from radioactive decay of naturally occurring isotopes in rock formations contains contributions by potassium (K^{40}) and daughter products (Bi^{214} and Tl^{208}) of uranium (U^{238}) and thorium (Th^{232}). Spectral analysis of borehole measurements identify the isotope abundance in the energy windows characteristic of each to produce a continuous well log of K (percent), U and Th (ppm), and total gamma energy (counts/s or API units). See *spectral gamma-ray log*.

natural remanent magnetism: See *remanent magnetism*.

Naudi method: *Inflection-tangent-intersection method* (q.v.).

navigation: Directing a craft from one point to another; determining (a) the location at a given moment and/or (b) the direction and distance to a desired location. See *positioning*.

Navstar satellite system: *Global Positioning System* (q.v.).

Navtrak: An acoustic locating or navigating system employing the traveltime of sonar waves. An Edo Western tradename.

NCN: A nitrocarbonitrate shothole explosive. Requires a primer to detonate.

near-dc: The commutated dc or low-frequency ac used in resistivity and IP surveying.

near field: The field near a source. Relationships near a source involve both effects which attenuate rapidly with distance as well as those which attenuate more slowly (such as spherical divergence). At large distances many near-field phenomena are relatively unimportant. If the distance from the source is R and the wavelength is λ, near field implies $R < \lambda$. In the near-field zone of an antenna, fields vary predominantly as the inverse cube of the distance. Compare *far field*. See Sheriff and Geldart, v. 1 (1982, p. 47-48).

near-surface corrections: Corrections applied to seismic reflection times to accommodate changes in elevation and in velocity within the first hundred feet or so. *Static corrections* (q.v.).

near-trace section: A seismic section which comprises only the data from the geophone group (or few groups) nearest the source. Also called **short offset section**.

Neel point: See *Curie point*.

negative area: An area subject to more-or-less continual subsidence.

negative frequency: The frequency of a sinusoidal wave train traveling in the negative direction.

negative IP effect: An IP decay voltage opposite in sign to that of the charging current, due to the geometric relationship of a shallow polarizable body and the measuring electrode array.

negative pole: See *magnetic pole*.

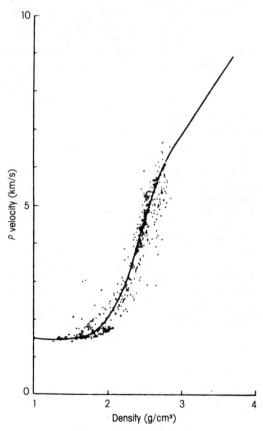

FIG. N-1. **Nafe-Drake** curve.

negative screening: An electromagnetic phenomenon in which a conducting overburden enhances the anomaly due to a buried conductor. The effect involves current flow between the overburden and the conductor.

negative standard polarity: See *polarity*.

neper: A unit for measuring ratios such as voltage ratios. N is the number of nepers if $N = \ln(V_1/V_2)$. It is thus similar to the *decibel* (q.v.). One neper = 8.686 dB.

neritic: Relating to water depths between low tide and 200 m.

nest: 1. An array of geophones, especially one containing many geophones close together; a **patch. 2.** To embed a subroutine or block of data in a larger routine or set of data.

nesting: A programming technique in which one or more iterative loops are included in a larger iterative loop.

network: 1. A set of points connected by communication channels. **2.** A system which converts an input function into an output function.

neutrally buoyant: Having the same buoyancy as the fluid in which it is immersed. Seismic streamers are nearly neutrally buoyant so that very little force is required to submerge or raise them.

neutral surface: The surface which separates compression and tension regions. When a layer is folded, the inside of the fold undergoes compression and the outside tension.

neutron activation log: A log of gamma rays of some characteristic energy. High-energy neutrons (about 14 MeV) bombard rocks and transmute elements to gamma-ray-emitting isotopes. From the gamma-ray energy calcium can be distinguished from silicon to permit lithology interpretation, or carbon from oxygen to distinguish oil from water. Sometimes called **calcium/silicon** and **carbon/oxygen logs,** respectively. See *induced gamma-ray spectroscopic log*.

neutron-lifetime log: A well log of the capture cross-section of thermal neutrons; similar to the thermal-decay-time log. A Van de Graaf neutron generator in the sonde periodically releases a burst of neutrons which enter the formation and begin to lose energy in collisions. At two discrete time intervals after a neutron burst, measurements are made of the thermal neutrons in the neutron-lifetime log. Measurements are made of the gamma rays which result from the capture of neutrons by nuclei in the **thermal-decay-time log.** The quantity plotted is sometimes (a) the reciprocal of the percentage which decay per unit of time, called the **thermal decay time** τ; (b) the time for the thermal neutron population to fall to half value, called the **neutron lifetime** L; and (c) the macroscopic **capture cross-section** Σ which is derivable from the foregoing ($\Sigma = 4.55/\tau = 3.15/L$). Thermal neutrons are captured mainly by the chlorine present and hence this log responds to the amount of salt in formation waters. Hydrocarbons result in longer decay times than salt water. Log readings are porosity-dependent and sensitive to clay content and permeability changes. This log is used in cased holes where resistivity logs cannot be run or to monitor reservoir changes to optimize production. It resembles a resistivity log with which it is

generally correlatable. Dresser Atlas tradename. See Figure N-2c. See also *pulsed neutron capture log*.

neutron log: A porosity well log which measures hydrogen density. Fast neutrons emitted by a source in the tool are slowed to thermal speed by collisions with (mainly) hydrogen atoms. The thermal neutrons are then captured by atomic nuclei of the surrounding material (mainly chlorine atoms) at which time a characteristic gamma ray of capture is given off. The neutron-log detector may record (a) the capture gamma rays ($n - \gamma$), (b) thermal neutrons ($n - n$), or (c) epithermal neutrons (those just above thermal speed). A low hydrogen density indicates low liquid-filled porosity. Porosity calculated from the neutron log is affected somewhat by the formation matrix and by the presence of gas. Neutron logs are used in crossplots to detect gas and determine lithology. Neutron logs are sometimes scaled in API units, sometimes in porosity units assuming a limestone matrix. The neutron log can be recorded in cased holes. See Figure N-2a. See also *sidewall neutron log*.

Newtonian potential: A potential associated with the inverse-square law, e.g., *gravitational potential* (q.v.).

Newton-Raphson technique: An iterative method of finding a numerical solution of an equation, $f(z) = 0$. A first value of z is tried. The next trial value is $z_{i+1} = z_i - f(z_i)/f'(z_i)$, where f' is the first derivative with respect to z. Where the process converges, trials are repeated until successive values come sufficiently close together.

Newton's laws: First law: A body does not change its state of motion unless acted on by external force. **Second law:** Acceleration equals unbalanced force divided by mass. **Third law:** If two bodies interact, the force exerted by the first on the second equals the force exerted by the second on the first. Enunciated by Sir Isaac Newton (1642-1727), English physicist.

n-factor: The exponent which expresses the change in amplitude of a field with distance. For an actual anomaly, n may be determined by a gradiometer arrangement or calculated from field measurements.

NG: No good.

nitrocarbonitrate: A class of mostly non-cap-sensitive explosives based on ammonium nitrate mixed with organic material; used as a seismic source.

Nitramon: An ammonia-gelatin explosive, not cap sensitive, which requires a primer to detonate it. Tradename of E.I. DuPont.

N-layered earth: A *layered earth* (q.v.) consisting of N-1 layers overlying a half-space (the Nth layer).

NML: *Nuclear-magnetism log* (q.v.).

NMO: *Normal moveout* (q.v.).

NMO velocity: Stacking velocity; see *velocity*.

NMR: Nuclear magnetic resonance. See *magnetic resonance*.

nodal plane: 1. A surface within a steady-state wave field which does not involve motion. **2.** The first motion from an earthquake may be either a push or pull, depending on the orientation of the station with respect to the epicenter and the direction of motion along the earthquake fault. A nodal plane separates the region where the first motion is a push from the region where it is a

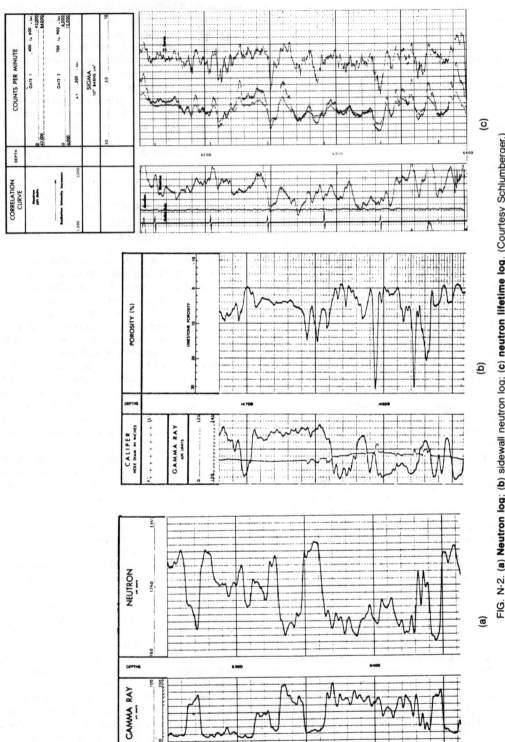

FIG. N-2. (a) **Neutron log**; (b) **sidewall neutron log**; (c) **neutron lifetime log**. (Courtesy Schlumberger.)

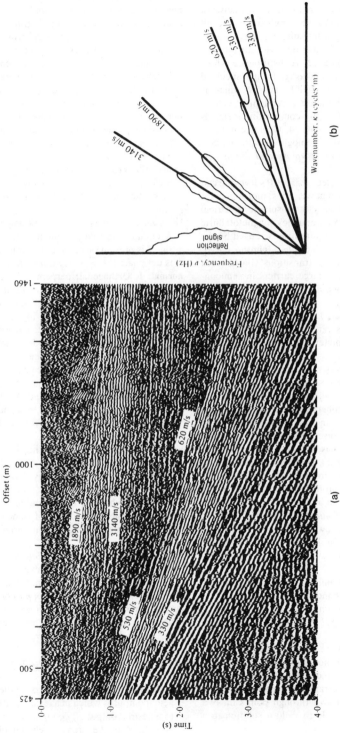

(a)

(b)

FIG. N-3. **Noise analysis** or walkaway. (**a**) Source was Vibroseis, geophones were spaced 1.5 m apart, offset to first phone was 425 m. The 1890 m/s arrival is a refraction from the base of the weathering; the 530 and 620 m/s arrivals are ground-roll modes; the 330 m/s arrival is an air wave. (Courtesy Chevron Oil Co.) (**b**) Frequency-wavenumber sketch for the data shown in part (a); see also Figure F-9.

pull. There are two such planes for first-energy *P*-waves, one of which is the fault plane.

node: 1. An end point of a branch in a network, or a junction common to two or more branches in a network or mesh. **2.** A point of rest in a vibrating system, a result of the interference of oppositely directed wave trains; e.g., one of the stationary points on a vibrating string.

noise: 1. Any unwanted signal. **2.** A disturbance which does not represent part of a message from a specified source. **3.** Sometimes restricted to energy which is random. To the extent the noise is random, it can be attenuated by a factor of $n^{1/2}$ by compositing n signals from independent measurements. **4.** Geologic noise is interference from unwanted geologic conditions. See also *noise (electrical)*; *noise (gravity and magnetic)*; *noise (seismic)*.

noise (electrical): 1. Noise in electrical or IP surveying can be due to interference from power lines, motor-generator or electronic components, atmospheric electrical discharges (sferics), or low-frequency magnetotelluric phenomena. **2.** Electrical circuit noise is caused by the randomness of conduction electrons (**Johnson noise**), the discreteness of magnetic transitions (**Barkhausen noise**), the discreteness of charge carriers in semiconductors (**shot noise**), modulation noise, and other causes.

noise (gravity and magnetic): Disturbances in observed data due to more-or-less random inhomogeneities in surface and near-surface material and errors in observation and reduction of data.

noise (seismic): Seismic energy other than primary reflections; includes microseisms, shot-generated noise, multiples, tape-modulation noise, harmonic distortion, etc. Sometimes divided into coherent noise (including nonreflection coherent events) and random noise (including wind noise, instrument noise, and other noncoherent energy). **Ambient seismic noise** refers to the background of random earth movements. Sometimes restricted to seismic energy not derived from the source.

noise analysis (seismic): A profile or set of profiles designed to gather data for an analysis of coherent noise trains. Usually consists of *microspreads* (q.v.) without any ground mixing so that low-velocity noise trains will be evident. See Figure N-3a. Results are often illustrated on a frequency-versus-wavenumber graph (see Figure N-3b).

noise survey: A mapping of ambient seismic noise levels within a given frequency band. A technique for detecting geothermal reservoirs which are often a source of short period seismic energy. Also called **ground noise survey**.

nominal time: The record of the time sequence of events, as the time scale on a seismogram, as contrasted with **real time**, the time when the events actually occurred.

nomogram: A set of scales arranged on a sheet of paper such that straight lines drawn through points on two scales intersect another scale to yield a solution to a mathematical expression.

nondipole field: See *magnetic field of the earth*.

nonfaradaic path: The virtual passage of current near an electrode as a result of reorientation of the ionic layers of the double layer. The process is analogous to charging a capacitor in that charge carriers are not transported across the interface.

nonimaging: See *remote sensing*.

nonnormal: 1. Not Gaussian; see *Gaussian distribution*. **2.** Not perpendicular. **3.** Different from the ordinary.

nonpolarizable electrodes: An electrode whose potential is not affected by the passage of current through it. Electrodes which are free of potentials caused by electrochemical action between the electrode and the ground. See *porous pot*.

nonsingular: Having an inverse. The determinant of a **nonsingular matrix** does not vanish so the matrix has an inverse. There is an inverse transformation for a **nonsingular transformation**.

nonvolatile memory: A type of computer memory which preserves data during power loss or system shutdown. Magnetic core read/write systems are typically nonvolatile.

NOR gate: The negative of an **OR gate** ("inclusive OR"). A circuit with multiple inputs which functions unless signal is present on any input. For inputs *A* and *B*, NOR is designated $\overline{A + B}$; see *gate* and Figure G-1.

normal: 1. Orthogonal; perpendicular to a surface or to another line. **2.** In the absence of an anomaly, as in a "normal time-distance curve." **3.** *Gaussian distribution* (q.v.). **4.** A resistivity well log in which a constant current is passed between a current eletrode in the sonde and a remote electrode (electrodes A and B) while the potential difference is measured between another electrode in the sonde and a reference electrode at the surface (electrodes M and N). The **spacing** is the distance between the A- and M-electrodes for the normal. A spacing of about 16 inches is used for the **short normal** and 64 inches for the **long normal**. See Figures E-7 and S-16.

normal correction: 1. Subtracting the normal magnetic field from magnetic data. **2.** Normalizing the ratio of successive Turam readings by dividing by the ratio of the primary fields. Differs from the free-air Turam correction wherever the ground is conductive.

normal dispersion: A decrease of velocity with frequency, the usual situation for a seismic surface-wave train. See *dispersion*.

normal distribution: See *Gaussian distribution*.

normal effect: An unwanted background IP effect due in part to membrane polarization, found to some extent in most rocks. See also *background polarization*.

normal equations: The set of linear simultaneous equations whose solution represents a least-squares fit (in particular, a *Wiener filter*, q.v.):

$$\phi_{zx}(\tau) = \Sigma f_t \, \phi_{xx}(\tau - t).$$

A digitized input x_t passed through the filter f_t yields the actual output y_t which in a least-squares sense is closest to a desired output z_t; $\phi_{xx}(\tau)$ is the autocorrelation of x_t and $\phi_{zx}(\tau)$ is the crosscorrelation of z_t with x_t as a function of a time shift τ. See *Levinson algorithm*.

normal fault: See *fault* and Figure F-2.

normal gravity: The value of gravity at sea level according to a formula which assumes the earth to have a simple, regular ellipsoidal shape. See *latitude correction*.

normal incidence: A raypath impinging on an interface at right angles. In isotropic media, equivalent to a wavefront striking an interface broadside, i.e., so that the angle between the wavefront and the interface (angle of incidence) is zero.

normalize: 1. Forming a ratio with respect to a standard (the normal). A normalized value usually is dimensionless. Normalizing often consists of scaling such that "something equals one". The "something" may be the rms value, the maximum value, etc. For example, an array response may be normalized by dividing each value by the rms value (or average energy) of the array. Hence for the array $X = (x_1, x_2, \cdots, x_n)$, the rms value $Y = [(x_1^2 + x_2^2 + \cdots + x_n^2)/n]^{1/2}$ and the normalized array is $(x_1/Y, x_2/Y, \cdots, x_n/Y)$. Autocorrelations are normalized by dividing by the value at zero time lag so that the maximum value of "one" indicates perfect correlation. Type curves (calculated effects for a model body) often are normalized so that the maximum effect is one. **2.** To adjust a floating point number so that the most significant bit (or digit) is held in the highest position of the mantissa, thereby permitting the maximum precision to be represented.

normalized apparent resistivity: Apparent resistivity divided by the resistivity of the upper layer. In constructing type curves, normalized apparent resistivity is plotted against normalized electrode interval (electrode interval divided by depth to the second layer). See *apparent resistivity curve*.

normal log: See *normal*.

normal magnetic field: 1. A smooth component of the Earth's magnetic field which is free of anomalies of exploration interest. Ordinarily computed from a low-order spherical harmonic expansion constrained by satellite measurements. The normal field of the earth varies slowly with time. Often identified with the *IGRF* (q.v.). **2.** The *magnetic field of the Earth* (q.v.) during an epoch when it is roughly aligned with the present-day field. Antonym: **reversed magnetic field**.

normal-mode propagation: The travel of waves trapped in a waveguide (**channel waves**). Seismic waves become trapped by total reflection such as occurs at a free surface or where the angle of incidence exceeds the critical angle. A surface water layer or a coal bed are two examples of possible channel-wave carriers. The mode of propagation is described by an **eigenfunction**. See Figure C-3 and Sheriff and Geldart, v. 1 (1982, p. 70-73).

normal moveout: NMO. The variation of reflection arrival time because of shotpoint-to-geophone distance (**offset**). The additional time required for energy to travel from a source to a flat, reflecting bed and back to a geophone at some distance from the source point compared with the time to return to a geophone at the source point.

normal-moveout correction: The time correction applied to reflection time because of normal moveout.

normal-moveout spectrum: The energy of a stacked trace as a function of arrival time and normal moveout. *Velocity analysis* (q.v.).

normal pressure: The state of a rock when its interstitial pressure is hydrostatic. See *abnormally high pressure*.

normal problem: A *direct problem* (q.v.).

normal ratio: Ratio between the readings of two coils in electromagnetic surveying, in the absence of conductive material. See *primary ratio*.

normal strains: See *strains*.

normal traveltime curve: A time-distance curve for a geologic section which does not contain anomalies of the type sought. Departures from the normal may indicate structures. Used in *fan shooting* (q.v.).

northing: Distance north of an east-west reference line. See *latitude*.

north-seeking pole: See *magnetic pole*.

nose: 1. A plunging anticline with structural closure in three out of four directions. **2.** An anomaly for which the contours do not close, as a "gravity nose" or "magnetics nose".

notch filter: A filter which is designed to remove a narrow band of frequencies. Often used to remove high-line effects; see *high line*.

NR: No reflection events. **1.** Denotes an event believed not to be a reflection. **2.** Denotes absence of reflections.

NRM: Natural *remanent magnetism* (q.v.)

NRZ: Nonreturn to zero. A method of digital recording on magnetic tape in which magnetization in one direction indicates a "0" and in the opposite direction indicates a "1". Compare *NRZI*.

NRZI: Nonreturn to zero invert, a system of encoding bits of information on magnetic tape, wherein a reversal of the magnetization polarity indicates a "1" and no change of polarity indicates a "0". Compare *NRZ*.

NS: Not shot; designates a scheduled shotpoint location (on a map) which has not been shot.

NSC: Necessary and sufficient conditions; a minimum complete set of conditions which are needed for a problem's solution to exist or for a situation to be true.

NTP: Normal temperature (0°C) and pressure (one atmosphere).

n-type semiconductor: A doped semiconductor with more electrons than holes available for carrying charges. Also called a **donor**.

nuclear cement log: A well log of scattered gamma rays, differing from the density log in that the gamma-ray source and detector are so spaced as to be sensitive to the density of material in the annulus. Used for distinguishing between cement and fluids behind casing. Can be run in an empty hole.

nuclear-magnetism log: A well log which is dependent on alignment of the magnetic moment of protons (hydrogen nuclei) with an impressed magnetic field. Abbreviated **NML** and also called **free-fluid log**. Protons tend to align themselves with an impressed magnetic field and when it is removed they precess in the earth's magnetic field and gradually return to their original state. The proton precession produces a radio-frequency signal whose amplitude is measured as the **free-fluid index (FFI)**. The rate of decay of the precession signal depends on interactions with neighboring atoms and hence on the nature of the molecule of which the proton is a part. The signal from the borehole fluid decays very

Decimal (Base 10)	Binary (Base 2)	Gray code	Quinary (Base 5)	Octal (Base 8)	BCD	BCDXS3	Biquinary	Hexadecimal
0	0	0	0	0	0000	0011	0 000	0
1	1	1	1	1	0001	0100	0 001	1
2	10	11	2	2	0010	0101	0 010	2
3	11	10	3	3	0011	0110	0 011	3
4	100	110	4	4	0100	0111	0 100	4
5	101	111	10	5	0101	1000	1 000	5
6	110	101	11	6	0110	1001	1 001	6
7	111	100	12	7	0111	1010	1 010	7
8	1000	1100	13	10	1000	1011	1 011	8
9	1001	1101	14	11	1001	1100	1 100	9
10	1010	1111	20	12	1 0000	100 0011	001 0 000	A
11	1011	1110	21	13	1 0001	100 0100	001 0 001	B
12	1100	1010	22	14	1 0010	100 0101	001 0 010	C
13	1101	1011	23	15	1 0011	100 0110	001 0 011	D
14	1110	1001	24	16	1 0100	100 0111	001 0 100	E
15	1111	1000	30	17	1 0101	100 1000	010 1 000	F
16	10000	11000	31	20	1 0110	100 1001	010 1 001	10
17	10001	11001	32	21	1 0111	100 1010	010 1 010	11
18	10010	11011	33	22	1 1000	100 1011	010 1 011	12
19	10011	11010	34	23	1 1001	100 1100	010 1 100	13
20	10100	11110	40	24	10 0000	101 0011	011 0 000	14

FIG. N-4. **Number systems**.

rapidly because of suppressant additives or disseminated iron (from steel worn from drill pipe and bits); by slightly delaying the time of measuring, the hole signal can be minimized. Fluids bound to surfaces (such as the water in shales) do not give appreciable response. Thus, the FFI indicates the free fluid (the hydrogen in free hydrocarbons and water). Gas gives a low reading because its hydrogen density is low. Sometimes **thermal relaxation time**, the rate of polarization buildup as a function of polarizing time, is measured to distinguish between water and oil. See also *proton-resonance magnetometer*.

nuclear-precession magnetometer: A magnetometer utilizing nuclear resonance; the resonance frequency is proportional to the absolute magnetic-field strength. See also *proton-resonance magnetometer* and *optically pumped magnetometer*.

null: Zero. A **null measurement** is one in which a balance is indicated by some quantity becoming zero, such as the current in one arm of a bridge circuit. The magnitude of the balancing "force" is then proportional to the quantity to be measured. For example, spring tension balances out the gravitational force in a gravimeter, the balance condition being indicated by null deflection.

number crunching: A computer program which involves a large amount of computation, particularly a repetitive operation on a large amount of data.

number system: A method of coding numbers for digital-computer storage and manipulation. Systems include decimal (base 10), binary (base 2), quinary (base 5), octal (base 8), *biquinary, binary-coded decimal* (BCD), *excess three code* (BCDXS3), *gray code*, duodecimal (base 12), hexadecimal (base 16), *one's complement, two's complement*, etc. See italicized entries for additional information and see Figure N-4.

numerical modeling: 1. Use of numerical techniques to calculate the theoretical response due to an assumed set of subsurface parameters (**forward modeling**). **2.** Use of direct or iterative methods for deducing subsurface parameters from geophysical data (**inverse modeling**).

Nusselt's number: Ratio of convective to conductive heat transfer.

nutation: Motion of the axis of rotation of a body about its mean position.

Nyquist frequency: A frequency associated with sampling which is equal to half the sampling frequency. Also called **folding frequency**. Frequencies greater than the Nyquist frequency alias as lower frequencies from which they are indistinguishable. See *alias*.

Nyquist theorem: *Sampling theorem* (q.v.).

O

object program: A computer program in machine language. Compare *source program*.

oblique configuration: A type of offlap reflection configuration, associated with high depositional energy. The top of the pattern indicates the wave base. See Figure R-8.

observed gravity: May refer to Bouguer, free-air, regional, or residual gravity fields. Sometimes means *raw gravity* (q.v.).

observer: 1. The person in charge of recording on a seismic crew. Sometimes the observer is also the field manager and sometimes is principally an electronic technician. **2.** The one who reads the gravimeter on a gravity crew.

Occam's razor: A dictum of scientific reasoning: The simplest explanation of observations is the most probable. "It is vain to do with more what can be done with fewer." Named for William of Occam (1300-1349), English philosopher.

occultation: An eclipse of a body by another body, as of a star by the moon. The observation of an occultation helps determine geodetic location.

oceanic layering: See Figure L-4.

oceanic trench: A long narrow depression of the ocean floor, commonly associated with subduction of an oceanic plate. See *plate tectonics*.

oceanography: See *geophysics*.

octal: A number system with a base of 8. See Figure N-4.

octave: The interval between two frequencies having a ratio of 2 (or 1/2). Filter rolloff is often given in decibels per octave.

odd function: A function which changes sign when its argument changes sign:

$$F(x) = -F(-x)$$

Also called **antisymmetric function**.

odograph: A time-distance graph.

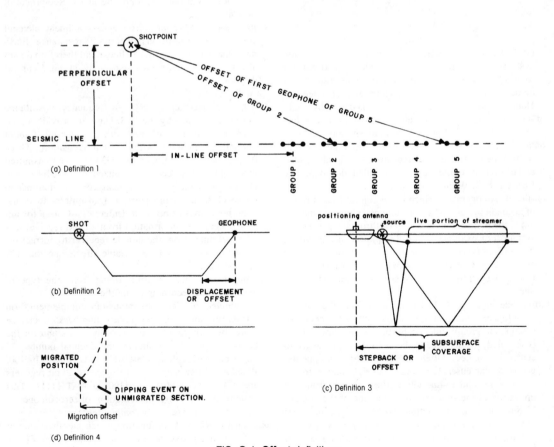

FIG. O-1. **Offset** definitions.

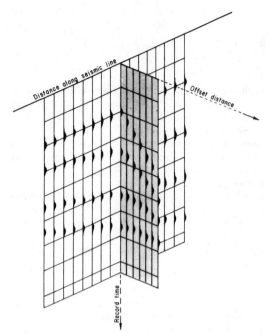

Distance along seismic line

Offset distance

Record time

FIG. O-2. **Offset space**. (Courtesy Compagnie Générale de Geophysique.)

oersted: A unit of magnetic-field intensity (in the cgs-emu system) in free space; the field which would exert a force of 1 dyne on a unit magnetic pole. Equal to 1 ampere turn/$(4\pi 10^3 m)$. See Figure M-1. Named for Hans Christian Oersted (1777-1851), Danish physicist.

off-end shooting: Having the seismic source located in-line and beyond the end of the geophone spread.

offlap: 1. Successive termination of strata farther seaward. When at the top of a depositional unit, also called **toplap**. When at the base, also called **downlap**. See Figure R-8. **2.** Without overlap.

off-line: 1. An operation which is not continuous with that of the main system, or an instrumental element which is not in the mainstream of data flow through a processing system. For example, an off-line plotter in a playback system is not directly connected to the main processing system. **2.** Referring to geophones or other instruments placed away from the line of shooting, opposite of in-line.

offset: See Figure O-1. **1.** The distance from the source-point to a geophone, or more commonly to the center of a geophone group; unless a particular geophone group is specified, the distance to the nearest geophone group center is implied. Often resolved into components: **perpendicular offset**, the distance at right angles to the spread line, and **in-line offset**, the distance from the projection of the shotpoint onto the line of the spread. **2.** Sometimes (in refraction work) the *displacement* (q.v.). **3.** In plotting marine data, the *stepback* (q.v.). **4.** The horizontal component of *migration* (q.v.). **5.** The horizontal component of fault displacement, measured

parallel to the strike of the fault. **6.** The distance between source and receiver in electromagnetic time-domain surveys.

offset frequency: The difference between an observed frequency and a reference frequency. See Figure S-1.

offset section: A display of traces having constant source-to-geophone distance (offset).

offset space: A way of thinking of a line of seismic reflection data in which the independent variables are arrival time, location along the seismic line, and offset. Events curve in the offset direction because of normal moveout. See Figure O-2.

offshore shooting: Marine seismic surveying.

off time: The time an IP pulse-type or an electromagnetic time-domain transmitter is off, during which the decay voltage is measured at the receiver.

ohm: A unit of electrical resistance or impedance. The potential drop across one ohm is one volt per ampere of current. Named for Georg Simon Ohm (1787-1854), German physicist.

ohmic: 1. A system which is electrically linear, i.e., obeys Ohm's law. **2.** The resistive component of an impedance as opposed to the reactive component.

ohmic contact: See *galvanic contact*.

ohm-meter: A unit of resistivity, also written ohm-meter^2/meter, being the resistance of a meter cube to the flow of current between opposite faces. Reciprocal of mho/m.

Ohm's law: The voltage drop across a linear element equals the current through it times its resistance. Earth materials are not necessarily linear and therefore do not always obey Ohm's law, especially at high current densities.

oil well: See *GOR*.

Omega: A long-range, very low-frequency positioning system developed by the U.S. Navy to provide world-wide all-weather positioning with an accuracy of about one mile (comparable with celestial navigation). Three frequencies (10.2, 13.6, and 11.333 kHz) are transmitted in a sequential pattern synchronized in phase by atomic clocks. The Omega receiver measures the difference in phase of signals from pairs of transmitters to define hyperbolic lines of position. **Differential-Omega (or micro-Omega)** uses information from a fixed receiver less than 100 miles from the mobile receiver to correct for sky-wave and other time-variable effects and thus improves accuracy.

omnitape: A device for transcribing from one type of magnetic-tape recording to another.

one's complement: The radix-minus-one complement form for representing negative binary numbers. It can be found by replacing all the ones by zeros and all the zeros by ones. For example, the decimal number 27 might be represented as 0011011, and -27 as 1100100. When a number is added to its negative all bit registers are full; e.g., $0011011 + 1100100 = 1111111$. This system contains two representations of zero; all ones or all zeros. Compare *two's complement*.

one-sided function: 1. A function which has the value of zero for all negative values of the argument; i.e., $F(x) = 0$ if $x < 0$ (or alternatively, zeros for all positive arguments). **2.** A function which is not defined for negative (or positive) values of the argument.

one-sigma: See *standard deviation*.

one-way time: Half the corrected traveltime for a reflection arrival. One-way time multiplied by average velocity gives reflector depth for a flat reflection and flat velocity layering.

onlap: 1. Successive landward termination of strata at the base of a depositional unit. **2.** A reflection termination at the base of a unit where the reflection is flat or dips away from the termination. See Figure R-8.

on-line: 1. A linear array of observation points, especially a seismic line. **2.** Equipment under the control of a central processing unit. **3.** A process (usually data output) which is concurrent with other operations.

onset: The beginning of a wave train. See *break*. In electric sounding, the start of a transient.

on time: 1. The time during which an IP transmitter is actually supplying current. **2.** The time during which the charging current or field from a pulse-type transmitter is observed at the receiver.

open chamber exploder: A marine seismic source which involves the detonation of an explosive mixture of gases in a chamber open to the water on the bottom side so the waste gasses are vented directly into the water.

open ended: 1. The situation where the addition of new elements does not disturb the prearranged system. **2.** Able to accommodate additional data.

open hole: A well bore which has not been cased where measurements are made.

open question: A question for which the answer is not known.

operand: A quantity participating in the execution of a computer instruction. An operand can be an argument, a result of computation, a parameter, an address, or the location of the next instruction to be executed.

operating system: An integrated system of routines for supervising the operation of a computer. Also called **executive.**

operation: A mathematical (or sometimes physical) process to be performed on data, usually indicated by a symbol. For example, a plus sign means the operation "add the number ahead of it to the number behind it". Differentiation, integration, convolution, Fourier transformation, crosscorrelation, etc., are likewise "operations". See also *operator*.

operational amplifier: A high-gain, high input-impedance amplifier requiring minimal current for operation. Ideally a voltage-controlled voltage source. External feedback components are used to obtain desired operations such as summing, integrating, differentiating, etc.

operator: 1. The specific thing involved in a particular operation. Thus, a filter operator is a specific filter expression involved in filtering (convolution). See also *operator length*. **2.** A symbol indicating an operation to be performed, and itself the subject of operations. **3.** That part of an instruction which tells the machine which function to perform: read, write, add, subtract, etc. **4.** An observer.

operator length: The time-domain length of the impulse response of a convolution operator. Often specified as a certain number of points; for example, a 56-point operator at 2-ms sample rate is 55 intervals (times 2 = 110 ms) long.

Opseis: A digital radiotelementry seismic recording system which stores data from each shot in remote units and on command transmits the data to the recording unit. A single UHF channel can handle the entire system requirements. Applied Automation tradename.

optical holography: See *holography* and Figure H-5.

optically pumped magnetometer: A magnetometer such as the cesium or rubidium-vapor magnetometer which involves nuclear magnetic resonance as a transfer mechanism between light and an RF field at the Larmor frequency. See Figure O-3. Such magnetometers can be made extremely sensitive. They measure absolutely the total magnetic field.

optimum: Best according to some criteria. A meaningless

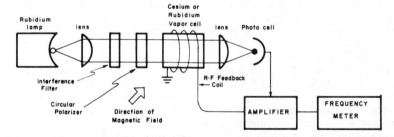

FIG. O-3. **Optically pumped magnetometer.** Atoms in a vapor cell precess about the steady magnetic field which is to be measured, which is at an angle of 20-70 degrees to the instrument axis. Monochromatic light which is circularly polarized in the plane perpendicular to the instrument axis has a component which can be absorbed by the precessing atoms. Once this absorption is complete, no further absorption can occur and then the vapor cell becomes transparent, a condition which causes an increase in the light to a photocell. The polarized precessing atoms have a component along the axis of a transverse RF field which permits the atoms to undergo transitions and hence become available to absorb more light. The precessing atoms thus become a transfer mechanism between the light and the transverse RF field when the field is at the Larmor frequency. The light intensity is used to monitor the precession and automatically adjust the RF frequency, whose frequency can be measured to give the steady magnetic-field intensity.

term unless the criteria are specified. Optimum in a least-squares sense means that the sum of the squares of all errors is minimized.

optimum damping: See *damping*.

optimum filter: A filter designed to maximize or minimize a certain performance measure. See l_p *fit*. Often means a **Wiener filter** in which the mean-square difference between actual and desired outputs is minimized.

optimum wide-band: A process of filtering and stacking which maximizes cancellation (according to certain mathematical criteria) of one type of event (a multiple in the case of horizontal stacking, a ghost in the case of uphole stacking) regardless of frequency content and at the same time reinforces another type of event (the primary). Application of the method requires precise knowledge as to the time differences between the events on the records to be stacked (i.e., precise differential normal-moveout information for multiple cancellation, precise uphole data for ghost cancellation). For optimum wide-band horizontal stacking, see Schneider et al. (1965). For optimum wide-band uphole stacking, see Schneider et al. (1964).

order: See *pole*.

order of magnitude: The nearest integer to $|\log_{10} X/S|$, where X and S are two quantities being compared. One order of magnitude indicates that one quantity is of the order of 10 (or a tenth) times the other value, two orders of magnitude 100 (or 1/100) times, etc. Used to make crude comparisons or to give the error or uncertainty of measurement.

order of a matrix: See *matrix*.

OR gate: A circuit with multiple inputs which functions when a signal is present at any input. Also called **inclusive OR gate**. For inputs A and B, signified by (AUB) or (A+B). An **EXCEPT gate** is called **exclusive OR** but the inclusive OR is intended unless "exclusive" is specifically stated. For the truth tables, see *gate* and Figure G-1.

O-ring: A rubber, Neoprene, Teflon, or other elastic, circular gasket with circular cross-section, used to effect a seal between parts of apparatus.

orogenic: Involving mountain-building by large-scale lateral forces. Often involves thrust faulting and folding. See *tectonic types*.

orthogonal: Normal or at right angles. Linear combinations of functions are orthogonal if they are linearly independent, i.e., if they cannot be expressed as combinations of each other. The nonvanishing of the determinant of coefficients is a test for the orthogonality of a set of equations. See also *Jacobian* and *matrix*.

orthometric correction: A correction to very precise land survey data because gravitational equipotential surfaces for different elevations are not parallel.

OS: *Operating system* (q.v.).

oscillograph: **1.** An instrument that renders visible a curve representing the time variations of electric phenomena. The recorded trace is an **oscillogram**. An example is the cathode-ray oscillograph. **2.** *Camera* (q.v.).

outer Helmholtz double layer: See *diffuse layer*.

outer product: *Cross product* (q.v.).

out-of-phase: The component of an electrical signal that has a 90-degree phase difference from the exciting or reference signal. Also called **quadrature**.

output: **1.** The power, current or voltage delivered by a circuit, system, or device. **2.** The terminals where the power, current, or voltage may be delivered. **3.** Data which have been processed.

output-energy filter: A filter which maximizes the energy of a signal while minimizing the energy of the filtered noise. See Treitel and Robinson (1969).

overburden: **1.** Material lying over an ore or valuable deposit. **2.** The section above a refractor or reflector. **3.** Loose, unconsolidated material above bedrock.

overdamped: See *damping*.

overdetermined: Having more equations to be satisfied than unknowns to be determined.

overflow: A condition occurring when a computer operation produces a result which has a magnitude exceeding the capacity of the computer's data-word size.

overlapping: The process of adding pairs of adjacent traces (mixing) and recording the sum as a single trace. The consequent section has one fewer trace than the input.

overlay: The technique of repeatedly using the same blocks of memory during different stages of a problem.

overload point: The input-signal amplitude for which the ratio of output to input first differs by 3 dB from the ratio within the linear operating range.

overpressured: Having larger interstitial fluid pressure than appropriate for the depth.

overshoot: **1.** After a step change, assuming too large (or too small) a value before settling down to the correct value. **2.** Amplitude which exceeds the gain permitted before clipping.

overvoltage: The extra potential (which in IP is proportional to impressed current density) due to an electrochemical and electrokinetic barrier set up at an electrode-to-electrode interface. **Activation overvoltage** is caused by current passage stimulating an electron-transfer reaction such that the electrode potential deviates from its reversible potential without appreciably changing the ion concentrations at the electrode surface. **Concentration overvoltage** is brought about by a depletion or accumulation of oxidized and reduced ion species at the electrode surface, causing a change in the reversible potential of the electrode. See *induced polarization*.

overvoltage method: Induced-polarization method; see *induced polarization*.

P

p: The *raypath parameter* (q.v.).

P&A: Plugged and abandoned. See *dry hole*.

Pacific margin: See *active margin*.

package program: A standard set of computer programs which are used as opposed to tailoring the processing to the specific needs of the data. Often implies back-to-back processing without intermediate decisions between stages in the processing.

packing: 1. The number of bytes of information per unit length of magnetic tape, often measured in bytes per inch (**bpi**). **2.** Increasing the density of stored data so that more data can be stored in the same space, as in placing more bits in a given length of magnetic tape. **3.** Arrangement of particles in a matrix, as of grains in sandstone.

pad: Sidewall pad; a footing on the end of an arm which presses against the borehole wall.

page back: To add a constant to values being plotted so that a wider range of values can be plotted on the same piece of paper. See Figure P-10.

paging: 1. The function in a printer routine which separates tabulations into separate pages. **2.** The function in a plotter routine which adds a fixed amount to a coordinate so that a graph being plotted will not run off the paper. Magnetometer records and seismic-profiler records often are "paged" so that a large plotting scale can be used without requiring excessively large paper. **3.** Division of data or program instructions into blocks called **pages**. Some of the pages may be stored in a storage device other than the computer's rapid-access memory and brought into the rapid-access memory only as needed. In this way a program can be larger than the rapid-access memory. Such storage is called **virtual memory**.

paleodatum: See *datum*.

paleomagnetism: Study of natural remanent magnetization of rocks and other materials in order to determine the intensity and direction of the Earth's field at the time the materials were magnetized. It has as adjuncts **archeomagnetism** (study of the Earth's magnetism during historical times) and **rock magnetism** (basic study of the magnetic properties of rocks and minerals). See *remanent magnetism*.

paleosection: A cross-section showing attitude of bedding and structure as it is assumed to have been at some past time. May refer to a seismic section on which one horizon has been flattened, assuming that this horizon was laid down flat and that the flattened section therefore shows the attitude of deeper structure at the time of deposition of the flattened horizon. Compaction because of overburden and other changes subsequent to deposition are often ignored. Also called a **palinspastic** or **restored section**.

palinspastic section: See *paleosection*.

pantograph: 1. A device for copying a drawing at a different scale. **2.** A device for mapping from one domain to another where there is a one-to-one correspondence between the domains. **3.** A device for plotting seismic events in their migrated position.

paradox of anisotropy: Relations between the apparent resistivity and true resistivity in a homogeneous transversely isotropic medium. If resistivities perpendicular and parallel to the bedding are ρ_T and ρ_L, respectively, the apparent resistivities measured in the transverse and longitudinal directions ρ_{aT} and ρ_{aL} are:

$$\rho_{aT} = \rho_L,$$

$$\rho_{aL} = (\rho_T \, \rho_L)^{1/2}.$$

Resistivity anisotropy coefficient is discussed under *anisotropy*.

parallax: A change in the apparent position of an object (such as a meter needle) with respect to a reference (such as the meter dial) which is in a different plane, when not viewed at right angles. Parallax error results when the observer is not correctly positioned for the reading.

parallel field: A uniform field in which current flow lines or equipotential surfaces are parallel.

parallel record: A test record made with all the amplifiers connected in parallel and activated by a single geophone. Also called a **bridle**. Used to check that all amplifier circuits perform similarly with respect to lead or lag, polarity, and phasing.

paramagnetic: Weakly magnetic with small positive susceptibility. The magnetic moments of individual atoms are uncoupled so that each atom behaves independently. Paramagnetism usually contributes only a few gammas to the magnetic field at the Earth's surface. Compare *diamagnetic* and *ferromagnetic*.

parameter: 1. A variable which can be changed independently and (usually) arbitrarily between calculations but which remains constant during any calculation. **2.** Quantities (each of which may represent a combination of quantities) which are sufficient to determine the response characteristic of a system.

parametric sounding: An electromagnetic depth sounding in which the frequency is varied while holding the geometry constant, as opposed to **geometric sounding** where the frequency is held constant and the geometry is varied. Used to resolve resistivity layering assuming the layering is horizontal. See *polarization ellipse*.

parasitic ferromagnetism: A weak ferromagnetism associated with imperfect antiferromagnetism in such substances as hematite.

paravane: A device with attached vanes, which is towed through the water and used to maintain equipment in a certain position relative to the towing vessel. The force of the flowing water on the vanes causes the device to dive, maintain a particular orientation, or move to the side. Used to tow a seismic streamer at depth or to tow sources to the side of a ship.

parity bit: One of the bits in a defined set which is dependent upon the other bits in such a way as to detect dropout. See *check*.

parity check: See *check*.

Parseval's theorem: For two aperiodic functions h_1 and h_2 with respective Fourier transforms H_1 and H_2,

$$\phi_{12}(0) = \int_{-\infty}^{\infty} h_1(t)\, h_2(t)\, dt$$

$$= \int_{-\infty}^{\infty} H_1(\nu)\, H_2(\nu)\, d\nu = P_{12}(0).$$

Thus the zero-lag value of the crosscorrelation $\phi_{12}(0)$, the two values on the left, equals the integral of the cross-product spectrum and the crosspower-spectral amplitude at zero frequency $P_{12}(0)$, the two terms on the right. Both equal the cross-energy in the time domain.

parsimonious deconvolution: A deconvolution technique which minimizes

$$(\Sigma e^p)^{1/2}/(\Sigma e^q)^{1/q},$$

where e is the prediction error and p is slightly larger than q. See Postic et al. (1980).

parsing: Breaking a unit into component parts, as is done in some computer operations.

partial fraction: One of a series of terms expressed as fractions involving roots of an expression; the sum of the series equals the expression. See Sheriff and Geldart. v. 2 (1983, p. 158).

partition gas chromatograph: A device for quantitative analysis of hydrocarbon constituents. A fixed quantity of sample is carried with a stream of sweep gas through a partition column packed with an inert solid coated with a nonvolatile organic liquid. The lighter fractions traverse the column faster than the heavier fractions so that the components appear separately at the column exit where their amounts can be measured.

party: The group working together to carry out a geophysical field project. Also called **crew** or **troop**.

party chief: The head of a geophysical party.

party manager: The person working under the party chief (or supervisor, if no party chief), who usually is responsible for the field work.

pascal: A unit of pressure, a newton per square meter. Named for Blaise Pascal (1623-1662), French mathematician.

Pascal: A high-level computer language.

pass: 1. A complete cycle through a computer involving input, processing, and output; a **machine run. 2.** The passage of a satellite from rise to set over the horizon.

passband: The range of frequencies which can pass through a band-pass filter without significant attenuation.

passive: 1. Having no source of energy. A "passive filter" involves no amplification, merely attenuating certain frequencies more than others. A **passive beacon** is a radar reflector which merely reflects radar energy, as opposed to an **active beacon** which transmits in response to a signal. **2.** Applied to a system which does not generate an output if there is no input. **3.** A positioning system which does not involve the transmission of a signal but only the observation and measurement of angle, amplitudes, phases, times, etc., in a system which exists independently of the craft being positioned.

passive seismic: Seismic techniques that do not use an artificial source of seismic energy. Such techniques are used for thermal exploration, studies of microearthquakes, amplitude spectra of ground noise, P- and S-wave delay studies, etc.

pass region: *Passband* (q.v.).

patch: 1. A large geophone group which feeds a single channel. Patch arrays occasionally are several hundred feet across containing several hundred geophones. Used in *transposed recording* (q.v.), especially with surface sources in areas of poor record quality. **2.** A jumper or a temporary connection, especially one which can be changed easily, as a connection on a "patch panel". **3.** A section of coding (or a subroutine) used to correct a mistake or alter a routine.

pattern: An *array* (q.v.).

pattern recognition: Analyzing data to see if subsets of the data contain arrangements which are distinctive of specific things.

pattern shooting: The firing of charges arranged in a definite *array (seismic)* (q.v.).

pay zone: The interval of rock in which an accumulation of oil or gas is present in commercial quantities.

P-band: Radar frequencies between 225 and 390 MHz; see Figure E-8. Used in remote sensing because it penetrates vegetation and shows a combination of vegetation and surface-soil effects.

PC: Continuous-type *micropulsations* (q.v.).

PC board: Printed circuit board.

PCM: Pulse-code modulation. See *modulation*.

PDMI: *Percent decrease in mutual impedance* (q.v.).

PDR: *Potential-drop ratio* (q.v.).

peak: The maximum upward (positive) excursion of a seismic wavelet; **crest**. Opposite of **trough**.

pearls: Continuous-type micropulsations "of the first kind" with periods from 0.2 to 5 s, amplitudes from 0.05 to 0.1 gamma. Strip-chart records of pearls look like amplitude-modulated sinusoidal waves (resembling a pearl necklace). See *micropulsations*.

pedestal effect: A time delay produced by absorption.

peel-off time: 1. A *static correction* (q.v.). **2.** The time above which data are to be removed. Used in making restored (palinspastic) sections.

peg-leg multiple: A multiple reflection involving successive reflection between different interfaces so that its travel path is nonsymmetric. See Figure M-12. Usually refers to short-path multiples (Type 1 in Figure M-11) within thin beds, which result in transferring energy from the front end of a wave train and adding it back later and thus is a mechanism for changing waveshape. See Sheriff and Geldart, v. 1 (1982, p. 106, 109).

pelagic deposits: Deep-sea sediments with little terrigenous material.

PEM: Pulse electromagnetic method, a *transient electromagnetic method* (q.v.).

penetration: 1. The greatest depth at which material properties significantly affect measurements. **2.** The greatest depth from which seismic reflections can be picked with reasonable certainty. Depends on the energy of the reflected wave, the presence of noise, and the processing to which the data are subjected as well as the recording system. **3.** See *skin effect*.

percent decrease in mutual impedance: The percent voltage change in a coupled circuit with respect to the low-frequency impedance. Used with reference to the amount of in-phase electromagnetic coupling.

percent distortion: See *distortion*.

percent frequency effect: PFE. 1. The basic polarization parameter measured in frequency-domain resistivity surveys, being the percent difference in resistivity measured at two frequencies:

$$PFE = 100\ (\rho_{dc} - \rho_{ac})/\rho_{ac},$$

where ρ_{dc} and ρ_{ac} are the low- and high-frequency resistivities. **2. Decade-normalized PFE** is multiplied by the log of the frequency ratio:

$$PFE = [100(\rho_{dc} - \rho_{ac})/\rho_{ac}]\ log_{10}(\nu_{ac}/\nu_{dc}).$$

3. Keller suggests that PFE be defined as:

$$PFE = 100(\rho_2 - \rho_1)/(\rho_2\rho_1)^{1/2},$$

where ρ_1 and ρ_2 are resistivities measured at two frequencies which are a factor of 10 apart. **4.** PFE is closely related to *chargeability* (q.v.) if defined as suggested by Brant:

$$PFE = 100\ (\rho_{dc} - \rho_{ac})/\rho_{dc}.$$

percent mineralization: In IP surveying, the volume-percentage of metallic-luster minerals in a rock. This value is usually about half the metallic-luster mineral content by weight.

perforate: To open holes through casing and into a formation so that fluids can flow through the formation into the borehole or vice versa.

perigee: Shortest distance from a satellite orbit to the Earth's center; see Figure E-9. The greatest distance is **apogee**.

period: 1. The time T for one cycle. The time for a wavecrest to traverse a distance equal to one wavelength, or the time for two successive wavecrests to pass a fixed point. For a monochromatic wave train,

$$T = 1/\nu = \lambda/V,$$

where ν = frequency, λ = wavelength, and V = phase velocity. See Figure W-2. **2.** A major standard geologic time unit; see Appendix L.

periodic function: A function that repeats after successive equal intervals of time; a **harmonic function**.

peripheral device: A general term designating machines which operate in conjunction with a computer or system but are not physically part of the system. Peripheral devices typically display, store, and return data to the computer on demand, prepare data for human use, or acquire and convert data to a form usable by the computer. Peripheral devices include printers, keyboards, graphic display terminals, paper tape reader/punches, analog-to-digital converters, disks, tape drives, etc.

permafrost: Permanently frozen soil or rock. Its effects on

seismic wave travel are discussed in Sheriff and Geldart, v. 2 (1983, p. 12-13). It usually has high seismic velocity and high electrical resistivity.

permeability 1. The ratio of the magnetic field **B** to the magnetizing force **H**. **2.** A measure of the ease with which a fluid can pass through the pore spaces of a formation. Measured in millidarcy (1/1000 darcy) units. The permeability constant k is expressed by **Darcy's law** as $\mu q/dp/dx$), where μ is fluid viscosity, q is linear rate of flow, and dp/dx is the hydraulic pressure gradient.

permeability of free space: See *magnetic field*.

permeance: The reciprocal of *reluctance* (q.v.).

permit: Permission from a landowner for a geophysical field party to work on his land.

permittivity: *Capacitivity* (q.v.) of a three-dimensional material, such as a dielectric. **Relative permittivity** is the dimensionless ratio of the permittivity of a material to that of free space; it is also called the **dielectric constant**.

perpendicular offset: See *offset*.

Peters' length: A measurement made on profiles across potential-field (especially magnetic) anomalies, the objective of which is to determine the depth to an anomalous mass (magnetized body). Peters' rule gives the depth as 5/8 of the horizontal distance between the points on the side of an anomaly, where the slope is half of the maximum slope, for a tabular body or dike model vertically polarized. See Peters (1949) and Figure D-6.

PFE: *Percent frequency effect* (q.v.).

PFN: Prompt fission neutron (log).

PGC: Preset or *programmed gain control* (q.v.).

phantom: A line on a seismic section drawn parallel to the dip of nearby reflection events. Phantoms are drawn and mapped where one cannot follow one event far enough to develop a map on that event alone.

phantom diffraction: A diffracted reflection, that is, an event diffracted because of a discontinuity (such as a velocity change at a fault) above the reflector. The arrival time is that of the associated reflection but the diffraction curvature is that appropriate to the diffracting point. See Sheriff (1982, p. 13).

phase: 1. The argument of a wave. If the representation of a wave is $\psi(x - vt)$, the argument $(x - vt)$ is the phase. **2.** The angle of lag or lead of a sine wave with respect to a reference; how far rotation, oscillation, or variation has advanced, considered in relation to a reference or assumed instant of starting. Commonly expressed in angular measure. Phase information, being the measure with respect to the instant of starting, carries the timing information of a seismogram and hence proper phase preservation is of utmost importance. See also *phase characteristics, phase response*, and compare *phasing*. **3.** In earthquake seismology, an event on a seismogram marking the arrival of a new group of waves, indicated by a change of period or amplitude or both.

phase angle: Tan^{-1} [quadrature component/in-phase component]. The phase angle is in quadrants 1 to 2 if the numerator is positive, in quadrants 1 or 4 if the denominator is positive. In induced polarization, phase angle is usually measured in milliradians.

phase array station: See *large aperture seismic array*.

phase characteristics: 1. Of all those wavelets or filters

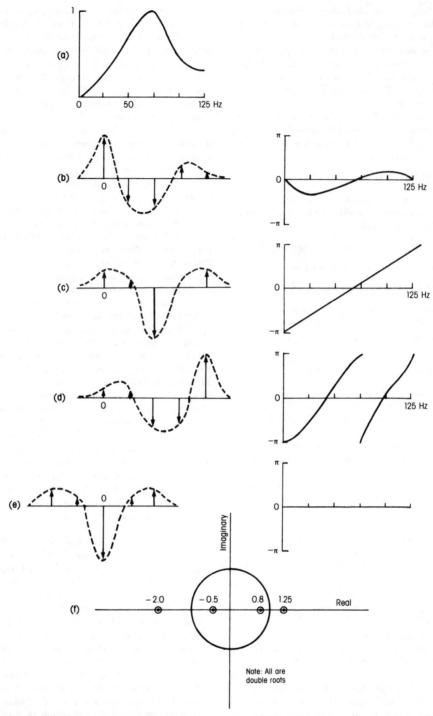

FIG. P-1. **Phase characteristics** of wavelets having the same amplitude spectrum (**a**). (**b**). Minimum-phase wavelet and its phase spectrum: $(1 - 0.8\,z)^2(1 + 0.5z)^2 = 1 - 0.6z - 0.71z^2 + 0.24z^3 + 0.16z^4$. (**c**) Linear phase: $(1 - 0.8z)(0.8 - z)(1 + 0.5z)(0.5 + z) = 0.4 + 0.18z - 1.25z^2 + 0.18z^3 + 0.4z^4$. (**d**) Maximum phase: $(0.8 - z)^2(0.5 + z)^2 = 0.16 + 0.24z - 0.71z^2 - 0.6z^3 + z^4$. (**e**) Zero phase: $0.4z^{-2} + 0.18z^{-1} - 1.25 + 0.18z + 0.4z^2$. A zero-phase wavelet is anticipatory, that is, begins before time zero. Phase curves depend on the location of the time reference. Other mixed-phase wavelets can also be made from these component doublets. (**f**) Z-plane plot of the roots of the autocorrelation function for the foregoing, all of which have the same autocorrelation: $\phi_{xx}(z) = (1 - 0.8z)^2(0.8 - z)^2(1 + 0.5z)^2(0.5 + z)^2$. In a more general case, roots may be complex.

with the same amplitude spectra or autocorrelation, particular members are characterized by their phase spectra (phase as a function of frequency). They can also be characterized in other ways, for example, by the location of their roots in the z-domain; see Figure P-1. The principal feature of **minimum phase** is that the energy arrives in a burst which is not followed by a larger burst. The phase of a minimum-phase wavelet is smaller and its energy builds up faster (i.e., it is **minimum-delay**) than that of any other causal wavelet with the same amplitude spectrum, or with the same autocorrelation. A two-term wavelet, **couplet** or **doublet** $[a,b]$ is minimum-phase or minimum-delay if $|a| > |b|$. Any wavelet may be represented as the convolution of couplets. A wavelet is minimum-phase if all the couplets which are its factors are minimum-phase. For example, the z-transform of a wavelet might be $(6 + z - z^2)$, which can be expressed as $(3 - z)(2 + z)$, each of which is minimum-phase; hence the wavelet is minimum-phase. Minimum-phase is sometimes expressed as having all roots outside the unit circle in the z-plane, or as having no zeros in the right half of the Laplace transform S-plane. A **maximum-phase** or **maximum-delay** couplet $[a, b]$ has $|a| < |b|$. Maximum-phase wavelets have all their roots inside the unit circle in the z-plane. For a **linear-phase** wavelet, the phase-frequency plot is linear; such a wavelet is symmetrical. A **zero-phase** wavelet has its phase identically zero; it is symmetrical about zero but is not causal. **2.** A multi-channel filter can be expressed as a matrix. It is minimum-phase if its determinant (which can also be expressed as the product of couplets) is minimum-phase. E.g., a multichannel response produced by impulsive inputs might be:

	Input channel 1	Input channel 2
Output channel 1	$(2 + z)$	(z)
Output channel 2	(1)	$(6 + z)$,

which has the determinant $12 + 7z + z^2 = (3 + z)(4 + z)$ which is minimum-phase; hence the multichannel response is minimum-phase. Statements similar to the preceding can be made for maximum-phase, linear-phase, or zero-phase by substituting those words everywhere for minimum-phase. See also Sheriff and Geldart, v. 2 (1983, p. 39 and 180-184).

phase coherence: The same phase relationship on adjacent traces; evidence for a reflection event.

phase comparison: A matching of the radio signals from two CW transmitters. Used to determine a line of position.

phase control: The process of rapid on-off switching which connects an ac supply to a load for a controlled fraction of each cycle.

phase-correction filtering: A filter which compensates for the nonlinear phase response of other components in a system.

phase curve: A phase curve of a seismic trace is a plot of the phase relationship of the component sinusoids de-

termined by harmonic analysis graphed versus frequency.

phase distortion: Change in waveshape because phase shift is not proportional to frequency. See *distortion*.

phase encoding: A method of recording on magnetic tape in which bits are indicated by changes in flux direction. Flux changes denoting a "1" bit are all in one direction while changes denoting a "0" bit are all the other direction. See *NRZ* and *NRZI*.

phase inversion: A change of 180 degrees in phase angle, mirror-imaging a trace about the zero-deflection position.

phase-lock: A technique in which a signal of almost constant frequency is generated within an instrument, which signal is brought to the same average phase as an external signal. Used as the reference signal in synchronous detectors to suppress noise.

phase response: A graph of phase-shift versus frequency, which illustrates the phase characteristics of a system or of a wave train. Filters with the same amplitude-frequency response but different phase characteristics affect the shape of pulses put through them differently. See *phase characteristics*.

phase reversal: A phase shift of 180 degrees, so that a peak becomes a trough and vice versa.

phase shift: The result of adding to or subtracting from a phase measurement. In the time domain, a phase shift of ϕ is equivalent to a time shift of $\phi/2\pi\nu$ where $\nu =$ frequency of the respective component being shifted. Phase shifts result in change of waveshape unless all components are shifted proportional to their frequencies.

phase spectrum: *Phase response* (q.v.).

phase splitting: Separation of a trough (or peak) into more than one trough (or peak). Refers to the appearance of an event on successive traces. Usually a consequence of interference between two or more events whose attitudes or strengths are changing laterally.

phase velocity: **1.** The velocity with which any given phase (such as a trough or a wave of single frequency) travels; may differ from group velocity because of *dispersion*. Sometimes called "trough" velocity or "peak" velocity. See Figure D-13. **2.** Sometimes, *apparent velocity* (q.v.).

phasing: A change in waveshape as a result of interference.

phasor diagram: A representation of a vector which rotates at some particular angular velocity. Characteristics can be interpreted by viewing the "frozen" position and spatial relationship of vectors. Master curves for the electromagnetic method are often presented as phasor diagrams.

phi units: A scale of particle size; $\phi = -\log_2 S$, where S is diameter in mm. See Figure W-10.

phone: A *geophone* (q.v.).

photoclinometer: A well-logging device which photographically records the angle and azimuth of borehole deviation from the vertical. Compare *poteclinometer*.

photoelectric absorption log: Measurement of induced gamma (gamma-gamma) radiation in two energy windows allows discrimination of the radiation resulting from Compton scattering (above 0.6 MeV) from that of photoelectric absorption (below 0.6 MeV). The photoelectric effect is strongly dependent on atomic number and hence lithology. Recorded with the compensated density log (CNL) to make a litho-density log (LDT); the P_e **index curve** (photoelectric absorption cross-section in barns/electron) is used to indicate lithology to aid in density-porosity determination.

photoelectric effect: The liberation of electrons because of the absorption of electromagnetic radiation (such as, but not restricted to, visible light) by a substance. Several phenomena may be involved.

photon log: A well log of scattered gamma rays, differing from a density log in that the sonde is not pressed against the borehole wall and hence the log is sensitive to changes in hole diameter and the density of the fluid in the borehole.

physically realizable: Satisfying two conditions: (a) not existing (having values of zero) before some initial time, and (b) containing finite energy (hence dying out toward infinity).

physical modeling: Subjecting an actual model to certain processes. In contrast to conceptual modeling where processes are imagined in a thought sequence, or computer or numerical modeling where processes are simulated by mathematical algorithms. **Analog modeling** or **scale modeling.**

P_i: Irregular-type *micropulsations* (q.v.).

pick: 1. To select an event on a seismic record, as to "pick" reflection events. **2.** An event or time on an event which has been selected. The arrival of an event signifying new energy should cause an increase in amplitude and should affect different channels in a systematic coherent way. Various statistical tests are used to make picking decisions, ranging from simple summing along possible coherent patterns to schemes

like semblance criteria. Some criteria search only for phase coherence, others look at the amplitude buildup, integrate over several half-cycles, equalize spectral variations, etc. Grading is intimately related to picking.

pickup: 1. *Geophone* (q.v.). **2.** Reception of a disturbance such as inductive or other input from an electric power line (high-line).

pico-: A prefix meaning 10^{-12}.

pi diagram: *Pole diagram* (q.v.).

Pie Slice: A fan-filter or *velocity-filter* (q.v.) process aimed at emphasizing a band of moveouts independent of frequency. Compare *butterfly filter.* Texas Instruments tradename.

piezoelectric: The property of a dielectric which generates a voltage across it in response to a stress, and vice versa. In a hydrophone the stress is produced by the pressure, and in an accelerometer the stress is produced by the inertia of the reaction mass. Piezoelectric transducers are commonly made of barium titanate or zirconate. Also called **electrostrictive.**

piezomagnetic: See *magnetostriction.*

piezoremanent magnetism: PRM. See *remanent magnetism.*

pigtail: The wire which connects a geophone to the seismic cable.

pigtail chart: See *dipmeter.*

pilot: An estimate for use as a basis of some analysis or process.

piloting: Determining location with respect to known geographical points. See *positioning.*

pilot trace: The seismic trace toward which other traces are adjusted. Used in time shifting for static corrections or in cross-equalization processes. The pilot trace may be composited from the traces being adjusted.

pinch out: The termination of a bed which thins gradually.

pinger: 1. A transponder or device which emits an acous-

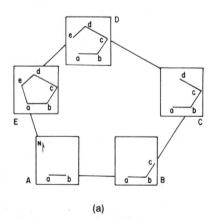

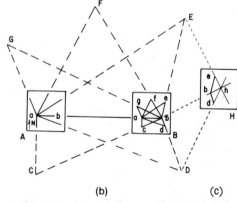

(a)   (b)   (c)

FIG. P-2. **Plane-table** methods. **(a) Plane-table traversing.** The plane table is set at A and oriented north; the rod at B is sighted with alidade and line ab is drawn along the alidade edge, the line length depending on the stadia reading. The table is set up at B, oriented by back-sighting on A, then the alidade is sighted on C and bc is drawn, and so on until the loop is closed at E by sighting on A. **(b) Plane-table intersection method.** The plane table is set at A and oriented; the rod at B is sighted with the alidade and line ab is drawn along the alidade edge with a length depending on the stadia reading. Points C, D, E, F, and G are sighted on and their directions indicated. The same points are then sighted on when the plane table is at B. **(c) Plane-table resection.** The plane table is set at an unknown location H and the alidade is sighted on known points e, b, and d. Point h is determined by the line intersections.

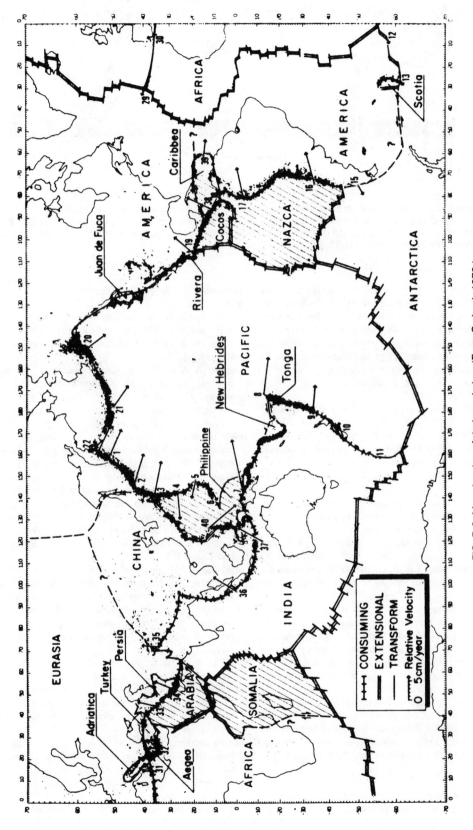

FIG. P-3. Map showing **plate** boundaries. (From Garland, 1979.)

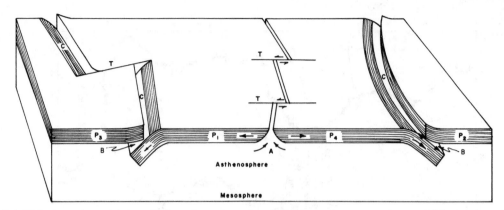

FIG. P-4. **Plate-tectonic** model. A = rift zone where plates P_1 and P_4 are moving apart, such as mid-Atlantic ridge where new crust is being created. C = ocean trench resulting from overriding of plate P_3 over P_1 (or P_2 over P_4); B = Benioff zone of earthquakes dipping along contact of plates P_1 and P_3 which are colliding. T = transform faults where plates are sliding by each other. (After Isaacs, 1968.)

tic signal upon being activated by sensing a coded acoustic signal. Pingers placed on the sea bottom or in anchored buoys can be interrogated by a ship transmitting a coded acoustic (sonar) signal and the distance to the pinger determined by traveltime measurements. **2.** A shallow-penetration high-power transducer used in marine engineering studies in soft-bottom areas.

pipe: 1. Production casing. **2.** Drill pipe.

piston corer: See *corer*.

pitch: 1. Angle between the major axis of polarization and the horizontal. Also called **dip** or **tilt**. See *polarization ellipse*. **2.** Rotational motion of a ship or aircraft about a horizontal axis perpendicular to the ship's course. Compare *roll*, *yaw*, and *trim*.

plane surveying: Surveying in which earth curvature is ignored.

pixel: A picture element, an addressable point in a raster image. A Landsat pixel represents an area of 1.2 acres (57 m east-west, 79 m north-south).

PL/1: Programming language one, a computer language used for both commercial and scientific applications.

plane polarized: Having all oscillation within one plane for a type of oscillation with more than one degree of freedom.

plane table: A survey instrument consisting of a drawing board which can be leveled on a tripod. An object is sighted through an alidade (Figure A-6) which rests on the table, allowing one to plot the line of survey directly from the observation by drawing a line along a ruler attached to the sighting telescope. Figure P-2 shows the use of the plane table.

plane wave: Having wavefronts which are planar (with no curvature), as might orginate from a very remote source. A common assumption in seismic and electromagnetic wave analyses which is not strictly true in actual situations. A plane wave can be expressed as

$$f(lx + my + nz \pm Vt),$$

where l, m, n are the direction cosines giving the wave direction, V is the velocity of the wave, and t is time.

plane-wave simulation: See *Simplan*.

plant: 1. The manner in which a geophone is placed on or in the earth. **2.** The coupling to the ground. The nature and quality of the plant affects the overall system response. **3.** To place a geophone in its proper place on the ground.

planter: A device which pushes geophones (or hydrophones) into soft marsh, perhaps 8–10 ft deep.

plasticity: The material property that allows a body to undergo permanent deformation without appreciable volume change, elastic rebound, and without rupture.

plate tectonics: A concept which envisions the Earth's crust divided into various plates (Figure P-3) which move slowly with respect to each other, being carried along by slow convection currents in the asthenosphere. Along major rifts (such as mid-ocean ridges) the plates are separating and new crust is being created. Elsewhere plates are overriding one another (in subduction zones) or sliding by one another (as along the San Andreas fault). See also *transform faults*, *Benioff zone* and Figure P-4.

playback: 1. To produce a new form of record from magnetic tapes (or other reproducible recording). Seismic playback may include filtering, gain adjustment, time shifting, mixing, stacking, migrating, etc. A possible playback program is shown in Figure P-9. **2.** The result of such processing, as opposed to the original recording.

plot: 1. A graph or plotted section. **2.** To draw lines representing events on a cross-section or map.

plot point: The location where a datum value is plotted. For symmetrical electrode arrays, the midpoint of the array; with asymmetric arrays, the convention may vary.

plotted section: Section on which seismic events are indicated by lines or sequences of points. The horizontal scale is usually distance along the seismic line and the vertical scale is usually either depth or reflection time. Data may or may not be migrated. Often called simply **seismic cross-section** or **cross-section**.

plotter: 1. A device for making a graphic display, frequently (but not necessarily) photographic. **2.** A device for graphing data, as an *X-Y* plotter. **3.** A person or device for drawing graphs, maps, or sections.

plough: A device for burying detonating cord (for use as a seismic source).

plugged and abandoned: P&A; see *dry hole*.

plumbing: Determining the point vertically over a survey point by dropping a weighted string (**plumb line**) to it.

plunge: The direction of the axis of a fold with a downward component. 2. To set the horizontal crosswire of a theodolite in the direction of a grade.

plus-minus method: A refraction interpretation method using reversed refraction profiles, also called **Hagedoorn method.** Let t_{AB} be the surface-to-surface time between A and B, and let t_A and t_B be arrival times at various intermediate locations from shots A and B, respectively. **"Minus"** values $t_A - t_B - t_{AB}$ are calculated for each location and plotted to give the velocity of the refractor. **"Plus"** values $t_A + t_B - t_{AB}$ are calculated for each location and plotted to give a picture of the refractor's depth. See Hagedoorn (1959) or Sheriff and Geldart, v. 1 (1982, p. 225).

pockmarks: Cone-shaped depressions in the seafloor, sometimes 5–10 m deep and 15–45 m in diameter, probably formed by ascending gas leakage.

point: 1. Refers to operator length, as a "56-point filter." See *convolution.* **2.** Source point.

pointing error: Systematic error in a ship's sense of direction such as might be caused by misalignment of sensors with the ship. Doppler-sonar pointing error shows as a fictitious cross-course velocity.

point mass: A mass theoretically concentrated at a point whose geophysical response is equivalent to some other mass distribution. In gravity, a uniform sphere can be treated as if its mass were concentrated at its center. Nonspherical masses at large distances can be approximated by point masses.

point sort: *Gather* (q.v.).

point source: 1. A source whose actual size is unimportant as far as the effects being observed are concerned. **2.** A single current electrode whose companion is a great distance away, such as the current pole of pole-pole or pole-dipole arrays.

poise: 100 *centipoise* (q.v.).

Poisson's equation: In space where the source density is ρ, the Laplacian of a potential U is

$$\nabla^2 U = 4\pi\rho K,$$

where ∇ is the operator del and K is a constant (gravitational constant in case of mass and gravitational potential). The constant 4π is deleted in some systems. In empty space where $\rho = 0$, this becomes Laplace's equation. Named for Simeon Denis Poisson (1781-1840), French mathematician.

Poisson's ratio: The ratio of the fractional transverse contraction to the fractional longitudinal extension when a rod is stretched; see *elastic constants* and Figure E-5.

Poisson's relation: For bodies having uniform susceptibil-

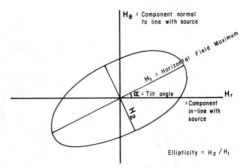

FIG. P-5. Polarization-ellipse relations.

ity and density contrast, the magnetic potential is directly proportional to the derivative of the gravity potential in the direction of magnetization.

Poisson solid: An isotropic elastic material for which the Lamé elastic constants λ and μ are equal. A Poisson solid has Poisson's ratio $\sigma = 1/4$.

polar form of complex number: Expressing a complex number $z = x + jy$ in the form $z = Ae^{j\theta}$; $A = $ **modulus** $= (x^2 + y^2)^{1/2}$ and $\theta = $ **phase** $= \tan^{-1}(y/x)$.

polarity: 1. The condition of power-supply terminals being electrically positive or negative. If opposite terminals are connected, electrons flow from the negative to the positive in the external connector. **2.** The "north" or "south" character of a magnetic pole.

polarity standard: SEG standards for seismic data specify that the onset of a compressional pulse is represented by a negative number. **Positive polarity** for a seismic waveshape relates to an increase in acoustic impedance or a positive reflection coefficient; for a zero-phase wavelet, a positive reflection coefficient is represented by a central peak, normally plotted black on a variable area or variable density display. The reverse situation is called **negative standard polarity.**

polarization: 1. Dipole moment per unit volume. In induced polarization, current dipole moment per unit volume. **2.** The polarity or potential near an electrode. **3.** A preferential direction of wave motion, as the component of S-waves whose motion is confined to a horizontal plane (SH). **4.** Magnetic orientation concerning only the vector direction and not the magnitude.

polarization ellipse: The locus of points in space described by the superposition of two fields having different directions and whose variations in time are of the form $A\cos(\omega t)$ and $B\cos(\omega t + \theta)$. In electromagnetic prospecting, neither the direction nor the phase of the primary and secondary fields are the same so the superposition of the two fields results in a polarization ellipse. For the ellipse shown in Figure P-5, the **modulus** of the magnetic wavetilt is given by $|H_1|/|H_2|$, the **tilt angle** or **pitch** by α, and the **ellipticity** by H_2/H_1. Ellipticity is positive or negative as the vector rotates clockwise or counterclockwise. See Smith and Ward (1974).

polarization filtering: 1. See *linear-phase filtering.* **2.** A technique for enhancing one mode of propagation with

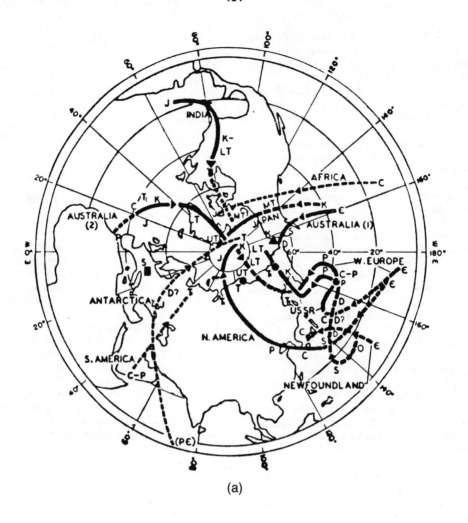

(a)

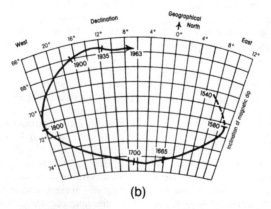

(b)

FIG. P-6. (a) **Polar wandering** curve showing location of the magnetic pole based on paleomagnetic data from different plates. The divergence of the curves backward in time indicates different plate movements for the different plates. Pc, Precambrian; c, Cambrian; O, Ordivician; S, Silurian; D, Devonian; C, Carboniferous; P, Permian; Tr, Triassic; J, Jurassic; LT, MT, UT, Lower, Middle and Upper Tertiary. (From Garland, 1979.) (b) Time variation in inclination and declination of the geomagnetic field at London. (After Parasnis, 1961.)

respect to another by combining the outputs of three-component recordings. Phase relationships for *P*-, *S*-, Rayleigh-waves, etc., are different at a free surface. See White (1964) and Sheriff and Geldart, v. 2 (1983, p. 59).

polarization resistance: The factor (RT/nFJ_0) employed in overvoltage theory, relating overvoltage η to current density *J*:

$$\eta = -(RT/nFJ_0)J.$$

R is the gas constant, *T* the absolute temperature, *n* the number of molar equivalents, *F* the faraday, and J_0 the exchange current density, all in cgs units. The units of polarization resistance are ohm-cm^2.

polarized: A connector is said to be polarized if the connector and its mate are so designed that they fit together in only one way. This prevents getting the wires connected incorrectly.

polarized electrode: See *ideal polarized electrode*.

polar projection: See *stereographic projection*.

polar wandering: Rocks become magnetized according to the direction of the Earth's field at the time of their formation, As the Earth's plates drift with time, the direction of the *remanent magnetism* (q.v.) changes. Changes in the apparent location of the magnetic pole with time is called ''polar wandering''. The polar wandering curves for different plates differ; see Figure P-6. There appears to be some movement of the magnetic pole with time in addition to the polar wandering because of plate movement, producing **secular changes.**

pole: 1. A singular point, where the value of a function becomes infinite. If a function has the factor $1/(x - a)^m$ and $m = 1$, then *a* is a **simple pole**. Otherwise, *a* is a pole of **order** *m*. **2.** A *magnetic pole* (q.v.). **3.** One electrode of a pair whose companion electrode (**infinite electrode**) is so far away that its location does not affect the measurements.

pole diagram: A *stereographic projection* (q.v.) on which the direction of lines is plotted. Also called a π **diagram.** A plane is represented by the direction of the line perpendicular to it. Compare *cyclographic diagram*.

pole-dipole array: See *array (electrical)* and Figure A-12.

pole-pole array: See *array (electrical)* and Figure A-12.

pole of spreading: Plate movement associated with sea floor spreading can be described as rotation of one plate with respect to the other about an axis through the Earth's center, the axis intersecting the Earth's surface at the pole of spreading.

pole strength: See *magnetic pole*.

polling: Calling stations to permit them to transmit information.

polyconic projection: See *map projection*.

poop shot: *Weathering shot* (q.v.)

population: The aggregate of a set of observations whose subaggregates have the same statistical properties. Where subaggregates have different statistical properties, they may be said to be ''of different populations.'' See *statistical measures*.

pore-pressure gradient: Interstitial pressure divided by the depth; usually measured in psi/ft. See *abnormally high pressure*.

porosity: Pore volume per unit gross volume. Often indicated by the symbol ø. Porosity is determined from cores, from sonic logs (see *Wyllie relationship*), from *density logs* (q.v.), from neutron logs, or from resistivity logs (see *Archie's formula*). See also *movable oil plot*. **Primary porosity** refers to the porosity remaining after the sediments have been compacted but without considering changes because of subsequent chemical action or flow of water through the sediments. **Secondary porosity** is additional porosity created by subsequent chemical changes, especially fissures, fractures, solution vugs, and porosity created by dolomitization. **Effective porosity** is the porosity available to free fluids, excluding unconnected porosity and space occupied by bound water and disseminated shale.

porosity overlay: A plot of porosity values calculated from different logs, plotted on top of each other. Compare *crossplot*.

porous pot: A nonpolarizable electrode which allows free ionic flow into the earth. A copper rod in a saturated copper-sulfate solution contained in a porous pot is such an electrode. Used in making voltage measurements where negligible current flows through it.

port: Connection point for an input or output device.

positioning: Determining the location of a survey ship or aircraft, usually with respect to geodetic coordinates but sometimes with respect to reference beacons whose geodetic locations may not be known. Positioning is sometimes divided into (a) **celestial navigation**, locating oneself by observing celestial bodies, which sometimes includes satellite navigation; (b) **piloting**, determining position with respect to geographical points, including many radio-navigation methods; and (c) **dead reckoning**, positioning by the extrapolation of track and direction from a previously known point of departure, including inertial positioning, Doppler-sonar, and Doppler-radar methods. Some modern positioning systems measure traveltime or differences in traveltime from reference stations, some measure the phase in standing-wave patterns set up by pairs of transmitter stations, some measure Doppler frequency shifts, some measure the direction of strongest signal. See Figure L-5. Features of some systems are listed in Figure P-7.

positive: 1. An anomalous area in which values are larger than expected or larger than in neighboring areas, as a ''gravity positive''. **2.** An area characterized by uplift. Positive often is used in a relative sense and might refer to an area which is subsiding less rapidly than surrounding areas.

positive pole: See *magnetic pole*.

positive polarity: See *polarity*.

positive separation: See *separation*.

posting: Marking data on a map or section at the appropriate location, often as a step prior to contouring.

postplot: Computation of locations which have been previously occupied, based on the best reconciling of all available data.

pot: 1. To fire a small charge in a hole to create space for loading a large charge. **2.** A *potentiometer* (q.v.). **3.** A *porous pot* (q.v.).

poteclinometer: A device for continuously measuring the

angle and direction of borehole deviation during a log run. A pendulum moves a variable-resistor arm so that the resistance is a measure of the angle with the vertical, and a compass needle moves another arm so that another resistance is a measure of the azimuth. Often run with dipmeters. Compare *photoclinometer*.

potential: 1. Electrical voltage with respect to a reference point. **2.** The amount of work required to position a unit charge, unit pole, or unit mass at a given position, usually with respect to infinity. Electric, magnetic, and gravitational fields are scalar potential fields. The gradient of a potential field is called the **field strength**, **field intensity**, or **flux density**; see *Gauss's theorem*. **3.** A function from which a quanlity can be determined by

Type	System	Frequency	Range (mf)	Accuracy (ft)	Principle	Remarks
Absolute reference	Celestial		Unlimited	15 000	Sighting on sun or stars	
	Transit satellite	150 MHz 400 MHz	Unlimited	150	Doppler shift	U.S. Navy operation Fix every 1–4 hours with 4–5 satellites
Piloting	Sightings		Near land		Sightings	Identify landmarks, buoys
	Bathymetry				Fathometer	Identify features in water bottom
	Pingers				Sonar transponder	Used in relocating a site such as a well head or in detail surveys
Low-frequency radio	Loran A	1900 kHz	600–750	25 000	Pulse timing	
	Loran C	100 kHz	1500	100–1500	Pulse + phase timing	Government maintained; hyperbolic
	Rho-rho mode		1500	100–500	Timing	Circular; atomic clock reference
	Omega	10–14 kHz	6000	1500–10 000	Pulse phase	Government maintained; hyperbolic
Medium-frequency radio	Decca	70–130 kHz	200–300	25–500	CW phase	Hyperbolic; 4 frequencies
	Hi-Fix	1.6–2 MHz	150–200	25–300	CW phase	Circular or hyperbolic
	Lambda	100–200 kHz	150–400	50–500	CW phase	Circular or hyperbolic; 4 frequencies
	Argo	1.6–2.0 MHz	200–700 km	5–20 m	Pulse phase	Circular or hyperbolic
	Lorac	1.6–2.5 MHz	150–250	25–300	CW phase	Hyperbolic; 2 frequencies
	Raydist DM	1.6–3.3 MHz	50–200	25–200	CW phase	Circular; 4 frequencies
	DR	1.5–5 MHz	50–200	25–200	CW phase	Circular; 2 frequencies
	N	1.5–3 MHz	150–200	25–300	CW phase	Hyperbolic; 2 frequencies
	Toran	1.6–3.8 MHz	150–250	25–300	CW phase	Hyperbolic or circular; 2 frequencies
	Toran O	100 kHz	600	50–300	Phase timing	Circular; atomic clock reference

FIG. P-7. **Positioning and navigation systems**.

Type	System	Frequency	Range (mf)	Accuracy (ft)	Principle	Remarks
High-frequency radio	Shoran	210–310 MHz	70	30–100	Pulse timing	Circular; 3 frequencies
	XR shoran	210–310 MHz	150–200	50–150	Pulse timing	Uses tropospheric scattered waves
	Syledis	420–450 MHz	60–300 km	5–10 m	Pulse PRN code	Circular or hyperbolic
	RPS	9300–9500 MHz	60	10–50	Pulse timing	Circular; uses transponders
	Autotape	2900–3100 MHz	60	10–20	Pulse phase	Circular; 3 frequencies
	Hydrodist	2800–3200 MHz	30	10–100	CW phase	Circular; 4 frequencies
	Trisponder	9500 MHz	15–50	10–20	Pulse timing	Circular; uses transponders
	Tellurometer	3000 MHz	15	5	Pulse timing	Circular; uses transponders
Dead reckoning	Doppler radar	8800 MHz			Doppler shift of radar waves	Used with aircraft
	Doppler sonar	300 kHz		600/hour	Doppler shift of sonar waves	
	Inertial				Accelerometers	Accumulates error with time

FIG P-7. **Positioning and navigation systems** (cont.).

specified mathematical operations, as a potential field from which seismic displacement, velocity, etc., can be ascertained by differentiation.

potential-drop ratio: An electrical-survey method which compares ratios of voltages between two adjacent, aligned pairs of potential electrodes.

potential electrode: The contact of an IP and/or resistivity receiver circuit with the ground, usually a porous-pot electrode.

potential field: A field which obeys Laplace's equations, such as gravity, magnetic, or electrical fields.

potential functions: Mathematical relations from which other relations can be derived by simple mathematical operations (such as differentiation). For example, the gradient of magnetic or gravity potential functions might give the magnetic or gravity field, or the divergence and curl of seismic potential functions might give the displacements involved in *P*- or *S*-waves. Potentials are used because they are often easier to describe than the relations which can be derived from them.

potentiometer: An electrical instrument for measuring low-level dc voltages without drawing current from the measured circuit, by using the unknown voltage as an arm in a direct-current bridge circuit.

pot resistance: The electrical resistance from a potential electrode to ground; the effective electrical resistance of a porous-pot potential electrode and adjacent region. Too high a pot resistance reduces sensitivity and increases susceptibility to noise.

Potsdam ellipsoid: The now obsolete international geodet-ic reference ellipsoid; see Figure G-3 and *Geodetic Reference System*.

Potsdam gravity: A former *gravity standard* (q.v.), the gravity at the Pendelsaal of the Geodetic Institute in Potsdam, East Germany.

Poulter method: See *air shooting*.

powder: Explosive.

powder factor: Pounds of explosives per ton of rock broken to the required size.

powderman: An individual licensed to handle explosives.

power series: An expression of the form $y = a + bx + cx^2 + \ldots$. See *Taylor series*.

power spectrum: 1. A power-density versus frequency relationship. The power spectrum $P(\nu)$ is the square of the amplitude-frequency response or the Fourier (cosine) transform of the autocorrelation function. 2. Occasionally implies cumulative power $\mathbf{P}(\nu')$, where $P(\nu)$ is the power density at the frequency ν and:

$$\mathbf{P}(\nu') = \int_0^{\nu'} P(\nu)\, d\nu.$$

power-transfer function: The function of frequency which represents the ratio of output-power density to input-power density.

ppm: Parts per million.

Pratt hypothesis: A model of compensation for *isostasy* (q.v.). See also Figure I-6.

preamplifier: An amplifier which precedes the main amplifier. Usually located near the signal source to improve

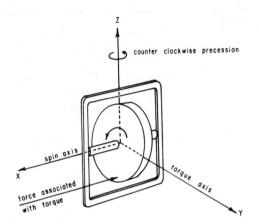

FIG. P-8. **Precession** rule: the spin axis precesses toward the torque axis.

signal-to-noise ratio. Often has a high input impedance to prevent loading and to give maximum signal transfer.

precession: The tendency of a gyroscope to turn when under the influence of a torque which tries to change the direction of its axis of spin. See Figure P-8.

precision: The repeatability of an instrument measured by the mean deviation of a set of measurements from the average value. Different from accuracy. See also Figure D-21.

precision index: See *error function*.

prediction error: The difference between a value predicted from observation of earlier measurements and the value actually observed. **Prediction-error filtering** involves use of a filter which minimizes some function of the errors. See l_p *fits* and *optimum filtering* and Sheriff and Geldart, v. 2 (1983, p. 191).

prediction lag: The time difference (often described by the number of data samples) between an input value and the prediction based on it. For example, multiples involving reverberation in a surface water layer are delayed by the two-way traveltime in the water layer, so the deconvolution operator may have a prediction lag of this much time; i.e., it does not begin functioning until after this lag time.

predictive deconvolution: Use of information from the earlier part of a seismic trace to predict and deconvolve the latter part of that trace. Some types of systematic noise, such as reverberations and multiples can be predicted. The difference between the predicted value and the actual value is called the **prediction error**; it is sensitive to new information such as primary reflections. Predictive deconvolution may also be used in a multitrace sense, where one tries to predict a trace from neighboring traces. See Sheriff and Geldart, v. 2 (1983, p. 43).

preemphasis: 1. Frequency filtering before processing (as in field recording) so as to emphasize certain frequencies compared to others. **2.** Recording so as to emphasize higher frequencies.

preliminary section: A seismic section made in an early

stage of processing before much of the processing has been carried out. As opposed to the final section.

preliminary waves: 1. The body waves of an earthquake, which arrive before the stronger surface waves. Usually includes both *P*-waves and *S*-waves. **2.** *P-wave* (q.v.).

preplot: 1. The locations planned to be occupied (usually in navigation-system coordinates) before the points are actually occupied. **2.** A list of programmed points in navigation-system coordinates. **3.** To calculate the navigation-system coordinates for programmed points.

preprocessor: A computer which operates on data prior to the main processing. Seismic preprocessing sometimes includes vertical stacking, reformatting, adding headers, editing, resampling, demultiplexing, etc.

preset gain control: *Programmed gain control* (q.v.).

pressure detector: *Hydrophone* (q.v.).

pressure remanent magnetism: PRM. See *remanent magnetism.*

pressure wave: *P-wave* (q.v.).

presuppression: *Initial suppression* (q.v.).

prewhitening: *Preemphasis* (q.v.) designed to make the spectral density more nearly constant.

prills: Pellets of ammonium nitrate used as a shothole explosive.

Primacord: *Detonating cord* (q.v). Ensign Bickford Co. tradename.

primary colors: The three colors which in combinations give any other color. The **additive primary colors** of light are red, green and blue; their sum gives white. The **subtractive colors** of magenta, yellow, and cyan apply to pigments; their sum in equal amounts gives black.

primary field: The electromagnetic field which would be generated if the source were in free space.

primary porosity: See *porosity*.

primary radar: Radar which relies on reflected energy to indicate targets, as opposed to using active transponders as targets.

primary ratio: The ratio of electromagnetic field readings at two locations in the absence of conducting material. Also called **normal ratio**.

primary reflection: Energy which has been reflected only once and hence is not a multiple. Usually includes the contribution of short-path multiples.

primary voltage: In IP surveying, the peak asymptotic charging voltage observed at a time-domain receiver.

primary wave: *P-wave* (q.v.).

prime: To prepare an explosive for firing, as to insert a cap in a stick of dynamite.

primer: An intermediate explosive which is set off by a cap and whose function is to detonate another explosive that is not cap-sensitive.

principal alias: A frequency between the Nyquist frequency and twice the Nyquist frequency.

principal alias lobe: See *directivity graph*.

principal diagonal: The matrix elements a_{ii}; that is, those elements lying along the diagonal line from a_{11} to a_{nn}, where n is the number of rows or columns (whichever is smaller) in the matrix.

principle of equivalence: Two conductive layers will carry nearly the same electrical current if their ratios of thickness to resistivity are the same. For resistive layers, it is their resistivity-thickness product which is important.

principle of least time: See *Fermat's principle*.

principle of reciprocity: The seismic concept that the same trace would result if source elements were replaced with receiver elements and vice versa. Similar concepts are involved with other methods; for example, with interchanging current and potential electrodes in electrical exploration.

principle of superposition: The concept that the result from two or more simultaneous causes can be obtained by summing the results of the individual causes. Implies linearity.

principle of suppression: Resistant layers sandwiched be-

tween conducting beds are electrically equivalent if the product of their thickness and resistivity is the same.

printed circuit: A thin laminated board on which electrical circuits are drawn and components mounted. Usually easily removed for testing and replacement.

private line: A communications channel dedicated to some exclusive use.

privileged: 1. A computer function whose use is restricted. For example, the ability to read a file of authorized passwords is a privileged function restricted to the system manager. In a computer system, privileged instructions may be executed only by the operating system when in the proper mode (e.g., supervisor state, kernal mode, etc.). **2.** A computer operation which has

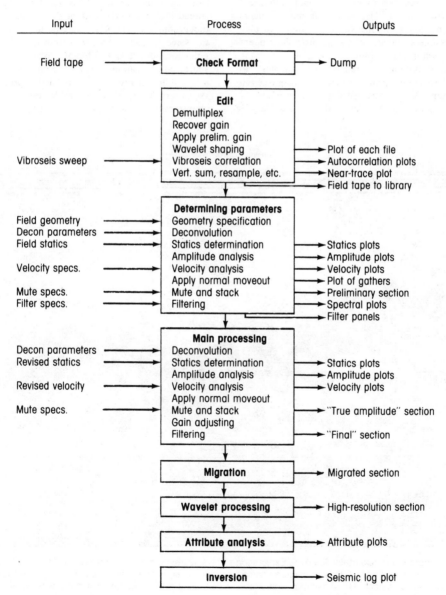

FIG. P-9. A possible **processing** flow chart for seismic data. After Sheriff (1978, p. 195).

priority over other operations. **3.** Data whose input has priority over other computer operations.

PRM: Pressure (or piezo) remanent magnetization. See *remanent magnetism*.

probabilistic: See *deterministic*.

probable error: The range within which half of a series of readings of the same quantity probably lie. For a Gaussian distribution, the probable error is 0.674 times the standard deviation. (The probable error is not an error which is more probable than others.) See *statistical measures*.

probing: *Sounding* (q.v.).

processing: Changing data, usually to improve the signal-to-noise ratio so as to facilitate interpretation. Processing operations include applying corrections for known perturbing causes, rearranging the data, filtering it according to some criteria, combining data elements, transforming, migrating, measuring attributes, display, etc. A possible sequence of seismic processing operations is shown in Figure P-9.

processor: 1. A program to translate programmer's instructions (**source program**) into machine language (**object program**). **2.** A device that does processing.

producibility-index log: A calculated well log showing

effective porosity and the percent (**q**) of the total porosity occupied by clay. Low clay content and high fluid-filled porosity suggest good permeability.

production log: A well log run inside tubing. Small-diameter sondes are used so that they can be lowered through 2-inch ID tubing. Devices include continuous flowmeter, packer flowmeter, gradiomanometer, manometer, densimeter, water-cut-meter, thermometer, radioactive tracer tools, through-tubing caliper, casing-collar locator, and fluid sampler.

profile: 1. A graph of a measured quantity against horizontal distance, as in a "gravity profile". **2.** A drawing showing a vertical section of the ground along a line. **3.** The series of measurements made from a single source point location into a recording spread. Additional shots from the same general source location into the same spread are considered part of the same profile (even though different shotholes may be used). However, if the same spread is shot into from a different source point location, it is a different profile, or if the same source point location is shot into a different spread, it is also a different profile. **4.** A **refraction profile** denotes the ensemble of individual profiles (as defined above) shot from the same shotpoint. Use of "profile" for both

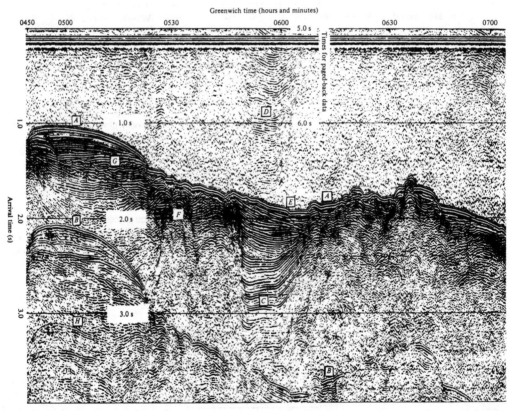

FIG. P-10. **Profiler** record. (Courtesy Teledyne Exploration.) The ship traveled 8.5 km between the 30 minute marks at the top of the record. The seafloor multiple B obscures primary reflections after its arrival. Events D result from multiples of C which have been paged-back. G indicates an unconformity truncating reflections below it and onlapped by reflections above it. F indicates diffractions, H a second-order seafloor multiple.

the component records and for the ensemble sometimes produces confusion. **Refraction set** is also used for the ensemble.

profile line: The line along which measurements are made.

profiler: A high-frequency marine seismic reflection system usually involving a low-energy source, used especially in oceanographic studies, as a reconnaissance tool in conjunction with gravity and other type surveys, in engineering studies, etc. Originally "profiler" meant use of a compact but weak source (often sparker) with only one or two hydrophone groups recorded by a single-channel plotter on electrosensitive paper. Profiler is now used for larger systems with stronger sources, multichannel streamers, and with data recorded on magnetic tape, so that the distinction versus conventional marine seismic work is often lost. See Figure P-10 and Sheriff and Geldart, v. 1 (1982, p. 190, 192-193).

profiling: A geophysical survey in which the measuring system is moved about an area (often along a line) with the objective of determining how measurements vary with location. Specifically, a resistivity, IP, or electromagnetic field method wherein a fixed electrode or antenna array is moved progressively along a traverse to create a horizontal profile of the apparent resistivity. Occasionally refers to *vertical profiling* or *sounding* (q.v.). See also *profiler*.

prograding: Deposition building successively toward the sea. A prograding pattern is shown in Figure R-8.

program: 1. The work schedule for a geophysical party; e.g., the lines to be shot on a seismic prospect. **2.** The instructions for processing data, as through a computer. **3.** The plan for gain control variations, as in programmed gain control **4.** To plan a program.

program flow chart: A display showing operations and decisions and their sequence within a program, to show how the job is done. Program flow charts are used as an aid to program development, as a guide to coding, and as documentation of a program. See Figure P-9.

programmed gain control: Predetermined gain for a seismic amplifier. The function describing amplifier gain with respect to time after the shot. Abbreviated **PGC**. See *gain control*.

programmer: One who develops the series of instructions required for a computer. He must know the operations available in the system with which he has to work and the procedure which must be translated step by step. He must consider (a) allocation of storage locations to data, instructions and related information; (b) conversion of input data; (c) availability of reference data such as tables and files; (d) requirements for accuracy and methods of checking; (e) ability to restart in case of unscheduled interruptions and error conditions; (f) automatic monitoring to ascertain that the required devices are operating; (g) housekeeping or setup procedures that preset switches and registers, type operator messages, check file labels, etc.; (h) format of output data; (i) availability of preexisting programs that may be used in this program; (j) editing of data; and (k) provision for exceptions not processable.

PROM: Programmable read-only memory.

propagation constant: In electromagnetic theory, the propagation constant k is given by

$$k^2 = \mu\epsilon\omega^2 + i\mu\sigma\omega$$

where μ is magnetic permeability, ϵ is dielectric permittivity, ω is angular frequency, and σ is electrical conductivity. Sometimes $\gamma = ik$ is called the propagation constant. Also sometimes called **wavenumber**.

propagation error: An error due to an unallowed-for change in velocity.

proper crossover: If the transmitter is west of the receiver, and if a counterclockwise rotation is recorded as a south dip angle, a proper crossover occurs where electromagnetic dip angles change from south on the south side of a point on the traverse to north on the north side. The axis of current concentration lies underneath the crossover point. Opposite of **backward crossover.**

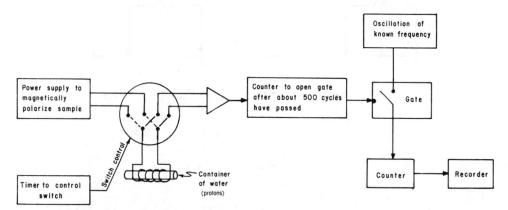

FIG. P-11. **Proton-resonance magnetometer**. Proton spin axes precess about a magnetic field. A polarizing field normal to the Earth's field is impressed for a short time to polarize the nuclei and is then removed. The nuclei, now oriented together, precess about the Earth's magnetic field at the Larmor frequency, inducing this frequency in a measuring coil. The induced frequency controls the length of time a gate is open, the time is determined by counting cycles of a standard frequency. The time duration is a measure of the Earth's field.

prospect: 1. An area characterized by a geologic or geophysical anomaly, especially one which is recommended for additional exploration. **2.** An area that is being investigated.

prospecting: Exploration of an area with the objective of locating oil, minerals, etc.

prospecting seismology: *Seismic exploration* (q.v.).

protection ratio: The ratio of transmission at a desired frequency to that at an undesired alias.

protocol: 1. A set of rules and priorities under which systems communicate with one another. **2.** A definition of the interface between systems or a system and its devices. Most frequently used in connection with telecommunication links.

proton-resonance magnetometer: The proton (hydrogen) nucleus has a magnetic moment because of its spin. The nuclei precess about the Earth's magnetic field **H** at a frequency ν_L (**Larmor frequency**):

$$\nu_L = \gamma|\mathbf{H}|/2\pi,$$

where γ = gyromagnetic ratio ($2\pi/23.4868$ Hz/gamma for protons). Precession of polarized nuclear-spins induces a voltage at the precession frequency in a measuring coil. The induced frequency is measured by a counting arrangement (Figure P-11) to determine the value of the Earth's magnetic field. For the normal Earth field of about 50 000 gammas, f_L = 2100 Hz. As the protons gradually relax into random orientation, the induced-field strength drops to zero. The drop rate depends on interatomic forces and hence on molecular structure. Compare *optically pumped magnetometer.*

proximity log: A microresistivity log similar to the microlaterolog but less sensitive to mud-cake thicknesses. Schlumberger tradename.

proximity survey: A survey to determine how far a well is from some feature. For example, shooting from the surface into a geophone in a deep well with the objec-

tive of determining the position of the flank of a salt dome. See also *ultra-long-spaced electric log.*

prune effect: When a surface has many local highs, methods of finding the maximum may find local maxima instead of the largest maximum if they begin with a bad initial guess.

pseudoanisotropy: Anisotropy other than the point property such as crystals possess. Most geophysical anisotropy concerns pseudoanisotropy. Also called **apparent** or **effective anisotropy**.

pseudogeometric factor: A coefficient used for estimating the response to a resistivity measurement R_a at different invasion depths:

$$R_a = R_{x0}J + R_t(1 - J),$$

where R_{x0} = flushed-zone resistivity, R_t = uncontaminated-zone resistivity, and J = pseudogeometric factor which is a function of invasion depth.

pseudogravity: The gravity field calculated from the magnetic-field measurements by means of *Poisson's relation* (q.v.).

pseudo-Rayleigh wave: See *Rayleigh wave.*

pseudosection: A plot of an electrical measurement or calculation, usually of apparent resistivity, as a function of position and electrode separation (which in turn is related to the *depth of investigation*, q.v.). Also called a **quasi-section**. For the dipole-dipole electrode configuration, the data are plotted (Figure P-12) beneath the point midway between the dipoles at a depth of half the distance between the dipole centers. For the Schlumberger array, the data are plotted beneath the array center at a depth of half the separation of the current electrodes; i.e., *AB*/2. Pseudosections are also used to portray induced-polarization data and are plotted in terms of percent frequency response, metal factor, or phase. Pseudosections may also be used to portray magnetotelluric data where apparent resistivity or some

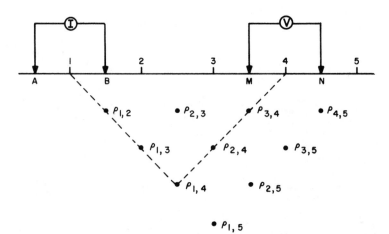

FIG. P-12. **Pseudosection**. The apparent resistivity $\rho_{1,4}$ (when the current electrodes A and B are centered at 1 and the potential electrodes M and N are centered at 4) is plotted 45 degrees below 1 and 4. Data for other pairs of measurements are plotted similarly. Apparent resistivities are then contoured to suggest variations with depth (but the pseudosection cannot be directly interpreted as showing resistivity variations with depth).

other parameter is plotted as a function of increasing period (decreasing frequency). A pseudosection conveys the variation of some parameter with location and penetration depth, but it can only be converted into a two-dimensional distribution by modeling.

pseudostatic SP: The "static SP" of a shaly sand. Strictly, static SP refers to a clean sand. See **SSP**. Abbreviated **PSP**. The ratio PSP/SSP is the **SP reduction factor** α.

pseudo-velocity log: *Seismic log* (q.v.).

P-shooter: A mechanical energy source which propels a 500 lb weight vertically toward a base plate. Employs a strong spring.

psi: Pounds per square inch, a unit of pressure. See Appendix J.

p-slowness graph: A graph of the *raypath parameter p* (q.v.) against the reciprocal of velocity (**slowness**).

PSP: *Pseudostatic SP* (q.v.).

p-tau mapping: See *tau-p mapping*.

p-type semiconductor: A doped semiconductor with more *holes* (q.v.) than electrons available for carrying charge; an **acceptor**.

pu: Porosity unit or 1 percent porosity.

pull-apart zone: A region subjected to extension, which often is accommodated by normal faulting. Implies a rift zone.

pull boat: A boat or raft on pontoons which moves by reeling in a line previously anchored ahead, using a power winch. Used in travel through swamps containing heavy timber.

pull-up: An apparent uplift produced by a local, shallower high-velocity region. Opposite of **push-down** which results from a local shallower low-velocity region.

pulse: A waveform whose duration is short compared to the time scale of interest and whose initial and final values are the same (usually zero). A seismic disturbance which travels like a wave but does not have the cyclic characteristics of a wave train.

pulse curve: *Beta curve* (q.v.).

Pulse-8: A commercial radio-positioning system operating around 100 kHz almost identical to Loran-C. Decca Survey tradename.

pulsed neutron-capture log: A neutron generator or accelerator produces short bursts of high-energy (about 14 MeV) fast neutrons which are slowed to thermal energy level (about 0.025 eV) where capture by nuclei (chlorine having the greatest capture cross-section of common elements) results in gamma-ray emissions. Two time-lapse measurements (minus a "background" count) of capture-gamma intensity from each neutron burst define the time-rate of thermal-neutron decrease (**decay lifetime, die-away**). Logs may depict count-rate curves and a curve (**sigma**) of macroscopic capture cross-section at a fixed time, or a curve (**tau**) of time to reach a "background" count rate. Such logs are used in cased oil wells where rock characteristics are known from previous logging and where mud filtration effects have dissipated, to determine oil saturation or changes in fluid saturation during oil production (as shown by successive logs). Dresser-Atlas tradename. See *neutron lifetime log*.

pulse electromagnetic method: *Transient electromagnetic method* (q.v.).

pulse method: Observation of the voltage decay after cessation of a transmitted current pulse. Also called the **time-domain IP method**, the **pulse-potential method**, and **dc-pulse method**.

pulse-position modulation: Similar to frequency modulation but uses pulses instead of full wave. See Figure M-10.

pulse shaping: To change the shape of a pulse into a more desired shape. Used to make square waves, to shape the time-break pulse so as to make the time break more definite, or to sharpen-up the effective onset of energy. See also *wavelet processing*.

pulse stabilization: Processing to ensure the same effective wavelet shape.

pulse stretching: The changing of a waveform because of applying different amounts of normal moveout to different parts of it.

pulse test: A seismic recording system test involving the application of a very short pulse to the system input.

pulse transient method: An *induced-polarization* (q.v.) method.

pulse-width modulation: A type of information encoding using a square carrier wave, the width of whose pulses are proportional to the amplitude of the modulating wave. See Figure M-10. Also called **ratio modulation**.

punch card: A card used for communication with computers. A standard card provides 80 vertical columns with 12 punching positions in each column but other types are also used. One or more punches in a single column represents a character.

punched paper tape: A medium for recording data in computer-readable form by means of punched holes precisely arranged along the length of a paper tape.

push-down: 1. See *pull-up*. **2.** To enter an object onto a last-in first-out stack.

push-pull wave: *P-wave* (q.v.).

PVC: Polyvinylchloride plastic, used for shothole casing, etc.

P-wave: An elastic body wave in which particle motion is in the direction of propagation. The type of seismic wave assumed in conventional seismic exploration. Also called **primary wave** (undae primae), **compressional wave**, **longitudinal wave**, **pressure wave**, **dilatational wave**, **rarefaction wave**, and **irrotational wave**. In an isotropic homogeneous solid, the *P*-wave velocity V_p can be expressed in terms of the elastic constants and the density (ρ):

$$V_p = [(\lambda + 2\mu)/\rho]^{1/2}$$
$$= [E(1 - \rho)/\rho(1 - 2\rho)(1 + \rho)]^{1/2},$$

where λ and μ are Lamé's constants, E is Young's modulus, and ρ is Poisson's ratio. See also *wave notation* and Sheriff and Geldart, v. 1 (1982, p. 44).

P-wave delay: The variation in *P*-wave traveltime from a reference traveltime value. Plotted to determine anomalously high or low velocity regions.

Q

Q: 1. The ratio of 2π times the peak energy to the energy dissipated in a cycle; the ratio of 2π times the power stored to the power dissipated. The Q of rocks is of the order of 50 to 300. Q is related to other measures of absorption:

$$1/Q = \alpha V/\pi\nu = \alpha\lambda/\pi = \delta/\pi = 2\,\Delta\nu/\nu_r.$$

where V, ν, and λ are, respectively, velocity, frequency, and wavelength. [See Sheriff and Geldart, v. 1 (1982, p. 55-56)]. The **absorption coefficient** α is the term for the exponential decrease of amplitude with distance because of absorption; the amplitude of plane harmonic waves is often written as $Ae^{-\alpha x}\sin 2\pi\nu(t - x/V)$, where x is the distance traveled. The **logarithmic decrement** δ is the natural log of the ratio of the amplitudes of two successive cycles. The last expression relates Q to the sharpness of a resonance condition; ν_r is the resonance frequency and $\Delta\nu$ is the change in frequency which reduces the amplitude by $1/\sqrt{2}$. **2.** The ratio of the reactance of a circuit to the resistance. **3.** A term to describe the sharpness of a filter; the ratio of the midpoint frequency to the band-pass width (often at 3 dB). **4.** A designation for *Love waves* (q.v.). **5.** Symbol for the *Koenigsberger ratio* (q.v.). **6.** See *Q-type section*.

q: The fraction of total porosity occupied by dispersed shale. See *producibility-index log*.

Q-band: Radar frequencies between 36 and 46 GHz; see Figure E-8.

QC: Quality control.

Q-factor: *Koenigsberger ratio* (q.v.).

Q-type section: A *three-layer resistivity model* in which the resistivities of the three layers decrease with depth.

quad: 10^{15} Btu, approximately equal to 10^9 ft^3 natural gas or approximately equal to 300×10^9 kW hr.

quadrant: Surveying angles are often measured in degrees clockwise from north; the northeast quadrant is called the first quadrant, the southeastern quadrant the second, the southwestern quadrant the third, and the northwestern the fourth. This is a different notation from that usually employed in mathematics. See Figure Q-1.

quadratic equation: The quadratic equation,

$$ax^2 + bx + c = 0,$$

has the solution

$$x = (1/2a)[-b \pm (b^2 - 4ac)^{1/2}].$$

quadratic spline: An interpolating operator whose slope ϕ at $(X_1 + \Delta X)$ is a linear combination of the slope at nearby points X_1 and X_2:

$$\phi = \phi_1 + (\phi_2 - \phi_1)\,\Delta X/(X_2 - X_1).$$

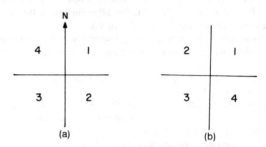

FIG. Q-1. **Quadrant**-numbering conventions. **(a)** As used in surveying; **(b)** as used in mathematics.

quadrature: 90 degrees out of phase. The **quadrature component** of a signal is the out-of-phase component; the part of an induced signal which is out of phase with the generating signal. See also *complex-trace analysis*.

quadrature filtering: *Polarization filtering* (q.v.).

quadrature spectrum: See *cross-spectrum*.

quadrature trace: See *complex trace analysis*.

quadric: A second-degree expression.

quarterboat: A boat or barge used by a geophysical crew as living quarters and/or base of operations.

quarterline: A line parallel to the boundary of a section of land (640 acres or one square mile) which bisects the section. The cross of quarterlines thus divides the section into **quarter sections** of 160 acres.

quasi-polynomials: Whereas polynomials are one-sided (i.e., involve increasing powers), quasi-polynomials are two-sided (i.e., involve both positive and negative powers). Thus $(a_0 + a_1 x + a_2 x^2 + \cdots)$ is a polynomial and $(\cdots + a_{-1}x^{-1} + a_0 + a_1 x + \cdots)$ is a quasi-polynomial.

quasi-section: *Pseudosection* (q.v.).

quasi-static: Varying very slowly, so that a solution which assumes constant values is an adequate approximation. In quasi-static electromagnetic problems, displacement currents are neglected. The quasi-static electromagnetic solution becomes poor where the loss tangent becomes less than 1, such as very resistive situations (on glaciers or deserts in the absence of groundwater) where displacement currents dominate over conduction currents.

quasi-transient method: See *transient electromagnetic method*.

quaternary gain: A gain control system in which amplification is changed only in discrete steps by factors of 4.

Compared with binary gain, fewer gain jumps are required.

quefrency domain: Fourier transform of the log of a function in the frequency domain. See *cepstrum*. A permutation of the letters in "frequency."

quenching: Blanking of sound reception in water caused by air bubbles; arises when a ship is undergoing excessive pitch and roll in bad weather.

Querwellen wave: *Love wave* (q.v.); also called *Q*-**wave**.

queue: A backlog of jobs awaiting action.

queuing: A system for handling random arrivals with minimum interference and delay. Rules for selection of items in the queue involve priorities and the demands of the arrivals on the capabilities of the system.

quiet: See *magnetically quiet*.

Q-**wave:** *Love wave* (q.v.).

R

R: Earthquake designation of a *Rayleigh wave* (q.v.). See also *wave notation*.

RA: Radiometric assay log.

radar: A system in which short electromagnetic waves are transmitted and the energy scattered back by reflecting objects is detected. Acronym for "radio detection and ranging." Ships use radar to help "see" other ships, buoys, shorelines, etc. Beacons sometimes provide distinctive targets. Radar is used in aircraft navigation (see *Doppler-radar*), in positioning, and in remote sensing. The radar spectrum is sometimes subdivided: **P-band**, 225-390 MHz; **L-band**, 390-1550 MHz; **S-band**, 1550 to 5200 MHz; **X-band**, 5200-11000 MHz; and **K-band**, 11-36 GHz; **Q-band**, 36-46 GHz; **V-band**, 46-56 GHz. See Figure E-8.

radar imagery: Mapping from an aircraft using short electromagnetic waves. A narrow radar beam which is transmitted perpendicularly to the aircraft flight line reflects from the ground to a receiver on the aircraft. The data display gives the appearance of an aerial photo.

radial array: See *azimuthal survey* and Figure A-13.

radial processing: RAMS; multichannel processing to remove seafloor multiples; the operator designed on a trace with offset x_1 is used to deconvolve the first multiple on the trace with offset $2x_1$, the second multiple on the trace with offset $3x_1$, etc. These all involve the same angle of incidence (see Figure R-1) and hence the same reflectivity.

radial refraction: 1. A pattern of shooting comparable with *fan shooting* (q.v.). **2.** Use of a detector deep in a borehole to receive and record seismic waves from sources located near the ground surface at different distances and azimuths. Used in determining salt-dome boundaries. Travel paths which are partly in salt show a lead which depends on the amount of salt travel. See Figure A-11.

radial survey: See *azimuthal survey*.

radian: A unit of angular measure such that the subtended arc equals the radius. Abbreviated **rad**. One radian = $180/\pi$ = 57.2958 degrees; one degree = 0.017453 radian.

radioactive-tracer log: A log involving the detection of radioactive materials dissolved in water or oil to determine the movement of fluids. A quantity of radioactive material (a **slug**) might be injected into the fluid and movement of the slug monitored to detect casing leaks or points of fluid entry or exit; or points where fluid enters formations might be shown by residual radioactivity at those levels.

radioactivity log: A well log of natural or induced radiation. Usually refers to a *gamma-ray log* (q.v.), but sometimes also to a *density log, neutron log, neutron-lifetime log* (q.v.), or other types of logs.

radioactivity survey: Measurements of variations in natural gamma radiation with the objective of mapping the

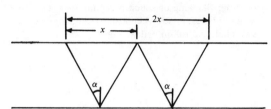

FIG. R-1. For horizontal reflectors, primary and its surface multiples involve the same reflection angle on traces selected such that the offset is proportional to the order of the multiple. This forms the basis of **radial processing**.

distribution of radioactive elements (usually K, U, and Th). Observations are also affected by radio nuclides, nuclear fallout, radon in the air, and cosmic radiation. See *gamma-ray surveying*.

radio-altimeter: An aircraft instrument in which the round-trip traveltime of radio waves which are beamed at the earth and reflected is measured to determine the height above ground.

radio Earth: A model of the Earth whose radius is 4/3 that of the Earth. The index of refraction for radio waves in the atmosphere decreases with altitude in a nearly linear manner, resulting in bending of radio waves toward the Earth. By assuming that the Earth's radius is 4/3 its actual radius, this refraction is roughly compensated and radio raypaths can be drawn as straight lines.

radio frequency: RF. A frequency above 3 kHz. Radio frequencies are subdivided into bands; see Figure E-8.

radiometer: A device which measures radiation, such as the infrared radiation used in thermal imaging.

radiometric survey: A survey of the amount of electromagnetic radiation emitted. Usually involves gamma ray spectrometry. See *gamma-ray surveys*. Compare *radioactivity survey*.

radio positioning: Determining position by electromagnetic (radio) wave measurement involving transmitters at fixed locations. Measurements are sometimes made of traveltimes, sometimes of the difference in arrival times of two radio signals, sometimes of the phase or phase difference in a standing-wave pattern resulting from the interference of two CW broadcasts, occasionally of a Doppler frequency shift. See Figure P-7.

radius of regionality: See *isostasy*.

radix: Number system base; the number whose power indicates the significance of different digit locations used to express a number. For example, 2 in the binary system, 10 in decimal.

radon method: Exploration for uranium by mapping radon.

raised-kernel function: The *kernel function* (q.v.) plus 1/2. Used in electrical exploration. See Koefoed (1968).

RAM: *Random access memory* (q.v.).

ramp: 1. To change in some continuous manner from one set of parameters to another as opposed to an abrupt step. Usually implies in a linear manner. E.g., the change between filter cutoffs for the early portion of a seismic record and those for the later portion may be "ramped" or distributed uniformly over the intervening transition portion. The amplitude of a Vibroseis sweep may increase gradually from zero to the amplitude which is held during most of the sweep. **2.** A method of terminating a data window. See Figure W-11. **3.** Motion between horizontal and vertical used to measure cross-coupling in testing shipboard gravimeters. **4.** The running integral of a unit step; a **triangular** function.

random: A relationship between two or more quantities where knowledge of one quantity does not help determine the other. A random process is called **stochastic**.

random access: Equal facility of access to any of the locations in a computer's memory.

random-access memory: A computer memory structured so that the time required to access any data item stored in the memory is independent of location. Abbreviated **RAM**.

random noise: Energy which does not exhibit correlation between distinct receiving channels; energy which is random with respect to the source parameters. By adding together n elements, random noise can be attenuated with respect to coherent signal by the factor $n^{1/2}$. Random seismic noise is attenuated by the use of many geophones per group or by stacking traces. In contrast to **coherent noise** such as unwanted modes of wave travel like ground roll, air wave, etc.

range: 1. Source-to-detector distance in refraction work; **offset. 2.** The distance to a positioning station, especially when measured directly. **3.** The extreme distance at which useful signals can be detected. **4.** The row of townships between successive meridian lines six miles apart. See Figure T-9. **5.** One of two measurements necessary to determine a location; see *drift*.

range line: The north-south boundary between townships.

range pole: A long pole which is sighted on when making long shots with a transit. Range poles are temporary survey markers and often have some distinguishing arrangement (perhaps survey flagging) on top of them to aid in locating and identifying them.

range-range determination: *Rho-rho determination* (q.v.).

ranging: Making a distance measurement as by sighting on a survey rod or with an electronic instrument. See also *positioning*.

rank: See *matrix*.

RAP: Reflection amplitude preservation; processing and display designed to preserve relative amplitudes of seismic reflections. Western Geophysical tradename.

rarefaction: Separation of molecules temporarily as a result of passage of a *P-wave*. (q.v.).

rarefaction wave: *P-wave* (q.v.).

raster: A method of representing a picture by a two-dimensional array. A pattern of scanning an area like the sweep of a beam of a television tube.

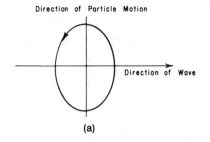

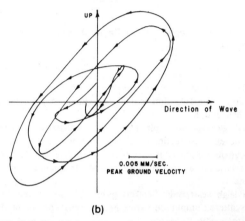

FIG. R-2. **Rayleigh-wave** motion. (**a**) Theoretical Rayleigh wave has retrograde elliptical motion at the surface. (**b**) Hodograph of particle velocity involved in ground roll. (From Howell, 1959, p. 80.)

rate: One of two radio-positioning range measurements necessary to determine a position. The other measurement is sometimes called **drift**. The distinction between rate and drift is often arbitrary.

ratiometer: An instrument for determining the ratio of two quantities. Ratiometers or compensators are used extensively in electromagnetic exploration equipment for measuring the ratio of two phasor voltages or currents. Bridge circuits are often used in ratiometers.

ratio modulation: *Pulse-width modulation* (q.v.).

rational number: A number which can be expressed as a quotient of integers.

raw gravity: Gravity measurements before applying latitude, terrain, and elevation corrections.

ray: See *raypath*.

Raydist: Medium-range radio positioning systems in which three or more transmitters emit continuous waves which differ by an audiofrequency. The position of a mobile station is determined by measuring the phase difference of the audiofrequency pattern which results from the interference of the waves emanating from transmitter pairs. See Figure P-7. Hastings-Raydist tradename.

Rayleigh distribution: A two-dimensional probability distribution which has the same variance in both direc-

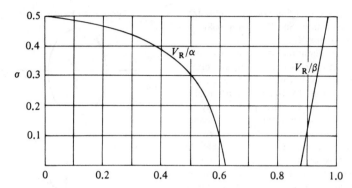

FIG. R-3. The dependance of **Rayleigh-wave** velocity V_R on Poisson's ratio σ. The P-wave velocity is α and the S-wave velocity β.

tions. The two-dimensional equivalent of a Gaussian distribution.

Rayleigh resolution limit: The minimum distance between successive reflections such that their individual entities can be recognized is $\lambda/4$ where λ = wavelength. Usually λ is taken as that of the dominant frequency component.

Rayleigh scattering: Scattering of radiation where the scattered amplitude varies as the fourth power of the frequency.

Rayleigh wave: 1. A type of seismic wave propagated along the free surface of a semiinfinite medium. Particle motion near the surface is elliptical and retrograde (i.e., the particle moves opposite to the direction of propagation at the top of its elliptical path) in the vertical plane containing the direction of propagation; see Figure R-2. Its amplitude decreases exponentially with depth, and the elastic properties to a depth of about one wavelength determine its velocity. For a Poisson ratio of σ = 1/4, the Rayleigh-wave velocity is 0.9194 times the S-wave velocity; see Figure R-3. A Rayleigh wave along a free surface can be thought of as a special case of a Stoneley wave (wave travel along an interface). Symbolized R-wave or L_R-wave. See Sheriff and Geldart, v. 1 (1982, p. 48-49). **2.** A similar type of wave where the medium is not semiinfinite; ground roll, such as encountered in seismic exploration, at times involves modes other than pure Rayleigh wave and may be called a **pseudo-Rayleigh** wave, although it is usually called simply a Rayleigh wave. Because the elastic constants change with depth in the real earth, long wavelengths depend on the elastic properties of different materials than short wavelengths and hence different wavelengths travel at different velocities. This dispersion can be used to calculate the thickness of surface layers. **3.** A surface wave in a borehole is sometimes called a Rayleigh wave; see *tube wave*. Named for John William Strutt, Lord Rayleigh (1842-1919), English physicist.

Rayleigh-Willis relation: The oscillation period T for the bubble effect varies as the cube root of the energy Q and inversely as the 5/6th power of the pressure. The relation is

$$T = 0.0450 \, Q^{1/3}/(D + 33)^{5/6},$$

where Q is the energy in joules and D is the depth in feet. See Figure R-4.

ray parameter: *Raypath parameter* (q.v.).

raypath: A line everywhere perpendicular to wavefronts (in isotropic media). A raypath is characterized by its direction at the surface, often expressed as stepout: θ = sin (Vp) (where θ = angle with vertical, V = instantaneous velocity and p = *raypath parameter* (q.v.). While seismic energy does not travel only along raypaths (i.e., seismic energy would reach points by diffraction even if the raypath were blocked), raypaths constitute a useful method of determining arrival time by ray tracing. Raypaths are usually shown on wavefront charts; see Figure W-3.

raypath curvature: Raypaths curve because of velocity changes, according to *Snell's law* (q.v.).

raypath parameter: 1. The quantity $p = dt/dx = (1/V)$ sin i, where dt/dx is the reciprocal of apparent velocity, V is instantaneous velocity, and i is the angle a raypath makes with the vertical. For horizontal velocity layering, i = angle of incidence on a horizontal reflector and p = constant specifies a raypath. See *tau-p mapping*. **2.** For a spherical velocity layering model (as used when considering long earthquake paths in the Earth), the raypath parameter is $p = (r/V)$ sin i where r is the Earth's radius.

ray tracing: Determining the arrival time at detector locations by following raypaths which obey Snell's law through a model for which the velocity distribution is known.

RBV: *Return-beam vidicon* (q.v.).

RDAU, RDU: *Remote data* (acquisition) *unit* (q.v.).

reactance: The opposition to alternating current flow offered by inductance or capacitance; the quadrature component of impedance.

read: To input data into a computer.

readability: The least discernible change in a readout device which can be readily estimated. Compare *sensitivity* and *resolution*.

read-after-write: Monitoring the recording of data by

reading a tape with a second magnetic head immediately after recording.

real: 1. The component of a complex number (vector) in the direction of the real axis, as opposed to the imaginary component. **2.** A component which is in-phase, as a component of an electromagnetic field which is in-phase with some reference signal (such as the input voltage). Compare quadrature component; see *quadrature*.

realizable: See *physically realizable*.

realizable filter: An electronic or digital filter which works in real time.

real memory: The main memory of a computer. Compare *virtual memory*.

real time: 1. Processing data at the time of detecting and recording them. **2.** Having the same time scale as actual time. **3.** Processing of data at the same rate as that at which they were recorded.

receiver: 1. A *geophone* (q.v.). **2.** As used in IP surveys, the part of an acquisition system which senses the information signal, often a sensitive, filterable ac or dc voltmeter with SP buckout controls. Generally a frequency-domain receiver is ac coupled and a time-domain voltmeter is dc coupled.

reciprocal method: A refraction method such as the *generalized reciprocal method* (q.v.).

reciprocal sonde: A sonde with the current and measuring electrodes interchanged. See *lateral*.

reciprocal time: The traveltime between common points

on reversed refraction profiles. Surface-to-surface time from a source point at A to a geophone at B must equal that from a sourcepoint at B to a geophone at A.

reciprocity principle: 1. The potential at a point M with respect to a current source at A is the same as if the points of measurement and source were reversed. Applied in electrical exploration. **2.** The seismic trace from a source at A to a geophone at B is the same as from a source at B to a geophone at A if sources and receivers are similarly coupled to the earth.

reconnaissance: 1. A general examination of a region to determine its main features, usually preliminary to a more detailed survey. **2.** A survey whose objective is (a) to ascertain regional geological structures, (b) to determine whether economically prospective features exist, or (c) to locate prospective features. As opposed to **detail** surveys which have the objective of mapping individual structures.

reconstitute: 1. To convert sampled data to a finer sample interval, e.g., to input 4 ms samples and output 2 ms samples. Opposite of **subsample** or **decimate**. **2.** To convert sampled data to continuous data, as with a digital-to-analog (D/A) converter.

record: 1. A recording of the energy picked up by a detector. **2.** A recording of the seismic data from one shot (or other type of energy release) picked up by a spread of geophones. See Figure R-5. **3.** A group of data handled by a computer as a single block of data. A number of records compose a file. **4.** To make a record.

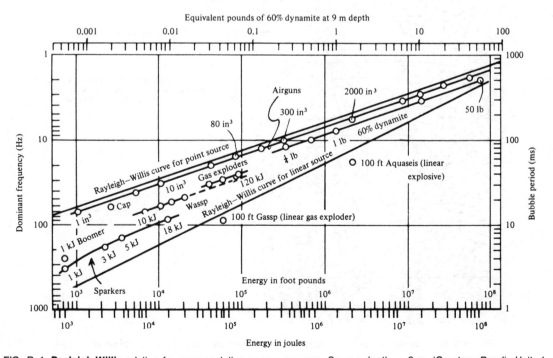

FIG. R-4. **Rayleigh-Willis** relation for representative energy sources. Source depth = 9 m. (Courtesy Bendix United Geophysical.)

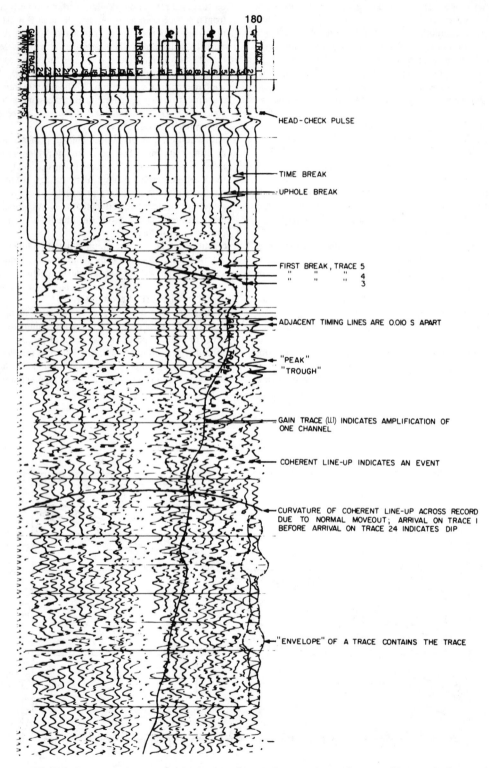

FIG. R-5. 24-trace split-spread seismic **record** in wiggle-trace form. (Courtesy Chevron Oil Co.)

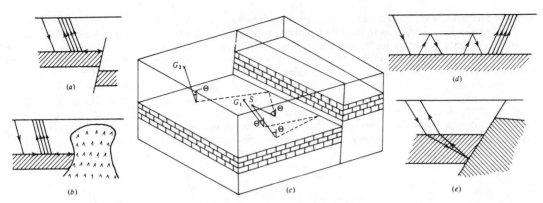

FIG. R-6. **Reflected refractions**. (a) Refraction reflected at a fault and (b) by a salt dome. (c) The travel paths involved are not always coplanar. (d) Multiply reflected refractions. (e) Reflected refractions do not always involve head waves.

recording truck: A vehicle which contains the amplifiers and recording gear for making a record in the field. The cab where the record is made is often called the **doghouse**. A doghouse is often mounted on small boats, pontoons, etc., for work where a truck cannot go.

record section: Display of seismic traces side-by-side. Originally made by splicing together individual seismic records but the entity of individual records now has been largely lost.

record time: 1. Time after the instant of energy release. **2.** Time after a certain reference.

recover: To relocate a preestablished survey location, as for use in tying a new survey to an older survey.

recovery: 1. The amount of core recovered compared to the amount cut. **2.** The amount of fluid in the drill pipe on a drill-stem test which did not produce enough fluid to fill the pipe and flow at the surface. **3.** To return to the same relative amplitude level as existed at the time of original recording, as in gain recovery.

rectify: 1. To allow current to pass in one direction only. **2.** To adjust a borehole log for true vertical depth. Logs in slant holes are usually recorded in distance measured along the hole, which makes them difficult to correlate with other logs. Also sometimes used for adjusting a log to what would be seen without formation dip. **3.** To replace an inverted image with an erect one. **4.** To eliminate time differences caused by weathering, elevation, normal-moveout or other differences.

rectilinear: A system of straight lines at right angles to each other, as with ordinary $x-y$ coordinate paper.

recursive filter: A filter for which the output depends on previous outputs as well as the input and the filter response. Part of the output is delayed and added to the input. Some types of filtering which require a long operator can be accomplished with a short operator in this way. Also called **feedback filter**. See Sheriff and Geldart, v. 2 (1983, p. 41, 186-187).

reduced latitude: See *geodetic latitude*.

reduced ratio: The ratio of electromagnetic field strength at two locations after being corrected for the **normal ratio** or **free air correction** (the electromagnetic-field gradient in the absence of perturbing conductive bodies). Used with the Turam method.

reduced refraction section: A display of refraction traces where x/V has been subtracted from the arrival time of each trace (x = offset distance) so that arrivals with the apparent velocity V are horizontal. If V is the velocity of the refractor, then the arrivals show the relief on the refractor. Reduced sections facilitate the picking of refraction arrivals even where V is not precisely correct.

reduced traveltime: 1. The difference t_r between the observed refraction traveltime and the traveltime which would have been observed if only one velocity (V) material had been present: $t_r = t - x/V$, where x = offset distance. The result yields the total delay time (shot delay time + geophone delay time) if the assumed velocity is the refractor velocity. An assumed velocity of 6 km/s is often used as the reduction velocity in crustal studies. **2.** Refraction records sometimes are plotted with the time reference advanced by an amount proportional to the offset so that arrivals with the refractor velocity appear flat. Thus relief on the refractor produces apparent relief on the event. Such a plot facilitates picking weak refraction arrivals (especially secondary arrivals). **3.** *Intercept time* (q.v.).

reduced vertical profile: A vertical seismic section display where arrival times are shifted down or up by the one-way first-arrival time. The result is to align horizontally either downward- or upward-traveling wave trains, which facilitates their study.

reduction: The process of substituting for observed values ones which have been corrected for known sources of error.

reduction to the pole: A method of removing the dependence of magnetic data on the angle of magnetic inclination. "Reduction to the pole" converts data which have been recorded in the inclined Earth's magnetic field to what the data would have looked like if the magnetic field had been vertical. Reduction to the pole removes anomaly asymmetry caused by inclination and locates anomalies above the causative bodies.

redundancy: A repetition of information, such as the same measurements made several times (perhaps in several ways). Redundancy permits the attentuation of some distorting effects. Successive measurements may differ from each other in that components other than the objective components differ; thus the objective data will be recorded the same each time and therefore may be distinguished from components which vary for the different measurements. For example, six-fold common-midpoint recording involves measuring the reflected energy from a given portion of the subsurface six times and hence has a redundancy of 6.

reef: 1. A local carbonate buildup. **2.** Sometimes restricted to a buildup produced by organisms such as coral. See Sheriff and Geldart, v. 2 (1983, p. 110-113) regarding evidences of reefs in seismic data.

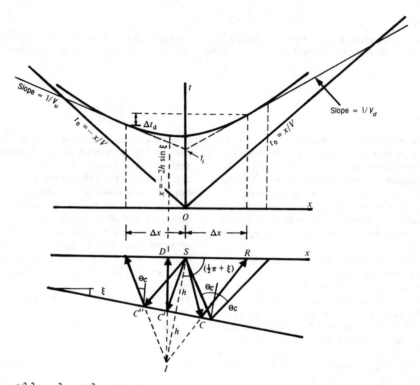

$$V^2 t^2 = x^2 + 4h^2 + 4hx \sin \xi;$$

$$t = \frac{2h}{V}\left(1 + \frac{x^2 + 4hx \sin \xi}{4h^2}\right)^{1/2} \approx t_0\left(1 + \frac{x^2 + 4hx \sin \xi}{8h^2}\right);$$

$$\sin \xi \approx \frac{V}{2}\left(\frac{\Delta t_d}{\Delta x}\right); \quad h = \tfrac{1}{2}V t_0.$$

Normal moveout $= \Delta t_n \approx (x^2/2V^2 t_0) \{1 - (x/4h)^2 + \ldots\}$, $\xi = 0$;
dip movement $= \Delta t_d/\Delta x$.

$$\sin \theta_c = V_1/V_2, \quad V_1 < V_2;$$
$$V_d = V_1/\sin (\theta_c + \xi); \quad V_u = V_1/\sin (\theta_c - \xi);$$
$$t_i = (2h/V_1) \cos \theta_c;$$
$$\theta_c = \tfrac{1}{2}\{\sin^{-1}(V_1/V_d) + \sin^{-1}(V_1/V_u)\}; \quad \xi = \tfrac{1}{2}\{\sin^{-1}(V_1/V_d) - \sin^{-1}(V_1/V_u)\}$$
$$V_2 = 2[(1/V_d) + (1/V_u)]^{-1} \approx \tfrac{1}{2}(V_d + V_u);$$
$$\xi \cos \theta_c \approx 1 - V_d/V_2 = V_u/V_2 - 1.$$

FIG. R-7. **Reflection** and **refraction** equations for constant velocity overburden.

reel: Winding apparatus for winding cables.

reel truck: A vehicle used for transporting cables and geophones.

reference plane: *Datum* (q.v.).

reference seismometer: A geophone which records successive shots to verify that they are similar. Used in well shooting to verify that they are similar. Used in well shooting to eliminate the possibility that source variability may produce error.

reflectance: *Reflectivity* (q.v.) or reflection coefficient.

reflected refraction: 1. Head wave energy which has been reflected or diffracted back from a discontinuity in the refractors, such as a fault. See Figures R-6a, b, c. **2.** A refraction multiple, involving head-wave energy which has been multiply reflected between reflectors. The extra travel may occur at any time during the travel along the refractor. The effect is to add cycles to the refraction arrival; see Figure R-6d. **3.** Any of a number of travel paths which involve reflection at an interface as well as travel through a high-velocity refractor; see Figure R-6e.

reflecting point: The point on a reflector where the angle of incidence equals the angle of reflection. Reflection actually involves a region surrounding the reflecting point; see *Fresnel zone.*

reflection: The energy or wave from a seismic source which has been reflected (returned) from an acoustic-impedance contrast (**reflector**) or series of contrasts within the Earth. The objective of most reflection-seismic work is to determine the location and attitude of reflectors from measurements of the arrival time of primary reflections and to infer the geologic structure and stratigraphy. The basic reflection equations are shown in Figure R-7. See also *reflectivity.*

reflection character analysis: Examination of waveshape variations to identify places where changes in stratigraphy or hydrocarbon content may occur and suggest the nature of the changes. See Sheriff (1980, chaps. 7-8).

reflection coefficient: 1. *Reflectivity* (q.v.). **2.** A ratio of resistivities ρ as derived from the method of images.

$$k = (\rho_2 - \rho_1)/(\rho_2 + \rho_1).$$

Used in describing type curves. Also called **resistivity-contrast factor** or **reflection factor**.

reflection configuration: A pattern of reflections; see Figure R-8.

reflection peak: 1. The maximum positive excursion produced by an event identified as a reflection. **2.** In electrical logging, an increase in resistivity reading as the upper (A) electrode of a lateral sonde (see Figure E-7) passes a thin high-resistivity formation.

reflection polarity: The direction of display of a reflection. See *polarity standard.*

reflection shooting: *Reflection survey.* (q.v.).

reflection strength: Amplitude of the envelope of a seismic wave; see *complex-trace analysis.*

reflection survey: A program to map geologic structure and/or stratigraphic features employing the seismic-reflection method. Measurements are made of the arrival times of events attributed to seismic waves which have been reflected from interfaces where changes in acoustic impedance occur, and of waveshape changes.

The objective usually is to map the depth, dip and strike of interfaces which usually are parallel to the bedding, and lateral changes in reflections. A second objective is to define stratigraphic variations from normal-moveout measurements or from the amplitude and waveshape of reflection events.

reflectivity: 1. Reflection coefficient; the ratio of the amplitude of the displacement of a reflected wave to that of the incident wave. The relationship is obtained by solving boundary condition equations which express the continuity of displacement and stress at the boundary. For normal incidence on an interface which separates media of densities ρ_1 and ρ_2 and velocities V_1 and V_2, the reflection coefficient for a plane wave incident from medium 1 is

$$R = (\rho_2 V_2 - \rho_1 V_1)/(\rho_2 V_2 + \rho_1 V_1).$$

This implies that displacement is measured with respect to the direction of wave travel so that a compression is reflected as a compression when the reflection coefficient is positive. A negative reflection coefficient implies phase inversion, that a compression is reflected as a rarefaction. Where displacement is measured with respect to a space-fixed coordinate system (e.g., positive means downward displacement), the signs are different from that given above. In the more general case of a plane-wave incident at an angle, both reflected P- and S-waves and transmitted P- and S-waves will be generated. The amplitude of each of these waves may be found from Knott's equations or *Zoeppritz's equations* (q.v.). **2.** The ratio of the reflected energy density to the incident energy density is the **energy reflectivity**. Such a usage gives the square of the values given by definition 1. **3.** *Reflectivity function* (q.v.). **4.** *Albedo* (q.v.).

reflectivity function: A time function or time series intended to represent reflecting interfaces and their reflection coefficients as nearly as possible, usually at normal incidence. The ability to determine this from seismic data is limited by information content (bandwidth), basic wavelet removal, and noise (data which are not primary reflections). Compare *structural section.*

reflectivity section: A display of the *reflectivity function* (q.v.) as a function of location along a seismic line, determined from seismic reflection data.

reflector: 1. A contrast in acoustic impedance, which gives rise to a seismic *reflection* (q.v.). **2.** A contrast in electrical properties which gives rise to an electromagnetic reflection.

reflector curvature effects: See Figure W-4, *buried-focus effect,* and Sheriff and Geldart, v. 1 (1982, p. 112-117).

reformat: To change the form of data from one *format* (q.v.) to another.

refraction: 1. The change in direction of a seismic ray upon passing into a medium with a different velocity. See *Snell's law.* **2.** Involving *head waves* (q.v.), i.e., involving a travel path in a high velocity medium parallel to the bedding. The basic equations involving head waves are shown in Figure R-7. See also Figure H-2.

refraction count: The difference between Doppler frequen-

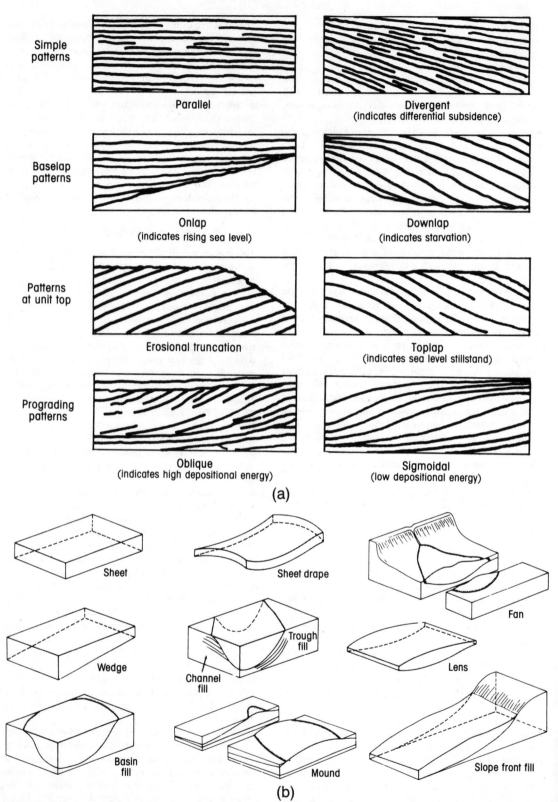

FIG. R-8. **Reflection configurations**. (a) Reflection patterns on seismic sections. (b) Three-dimensional shapes of seismic facies units. After Sangree and Widmier (1979).

cy counts from a navigation satellite for 400 MHz and 150 MHz radio waves. This difference is usually within ±25 Hz. Ionospheric refraction (which is inversely proportional to frequency) can be determined from the difference, allowing correction for refraction above elevation angles of about 7.5 degrees.

refraction marker: *Refractor* (q.v.).

refraction method: See *refraction survey*.

refraction profile: See *profile*.

refraction shooting: *Refraction survey* (q.v.).

refraction survey: 1. A program to map geologic structure by using head waves. Head waves involve energy which enters a high-velocity medium (refractor) near the critical angle and travels in the high-velocity medium nearly parallel to the refractor surface. The objective is to determine the arrival times of the head waves to map the depth to the refractors in which they traveled. For a summary of refraction interpretation methods, see Sheriff and Geldart, v. 1 (1982, p. 218-229). **2.** Refraction surveying also includes the detection and mapping of high-velocity masses such as salt domes (see *fan shooting*) and the delineation of such masses by shooting into seismometers located in deep wells (see Figures A-11 and T-6).

refraction test: *Depth probe* (q.v.).

refraction wave: A *head wave* (q.v.).

refractive index: The ratio of the velocity of light in a vacuum to that in a given medium.

refractor: A layer of higher velocity than overlying layers, through which a *head wave* (q.v.) can travel. In order to be useful for mapping, refractors must be (a) sufficiently thick ($> 1/10$ of a wavelength) that the head wave can carry energy over an applicable distance, (b) sufficiently extensive that the same refractor is mappable over an appreciable area, (c) sufficiently distinctive in velocity that the head wave can be distinguished from the waves carried in other layers, and (d) not be "hidden" by a shallower refractor of higher velocity.

regional: The general attitude or configuration disregarding features smaller than a given size. **Regional dip** is the general dip attitude ignoring local structure. **Regional gravity** is the gravity field produced by large-scale variations ignoring anomalies of smaller size, often the field produced by density variations within or below basement. See *residualize*.

register: A temporary storage for data being processed. Typically, a register stores a single computer word.

regression: A seaward movement of the shoreline with time. Opposite of **transgression**.

regression analysis: A method of finding the statistical dependence of one quantity on other quantities. Often implies the same as *factor analysis* (q.v.) or **multivariant analysis**.

reject region: The range over which a filter exercises considerable attenuation. May refer to those frequencies, dips, or apparent velocities which filters attenuate.

relative apparent resistivity: The ratio of apparent resistivity to the true resistivity of a portion of a model; for example, apparent resistivity divided by the resistivity of the upper layer in a simple two-layer case. Such dimensionless ratios are used in resistivity type curves.

relative bearing: Azimuth of hole deviation with respect to the reference on a dipmeter sonde (such as the no. 1 electrode).

relative ellipse area: In telluric surveying, electric-field vectors are observed at a field station ($\mathbf{E}_u$ and $\mathbf{E}_v$) and also simultaneously at a base ($\mathbf{E}_x$ and $\mathbf{E}_y$). The vectors are related to each other in linear combinations:

$$\mathbf{E}_u = a\mathbf{E}_x + b\mathbf{E}_y$$
$$\mathbf{E}_v = c\mathbf{E}_x + d\mathbf{E}_y.$$

The determinant of this transformation, $J = ad - bc$, called the relative ellipse area, is a measure of the electrical properties at the station relative to the base.

relative permeability: 1. The ratio of the permeability for a given phase to its permeability when other phases are not present; the ability of a porous medium to permit fluid flow through it when two or three phases are present in the pore space. The relative permeability depends on the fraction of the pore space occupied by that phase. Since various phases inhibit the flow of each other, the sum of the relative permeabilities of all phases present is always less than unity. **2.** Relative *magnetic permeability* (q.v.).

relative permittivity: Dielectric constant normalized by dividing by the permittivity of free space (8.85×10^{-12} farad per meter) so as to give a dimensionless quantity.

relative thickness: The ratio of thickness of a layer to the electrode interval. A dimensionless ratio used in drawing apparent-resistivity curves.

relaxation time: See *time constant*.

relay: A switching device, usually controlled by a separate electrical circuit. Mercury relays are sometimes used to switch current in IP transmitters.

release time: See *AGC time constant*.

relief: 1. The difference between the highest and lowest elevations in an area. **2.** The range of values over an anomaly or within an area. While elevation differences are usually meant, one also speaks of "gravity relief" for the magnitude of a gravity anomaly, etc.

reluctance: Magnetic flux per unit magnetomotive force.

remanence: See *remanent magnetism*.

remanent magnetism: Remanence: the magnetization remaining in the absence of an applied magnetic field. **(a) Natural remanent magnetization** is the residual magnetization possessed by rocks and other materials in situ; unless otherwise qualified, this is the meaning implied. **(b) Thermoremanent magnetization (TRM)** remains after a sample has been cooled from a temperature above the Curie point in a magnetic field. **(c) Chemical remanent magnetization (CRM)** is acquired when a magnetic substance is chemically formed or crystallized in a magnetic field at temperature below the Curie point. **(d) Depositional** (or **detrital**) **remanent magnetization (DRM)** is acquired in sediments when magnetic mineral particles are preferentially aligned during deposition (usually by settling through water) in response to the effect of the ambient magnetic field. **(e) Isothermal remanent magnetization (IRM)** is remanent magnetization in the ordinary sense, i.e., the magnetization after application and subsequent removal of a magnetic field; it is not involved in paleomagnetism. (f) **Pressure** (or **piezo**) **remanent magnetization** (**PRM**) is remanence acquired as a result of the application of stress; the effects generally become more pronounced as the strain proceeds from elastic to plastic deformation. See also *viscous magnetization*.

remote data unit: RDU. A portable unit which receives signals from several (often 4) geophone groups, digitizes them, stores the data temporarily, and transmits the data to the recording system upon command. Also **RDAU, remote data acquisition unit.**

remote electrode: *Infinite electrode* (q.v.).

remote job entry: Input of data-processing jobs from a terminal connected to the computer by a communication line. Abbreviated **RJE.**

remote (reference) magnetotelluric method: Magnetotelluric survey conducted with either an electric field or magnetic field reference located a few kilometers from the point of measurement. The reference is used to obtain a better estimate of the true impedance tensor Z by using average crosspowers between the reference fields and the electric and magnetic fields at the sounding location. In contrast to conventional magnetotellurics, the impedance estimate will be unbiased by noise power, provided the noise in the reference signal is uncorrelated with noise in the electric and magnetic channels. (See *magnetotelluric method* and Gamble et al (1979).

remote sensing: Measurements made from large distances, as from high flying aircraft or Earth satellites. Especially refers to measurements of either natural radiation or radiation from a source in the sensor which has been reflected back from the Earth. Several portions of the electromagnetic spectrum (ultraviolet, visible, infrared, microwave, or radar) are used. Both **passive methods** (measuring natural radiation) and **active methods** (beaming from a source on the aircraft and measuring the reflected energy) are used. Displays sometimes use **imaging** (so as to produce a picture of the radiation, as in a photograph), sometimes are **nonimaging** (so as to produce a profile of the variation of radiation along the flight path). May also include measurements of the magnetic field or other nonradiation measurements.

remote triggering: A method of controlling "on" and "off" voltage switching of IP receiving equipment to record the decay signal. A synchronous-detection method which uses the ground signal for a timing channel.

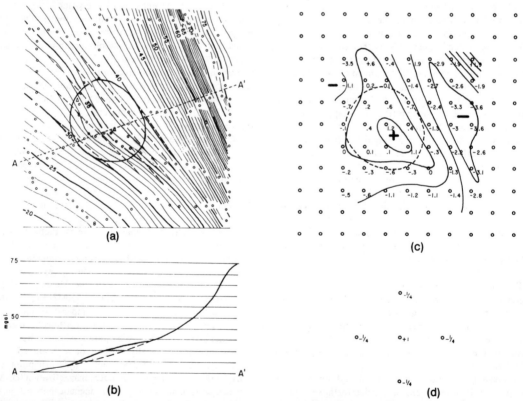

FIG. R-9. Residualizing methods. (**a**) Graphical smoothing of map contours. The difference between smoothed and actual contours localizes an anomaly. (**b**) Graphical smoothing of profile. The profile should be drawn where the contours are controlled by the data, and different profiles must be consistent as to the smoothed regional. (**c**) Reading values on a regular grid and convolving with a template. Margins are lost in the process so the residual covers a smaller area than the data. Template and weighting can be varied to yield second derivative and other types of maps. (**d**) Template used to make map shown in (c) above.

repeatability: The deviation from the average of corresponding data points taken from repeated tests under ideal conditions.

replacement velocity: The velocity used in static corrections to compensate for low-velocity near-surface materials.

replicate: To reproduce a function, as by convolving with a series of impulses.

replicating function: *Comb* (q.v.).

reproducible recording: Recording on a medium which readily permits automatic reading, such as on magnetic tape, but excluding an ordinary print on paper.

resample: To change the sampling frequency (or the interval between adjacent samples). To decrease the number of samples is to **decimate**; for example, a data set at 2 ms intervals can be "decimated" to a data set at 4 ms intervals by dropping every other sample. To increase the number of samples is to **reconstitute**; for example, to obtain samples interpolated between the values of a data set at 2 ms intervals to yield a data set at 1 ms intervals.

resection: Determination of the horizontal location of a survey station by the intersection of lines indicating the direction from other stations. See Figure P-2.

resident programs: Programs stored within a computer's memory which do not need to be specifically loaded in order to use them.

residual: 1. The difference between observed data and the regional, as in gravity and magnetics. What is left after the regional has been removed. See *residualize*. **2.** Sometimes an anomaly, the difference between a measurement of a quantity and the expected value of that quantity; what is not accounted for. **3.** *Salt residual* (q.v.).

residual amplitude: The area of amplitude excursions which exceed the average amplitude, divided by the number of cycles. Used to indicate the degree of interbedding.

residual amplitude section: A display of only the portion of peaks which exceed the average amplitude.

residual disturbance: Disturbances that persist during quiet days following a magnetic storm.

residualize: 1. The process of separating a curve or a surface into its low-frequency parts (called the **regional**) and its high-frequency parts (called the **residual**). Residualizing attempts to predict regional effects and find local anomalies by subtracting the regional effects. This separation is not unique. The regional may be drawn graphically (Figure R-9a,b), by gridding methods (Fig-

ure R-9c,d), by surface fitting, by Fourier analysis and filtering, and by other methods, most of which can be thought of as two-dimensional convolution operations (**map convolution**). Some of these methods produce halo effects about local anomalies. Residualizing sometimes is done in steps. **2.** The process of determining what is not accounted for by a particular model. The effects of the model are calculated and subtracted from the observed field, the residual being those effects still unaccounted for. See *salt residual*.

residual normal moveout: A small amount of normal moveout which remains because of incomplete normal-moveout removal. Compare *differential normal moveout*.

resistance: Opposition to the flow of direct current. Compare *impedance* and *reactance* and see *Ohm's law*.

resistive coupling: See *coupling*.

resistivity: The property of a material which resists the flow of electrical current. Also called **specific resistance**. The ratio of electric-field intensity to current density. The reciprocal of resistivity is **conductivity**. In nonisotropic material the resistivity is a tensor. See *ohmmeter* and *apparent resistivity*.

resistivity anisotropy coefficient: See *anisotropy*.

resistivity-contrast factor: The ratio, $(\rho_2 - \rho_1)/(\rho_2 + \rho_1)$, which appears in the analysis of resistivity relationships between materials of resistivity ρ_1 and ρ_2. Also called **reflection coefficient**.

resistivity index: The ratio of the resistivity of a formation bearing hydrocarbons to the resistivity if 100 percent saturated with formation water. See *Archie's formulas*.

resistivity logs: 1. Well logs which depend on electrical resistivity: *normal, lateral, laterolog,* and *induction log* (q.v.). Most resistivity logs derive their readings from 10 to 100 ft^3 of material about the sonde. *Microresistivity logs* (q.v.), on the other hand, derive their reading from a few cubic inches of material near the borehole wall. **2.** Records of surface *resistivity methods* (q.v.).

resistivity method: 1. Observation of electric fields caused by current introduced into the ground as a means for studying earth resistivity. The term normally includes only those methods in which a very low frequency or direct current is used to measure the apparent resistivity. Includes *electric profiling* and *electric sounding* (q.v.). Various *arrays* (q.v.) are used. See also *Gish-Rooney method* and *Lee partitioning method*. **2.** Sometimes also includes *induced-polarization* and *electromagnetic methods* (q.v.).

resistivity spectrum: The resistivity of a polarizable mate-

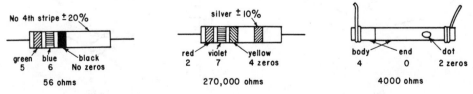

FIG. R-10. **Resistor** identification code. The color code is: brown = 1, red = 2, orange = 3, yellow = 4, green = 5, blue = 6, violet = 7, gray = 8, white = 9, black = 0, gold = 0.1 (or 5 percent if tolerance), silver = 0.01 (or 10 percent if tolerance).

rial, measured at successive frequencies and plotted against frequency. This spectrum can be transformed into an IP decay curve and vice versa. The IP resistivity spectrum of a polarizable material appears to be distinctive of the nature of some kinds of substances (type of mineralization).

resistor: An electrical circuit element possessing a certain resistance. Resistor color codes are shown in Figure R-10.

resolution: 1. The ability to separate two features which are very close together. The minimum separation of two bodies before their individual identities are lost on the resultant map or cross-section. See Sheriff and Geldart, v. 1 (1982, p. 117-122). **2.** The smallest change in input that will produce a detectable change in output. **3.** The ability to localize an event seen through a window, usually taken as the half-width of the major lobe.

resolvable limit: For discrete seismic reflectors, the minimum separation so that one can ascertain that more than one interface is involved. The value depends on the criteria for ascertaining. The **Rayleigh resolution limit** is λ/4 where λ is the dominant wavelength, the **Widess limit** is λ/8. Compare *detectable limit*. Horizontal resolution on unmigrated seismic sections is often taken as the width of the first *Fresnel zone* (q.v.).

resolve: To separate into parts. **1.** To determine the component orthogonal vectors which add together to form a given vector. The direction of the component vectors has to be prescribed. Usually implies finding the components in the directions of an orthogonal coordinate system. **2.** To show two features as separate rather than blended together.

resolved bands: The **number of resolved bands** is the ratio of the Nyquist frequency to the resolution.

resonance: Buildup of amplitude in a system having a natural frequency to a stimulus of nearly the same frequency.

resonant frequency: *Natural frequency* (q.v.)

response time: The time between the initiation of an operation and the receipt of results. Response time includes transmission of data to the computer, processing, file access, and transmission of results to the output terminal.

restore: A write-after-read computer operation.

restored section: *Paleosection* (q.v.).

retarded potential: A potential function whose argument is *retarded time* (q.v.), from which the response can be determined.

retarded time: The response at a point to a distant cause is delayed by the traveltime to the point. This time delay can be accommodated if the time scales at the observing point differ from that at the cause, that is, if observations are made in "retarded time". See Sheriff and Geldart, v. 1 (1982, p. 39).

retrocorrelation: Correlation of a function with a reversed version of itself, or the equivalent, convolution of a function with itself; **autoconvolution**. Retrocorrelation of a seismic trace yields a **retrocorrelogram**, which contains multiples which involve the surface.

retrograde: 1. Rotational motion opposite to the usual direction. Rayleigh waves are sometimes called **retrograde waves**, because motion near the surface is in elliptical orbits such that the particle is traveling opposite to the direction of the wave while at the top of the ellipse. See Figure R-2. **2.** A branch of a time-distance curve resulting from very steep velocity gradients. See Figure D-15b. Also called **reverse branch**.

return-beam vidicon: A TV camera which is scanned by an electron beam; used on Landsat satellites (band 3) to form images. **RBV.**

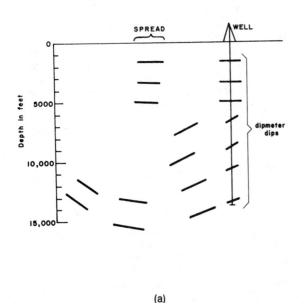

(a)

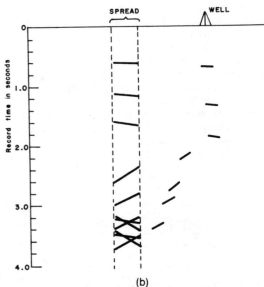

(b)

FIG. R-11. **Reverse migration.** (**a**) Depth section showing migrated seismic events. (**b**) Time section showing reverse-migrated dipmeter data. Structural models may be reverse migrated to find where their seismic evidences should be sought.

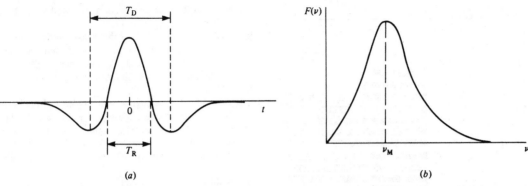

FIG R-12. **Ricker wavelet.** (a) Time-domain and (b) frequency-domain representations. Expressions for the Ricker wavelet are: $f(t) = (1 - 2\pi^2 v_M^2 t^2) \exp[-\pi^2 v_M^2 t^2] \leftrightarrow (2\pi^{-1/2})(v^3/v_M^2) \exp[-v^2/v_M^2] = F(v)$. Also, $T_D = (6/\pi)^{1/2}/v_M$ and $T_R = 3^{-1/2} T_D$.

reverberation: 1. Multiple reflection in a layer, usually the water layer in marine work; **ringing; singing.** Sometimes the term is reserved for the case where the successive multiples blend together into a more or less steady oscillation; occasionally it includes also the situation where the water is so deep that the successive multiples are discrete. Reverberation sometimes occurs on land records also. Removing reverberation effects may be the objective of deconvolution. See Sheriff and Geldart, v. 2 (1983, p. 32-33). **2.** The effect of a long oscillatory source such as sometimes occurs with explosive or air gun (or other) sources.

reversal: A local change of approximately 180 degrees in the direction of the dip along a line. Often refers specifically to a dip reversal which denotes a structural lead.

reverse branch: A seismic event where right-to-left orientation on the event corresponds to the reverse orientation (left-to-right) on the reflector. This is a consequence of a buried focus such as can be produced by sharp synclinal curvature or focusing by a velocity lens. For zero offset and constant velocity this occurs where the radius of curvature for a concave-upward reflector is less than the reflector depth. See Figure B-8 and *buried focus*. The reverse branch involves a phase shift; see Sheriff and Geldart, v. 1 (1982, p. 116-117). Compare *backward branch*. Also called **retrograde branch.**

reverse control: Profiles shot from opposite directions; **two-way control. 1.** The portions of a refractor for which there is overlap of relative delay-time data resulting from shooting in opposite directions (i.e., the subsurface coverage from which arrivals are seen in opposite directions). **2.** Sometimes merely the portion of a refraction profile which has been shot from opposite directions. **3.** Reflection shooting where dip can be verified by data shot in opposite directions.

reversed: A spread or refraction profile shot from opposite directions.

reversed magnetic field: See *normal magnetic field*.

reversed polarity: Having polarity opposite to "normal". Seismic sections are often plotted with both normal and reverse polarity, all amplitude values in the former being multiplied by −1 to achieve the latter. An attempt

is usually made to have positive reflectivity show as a central peak which is shaded black on a normal polarity plot but achieving absolute phase significance is often elusive.

reverse fault: See *fault* and Figure F-2.

reverse migration: Determining where the reflection event from a given portion of reflector would be observed at normal incidence. See Figure R-11.

reverse Polish: 1. The scheme for entering data into Hewlett-Packard hand computers; the sequence for A + B would be: "A, enter, B, add." As opposed to the sequence used with Texas Instruments computers, where the sequence would be: "A, add, B." **2.** A method used in parsing computer language statements into operator and operand stacks (last in, first out buffers). Operators and operands are pushed onto their respective stacks in reverse order of execution so they can be popped off the stacks in correct order.

reverse SP: Where the mud is more saline than the formation water, SP voltages are reversed with respect to the usual situation.

reverse symmetric: A *matrix* (q.v.) for which

$$A_{xx}(t) = A_{xx}^T(-t).$$

reverse-wound geophone: See *humbucking*.

reversible process: A physical or chemical change which can be caused to proceed in either direction by small alteration in one of the controlling equilibrium conditions such as concentration, temperature, or pressure. If a small change in current shifts the equilibrium position, a change back to the original value restores the equilibrium to its original position. See *equilibrium conditions*.

review: To reinterpret data or to rework.

Reynolds number: The ratio of inertial force to viscous force for viscous fluid flow. In the bubble effect, $R = 2uap/\mu$, where u = radial water velocity, a = bubble radius, ρ = water density, and μ = water viscosity.

RF: *Radio frequency* (q.v.).

R_g **wave:** A short-period Rayleigh wave that travels as a guided wave in the crust. The subscript "g" refers to the granitic layer.

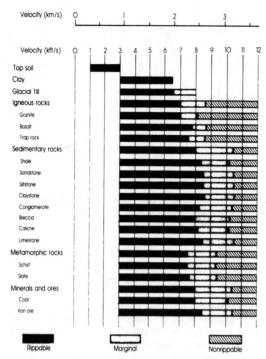

Velocity (km/s)

Velocity (kft/s)

Top soil
Clay
Glacial till
Igneous rocks
  Granite
  Basalt
  Trap rock
Sedimentary rocks
  Shale
  Sandstone
  Siltstone
  Claystone
  Conglomerate
  Breccia
  Caliche
  Limestone
Metamorphic rocks
  Schist
  Slate
Minerals and ores
  Coal
  Iron ore

Rippable   Marginal   Nonrippable

FIG. R-13. **Rippability**. (Courtesy Caterpillar Tractor Co.)

rho-rho determination: *Positioning* (q.v.) in which the distance from two known points is used to determine the location. Also called **range-range**. Other ways of determining location include rho-theta determinations, measurement of differences in distances, use of azimuthal systems, dead reckoning, etc.

rho-theta determination: *Positioning* (q.v.) in which the distance and direction from a fixed point are used to determine location.

rhumb line: A line on the surface of the earth making the same angle with all meridians, thus having constant azimuth (constant direction with respect to geographic north); **loxodrome**. A straight line on a Mercator projection.

ribbon: A very thin sheet used as a magnetic model; a very thin dike model.

Richter scale: A scale for measuring the *magnitude of* an *earthquake* (q.v.). The logarithmic scale was devised in 1935 by Charles Francis Richter (1900—), American seismologist.

Ricker wavelet: A zero-phase wavelet, the second derivative of the error function. See Figure R-12. Named for Norman H. Ricker (1896-1980), American geophysicist.

Rieberize: To time-shift adjacent traces and sum, thus emphasizing energy which comes from a particular direction relative to energy from other directions. By varying the amount of time shift, various directions can be searched to find a maximum for a particular arrival, thus ascertaining the raypath direction associated with it. Interfering events from different directions can be sorted out in this way. The display as a function of

arrival time for various directions is called a **sonograph** or **sonogram**. See Rieber (1936).

rift: 1. A long graben such as the mid-Atlantic rift or the rift valleys of East Africa. **2.** A long graben associated with a pull-apart zone. **3.** A fault; especially a long strike-slip fault such as the San Andreas fault.

rig: *Drill* (q.v.).

right ascension: 1. The angular distance east of the vernal equinox line (first line of Aries) on the celestial sphere to a celestial body. **2.** The angle between the vernal equinox line and the intersection of the plane of the orbit of a satellite with the plane of the celestial equator. See Figure K-1.

right-hand rule: A rule which gives the direction of the force on a current in a magnetic field. See Figure I-2.

right lateral fault: See Figure F-2.

rigidity modulus: See *elastic constants*.

rim syncline: Ring of depressed sediments surrounding a salt dome, caused by collapse following salt withdrawal.

ringing: 1. *Reverberation* (q.v.). **2.** The oscillatory effect produced by a narrow-band filter.

ringy: Oscillatory.

rippable: Material which can be excavated with relative ease using a ripper mounted on a tractor. Can be roughly related to *P*-wave velocity as shown in Figure R-13.

rise-time error: A delay in measuring the onset of an event because of the time required to reach detectable amplitude level.

RJE: *Remote job entry* (q.v.).

rms: *Root-mean-square* (q.v.).

rms error: The square root of the average of the squares of the differences between a series of n measurements m_i and their mean $\overline{m}$. The rms error

$$\sigma = [(1/n) \ \Sigma \ (m_i - \overline{m})^2]^{1/2};$$

also called the **standard deviation**. For a normal distribution, 0.683 of the population has less than the rms error. See *statistical measures*.

rms positional error: The circle whose radius is the square root of the mean of the sum of the squares of the distances of measured positions from a point. If the uncertainties in x and y directions are equal, the probability of a measurement lying within the rms positional-error circle is 41 percent. Called **one-sigma circle** because this radius is equal to the standard deviation when taken about the mean of the observations. Compare *CEP*.

rms velocity: Root-mean-square velocity. For a series of parallel layers of velocity V_i, where the traveltime for seismic energy perpendicularly through each is t_i, the rms velocity is

$$V_{rms} = [(\Sigma V_i^2 t_i)/(\Sigma t_i)]^{1/2}.$$

The average velocity V_{av}, on the other hand, is

$$V_{av} = (\Sigma V_i t_i)/(\Sigma t_i).$$

The velocity determined from velocity analysis based upon normal-moveout measurements (stacking veloci-

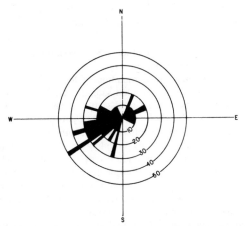

FIG. R-14. **Rose** showing frequency of observations in various directions.

ty) where velocity layering and reflectors are parallel and horizontal, approaches V_{rms} as the offset approaches zero.

robust: A computer program or process that can tolerate poorly conditioned data.

rockbit: Roller bit. See *bit* and Figure B-3.

rod: A graduated pole used as a target in surveying. See *stadia*.

rodman: A surveyor's assistant, one of whose jobs is holding the stadia rod.

roll: Rotational motion of a ship about the axis of principal direction of motion. As opposed to average tilt to starboard or port, which is called **list**. Compare *pitch* and *yaw*.

roll along: The field method for common-midpoint recording; see *common-midpoint stack*.

roll-along switch: A switch which permits connecting different geophone groups to the recording instruments, used in common-midpoint recording.

roller bit: A type of drill *bit* (q.v.). See Figure B-3.

roll off: 1. The frequency beyond which a filter produces significant attenuation. Usually taken as the point where the response is down by 3 dB, sometimes by 6 dB. **2.** Attenuation produced by a filter, often given in decibels per octave.

roll over: 1. Reversal of dip direction such as produced by rotation of a block resulting from sliding along a curved fault plane (**listric fault**). Usually associated with gravity faulting contemporaneous with deposition (**growth fault**). **2.** Cable roll over.

ROM: Read-only memory.

roentgen: The radiation which will produce 2.083×10^{15} pairs of ions per m^3 of air at STP. Named for Wilhelm Konrad von Roentgen (1845-1923), German physicist.

root: 1. Relatively light crustal material projecting down into the mantle and thus balancing topographically high areas (mountains). See Figure I-6. **2.** The values of x which satisfy the equation $F(x) = 0$.

root-mean-square: The square root of the average of the squares of a series of measurements. Abbreviated **rms**.

The autocorrelation value (without normalizing) for zero lag is the mean square value. For a sine wave, the rms value is $(1/2)\sqrt{2}$ times the peak amplitude.

rose: A polar diagram in which radial length indicates the relative frequency of observation at a certain angle. Also called *rosette*. Used to illustrate the direction probabilities of wind, dip direction from dipmeter measurements, lineations, faults, etc. See Figure R-14.

Rossi-Forel intensity scale: A scale of earthquake intensity devised in 1878 by M. S. de Rossi and F. A. Forel. Compare *Mercalli scale*.

rotary drill: See *drill*.

rotary-field electromagnetic method: A method using circularly polarized electromagnetic energy. See Sinha (1970).

rotary table: See *drill*.

rotating dipole: A rotating dipole source consists of two fixed dipoles, either magnetic or grounded electric current dipoles, oriented with their axes mutually perpendicular and energized by sinusoidal current with a 90-degree phase shift between sources so that the resultant magnetic and electric fields appear to rotate in space.

rotational latency: See *latency*.

rotational wave: *S-wave* (q.v.).

round-off error: The error created when a decimal number is approximated by a number with fewer digits.

row binary: A method of recording binary information on cards. Each punched hole is a binary one and lack of a punch indicates zero. One 36-bit word can be punched in any half row, so 24 words may be put on a card.

row matrix: See *matrix*.

row vector: See *matrix*.

RPS: 1. Motorola tradename for a high-frequency short-range radio-positioning system. **2.** Rotational position sensing, a feature of a rotating memory (usually disk) wherein the device informs the host (by interrupt) that a specified position has been reached.

RTU: Remote telemetry unit. See *remote data unit*.

rubidium-vapor magnetometer: A type of *optically pumped magnetometer*. (q.v.).

rugosity: Roughness; irregularity of a borehole wall. The readings of logging tools which have to be held in contact with the hole wall (such as density and microresistivity logs) are affected by rugosity.

Runge-Kutta method: A numerical method of solving differential equations. See Sheriff and Geldart, v. 2 (1983, p. 157).

run time: 1. The time required to complete a single, continuous execution of an object program. Usually measured in CPU time rather than elapsed time. **2.** Occurring during a run, e.g., a "run-time" error.

running window: A window whose position moves as a function of time.

R_{wa}-**analysis log:** *Formation analysis log* (q.v.).

R-**wave:** *Rayleigh wave* (q.v.).

RZ: Return to zero; a method of digital recording on magnetic tape in which a one is indicated by magnetization in one direction whereas a zero is indicated by magnetization in the opposite direction. Compare *NRZ* and *NRZI*.

S

S: Admittance; see *S-rule*.

sabkha: A supratidal to semi-arid environment, often characterized by evaporite-salt, tidal-flood, eolian deposits.

Saile: Seismic acoustic impedance log estimation; see *seismic log*. A Conoco tradename. Sometimes spelled **Sail**.

salinity: Total salts in solution (sodium, potassium, chloride, sulfate, etc.). Normal sea water has around 35 000 ppm (parts per million) salinity; less than 2000 ppm often is regarded as **fresh**.

salt lead: An earlier-than-normal arrival time for a head wave attributed to a salt dome intervening between source and detector.

salt residual: A gravity map from which the calculated effects of a salt-dome model have been subtracted, thus showing the gravity effects for which the model does not account.

sample interval: The interval between readings, such as the time between successive samples of a digital seismic trace or the distance between gridded gravity values. Also called **sample period**.

sample log: A log depicting the sequence of lithologies penetrated in drilling a well, usually compiled by a geologist from low-power (12 to 20X) binocular microscopic examination of drill cuttings and cores recovered at the well site. Also called **strip log**.

sample rate: *Sample interval* (q.v.).

sampling function: A *comb* (q.v.); an infinite sequence of impulses occurring at equal time intervals. Multiplying a waveform by the sampling function produces the set of sample values which represents the digitized waveform. See Figure C-7.

sampling skew: A small systematic channel-to-channel time delay in seismic digital recording, produced because the multiplexer samples adjacent channels successively in order to present them to a single channel A/D digitizer.

sampling theorem: Band-limited functions can be reconstructed from equispaced data if there are two or more samples per cycle for the highest frequency present. Also called **cardinal theorem** and **Nyquist theorem**. See *alias* and Sheriff and Geldart, v. 2 (1983, p. 29-32 and 177-178).

sand count: 1. The total effective thickness of permeable section excluding shale streaks or other impermeable zones. Often determined from electrical logs (SP or microlog). 2. The number of separate permeable sands separated by impermeable zones.

sand line: 1. A line that can be drawn through the maximum deflections for thick, clean sands on an SP log in a section where the formation water is of constant salinity. Compare *shale baseline*. See Figure S-16. 2. A wire line on a drilling rig often used to run or recover tools inside the drill pipe.

saphe: The cepstrum-domain equivalent of phase in the frequency domain. A permutation of the letters in "phase."

SAR: Synthetic aperture radar. See *synthetic aperture*. Aero Service tradename.

satellite navigation: Determination of location based on observations of a navigation satellite. Transit satellites in polar orbits transmit ultra-stable carrier frequencies of 150 and 400 MHz (accurate to one part in 10^{11}). Each successive 2 minute transmission begins precisely on every even minute, thus providing timing. Information about the satellite's position is carried by phase modulation (doublet modulation) of the carriers; see Figure K-1. This position information is updated by ground tracking stations and reinjected to the satellite every 12 hours. The distance to the nearest point of the satellite's orbit can be determined from frequency measurements; higher frequency is seen as the satellite approaches because of the Doppler effect and lower frequency after the satellite has passed (Figure S-1). In the **long-count mode**, the number of cycles in a two-minute observation period are counted; 8 such counts are possible on a satellite pass. In the **short-count mode**, observation periods of multiples of 4.6 s are used. Three or more long counts (including one before and one after closest approach) or 15 short counts are adequate for a fix. The accuracy of satellite fixes depends on the accuracy with which the velocity of the observing station is known (both speed and direction). Accuracy can be improved by *translocation* (q.v.). Satellite fixes are commonly combined with other navigation tools such as Doppler sonar or Loran C, which give locations between satellite fixes and determine the velocity of the observing station. See also *refraction count* and Sheriff and Geldart, v. 1 (1982, p. 189-190). Abbreviated **satnav**.

satellite pass: The transit of a navigation satellite which can be used for a determination of position.

satnav: *Satellite navigation* (q.v.).

saturable system: A system which can be used by only one (or a few) users at a time.

saturation: 1. Apparent nonlinear resistivity or IP behavior due to large contrasts in electrical properties and extreme values of resistivity contrast. These conditions make it difficult to evaluate the true resistivity and IP effects of a body but easier to find its depth. 2. In induced-polarization measurements, the IP response sometimes varies nonlinearly with charging current, probably due to exceeding the current-density limit of the polarizable body. 3. The limiting value of a nonlinear variable. 4. The maximum magnetization as the magnetic field is increased. 5. The extent to which one **hue** is dominant in a color. The deviation of a color from the gray axis on a color chart.

saturation exponent: See *Archie's formula*.

saturation prospecting: The blanket use of several exploration methods over an area.

sausage powder: A long plastic tube containing low-velocity explosive, used to provide a directional charge.

sawtooth pattern: A method of moving thumper or vibrator trucks involving zig-zagging across the seismic line.

sawtooth SP: A jagged SP curve. When a very permeable

salt-water sand containing a shaly streak is invaded by fresh-mud filtrate, the filtrate tends to accumulate just below the shaly streak, setting up an electrochemical cell which causes the SP to develop a sawtooth appearance.

S-band: Radar frequencies between 1550 and 5200 MHz; see Figure E-8.

scalar: 1. A number which is not associated with a direction, as opposed to a *vector* (q.v.). Compare *scaler*. **2.** A single data element as opposed to a set; compare *vector*.

scalar magnetotelluric method: A method in which one electric component and one orthogonal magnetic component are used to define the apparent resistivity sounding curve. Applicable only to isotropic horizontal layering, such as in some basin environments. In areas of more complex geology, the *tensor magnetotelluric method* (q.v.) is used.

scalar wave equation: See *wave equation*.

scale modeling: *Physical modeling* (q.v.).

scaler: A constant expressing a proportionality; a scale factor. Compare *scalar*.

scaling: Changing the amplitude by multiplying by a constant. See *linear system*.

scan: 1. One complete sequence of events, such as sampling all record channels or sonogramming for all apparent dips. **2.** To examine in a systematic way, as to look through a data set to see the effects of a parameter or parameter change. **3.** To traverse systematically as is done by an electron beam in a television tube.

scatter diagram: A diagram on which points are plotted against two (or more) variables to demonstrate the relationship between the variables.

scattergram: Diagram showing the geographical distribution of midpoints in a crooked-line or 3-D survey. See Figure C-16.

scattering: The irregular and diffuse dispersion of energy caused by inhomogeneities in the medium through which the energy is traveling.

schlieren: 1. Streaks of density and index-of-refraction differences in a turbulent fluid. Such streaks permit photographic detection of density variations. **2.** A seismic experimental method with physical models using spark sources and high-speed camera. From the German for "streak."

Schlumberger electrode array: Electrode arrangement used in surface resistivity surveying consisting of four collinear electrodes, with the outer two serving as current sources and the inner two (which are closely spaced about the midpoint of the outer pair) serving as

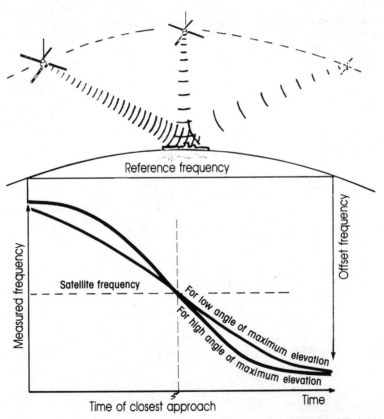

FIG. S-1. **Satellite** pass. For satellites in polar orbits, the time of closest approach gives the latitude, and the rate of change of frequency gives the longitude, both with respect to the satellite. Often the quantity measured is the beat frequency (offset frequency) between the satellite frequency and a reference frequency.

potential-measuring points. See *array (electrical)* and Figure A-12. Named for Conrad (1878-1936) and Marcel (1884-1953) Schlumberger, French geophysicists.

Schmidt diagram: A polar plot where the angle indicates dip or drift direction and the distance from the origin indicates the dip or drift magnitude. In the **modified Schmidt diagram** used for plotting low dips, zero dip is on the outside and dips become larger toward the diagram center. See Figure S-2.

Schmidt field balance: A magnetometer which consists of a permanent magnet pivoted on a horizontal knife edge. The torque of the magnet trying to align with the earth's field is balanced by a gravitational torque because the magnetometer is not pivoted at its center of gravity.

Schmidt net: An equal-area plot of latitude and longitude, used in plotting geologic data such as the direction of structural features. The Schmidt net is the same as Lambert's azimuthal equal-area projection (but different from Lambert's conic projection).

Schuler period: The time equal to $2\pi(R/g)^{1/2}$, where R is the Earth's radius and g is the acceleration of gravity; equal to 84 minutes. The Schuler period is a natural precession rate for gyrocompasses.

Schumann resonance: An electromagnetic waveguide phenomenon between the Earth and the ionosphere; the space between acts as a cavity resonator. Frequencies of the lowest-order modes are about 8, 14, 20, and 26 Hz.

scintillometer: An instrument for measuring radioactive radiation, especially from gamma rays. Gamma radiation impinging on a sensitive phosphor causes it to emit light (scintillations) which is measured by a photomultiplier tube.

scissors fault: See Figure F-2.

scratch-pad memory: Any memory space used for the temporary storage of data. Typically, scratch-pad memories are high-speed integrated circuits which are addressed as internal registers.

sea chest: A fitting in a ship's hull below the water line, such as used to mount sonar transducers. See Figure S-3.

seamount: Steep-sided peak rising from the ocean floor, with its top below sea level; **guyot**. Usually of volcanic origin.

sea state: A scale of the height of sea waves. The Douglas sea state scale is shown in Figure B-1.

secant projection: See *map projection*.

second arrival: 1. An energy wave train, especially a refraction event, which is not the first arrival. **2.** A ghost or simple multiple.

secondary: *Second arrival* (q.v.).

secondary field: 1. The electromagnetic fields which result from the induction of currents in a medium by a primary field **2.** In methods where galvanic energization of the earth is used, the secondary field is that resulting from the difference between the actual current system in the earth and the system which would exist if the earth were homogeneous.

secondary lobe: A pass region other than the principal. Applies especially to the *directivity graph* (q.v.) of arrays, the mixing of traces in velocity filtering, etc.

secondary patterns: The use of the sum and difference frequencies ("red + green" and "red − green") to yield "coarse" networks which can be used to remove lane ambiguities. Used with medium-frequency radio-positioning systems.

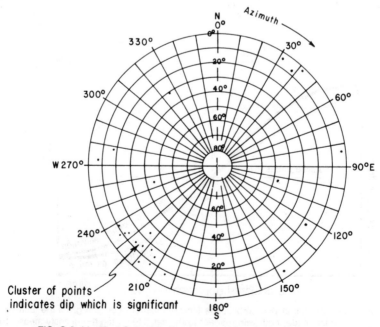

FIG. S-2. Modified **Schmidt diagram** used for plotting dipmeter data.

secondary porosity: *Porosity* (q.v.) resulting from the alteration of formations, such as by fractures, vugs, solution channels, dolomitization, etc.

secondary-porosity index: A measure of the secondary porosity, usually attributed to fractures or vugs, calculated from sonic-log values in conjunction with either density-log or neutron-log values. Symbol: **SPI**. If ϕ_D is the porosity calculated from a density (or neutron) log and ϕ_{sonic} the porosity calculated from a sonic log, SPI is sometimes defined as $(\phi_D - \phi_{sonic})$, usually as $(1 - \phi_{sonic}/\phi_D)$.

secondary recovery: Injection of water or gas into a reservoir to help flush more hydrocarbons from the reservoir. Compare *enhanced oil recovery*.

secondary reflection: *Multiple* (q.v.).

secondary voltage: In IP surveying, the polarization voltage observed at a time-domain receiver immediately after the primary current is turned off. Sometimes called **initial transient** or **initial decay voltage**.

secondary wave: *S-wave* (q.v.).

second critical angle: For an incident *P*-wave, the angle $\theta = \sin^{-1}(V_{S2}/V_{P1})$ where V_{P1} is the *P*-wave velocity in the incident medium, and V_{S2} is the *S*-wave velocity in the second medium (which is to support the *S* head wave).

second-derivative map: A map of the second vertical derivative of a potential field, such as gravity. Such maps tend to emphasize local anomalies and isolate them from a regional background. Often made using Laplace's equation relating the second vertical derivative to second horizontal derivatives, which can be approximated from differences in the values near the point. See also *grid residual*.

second-order correction: 1. A correction which is significantly smaller than first-order corrections. 2. Frequently refers to refinements to correct for slightly incorrect normal moveout or static corrections.

second-order triangulation: See *triangulation*.

section: 1. What might be seen by slicing through a solid object, such as a slice through the Earth. 2. A profile showing the geologic formations which would be exposed in a vertical cut, or some physical property of what would be so exposed. 3. A plot of seismic events, as a record section. 4. Geologic formation, as "the section in this area is Mesozoic." 5. A square mile = 640 acres. Section numbering is shown in Figure T-9.

sectional correlogram: The autocorrelation of successive traces, displayed like a record section.

section gauge: *Caliper log* (q.v.).

secular variation: Time variations whose periods are measured in decades, as "the secular variation of the Earth's magnetic field" with periods of 30 to 300 years.

seek latency: See *latency*.

SEG A, SEG B, SEG C, SEG Y: SEG standard magnetic tape formats recognized by the Society of Exploration Geophysicists. See SEG (1980). SEG X is no longer supported as a standard.

seiche: Free oscillation of an enclosed body of water.

seis: Seismometer or *geophone* (q.v.).

Seiscrop: GSI tradename for a *time-slice map* (q.v.).

Seislog: Teknika tradename for a *seismic log* (q.v.).

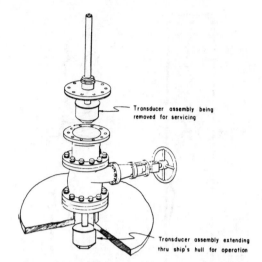

Transducer assembly being removed for servicing

Transducer assembly extending thru ship's hull for operation

FIG. S-3. **Sea chest** opening in ship's hull for Doppler-sonar transducer.

Seisloop: GSI tradename for a three-dimensional survey arrangement. See Figure T-3.

seismic: 1. Having to do with elastic waves. Energy may be transmitted through the body of an elastic solid by body waves of two kinds: *P*-waves (compressional waves) or *S*-waves (shear waves) (see *P-waves*, *S-waves*), or along boundaries between media of different elastic properties by *surface waves* (q.v.). Often equated with "acoustic," "elastic," and "sonic." 2. Having to do with natural earthquakes. Derived from the Greek "seismos" meaning "shock".

seismic cap: See *cap*.

seismic constant: In building codes dealing with earthquake hazards, an arbitrary amount of horizontal acceleration that a building must be able to withstand.

seismic datum: See *datum*.

seismic discontinuity: 1. Any discontinuity in elastic properties and/or density at which seismic velocity and/or acoustic impedance change abruptly. 2. Specifically, the Moho discontinuity between the Earth's crust and mantle, the Gutenberg discontinuity between the mantle and the core, and the gradational change between the outer and the inner core.

seismic-electric effect: A voltage between two electrodes in the ground caused by passage of a seismic wave.

seismic event: Arrival of a new seismic wave, usually ascertained by a phase change or an increase in amplitude on a seismic record. It may be a reflection, refraction, diffraction, surface wave, random signal, etc.

seismic exploration: The use of seismic techniques to map subsurface geologic structure and stratigraphic features with the aim of locating deposits of oil, gas, or minerals; Synonyms: **prospecting seismology, exploration seismology, applied seismology**. Includes use of both reflection and refraction surveys, although often only the former is meant.

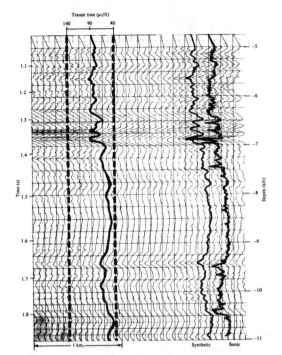

FIG. S-4. Seismic log (synthetic sonic log) traces, each derived by inversion from one seismic traces, are displayed side-by-side. The scale is shown for one trace (heavied) and one trace has been heavied for comparison to an actual sonic log filtered the same as the sonic trace. (Courtesy Teknica.)

seismic facies: The character of a group of reflections involving the general amplitude, abundance, continuity, and configuration of reflections. See Figure R-8 and Sheriff (1980, chap. 5).

seismic gap: A region in which earthquakes waves are not observed with the amplitudes and arrival times expected by interpolating observations. Usually indicates a velocity anomaly.

Seismic Immunity Group: A group of companies who agreed to cross-license each other's patents. Abandoned in 1950s.

seismicity: 1. The likelihood of an area being subject to natural earthquakes. **2.** The relative frequency, intensity, magnitude, and kind of natural earthquakes.

seismic log: Log of acoustic impedance, velocity or reciprocal of velocity (**transit time**), derived by inversion of seismic data and assumptions as to density values. The vertical scale may be either time or depth and the data may or may not have been migrated before inversion. Also called **synthetic sonic log, instantaneous velocity section, G-log, Seislog, synthetic acoustic impedance section, Saile.** See Figure S-4 and Lindseth (1979).

seismic map: A contour map constructed from seismic data. Values may be in either time or depth, unmigrated or migrated, with respect to a datum or with respect to another reflector (in which case it is called an **isopach**).

seismic method: See *seismic survey.*

seismic profile: See *profile.*

seismic pulse: The signal generated by an impulsive seismic energy source (explosive, thumper, air gun, sparker, etc.). Sometimes "wavelet" is used as a synonym. A correlated Vibroseis sweep signal is sometimes included.

seismic record: A plot of the seismic traces from a single source point; a **seismogram.**

seismic refraction method: See *refraction survey.*

seismic section: A plot of seismic data along a line. The vertical scale is usually arrival time but sometimes depth. Sometimes the data have been migrated.

seismic sequence analysis: Defining separate depositional sequences by locating their boundaries, usually by seismic evidences of unconformities. See Sheriff (1980, chap. 3 and 4).

seismic stratigraphy: Methods to determine the nature and geologic history of sedimentary rocks and their depositional environment from seismic evidence. See Sheriff (1980).

seismic survey: A program for mapping geologic structure by observation of seismic waves, especially by creating seismic waves with artificial sources and observing the arrival time of the waves reflected from acoustic-impedance contrasts or refracted through high-velocity members. See *passive seismic* survey, *reflection survey, refraction survey.*

seismic velocity: See *velocity.*

seismic wave: An elastic disturbance which is propagated from point to point through a medium. Seismic waves are of several types: (a) Two types of *body waves* (or *preliminary waves:*) *P-wave* and *S-waves;* (b) Several types of boundary waves or *surface waves: Rayleigh waves,* pseudo-Rayleigh waves or *ground roll, Love waves, Stoneley waves, tube waves;* (c) *channel waves;* (d) *air waves* or shock waves (= Mach effect); (e) *standing waves* = stationary waves. See individual *italicized* entries.

seismogram: A *seismic record* (q.v.).

seismograph: 1. A seismic recording instrument or system. **2.** A *geophone* (q.v.).

seismologist: 1. One versed in seismic principles such as required in oil exploration or earthquake analysis. **2.** An individual occupying a position on a seismic field party which requires knowledge of seismic principles.

seismology: The study of *seismic waves* (q.v.), a branch of geophysics. Especially refers to studies of *earthquakes* (q.v.) or to *seismic exploration* (q.v.) for oil, gas, minerals, engineering information, etc.

seismometer: *Geophone* (q.v).

seismoscope: An instrument that merely indicates the occurrence of an earthquake; see Figure S-5.

Seisviewer: *Borehole televiewer* (q.v.). Birdwell tradename.

selective stacking: A way of stacking several time series in which extrema values (those which fall outside the standard deviation) are eliminated. Used to stack time-domain electromagnetic soundings.

self-potential: Spontaneous-potential or *SP* (q.v.).

self-potential method: Observation of the static natural voltage existing between sets of points on the ground, sometimes caused by the electrochemical effects of ore bodies. Used in mining exploration, especially for shallow sulfide bodies. Compare *telluric current method* and see also *SP*.

self-resistance: See *electrode resistance*.

semblance: A measure of multichannel coherence; the energy of a sum trace divided by the mean energy of the components of the sum. If f_{ij} is the jth sample of the ith trace, then the semblance coefficient S_c is

$$S_c(k) = \frac{\displaystyle\sum_{j=k-N/2}^{k+N/2} \left[\sum_{i=1}^{M} f_{ij}\right]^2}{\dfrac{1}{M}\displaystyle\sum_{j=k-N/2}^{k+N/2} \sum_{i=1}^{M} (f_{ij})^2},$$

where M channels are summed; the coefficient is evaluated for a window of width N centered at k. It is basically the energy of the stack normalized by the mean energy of the components of the stack. This is equivalent to the zero-lag value of the autocorrelation of the sum trace divided by the mean of the zero-lag values of the autocorrelations of the component traces.

semiconductor: A substance such as germanium or silicon whose electrical conductivity at normal temperature is usually intermediate between that of a metal and an insulator, and whose conductivity is anisotropic. Its concentration of charge carriers increases with temperatures over a given range. Current flow may be by the transfer of positive holes or missing electrons (**p-type**) as well as by movement of electrons (**n-type**). Many common metallic sulfides and oxides are semiconductors. See *transistors*.

semiinfinite: Extending so far that some of the boundaries have no effect. Thus a **semiinfinite slab** is a horizontal unit of finite thickness bounded by a vertical plane on one side but extending so far in other directions that the boundaries in those directions create no measurable effects. A **semiinfinite prism** is a vertical prism with a bottom so remote as to not affect measurements.

sender: A current waveform generator for IP or resistivity surveying; **transmitter**.

sensitivity: The least change in a quantity which a detector is able to perceive. An instrument can have excellent sensitivity and yet poor *accuracy* (q.v.). Compare *readability*.

separation: 1. A difference in values between two measurements, especially well-log based measurements. 2. The difference in resistivity readings from two logging tools which have different depths of investigation (see Figures I-3 and M-6). Low-resistivity mud cake (which becomes thicker where formations are more permeable) causes the apparent resistivity of a shallow measurement to be lower than that of a deeper measurement, a situation called **positive separation**. See *microlog* and *movable-oil plot*. 3. Apparent displacement on a fault. 4. Displacement between source and receiver.

sequence analysis: The procedure of picking unconformities and correlative conformities on seismic sections so as to separate out the packages involved with different

FIG. S-5. **Seismoscope**. This Chinese seismoscope dates from about 100 A.D.. An earthquake would upset a pendulum fastened to the base of the vase and knock a ball from the dragon's mouth into the toad's mouth to indicate the direction from which the temblor came.

time depositional units. See Sheriff (1980, chap. 3 and 4).

setup: A particular arrangement of cables, geophones, shotholes, etc., for making a recording in the field. Several records are sometimes made at one setup.

seven-bit alphameric code: A computer code in which numeric, alphabetic and special characters are represented by seven binary positions: one check position, two zone positions, and four numeric positions. This is an even-parity code.

sextant: A doubly reflecting instrument for measuring angles, particularly the altitude angle of a celestial body above the horizon.

Sezawa M_2 wave: *Hydrodynamic wave* (q.v.).

sferics: Natural ''atmospheric'' fluctuations of the electromagnetic field, generally at frequencies from 1 to 10^5 Hz, caused by lightning discharges. See *Schumann resonance*. Also spelled **spherics**.

SFL: *Spherically focused log* (q.v.).

SH: See *SH-wave*.

Shadcon: A color display of seismic attributes. Western Geophysical tradename.

shadow zone: 1. An area in which there is little penetration of waves, usually because of the velocity distribution. 2. A portion of the subsurface from which reflections are unobservable because their raypaths do not reach the surface. The overlying beds may have such dips and velocity contrasts that raypaths to or from reflectors within the shadow zone become refracted or undergo total reflection. 3. A portion of the subsurface which does not evidence itself on refraction profiles, such as beds whose velocity is lower than an overlying refractor. See also *blind zone, hidden layer, channel wave*. 4.

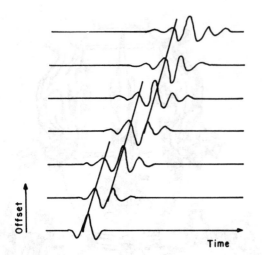

FIG. S-6. **Shingling** resulting from shift of energy in a wave train. Compare Figure D-13.

A region 100-140 degrees from the epicenter of an earthquake in which there is no direct penetration of seismic waves due to refraction from the low-velocity zone at the core boundary. **5.** A region of weakened amplitude, such as sometimes appears under a hydrocarbon accumulation.

shah: A *comb* (q.v.).

shake table: A test instrument on which geophones can be mounted and driven with known frequencies and amplitudes so that geophone characteristics can be determined.

shale baseline: 1. A line drawn through the minimum deflections characteristic of impermeable shales on an SP log (see Figure S-16), which is used as the reference in making measurements to determine the characteristics of sands and their formation waters. Compare *sand line* and see also *SP* and *baseline shift*. **2.** A line drawn through the characteristic of thick shales on the gamma-ray log.

shale potential: A part of the *electrochemical SP* (q.v.).

shaliness: The content of shale (or clay) in a dominantly nonshale formation; the degree to which ion-exchange processes contribute to resistivity measurements. Electrical conduction in shales is an ion-exchange process whereby electrons move between exchange sites on the surface of clay particles. See *dirty* and *pseudostatic SP*.

shallow-focus earthquake: An earthquake whose focus occurs at a depth of less than 50 to 70 km. Most earthquakes are of this type.

shallow-water survey: A geophysical survey in waters where conventional marine survey ships cannot operate easily because of insufficient water depth, obstructions such as reefs, etc.

shape anisotropy: Anisotropy resulting from preferred orientation of nonspherical particles which are themselves isotropic.

shaped charges: Explosives so designed that the explosive effect is concentrated in one direction. Their effect on seismic waves is usually minimal.

shaping deconvolution: Wiener *deconvolution* (q.v.) wherein the desired wavelet shape is specified. A zero-phase waveshape is often specified (see **phase characteristics**).

Sharpe's equation: Relation for the wave generated by a spherically symmetric source. See Sharpe (1942) or Sheriff and Geldart, v. 1 (1982, p. 47-48).

shear modulus: See *elastic constants*.

shear wave: An *S-wave* (q.v.).

shielding: 1. Enclosing electrical wires or components in electrical or magnetic conductors in order to reduce the effects of noise and electrostatic, magnetic, or electromagnetic coupling. **2.** The encircling conductors which shield the interior wires or components.

shingling: Forming an echelon pattern; see Figure E-4. **1.** On refraction recordings, loss of visibility of early cycles with increase in range; see Figure S-6. **2.** Incorrect migration of segments of a continuous reflection event so that the migrated segments do not join to form a continous event. **3.** A seismic facies pattern indicating prograding; see Figure R-8.

shipboard gravimeter: An instrument or system for measuring the acceleration of gravity from a ship in motion. Complex arrangements are used to insulate the meter from the many accelerations to which the ship is subject and to correct the data for the effects of measuring while moving (see *Eötvös effect*).

shock wave: A high-amplitude wave which propagates at greater than seismic (sonic) velocity. In contrast to an elastic wave.

shoot: 1. To fire an explosive. **2.** To generate seismic energy by means other than the detonation of explosives. **3.** To carry out a seismic survey, as "to shoot a prospect."

shootback method: An electromagnetic-surveying method which employs two tilted coils, each of which serves both as a transmitter and a receiver. At every station, readings are taken with each coil serving as transmitter for one and as receiver for the other. By averaging the readings, errors due to misalignment or topographic differences are essentially eliminated. Also called **Crone shootback**.

shooter: The person on a seismic party in charge of detonating the explosive.

shooting a well: The procedure of measuring directly the traveltime from a source on the surface to a geophone positioned in a well. Compare *vertical seismic profiling*.

shooting under: See *undershooting*.

shoot on paper: Thinking through a field program before starting it; calculating the results which may be expected, trying to anticipate problems which may occur and ambiguities which are likely to arise, and deciding whether the desired objectives are likely to be achieved.

shoran: Short-range navigation; a positioning system where the distances from reference transponders are determined by measuring the traveltime of pulsed radio waves to and from the station. See Figure S-7. Generally range is line-of-sight limited. See also *extended-range shoran*.

short Doppler count: See *satellite-navigation*.

short lateral: see *lateral*.

short normal: A normal resistivity log made with the A and M electrodes in the sonde about 16 inches apart. See *normal*.

short-path multiple: A multiple reflection in which energy is reflected back and forth over a small portion of the section so as to blend with the primary pulse, changing its waveshape and adding a tail. See Figure M-12.

short shot: *Weathering shot* (q.v.).

short-trace section: A record section composed of one trace from each successive record, the trace for the minimum source-geophone distance.

shot: 1. The detonation of an explosive. **2.** Any impulsive source of seismic energy. **3.** Any source of seismic energy. **4.** A measurement through an alidade or transit sighting on a stadia rod.

shot bounce: Noise on a seismic record caused by mechanical motion of the recording truck.

shotbox: *Blaster* (q.v.).

shot break: *Time break* (q.v.). **1.** The instant of the explosion. **2.** Initiation of a seismic wave by a nonexplosive source.

shot depth: The distance down the hole from the surface to the top (usually) of the explosive charge, often measured with loading poles. With large charges the distance to both top and bottom of the column of explosives is usually given.

shot elevation: Elevation of the top of the explosive charge in the shot hole.

shot hole: The borehole in which an explosive is placed for blasting.

shot-hole elevation: The elevation of the ground at the top of the shot hole.

shot-hole fatigue: See *hole fatigue*.

shot-hole log: The driller's record of the depth and lithologic characteristics of the formations encountered in a seismic shot hole.

shot instant: Time break, the time at which a shot is detonated.

shot-moment line: A wire wrapped around an explosive charge which is ruptured when the charge is detonated, a primitive way to record the time break.

shot noise: 1. *Hole noise* (q.v.). **2.** Noise in a semiconductor because of the discreteness of current carriers. Such rms current noise is proportional to the square root of the product of current and bandwidth.

shotpoint: Source point; abbreviated **SP. 1.** The location where seismic energy is released. Originally meant where the explosive charge is detonated but now refers to the location of any source of seismic energy, such as Thumper drops, Dinoseis pops, Vibroseis excitations, etc. Where patterns of sources are used, usually refers to the center of the pattern. **2.** The area surrounding the shotholes.

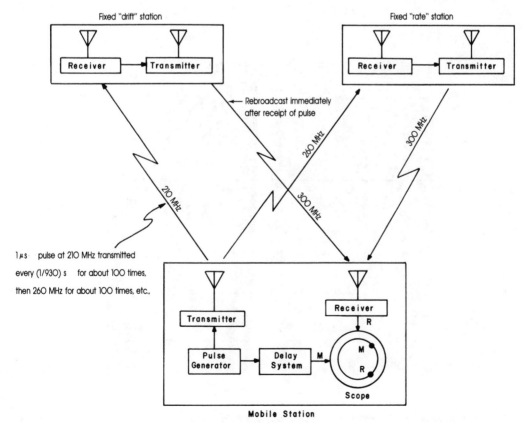

FIG. S-7. **Shoran** system.

shotpoint gap: The greater distance between geophone groups on opposite sides of the shotpoint in an otherwise uniform spread. Used so that the geophone groups nearest the shotpoint will be far enough away that source-associated noise will have little effect.

shotpoint seis: *Uphole geophone* (q.v.).

shoulder-bed effect: Effect of adjacent beds on a log reading. Also called the **adjacent-bed effect**. For example, high-resistivity beds adjacent to a low-resistivity bed may result in more current flowing in the low-resistivity bed than if the high-resistivity bed were not present, thus changing the apparent resistivity of the low-resistivity bed.

Shover: A horizontal vibrator used for generating *S*-waves. Prakla-Seismos tradename.

show (of oil or gas): A small amount of oil or gas in a well or a rock sample. ''Show'' usually signifies that it is not present in commercial quantity.

shuttle: The moving part of an air gun; compressed air is confined in the chamber when the shuttle is closed but released into the water when the gun is fired by opening the shuttle. See Figure A-4.

SH-**wave: 1.** An *S*-wave which has only a horizontal component of motion. **2.** The horizontal component of any *S*-wave. **3.** A surface *Love wave* (q.v.) is often mistaken for (and called) an *SH*-bodywave.

SI: International System of units, very similar to the mks or mksa system. See Appendix G and Figure M-1.

sial: Granitic, generally acidic continental crust, plus the overlying sediments. Composition is dominantly silica-aluminia composition and specific gravity about 2.7. The name comes from the chemical symbols Si, Al. Compare *sima*. The concept that the crust is made up of a sial layer over sima is now regarded as overly simplistic.

side lobe: A subsidiary passband outside of the main

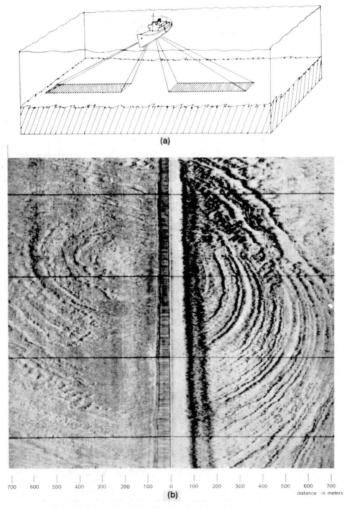

(a)

(b)

distance in meters

FIG. S-8. **Side-scan sonar.** (a) Schematic diagram; (b) record showing reflections from sea-floor relief. There is a blind zone under the ship's track. (Courtesy Compagnie Générale de Géophysique.)

passband. Can refer to filter curves, array directivity patterns, multichannel mixing arrangements, velocity filters, etc. See Figure D-12.

sideral hour angle: See *hour angle*.

side-scan sonar: A method of locating irregularities on the ocean bottom. A pulse of sonar energy (typically 120 kHz) is emitted from a fish which is towed 50 to 500 ft above the bottom, depending on the range and resolution sought. The sonar beam is narrow in the direction of traverse because the source consists of a line array of elements. Bottom irregularities (rock outcrops, pipelines, shipwrecks, boulders) and variations in bottom sediments produce changes in the amount of energy return; see Figure S-8. The arrival time measures the distance from the fish to the reflecting object. Other names include: asdic, basdic, sideways asdic, sideways-looking sonar, sideways sonar, echo-ranger, horizontal echo-sounder, and lateral echo-sounder.

side shot: A reading or measurement from a survey station to locate a point which is not intended as a base for an extension of the survey. Usually made to determine the position of some object to be shown on the map (such as a seismic source point).

sideswipe: Evidence of a structural feature which lies off to the side.

sidewall core: A formation sample obtained with a wireline tool from which a hollow cylindrical bullet is fired into the formation and retrieved by a cable attached to the bullet. The type of bullet and size of charge is varied to optimize recovery in different formations.

sidewall neutron log: An epithermal neutron log made with a skid which is pressed against the borehole wall and which may cut into the mud cake to minimize borehole effects. Abbreviated **SNP**. SNP is a Schlumberger tradename.

sidewall pad: A measuring device which is pressed against the side of a borehole, such as used with microresistivity logs, density logs, many radioactivity logs, etc.

sidewall sampler: A wireline device for taking *sidewall cores* (q.v.).

siemen: A unit of electrical conductivity; the reciprocal of ohm. Also called **mho**. Named for Werner (1816-1892) and Wilhelm (1823-1883) Siemens, German inventors who pioneered in electricity applications.

sight: 1. A bearing or angle measured with a compass, transit, or alidade. 2. Any established point on a survey. 3. An opening through which an object can be seen; used to determine the direction and/or distance to the object.

sigma: 1. *Standard deviation* (q.v.). 2. *Poisson's ratio* (q.v.). 3. A *sigma unit* (q.v.).

sigma unit: A unit of measure of capture cross-section. Also called *capture unit* (q.v.). Abbreviated **su**.

sigmoid configuration: A type of offlap *reflection configuration* (q.v.) signifying quiet-water deposition; see Figure R-8.

signal: That which is sought, which carries desired information. As opposed to **noise**. See *signal-to-noise ratio*. Sometimes ''message'' is used for the desired information and ''signal'' is used to include both message and noise.

signal averager: An electronic device used to stack a repetitive signal many times to improve the ratio of signal-to-random noise.

signal compression: A process whose objective is to shorten the effective length of the *embedded wavelet* (q.v.), usually by wavelet processing (deconvolution).

signal correction: A correction for differences in timing or waveshape resulting from changes in the outgoing signal at various source locations.

signal enhancement: Vertical stacking, that is, adding successive waveforms from the same source point and thereby discriminating against random noise. Especially used with seismic recorders for engineering work.

signal/noise ratio: See *signal-to-noise ratio*.

signal theory: The concept that a relatively pure signal is transmitted from a source through some medium, is received at a receiving station together with superfluous information called noise, and that the problem is to separate the signal result from the noise so that the final result approximates as closely as possible the original signal.

signal-to-noise ratio: The energy (or sometimes amplitude) divided by all remaining energy (noise) at the time; abbreviated **S/N**. Sometimes the denominator is the total energy, that is, $S/(S + N)$. Signal-to-noise ratio is difficult to determine in practice because of the difficulty in separating out the signal (the desired portion).

signature: 1. The aspect of a waveshape which makes it distinctive; **character**. 2. A waveshape which distinguishes a particular source, transmission path, or reflecting sequence.

signature deconvolution: A wavelet processing or deconvolution operation in which one attempts either to remove trace-to-trace variations in the *embedded wavelet* (q.v.) or replace the embedded wavelet with some other wavelet shape.

signature log: See *full-waveform log*.

sign bit: The bit which indicates the algebraic sign (plus or minus) of the number.

sign-bit recording: 1. Recording for a series of measurements only the information as to which samples were positive and which negative. 2. Recording of zero-crossings only, that is, when values changed from positive to negative or vice versa.

significance: The ratio of the smallest change which can be detected to the magnitude of a quantity at that time, such as the number of significant bits in a quantity. See Figure D-21 and *dynamic range*.

significance level: The complement of probability; e.g., 10 percent significance = 90 percent probability.

sima: Basaltic, peridotitic, basic oceanic crust, of specific gravity 3.0 to 3.3. Name derives from its silicamagnesium composition. Compare *sial*.

similar fold: See *folding*.

Simplan: A method to determine the response of the Earth to plane or cylindrical waves by summation of observations with spherical waves. Acronym for simulated plane waves. A Seiscom Delta United tradename. See Sheriff and Geldart, v. 2 (1983, p. 55-56).

Simplan stack: Stack of a common-source gather without applying normal-moveout corrections. See *Simplan*.

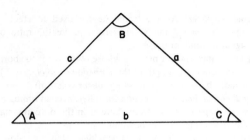

Law of sines:

$$\frac{a}{\sin A} = \frac{b}{\sin B} = \frac{c}{\sin C}$$

Law of cosines:

$$c^2 = a^2 + b^2 - 2ab\cos C$$

FIG. S-9. Laws of **sines** and **cosines**. The law of sines is: $a/\sin A = b/\sin B = c/\sin C$. The law of cosines is: $c^2 = a^2 + b^2 - 2ab\cos C$.

simple multiple: A long-path multiple which has undergone only three reflections (that is, twice reflected from a deep interface and once from a shallow interface, usually that at the base of the weathering or at the surface). Most multiples are more complex than this.

simple pole: See *pole*.

simplex: A one-way circuit; *half duplex* (q.v.).

simulator: A device or a computer program that simulates the operation of another device or computer.

sinc x: $(\sin x)/x$, also called the **diffraction function**. This function is the Fourier transform of a unit *boxcar* (q.v.) and is extensively used in seismic data processing. Sometimes sinc $x = (\sin \pi x)/\pi x$.

sines, law of: In any plane triangle (Figure S-9), the ratio of the lengths of any two sides is the same as the ratio of the sines of the angles opposite the sides:

$$a/(\sin A) = b/(\sin B) = c/(\sin C).$$

sine tranform: The *Fourier transform* (q.v.) of the odd or antisymmetrical part of a function. The **cosine transform** involves the even or symmetrical part.

sing around: A system where the echo from one transmitted pulse triggers the next pulse; used in velocimeters. The sing-around frequency is related to the path length and the propagation velocity.

singing: Ringing or *reverberation* (q.v.) produced by short-path multiples in a water layer.

single-ended spread: A reflection profile which is shot from one end. Also called **end-on spread**. See Figure S-17 and compare *off-end shooting*.

single-shot tool: A device to obtain one measurement of the direction of a borehole at a particular depth. See *directional survey*.

singularity: 1. A point where a function is not differentiable. **2.** A value of a variable for which a function becomes infinite. Also called a **pole**.

singy: Having an oscillatory character; **ringy**.

sinistral: Rotational to the left or counter-clockwise. A "sinistral" wrench fault is **left lateral**. Opposite to **dextral**. See Figure F-2.

sink: A singularity or pole to which lines (of force) converge, the opposite of a **source**.

sinusoid: A sine or cosine curve.

site damping: See *damping*.

six-fold: Having a redundancy of six. "Six-fold common-midpoint shooting" involves recording such that six different traces have the same midpoint.

skew: 1. A condition of the bits not being written straight across a magnetic tape. It occurs when a magnetic tape is not mounted properly with respect to the magnetic heads. The tracks may deviate from their proper position and produce crossfeed, time displacement between channels, parity errors, etc. **2.** In magnetotellurics, an invariant of the tensor impedance which indicates the amount of three-dimensionality. It is zero for one- and two-dimensional earths with noise-free data. **3.** See *skewness*.

skew box: A hardwired device for automatically correcting for skew when reading a digital tape.

skewness: Asymmetry in a distribution. See *statistical measures*.

skew-symmetric matrix: See *matrix*.

skid: 1. A mounting for a borehole sonde which cuts into the mud cake and presses the sonde against the borehole wall to minimize borehole effects. See Figure D-3. **2.** A sled on pontoons on which geophysical equipment is carried and which is dragged from one location to another. Used in marsh work.

skidded shot: A shotpoint which has been moved a short distance from its normal location, usually because of access difficulties.

skin depth: The effective depth of penetration of electromagnetic energy in a conducting medium when displacement currents can be neglected. The depth at which the amplitude of a plane wave has been attenuated to $1/e$ (or 37 percent):

$$\text{Skin depth} = \delta = (2/\sigma\mu\omega)^{1/2} \text{ meters,}$$

where σ = conductivity in mhos/meter, μ = permeability in henrys/meter, and ω = angular frequency in radians/second. Also called **effective depth**.

skin effect: 1. The tendency of alternating currents to flow near the surface of a conductor. **2.** A reduction in apparent conductivity. The propagation of an electromagnetic wave through a conductive formation results in a phase shift. In induction logging in high-conductivity formations, a correction is made for this effect. See *skin depth*.

skip: 1. See *cycle skip*. **2.** A portion of a seismic section where data are not available. May refer to a station which was not occupied for some reason or to stations near the source point which are not being used because they are too noisy. **3.** A local loss of information on a continuous profile.

skip distance: The minimum distance from a transmitting

antenna at which a *sky wave* (q.v.) can normally be received.

skip mixing: The mixing of data from alternate (nonadjacent) channels. Consequently, adjacent output channels do not contain common input information.

sky wave: Electromagnetic (radio) waves which reflect from ionized layers in the ionosphere. Involved in sky-wave interference and in making radio waves receivable beyond the line-of-sight horizon. Compare *tropospheric scatter*.

sky-wave interference: Interference between the direct (or ground) radio wave and waves reflected from ionized layers in the ionosphere. The ionization results from sunlight and the ionized layers vary around sunrise and sunset, so sky-wave interference is especially variable at these times. Sky-wave interference degrades the accuracy and range of radio-positioning systems.

Slalom: *Crooked line* (q.v.). CGG tradename.

slant path correction: A correction to side-scan sonar data to yield a display with linear scales.

slant range: A distance measurement which involves both horizontal and vertical components, such as the distance from an observing station to a navigation satellite.

slant stack: Time-shifting traces proportional to their distance from some reference point and then stacking; the effect is to emphasize events with certain dips, that is, to *beam steer*. Also used with *tau-p mapping* (q.v.).

SLAR: Side-looking airborne radar. See *SLR*. Aero Service tradename.

slave drum: A recording drum which is kept synchronous with another drum.

slave station: A transmitting station used to retransmit signals from another station so that the two transmissions are synchronous and thus will set up standing wave patterns. Used in phase-measurement positioning systems. The slave station may or may not be under the control of the master station.

sleeper: An explosive charge loaded into the hole an appreciable time (up to several days) before it is to be used.

Sleeve exploder: A seismic marine energy source in which a gas (propane or butane) is exploded in a thick rubber bag (the **sleeve**). Waste gases are vented directly to the air rather than into the water, thus reducing the bubble effect. See Figure S-10. Esso Production Research tradename. Also called **Aquapulse**, a Western Geophysical tradename.

slew rate: 1. The highest speed at which a digitizer input device can be moved without affecting the accuracy. **2.** Reciprocal of the time delay between the sampling of successive channels in a multiplexing operation. Compare *slue*.

S-line: See *S-rule*.

slingram: A moving-source electromagnetic profiling method in which the mutual coupling between two loops is measured. Primary field effects are nulled by a cable link between the coils. Sometimes called *horizontal-loop method* (q.v.).

slip-vector analysis: The first motion from strike-slip earthquakes appears as a push or pull depending on the location of the observing station with respect to the

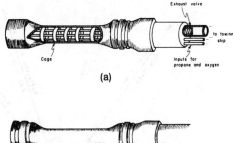

(a)

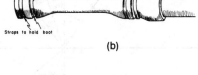

(b)

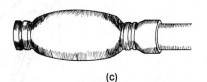

(c)

FIG. S-10. **Sleeve exploder.** (**a**) Metal framework made of pipes over which thick rubber boot is placed. Water passing through the pipes cools the unit. (**b**) Boot over cage prior to explosion. (**c**) Boot expands because of pressure of exploding gases. The framework prevents complete implosive collapse after the explosion, thus lessening the bubble effects. (Courtesy Exxon Production Research.)

epicenter and the direction of first motion involved in the earthquake. Objective of analysis is determining the fault motion involved. See *nodal plane*.

slope-distance rule: See *depth rule*.

Slotnick method: A graphical refraction interpretation method applicable for plane multilayer refractors. See Slotnick (1950).

slowness: The reciprocal of velocity.

SLR: Side-looking radar. A remote-sensing method which involves sweeping the earth to either side of an aircraft with a radar beam and recording the reflected signals, which are displayed so as to give the appearance of an aerial photograph. An aircraft at 20,000 feet can map a strip about 12 miles wide. Also **SLAR**.

slue: To turn about its own axis. The sluing characteristics of a gyrocompass enable it to follow faithfully variations of the ship's heading. Also spelled **slew**.

slug: See *radioactive-tracer log*.

slurry explosive: A bulk-type explosive which can be poured into boreholes. Not cap sensitive and requires a primer charge to detonate.

slush pit: The pit used in rotary drilling for storage of water or mud for circulation through the hole. Muds can be mixed in the pit. Sometimes a pit is dug in the ground, sometimes a portable sheet-iron pit is used.

smart stacking: *Selective stacking* (q.v.).

smear: 1. To mix data not in complete register. **2.** To average seismic data originating from sources at different locations, or recorded by geophones at different

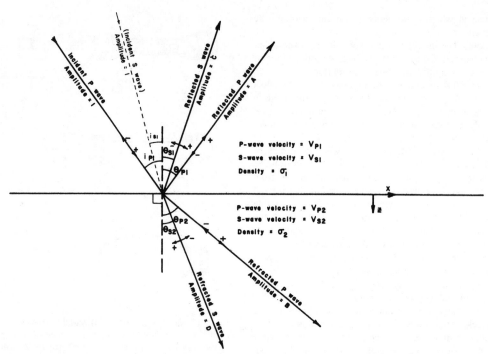

FIG. S-11. **Snell's law** relations for either incident *P*-wave or *S*-wave: $V_1/\sin i = V_{P1}/\sin \Theta_{P1} = V_{S1}/\sin \Theta_{S1} = V_{P2}/\sin \Theta_{P2} = V_{S2}/\sin \Theta_{S2} = p$.

locations, or both; a consequence of ground mixing. The data from several shots or other source impulses often are combined without adjustment by vertical stacking, especially with surface sources. The amount of smear is the distance over which the sources are placed plus the distance over which the geophones feeding one channel are planted. **3.** The effect of stacking common-midpoint traces for a dipping reflector, because the reflecting point is not common. See Figure C-8.

smile: A wavefront-shaped event on a migrated seismic section, the result of the migration of a noise burst. Also produced by data truncation.

smoothing: 1. Averaging adjacent values according to some scheme; involves filtering out higher frequencies. Often accomplished by use of a *running window* (q.v.). **2.** Straightening a good, shallow reflection and using it as a reference to remove undesirable lateral velocity and topography variations. Also called **datuming**.

S/N: *Signal-to-noise ratio* (q.v.).

Snell's law: When a wave crosses a boundary between two isotropic media, the wave changes direction such that

$$(\sin i)/V_i = (\sin \theta_{P1})/V_{P1} = (\sin \theta_{S1})/V_{S1}$$
$$= (\sin \theta_{P2})/V_{P2} = (\sin \theta_{S2})/V_{S2} = p,$$

where *i* is the incident wave with a velocity $V_i = V_{Pi}$ if a *P*-wave, or $V_i = V_{Si}$ if an *S*-wave; θ_{P1} and θ_{S1} are the angles of reflection of the *P*- or *S*-waves in medium 1, which have velocities V_{P1} and V_{S1}, respectively; θ_{P2}

and θ_{S2} are the angles of refraction of the *P*- and *S*-waves in medium 2 which have velocities V_{P2} and V_{S2}, respectively; *p* is the **raypath parameter**. The respective angles are shown in Figure S-11. If sin θ_{P2} or sin θ_{S2} exceed 1 as given by this equation, total reflection occurs. See also *Zoeppritz's equations*. The relationships become complicated if anisotropy is present; then additional shear wavefronts are generated and the wavefronts are not necessarily orthogonal to the direction of wave propagation. Snell's law is also called **Descartes' law**. Named for Willebrord Snell (1591-1626), Dutch mathematician.

sniffer: 1. A device which collects samples of sea water and analyzes them for hydrocarbon content by determining rate-of-flow through a diffusion column [*colorometric* (q.v.) technique]. **2.** A device which collects gas samples and analyzes them for radioactive content.

snorkel: A tube to the surface from an underwater energy source through which waste gases escape.

SNP: *Sidewall neutron* (porosity) *log* (q.v.).

soak: To allow a system to come to equilibrium with its surroundings. For example, a borehole gravimeter may have to rest at the bottom of a deep well bore for a period of time to allow it to adjust to the temperature and pressure regimen before meaningful readings can be obtained from it.

SOFAR: A depth region in the oceans of low velocity which carries *channel waves* (q.v.). See Figures C-3 and W-1.

soft error: A read error that can be recovered from by one or more rereads.

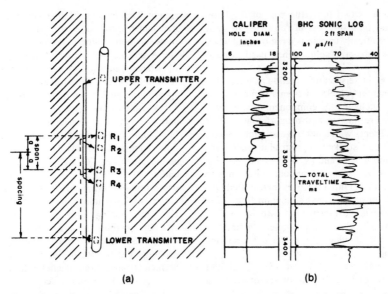

FIG. S-12. **Sonic log**. (a) Schematic compensated sonic logging sonde. (b) Sonic log. (Courtesy Schlumberger.)

soft formation: A poorly consolidated sand-shale sequence.

soft magnetism: That component or portion of remanent magnetization which has relatively lower coercive force. In paleomagnetism studies this softer magnetization is removed by alternating field demagnetization in order to isolate harder remanent magnetism.

soft mantle: Upper mantle which has lower than normal velocity; occurs under plate boundaries.

soft rock: Sedimentary rock. Used to distinguish between metal mining (**hard rock**) and petroleum (**soft rock**) objectives.

soft spring: A spring with very low natural frequency, used to insulate from high-frequency noise.

software: Programs for data processing, including those that control the internal operation of the processing system itself.

Sokolov rule: See *depth rule* and Figure D-6.

solar wind: Ionized particles flowing radially outward from the sun. Transient magnetic disturbances (see *K-index*) are correlated with solar wind variations.

sole: The lowest movement surface of a fault, especially where a gravity listric fault or a thrust fault becomes a bedding-plane fault. Also called the **fault baseplate**.

solid-state circuitry: The use of semiconductor elements such as transistors, integrated circuits, etc., which do not require much space or power.

sonar: Sonic (acoustic) waves in the water. Used for navigation, positioning, and communications. See *Doppler sonar, acoustic positioning,* and *side-scan sonar.*

sonar reference intensity: For a plane wave, an rms pressure of 1μPa.

sonde: A logging tool such as is lowered into a borehole to record (while being withdrawn) resistivity, sonic, radioactivity or other types of well logs.

sonic: Pertaining to acoustic or *P*-waves in fluids. Sometimes includes other wave modes and hence becomes synonomous with seismic and elastic.

sonic log: A well log of the traveltime (transit time) for acoustic waves per unit distance, the reciprocal of the *P*-wave velocity. Also called **acoustic velocity log** and **continuous velocity log**. Usually measured in microseconds per foot. Used for porosity determination by the *time-average equation* (q.v.). The interval transit time is integrated down the borehole to give total traveltime. For the compensated sonic log, two transmitters are pulsed alternately and measurements are averaged to cancel errors due to sonde tilt or changes in hole size. See Figure S-12 and also *cycle skip, three-D log, full-waveform log, cement bond log,* and *fracture log.*

sonic wave: *Acoustic wave* (q.v.).

sonobuoy: 1. A device used in marine refraction surveys for detecting energy from a distant shot and radioing the information to the recording ship; see Figure S-13. Used in marine refraction and extended profile work. The sonobuoy is a free-floating buoy which usually is simply thrown off the shooting ship. Once the sonobuoy is in the water, sea water activates the buoy's batteries, one or more hydrophones are dropped, and a radio antenna is extended upward into the air. As the ship travels away from the buoy, firing charges (or other energy source) as it goes, the seismic arrivals are detected by the hydrophones and transmitted to the ship where they are recorded and timed. The distance from the energy source to the sonobuoy can be determined by the arrival time of the wave which travels directly through the water. The buoy is expendable and sinks itself after a certain time, the cost of the buoy usually being less than the cost of retrieving it. **2.** A buoy which automatically transmits a radio signal when triggered by a water-borne sonic signal; used in positioning.

sonogram: A display of seismic information as a function of dip or the apparent velocity of events. An ensemble of input traces such as the component traces of a reflection profile are time shifted proportional to their position so that events with a certain apparent velocity line up; they are then stacked to produce a single trace on the sonogram. Large energy is present where a coherent event occurs whereas random noise tends to cancel. The ensemble of input traces is then shifted by a different amount to make a next sonogram trace. The complete sonogram is the transform of the data from the space-time domain to the moveout-time domain. See Figure S-14. Sonogramming is also called **Rieberizing** and **sonographing**.

sonograph: 1. *Sonogram* (q.v.). **2.** An acoustic picture obtained under water by side-scan or sector-scanning sonar.

Sonoprobe: A marine echo sounder that generates sound waves and records their reflections. Usually has more power and penetration than a fathometer but less than a sparker or gas exploder. Mobil Oil tradename.

sophisticated: Complex or intricate. Often refers to methods which were not feasible before the use of high-speed digital computers.

sorption: The binding of one substance to another by mechanisms such as adsorption (holding on the surface) or absorption (taking in completely).

Sosie: A seismic method which employs a random series of seismic impulses to generate seismic waves. The recorded data can be correlated with the random series to produce an interpretable result. Societe Nationale Elf-Aquitaine tradename.

sound: 1. To measure the depth of water. **2.** To determine how some quantity varies with depth. **3.** *P*-waves in fluids such as air or liquids.

sound channel: See *SOFAR*.

sounding: Measuring a property as a function of depth; a **depth probe** or **expander**. Especially a series of electrical resistivity readings made with successively greater electrode spacing while maintaining one point in the array fixed, thus giving resistivity-versus-depth information (assuming horizontal layering); **electric drilling**, **probing**, **VES** (vertical electric sounding). Also used with electromagnetic, magnetotelluric, and other types of geophysical surveying. See also *geometric sounding* and *parametric sounding*. As opposed to **profiling** where the objective is to ascertain lateral rather than vertical variations in a quantity.

sound wave: *Acoustic wave* (q.v.).

source: 1. A device that releases energy, such as an explosion or an air gun release for seismic energy. Some seismic source waveforms are shown in Figure S-15. **2.** The point from which lines of force in a vector

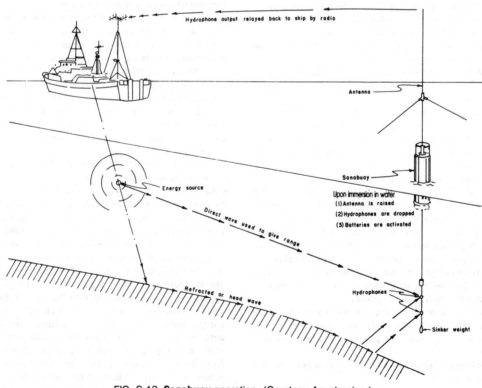

FIG. S-13. **Sonobuoy** operation. (Courtesy Aquatronics.)

field originate; a mass for the gravitational field. Opposite of **sink**.

source code: See *source program*.

source pattern: See *array (seismic)*.

source point: The location of a seismic source. Abbreviated SP.

source program: A computer program prior to machine decoding; e.g., a program in some symbolic language such as Fortran.

source-receiver product: The number of separate raypaths mixed together to produce the final display; also called **effort**. The product of the number of holes per shot (or impulses per record), the number of geophones per group, the number of records stacked, and (for Vibroseis) the duration of the sweep.

southing: See *latitude*.

south-seeking pole: See *magnetic pole*.

SP: Source point or shotpoint.

SP: Spontaneous-potential or **self-potential. 1.** A well log of the difference between the potential of a movable electrode in the borehole and a fixed reference electrode at the surface. The SP results from *electrochemical SP* and *electrokinetic potentials* (q.v.) which are present at the interface between permeable beds adjacent to shale. In impermeable shales, the SP is fairly

constant at the **shale-baseline** value (see Figure S-16). In permeable formations the deflection depends on the contrast between the ion content of the formation water and the drilling fluid, the clay content, the bed thickness, invasion, bed-boundary effects, etc. In thick, permeable, clean nonshale formations, the SP has the fairly constant **sand-line** value, which changes if the salinity of the formation water changes. In sands containing disseminated clay (shale), the SP will not reach the sand line and a **pseudostatic** SP value will be recorded. The SP is positive with respect to the shale

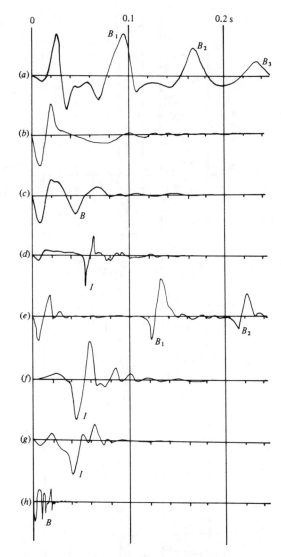

FIG. S-15. Far-field waveforms generated by marine seismic **sources**. (**a**) Single air gun; (**b**) array of air guns; (**c**) sleeve exploder; (**d**) Vaporchoc; (**e**) Maxipulse; (**f**) Flexichoc; (**g**) water gun; (**h**) sparker. Amplitudes are not to scale. *B* indicates bubble effects, I indicates implosion. (From Sheriff and Geldart, 1982, p. 180.)

FIG. S-14. **Sonogram.** (**a**) Hypothetical seismogram showing two events and a noise burst on one channel. (**b**) Sonogram of these events. (From Trorey, 1961.)

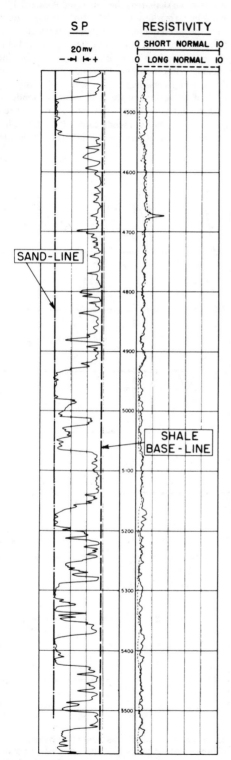

SP

20 mv
—•→| |←•→

RESISTIVITY

0 SHORT NORMAL 10

0 LONG NORMAL 10

SAND-LINE

SHALE
BASE-LINE

FIG. S-16. **SP log** in sand-shale series with fresh mud in the borehole. (Courtesy Schlumberger.)

baseline in sands filled with fluids fresher than the borehole fluid. See also **SSP. 2.** The natural ground voltage observed between nearby nonpolarizing electrodes in field surveying. In many mineralized areas this is caused by electrochemical reaction at an electrically conducting sulfide body. In geothermal areas, SP can be caused by the motion of ions (streaming potential) or from contrasts in temperature. Compare *induced polarization.*

space filtering: See *apparent velocity filtering.*

space-frequency domain: The two-dimensional transform of a seismic section from location versus time to wavenumber versus frequency.

spacer section: See *streamer.*

space-time filter: An *apparent-velocity filter* (q.v.) or frequency-wavenumber (f,k) filter. Also called **beam pointing**.

spacing: The separation of certain electrodes or sensors on logging sondes. In nuclear devices, usually the distance from the source to the detector. See Figures D-3, E-7, and S-12, and compare *span.*

span: The separation of certain sensors on logging sondes. On the sonic sonde, span is the distance between two receivers of a pair whereas **spacing** is the distance from transmitter to the midpoint of the corresponding receiver pair; see Figure S-12.

span adjustment: Calculation of a log which would have resulted from the use of a span or spacing different from the one actually used.

sparker: A seismic source in which an electrical discharge in water is the energy source. The discharge is between two electrodes in the salt water; the heat generated by the discharge vaporizes the water, creating an effect equivalent to a small explosion. See also *exploding wire.*

spatial aliasing: Aliasing resulting from spatial sampling; see *alias.*

spatial frequency: *Wavenumber* (q.v.), the number of wave cycles per unit of distance in a given direction (direction of the spread).

spatial sampling: Making measurements only at discrete locations. The sampling potentially involves spatial aliasing problems (see *alias*).

SP buckout: A variable voltage-compensation circuit in series with the input terminals of an IP, resistivity, or SP receiver. Used to match the input voltage level of the voltmeter-receiver to that of the ground. The buckout voltage is the dc self-potential.

specific: Refers to normalized dimensional or volume properties of a material.

specific acoustic impedance: Acoustic impedance divided by the acoustic impedance of water.

specific capacity: A polarization parameter in the time domain similar to the metal factor in the frequency domain. Long-time chargeability (area under decay curve) divided by resistivity. Also called **static capacity.** Dimensions are farad per meter.

specific conductance: See *conductivity.*

specific factor: See *factor analysis.*

specific impedance: See *resistivity.*

specific resistance: See *resistivity.*

spectral density: The square of the amplitude spectrum; see **Fourier transform**.

spectral gamma-ray log: A log of natural gamma-radiation intensity within discrete energy bands which are characteristic of specific radioactive series (uranium-radium, thorium) or of potassium-40. Differs from the conventional gamma-ray log which measures broad-spectrum, undifferentiated gamma energy. Useful for correlation where other methods fail. Also useful for uranium exploration where thorium series or potassium minerals contribute significantly to total gamma radiation. See also *induced gamma-ray spectroscopy log* and *natural gamma-ray spectroscopy log*.

spectrum: **1.** Amplitude and phase characteristics as a function of frequency for the components of a wave train or wavelet. See *Fourier analysis.* **2.** Filter response characteristics. See *transfer function.* **3.** Other ''quantities'' displayed so as to show the relative content of various components. Thus a **velocity spectrum** (or **normal-moveout spectrum**) shows the amount of coherent energy which appears to have various amounts of normal moveout as a function of arrival time. A **dip spectrum** (or **sonograph**) shows the amount of coherent energy which appears to have various amounts of dip moveout or apparent velocity.

specular: Mirror-like.

spheric: *Sferics* (q.v.).

spherical coordinates: A system of three-dimensional coordinates defined by a radius and two angles (like latitude and longitude). See Figure C-13.

spherical divergence: **1.** The decrease in wave strength (energy per unit area of wavefront) with distance as a result of geometric spreading. A spherical wave traveling through the body of a medium continually spreads out so that the energy density decreases. For a point source the energy density decreases inversely as the square of the distance the wave has traveled. For energy which travels along a surface, the analogous term is **cylindrical divergence**, which varies inversely as the distance. Other mechanisms by which a seismic wave loses energy involve absorption and loss at interfaces by reflection (including diffraction, mode conversion, and scattering). **2.** The decrease in field strength (flux density) for gravity, magnetic, and similar fields where the intensity decreases as the square of the distance.

spherical excess: The amount by which the sum of the angles of a triangle on the surface of a sphere exceeds 180 degrees. For a sphere, this excess is the area of the triangle. For a spherical triangle on an ellipsoid such as the earth approximates, the spherical excess ϵ is approximately:

$$\epsilon = mbc \sin \alpha$$

where b and c are two adjacent sides of the triangle which intersect at the angle α, m = latitude function = $\rho/2RN$, ρ = number of seconds of arc/radian = 206264.8, R = radius of curvature in the meridian, and N = radius of curvature in the prime vertical. Values of m for various ellipsoids are obtained from tables; the

value for the center of the triangle to the nearest half-degree is usually used. For greater precision a correction factor is often applied in iterative fashion.

spherical harmonic: Solution of Laplace's equation in spherical coordinates.

spherically focused log: A short-normal resistivity device to which current-focusing electrodes have been added to maintain approximately spherical distribution of the measuring current for better thin-bed resolution and response even with high formation-to-mud resistivity contrast. Acts to prevent current travel in the borehole. SFL is a Schlumberger tradename.

spherical triangle: The triangle formed on the surface of a sphere by the intersections of three arcs of great circles.

spherical wave: A wave generated by a point source. In the case of constant velocity V, a spherical wave is any function

$$(1/r) f(r \pm Vt)$$

where r is distance from the source.

spherical wavefront: Surface which a given phase of a seismic impulse generated by a point source occupies at any particular time. The surface is not necessarily spherical if the velocity varies with location.

spheroid: The oblate ellipsoid of revolution used to approximate the Earth's shape. The Earth's shape can now be determined by radar ranging from satellites. See Figures G-2 and G-3 and *Geodetic Reference System*.

SPI: *Secondary-porosity index* (q.v.).

spike: **1.** To work away from the crew's normal area; **hot shot**. **2.** An *impulse* (q.v.).

spiking deconvolution: Deconvolution in which the desired wavelet is a spike or impulse. Also called **whitening deconvolution.**

spill point: The lowest contour on a hydrocarbon trap capable of holding hydrocarbons under gravitational equilibrium if the formation is permeable.

spinel: A mineral with the general formula AB_2O_4. Some iron minerals are spinels and this crystal structure is important in magnetic and electrical properties.

spinner magnetometer: A device which spins a sample and measures the induced ac voltage to determine the strength and direction of the sample's magnetic field.

spinner survey: A log of the rate of flow of fluid in the wellbore, casing, or tubing. See *flowmeter*.

S-P interval: In earthquake seismology, the time interval between the first arrivals of P- and S-waves, one measure of the distance from the earthquake hypocenter.

spirit leveling: Determining relative elevation by sighting on the rod when the sighting level is horizontal. If the elevation difference exceeds 1-2 m, the process has to be carried out in steps.

spit-out: To print out detailed data. Sometimes, to *dump* (q.v.).

S-plane: The Laplace transform domain σ versus ω, where $s = \sigma + i\omega$. See *Laplace transform*.

spline: **1.** A long flexible strip used in drawing a smooth curve. **2.** A digital-to-analog conversion method employing a curve (surface) fitting scheme which assures a

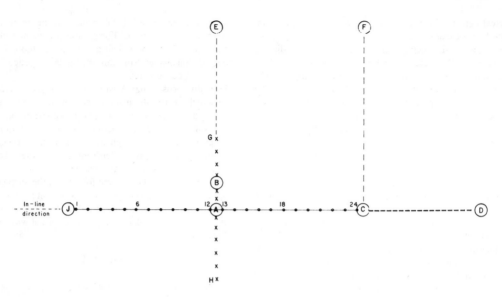

FIG. S-17. **Spread** arrangements for 24 in-line geophone groups. **Split spread** or **split dip** if the shotpoint is at A; **offset split dip** if the shotpoint is at B; **end-on** if the shotpoint is at C; **in-line offset** if the shotpoint is at D; **broadside T** if the shotpoint is at E; **broadside L** if the shotpoint is at F; and **cross** if some geophone groups are at right angles (G − H).

desired degree of smoothness to the fitted data. An *n*th order spline has all derivatives up to *n*-1 continuous. Thus a quadratic spline has a continuous first derivative and a cubic spline has both first and second derivatives continuous. **3.** Both the long flexible strip and analytic splines are sometimes used in residualizing, the smooth curve representing the regional and the difference between the smooth curve and the gravity profile representing the residual. By extension, a smooth surface used to represent a regional gravity field.

split: *Split spread* (q.v.).

split-dip shooting: See *split spread*.

split spread: A method of reflection shooting in which the source point is at (or perpendicularly offset from) the center of the geophone spread. Also called a **straddle spread**. See Figure S-17. A split-spread record is shown in Figure R-5.

spontaneous-potential: Self-potential or *SP* (q.v.).

spot correlation: Correlation of reflections on nonadjacent seismic records based on character or intervals between events.

spread: 1. Arrangement of geophone groups in relation to the source point. Various arrangements are used; see Figure S-17. See also Figure F-1 for fan shooting, Figure T-3 for 3-D spreads. Spreads are **interlocking** if the geophone group location and the source point for one profile are located at the source point and geophone group location (respectively) for another profile (for example, source at A into geophone 24 and source at C into geophone at 13 in Figure S-17). Spreads are **reversed** if the same array of geophones is shot into from shotpoints in opposite directions in-line (for example, shot into spread from 1 to 24 from both shotpoints at C and J in Figure S-17). A **microspread** has very small

geophone group intervals (2-15 ft). **2.** The layout of electrodes or antennas in resistivity or electromagnetic surveying. See *array (seismic)*.

spread correction: 1. *NMO* or *normal moveout correction* (q.v.). **2.** Correction applied to refraction data to make *reduced traveltime* (q.v.) curves.

spreading: Divergence; loss of amplitude because of geometrical spreading; spherical divergence (for body waves) or cylindrical divergence (for surface waves).

SP reduction factor: The ratio of actual *SP* to *SSP* (q.v.). See *pseudostatic SP*.

SPS: Shotpoint seismometer or *uphole geophone* (q.v.).

spudder: A drill used for making holes in hard rock. The bit is raised and dropped and the resulting cuttings are removed by a bail, a pipe with a flap valve at the bottom. Also used to pound casing into gravel or formations with boulders such as glacial drift.

spur: See *trace*.

square-wave: A **full square-wave** is a waveform consisting of alternating equal magnitude "positive on" and "negative on" portions. A **half square-wave** is switched on and off. A **pulsed square-wave** has portions which are "positive on, off, negative on, and off." See Figure S-18.

squash plot: See *compressed section*.

squeeze camera: A camera or printer that changes scale in one direction (usually horizontally) without altering the scale in the orthogonal direction.

squeeze section: A section with a highly reduced horizontal scale; see *compressed section*.

squelch circuit: A control used in uphole geophone amplifier circuitry which permits the uphole geophone signal to be recorded on one of the ordinary geophone channels prior to the first breaks. It then blocks or disconnects the uphole geophone signal after the uphole break has been recorded, so that the uphole geophone output will not interfere with the subsequent record.

Squid magnetometer: A sensitive magnetometer which detects magnetic field changes by means of a superconducting loop containing one or two Josephson junctions. Acronym for "superconducting quantum interference device". A Squid carries supercurrent up to a certain critical value, beyond which a finite resistance appears in the loop. The value of this critical current depends upon the external flux as well as the geometry. In the **rf-Squid magnetometer**, a loop with one Josephson junction is driven inductively by a high-frequency (typically 30 MHz) alternating current which periodically exceeds the critical current. The voltage appearing depends upon the value of the critical current which in turn depends upon the external flux. In the **dc-Squid magnetometer** a dc-current slightly greater than the critical current is fed into a loop containing two Josephson junctions. This produces high-frequency oscillations in the loop due to the ac-Josephson effect. Thus the current periodically exceeds the critical current and a voltage appears which depends upon the external flux. Both the rf- and the dc-Squids are usually incorporated in a negative feedback circuit which detects and nulls the flux. The output of the negative feedback circuit is proportional to the output of the Squid. Squid magnetometers are capable of detecting fields on the order of 10^{-5} gamma, and are used in magnetotelluric and controlled-source electromagnetic field techniques. See Clarke (1974) and Weinstock (1981).

squiggle: A **wiggle trace** or trace of galvanometer deflection versus time. See Figure D-14.

S-rule: An application of the equivalence principle used in resistivity sounding. Sounding graphs over a series of strata with resistivities ρ_i and thicknesses h_i above a highly resistive substratum possess the same asymptote if the sum of longitudinal conductance (h_i/ρ_i) is constant, i.e., if

$$S = \sum (h_i/\rho_i) = \text{constant},$$

S is the **admittance**.

S/(S + N) filter: See *Wiener filter*.

SSP: Static self-potential; the maximum SP that would be recorded when the borehole logging sonde passes from a position well inside a very thick, porous, permeable, clean sand to a point well within a thick shale. The **electrochemical SSP** (the electrokinetic SP is often neglected) is approximately given by:

$$\text{SSP} = -K \log_{10}(a_w/a_{mf}),$$

where a_w is the activity of the formation water and a_{mf} that of the mud filtrate. Because of the inverse relationship between activity and equivalent resistivity, this equation can be written:

$$\text{SSP} = -K \log_{10}(R_{mfe}/R_{we}).$$

For NaCl muds that are not too saline, $R_{mfe} = R_{mf}$; for other muds an activity correction should be made. In

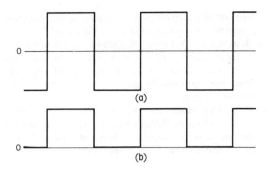

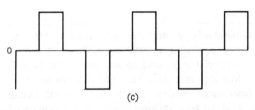

FIG. S-18. **Square waves**. (a) Full square wave; (b) half square wave; (c) pulsed square wave.

these equations $K = 60 + 0.133\ T$, where $T =$ Fahrenheit temperature. See also *SP* and *pseudostatic SP*.

stability (of a filter): A filter is stable if the energy of its impulse response is finite. Stable minimum-phase filters have stable inverses. Maximum-phase wavelets do not have stable inverse memory functions but inverse filtering can be accomplished by stable anticipation functions. The inverse of a mixed-delay wavelet requires both a stable memory function and a stable anticipation function.

stabilized platform: A platform on which instruments (such as gravimeters) can be mounted where they will remain nearly level despite tilt of the platform support. The platform, mounted on gimbals, is controlled by a gyroscope coupled to an accelerometer-controlled servo system on each gimbal axis. Used in measuring gravity on a ship in motion and in mounting inertial navigation sensors.

stack: 1. A composite record made by combining traces from different records. Some of the types of stacking are listed in Figure S-19. See *common-midpoint stack, diversity stack, uphole stack,* and *vertical stack*. Stacking involves filtering because of timing errors or waveshape differences among the elements being stacked; see Figure S-20. **2.** A computer buffer operated on a last-in first-out basis.

stacking chart: A diagram showing the interrelationships among the traces from common-midpoint shooting; a graph of source point location s versus geophone group location g for a **surface stacking chart**, of s versus $(s + g)/2$ for a **subsurface stacking chart**. Used to determine the proper traces for stacking and for determining parameters for shifting traces (as in making static corrections). Components along various alignments represent common midpoint, common source, common geophone, or common offset. See Figure S-21.

stacking velocity: Velocity calculated from normal-moveout measurements and a constant-velocity model. Used to maximize events in common-midpoint stacking. Sometimes erroneously called "rms velocity". See *velocity analysis*.

stadia: 1. An instrument for measuring distances, consisting of a telescope through which a vertical graduated rod can be seen, overlain by horizontal parallel cross hairs (located in the focal plane of the telescope eye-

---

Ground mixing combines outputs of several geophones and/or sources located over a limited area of the ground to attenuate events with low apparent velocities and random noise.

Instrument mixing combines adjacent traces without time shifting to attenuate events with low apparent velocity and random noise.

Vertical stacking combines traces which have nearly identical source-group positions without time shifting; done to increase signal strength and attenuate random noise.

Uphole stacking combines data from different shot depths after static shifting based on uphole-time measurements, to attenuate ghosts.

Common-midpoint (CMP) stacking combines traces for which midpoint between source and receiver is the same after normal-moveout correction.

Common-offset stacking combines data from a limited area for which the offset is the same or nearly the same.

Simplan stacking combines common receiver (or common source) gather to simulate effect of having a plane wave.

Slant stacking combines data along various apparent velocity lines to accomplish tau-*p* transform.

Diversity stacking combines data elements excluding portions where amplitudes exceed a threshold, to attenuate large nonsource-generated noises.

Coherency filtering combines data which satisfy certain trace-to-trace coherency criteria.

Velocity or f-k filtering often has same effect as mixing along certain dip attitudes.

Automatic migration can be thought of as combining data along hyperbolic diffraction curves, also called **diffraction stack**.

FIG. S-19. **Stacking** types.

piece); the amount of rod seen between the cross hairs allows one to determine the distance to the rod. **2.** The rod alone.

stadia factor: The ratio of the distance from a rod to the portion of the rod subtended between the stadia cross hairs. Often a value of 100.

stadia tables: Tables giving values of the quantities (sin 2α)/2, $\cos^2\alpha$, and/or $\sin^2\alpha$ as functions of α. Used in calculating horizontal (H) and vertical (V) distances from a transit station to a stadia rod. If α is the angle which the line of sight makes with the horizontal, F is the stadia interval factor (often 100), and X is the distance on the rod between the cross hairs, then $H = FX \cos^2\alpha = FX - FX \sin^2\alpha$, and $V = FX$ (sin 2α)/2.

Stagarray: A multi-air gun array. Petty Ray tradename.

stake: 1. A marker used by field parties to locate gravity stations, shotpoints, geophone locations, survey locations, etc. Usually indicates a temporary location as opposed to a **monument** which is a permanent location. **2.** To locate the site for a well. **3.** An electrode, such as might be used in electrical exploration or to ground a seismic truck. **4.** To mark the boundaries of a mineral claim.

stake resistance: The electrical resistance between a current electrode and the ground.

stand alone: Complete in itself rather than as a part of a larger system.

standard curve: *Type curve* (q.v.).

standard deviation: The standard deviation σ of n measurements of a quantity X_i with respect to the mean $\overline{X}$, is

$$\sigma = [(1/n) \ \Sigma \ (X_i - \overline{X})^2]^{1/2}.$$

With a normal or Gaussian distribution of data, 68.3 percent of the data fall within a standard deviation about the mean. The square of the standard deviation is the **variance**. See *statistical measures*. For two degrees of freedom, measurements (X_i, Y_i) with respect to the means ($\overline{X}, \overline{Y}$), σ is

$$\sigma = [(1/h) \ \Sigma(X_i - \overline{X})^2 + (Y_i - \overline{Y})^2]^{1/2}.$$

For a Rayleigh distribution of data, 40.5 percent fall within a circle of radius σ (called **one sigma**).

standard Earth: An Earth model with spherical shells of seismic velocity which contain the same volume as the corresponding layers of equal velocity in the actual Earth.

standard error: *Standard deviation* (q.v.).

standard format: For seismic formats, see SEG (1980).

standard lines: See *map projection*.

standard meridian: See *map projection*.

standard parallel: See *map projection*.

standard polarity: See *polarity*.

standard section: A diagram showing all the stratigraphic units in an area in their sequence of deposition; used as a standard for correlation. Often shows the maximum thickness of units.

standoff: 1. The distance a sonde is from the wall of the borehole. 2. A device for keeping the sonde from lying against the borehole wall.

standout: The amount by which the amplitude of an event exceeds the mean amplitude.

standing wave: A phenomenom produced by the interference of two continuous wave trains. A standing wave may result from the interference of a continuous wave train from a source and one resulting from reflection, or it may appear between two reflections. After being excited the wave will decay exponentially. Standing-wave patterns show amplitude nodes and antinodes alternating at 1/4 wavelength intervals. Continuous-wave radio-navigation systems set up standing-wave patterns.

star: A configuration with a center and points in various directions from the center. See *array (seismic)* and *triad.*

starved: Having low availability of sediments for deposition. In a **starved basin,** the subsidence rate exceeds the sedimentation rate.

state variable: One of the sets of variables which completely describe a system at any time. A state variable may represent a derivative of a quantity which is itself a state variable, allowing differential equations to be expressed as a set of linear simultaneous equations. For example, the voltage drop around an electrical circuit which includes capacitance, inductance, and resistance may be expressed by the differential equation:

$$E(t) = C \int I \, dt + RI + L \, dI/dt.$$

Using state variables of I, $Q = \int I \, dt$, and $P = dI/dt$ permit this to be written as a set of three simultaneous equations:

$$E(t) = CQ + RI + LP,$$

$$dQ/dt = I,$$

and

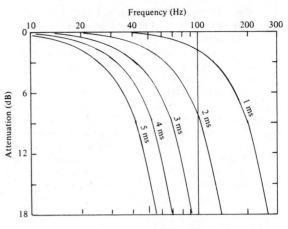

FIG. S-20. Filter effect of timing errors in **stacking**. The numbers on the curves are standard deviations of the timing differences among the traces stacked. (From Sheriff and Geldart, v.1, 1982, p. 151.)

$$dI/dt = P.$$

See also *parameter.*

static capacity: *Specific capacity* (q.v.).

static corrections, often shortened to **statics:** Corrections applied to seismic data to compensate for the effects of variations in elevation, weathering thickness, weathering velocity, or reference to a datum. The objective is to determine the reflection arrival times which would have been observed if all measurements had been made on a (usually) flat plane with no weathering or low-velocity material present. These corrections are based on uphole

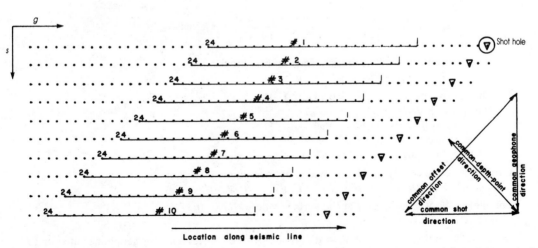

FIG. S-21. Surface **stacking chart**. Each horizontal line shows the location of geophone groups and shotpoint for a single record. To the right is shown the directions for finding all traces with various elements in common. s = source coordinate, g = geophone coordinate.

data, refraction first-breaks (see Figure S-22), and/or event smoothing. (a) **Uphole-based statics** involve the direct measurement of vertical traveltimes from a buried source; see *uphole shooting*. This is usually the best method where feasible. (b) **First-break based statics** are the most common method of making field (or first-estimate) static corrections, especially when using surface sources. The *ABC method* (q.v.) and variations for more complex assumptions are used for this determination. (c) **Data-smoothing statics methods** assume that patterns of irregularity which most events have in common result from near-surface variations and hence static-correction trace shifts should be such as to minimize such irregularities. Most automatic statics-determination programs employ statistical methods to achieve the minimization. (d) Underlying the concept of static corrections is the assumption that a simple time shift of an entire seismic trace will yield the seismic record which would have been observed if the geophones had been displaced vertically downward to the reference datum, an assumption not strictly true. See Sheriff and Geldart (v. 1, 1982, p. 193-196; v. 2, 1983, p. 47-49 and 75-76).

static SP: Static spontaneous or self-potential or *SSP* (q.v.); compare *pseudostatic SP*.

station: 1. A ground position at which a geophysical instrument (gravimeter, geophone, etc.) is set up for an observation. **2.** An input and/or output point on a communications system.

stationary: Having statistical properties which do not

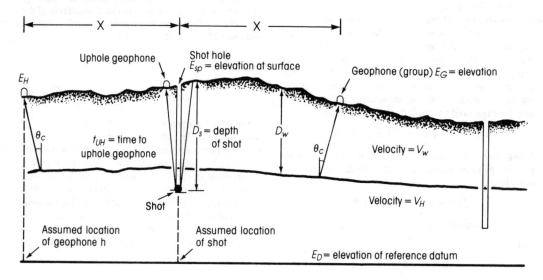

Excess time = vertical time through weathering − time at subweathering velocity

$$t_{xs} = D_W(1/V_W - 1/V_H) = t_i \sqrt{\frac{V_H - V_W}{V_H + V_W}} = t_i K, \text{ where}$$

t_i = intercept time for refraction first breaks, and
$K = [(V_H - V_W)/(V_H + V_W)]^{1/2}$.

For shot below the datum (see also Figure U-1),
$$D_W = (t_{UH} - D_S/V_H)/(1/V_W - 1/V_H)$$
$$= V_W(t_{UH}V_H - D_S)(V_H - V_W)$$
$$= t_i V_W/\cos\theta_c, \text{ where}$$
θ_c = critical angle = $\sin^{-1}(V_W/V_H)$.

Correction to effectively place shot and geophone on datum = t_c:
$$t_c = (E_{SP} - D_S - E_D)V_H + (E_G - E_D)/V_H - t_{xs}.$$

For shot in the weathering (or at the surface where $t_{UH} = 0$),
$$D_W = \tfrac{1}{2}V_W(t_i/\cos\theta + t_{UH}).$$

Differential weathering correction = difference between corrections for geophones at G and H:
$$t_{DWC} = (t_H - t_G)(1 - V_W/V_H) + (E_H - E_G)/V_W.$$

FIG. S-22. **Statics** corrections based on first-break intercept time.

change with time and/or position. The statistics are the same if the time origin is changed.

stationary filter: A filter which is not time variant.

stationary mass: A weight that tends to remain quiescent during the passage of seismic waves.

stationary wave: A *standing wave* (q.v.).

statistical measures: Many "measures" of a distribution of data are used. The most common for a set of n values X_i are

$$\text{mean} = \overline{X} = (1/n) \sum_{i=1}^{n} X_i;$$

$$\text{rms value} = \left[(1/n) \sum X_i^2 \right]^{1/2};$$

$$\text{mean deviation} = (1/n) \sum_{i=1}^{n} |X_i - \overline{X}|;$$

$$\text{standard deviation} = \sigma = \left[(1/n) \sum_{i=1}^{n} (X_i - \overline{X})^2 \right]^{1/2};$$

$$\text{variance} = \sigma^2;$$

mode = value associated with the highest associated frequency;

$$\text{coefficient of variation} = \sigma/\overline{X};$$

$$\text{skewness} = \gamma = (1/n) \sum_{i=1}^{n} [(X_i - \overline{X})/\sigma]^3;$$

$$k\text{th moment about } A = (1/n) \sum (X_i - A)^k;$$

$$\text{dispersion} = (X_{\max} - X_{\min})/X.$$

If X_i is a Gaussian distribution, then the **probable error** is 0.6745 σ, σ^2 is a **chi-square distribution**, and γ is approximately Gaussian.

statistical stacking: *Selective stacking* (q.v.).

steady mass: *Stationary mass* (q.v.).

steady state: Equilibrium conditions observed when variations over the short time are absent.

steam gun: *Vaporchoc* (q.v.).

steam quality: 1. In geothermal development, the quality of steam produced from underground is measured in terms of the weight of steam required to generate one kilowatt-hour of electrical energy. 2. The mass fraction of steam divided by total mass.

Steenland-Vacquier rule: See *depth rule*.

steepest descent: 1. An iterative method of approaching a minimum by taking an increment along the steepest gradient to arrive at the next approximation, the step length often being proportional to the magnitude of the gradient. Provision can be made to speed-up convergence onto the minimum and prevent oscillation about the minimum. It is assumed that the function is continuous and that the initial estimate is near enough to the correct minimum, in the event that the function has more than one minimum. Sometimes called **steepest ascent** when used to approach a maximum. 2. A method to compute the asymptotic behavior of an integral, also called the "saddle-point method;" see Morse and Feshbach (1967, p. 437).

steer: To introduce a time shift into an ensemble of traces so that energy approaching from a given direction appears at the same time on all traces. Used in sonogramming, in studying earthquakes with large arrays, etc. Variations include weighting the components and/or filtering before summing.

Stefanesco function: *Kernel function* (q.v.).

stepback: The correction applied to a location (such as the location of a seismic ship determined by radio methods) to yield the midpoint for seismic data, allowing for the positions of the streamer and source with respect to the navigation antenna. See Figure O-1.

step function: An abrupt increase or decrease from one constant value to another (often from zero to one, or vice versa). The first derivative of a step function is an impulse. Also called **Heaviside function**.

step-function response: Output of a system when the input consists of a step function. For a linear system it is the integral of the impulse response.

stepout: Differences in arrival time because of dip; **moveout**. **Normal moveout** is used for differences because of offset (source-to-geophone distance).

stepped-gain amplifier: An amplifier whose gain is variable in discrete steps. Gain changes may be programmed, i.e., predetermined by the user, or they may be controlled by the magnitude of the signal being amplified.

stepping method: A surveying method for determining the elevation of the **stadia** rod with respect to the transit. A point level with the transit is sighted through the telescope which is then tilted so that the bottom cross hair is aligned with the point. Another point in the line of the top cross hair is sighted and the procedure is repeated until the stadia rod comes into the view of the telescope. The number of steps or times of retilting the telescope is counted and converted to Beamans, the full intercept between the stadia wires being one **Beaman**.

steradian: A unit of measure of solid angle. A sphere = 4π steradians.

stereographic projection: 1. A representation of directional information used in three-dimensional structural problems. A **stereonet** or **Wulff net** is used if angular relations are to be preserved and a **Schmidt** or **Lambert net** if areas are to be preserved. Lines are represented as points indicating their direction, planes as either great circles (**cyclographic projection**) or points for axial lines perpendicular to the plane (**polar projection**). See Figure S23. 2. A projection used to map the Earth; see Figure M-3.

stick: A time-domain representation of amplitude at a particular time. A scaled impulse. See *stickogram* and compare *stick plot*.

stickogram: 1. Graph of reflection coefficients as a function of depth, often made from sonic-log data as an intermediate step in *synthetic-seismogram* (q.v.) preparation; see Figure S-28. Stickograms may or may not include sticks which represent multiples. 2. A time-

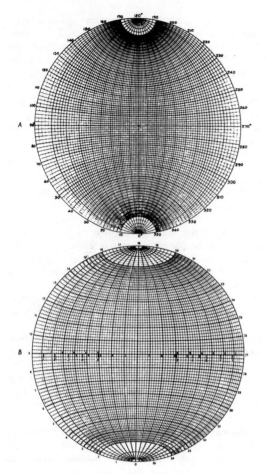

FIG. S-23. **Sterographic** net. (a) Wulff net; (b) Schmidt net or Lambert equal-area plot.

domain diagram of a sampled waveform or filter showing the successive quantized values of the waveform or of the impulse response of the filter.

stick plot: A presentation of dipmeter results where the well bore is represented by a line according to the projection of the well onto a vertical plane, and the components of dip in this plane are indicated by short line segments.

stiffness: The ratio of stress acting on an elastic medium to the strain, an elastic constant.

stillstand: A period of time during which there is not much variation in the level of the land with respect to sea level.

stochastic: Random; a value determined from a specified distribution by chance. Opposite of *deterministic*. Compare *Markovian variable*.

Stoneley wave: 1. A type of seismic wave propagated along an interface. Such a wave is always possible at solid-fluid interfaces and under very restricted conditions at solid-solid interfaces. See Sheriff and Geldart, v. 1 (1982, p. 51). **2.** A surface wave in a borehole; see *acoustic wave*.

stone-slab correction: *Bouguer correction* (q.v.).

storage: A computer memory system; a device where data can be stored and from which it can be retrieved. Storage is divided into locations, each with an assigned address; each location holds a specific unit of data (a digit, a word, or a complete record, depending on the system).

storm: A temporary, considerable disturbance of a geophysical field, e.g., a *magnetic storm* (q.v.).

STP: Standard temperature (0°C) and pressure (one atmosphere).

straddle spread: *Split spread* (q.v.).

straight: Not mixed; see *mixing*.

straight slope measurement: The horizontal distance over which a magnetic anomaly is nearly linear at the maximum slope; used in depth determination. See Figure D-6.

strain: The change of dimensions or shape produced by a stress. Strain is usually expressed in dimensionless units such as change of length per unit of length, angle of twist, change of volume per unit of volume. Rotation or translation without change of shape is not strain. See *elastic constants* and Sheriff and Geldart, v. 1 (1982, p. 34-35). If u,v,w are the stress-produced displacements in the x,y,z directions of a point in a body, the strains are:

normal strains: $\epsilon_{xx} = \partial u/\partial x,$

$$\epsilon_{yy} = \partial v/\partial y,$$

$$\epsilon_{zz} = \partial w/\partial z.$$

shearing strains: $\epsilon_{xy} = \epsilon_{yx} = \partial v/\partial x + \partial u/\partial y,$

$$\epsilon_{yz} = \epsilon_{zy} = \partial w/dy + \partial v/dz,$$

$$\epsilon_{zx} = \epsilon_{xz} = \partial u/\partial z + \partial w/\partial x.$$

The fractional change in volume (**dilatation**) Δ is

$$\Delta = \epsilon_{xx} + \epsilon_{yy} + \epsilon_{zz}.$$

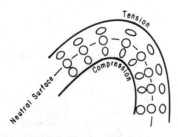

FIG. S-24. **Strain** ellipsoid.

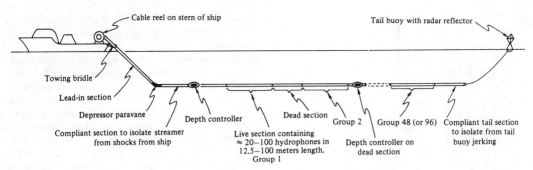

FIG. S-25. **Streamer**.

strain ellipsoid: A representation of strain by showing the ellipse into which a circle would be distorted if subject to the same strain. See Figure S-24.

strain energy: The work involved in straining a body; for a conservative system, the potential energy stored in a strain. If the body is elastic, the work is

$$E = (1/2) \, \Sigma \, \Sigma \, \sigma_{ij} \, \epsilon_{ij},$$

where σ_{ij} is the stress in the i-direction on a face perpendicular to the j-direction, and ϵ_{ij} is the rate of change of i-direction displacement in the j-direction.

strain seismometer: A seismometer that is designed to detect deformation of the ground by measuring relative displacement of two points.

stratal surface: The surface of a sheet-like rock unit visibly distinguishable from units above and below it. Represents the surface of the solid Earth at some time. Seismic reflections parallel stratal surfaces.

stratigraphic interpretation: Prediction of lithology, depositional environment, and/or interstitial fluid based on seismic measurements, especially reflection patterns. See Sheriff (1980).

streamer: A marine cable incorporating pressure hydrophones, designed for continuous towing through the water. A marine streamer (Figure S-25) is typically made up of 96 or more **active** or **live sections** which contain hydrophone arrays separated by **spacer** or **dead sections**. Usually the streamer is nearly neutrally buoyant and depressors or *depth controllers* (q.v.) are attached to depress the streamer to the proper towing depth. The entire streamer may be 3-4 km in length.

streamer feathering: Drift of a marine streamer to one side because of a crosscurrent. See Figure T-3.

streaming potential: See *electrokinetic potential*.

strength: The limiting stress before failure.

stress: The intensity of force acting on a body, in terms of force per unit area.

stretch: The change in wavelet shape produced by applying a normal-moveout correction.

stretch modulus: Young's modulus; see *elastic constant*.

stretch section: A portion of a marine seismic streamer designed to isolate the sensitive portion of the streamer from shocks to the towing ship.

strike: 1. The direction of the intersection of a surface and a horizontal plane; the horizontal direction at right angles to the dip. As in the "strike of a bed". **2.** The projection on the horizontal of the major axis of the ellipse of polarization.

strike-slip fault: A fault across which motion has been predominantly horizontal. See Figure F-2.

string: 1. Several geophones which are permanently connected together; a *flyer*. **2.** A **computer string** is a sequence of elements (such as bits or characters).

stringer: A thin high-speed layer (whose presence may be erratic) within low-speed formations. A stringer is too thin or not continuous enough to carry refracted energy very far.

string galvanometer: An electrical wire in a magnetic field so arranged that the wire is deflected proportional to the current flowing through the wire. The shadow of the wire is projected onto photographic film to make a record of the current variations. Refers to a type of seismic camera in use up to the 1950s.

strip log: A *sample log* (q.v.).

stripping: A procedure which successively removes the effects of upper layers. **1.** Making corrections which effectively place seismic sources and receivers at the base of the stripped layers. **2.** Removing the calculated effects of layers successively. Stripping is sometimes used in gravity interpretation. Synonym: **layer stripping**.

strobe: To read or measure at discrete time intervals.

strong-motion accelerograph: A self-actuating, triaxial earthquake recorder designed to provide acceleration data on strong, local earthquakes. Used in engineering studies for dams, power plants, etc.

structure: 1. The general disposition, attitude, arrangement, or relative position of the rock masses of an area; the sum total of the structural features of an area, consequent upon such deformational processes as faulting, folding, and igneous intrusion. **2.** Any physical arrangement of rocks (such as an anticline, fault, or dome) that may involve the accumulation of oil or gas. **3.** A subsurface area characterized by folding and/or faulting. **4.** Seismic anomaly, usually a closed high.

structural section: A display of seismic reflections which attempts to portray the attitude of bedding.

structural style: The structural features which result from a certain stress history. Structural style considerations

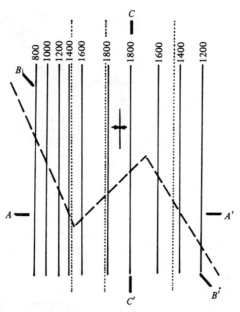

FIG. S-26. **Subsurface** trace. The straight vertical lines are contours on a north-south cylindrical syncline. For seismic line AA′ (perpendicular to strike) the subsurface trace is vertically under the line; for line BB′ it is as shown by the dashed line; for line CC′ (parallel to strike) there are three parallel subsurface traces as indicated by the dotted lines. (From Sheriff and Geldart, v. 1, 1982, p. 123.)

can assist in seismic interpretation. See Sheriff and Geldart, v. 2 (1983, p. 93-97).

structure-sensitive conductivity: See *extrinsic conduction*.

strum: See *cable strum*.

stunt box: A device which controls the nonprinting functions of an output device upon receiving orders.

su: Sigma unit or *capture unit* (q.v.), a unit for measuring capture cross-section.

subbottom profiler: An instrument which produces a cross-section-like record of sediments below the sea floor.

subcarrier: A carrier which is applied as a modulating wave to another carrier.

subduction zone: The zone where one plate plunges beneath another plate. See Figure P-4. **B-type (Benioff) subduction** involves an oceanic plate plunging underneath another plate, **A-type (alpine)** involves a continent-continent collision.

subpoint: The location of one of the elements in an array.

subroutine: A computer program called for as part of a larger program.

subsample: To resample digitized data at a longer interval than formerly used. For example, to subsample from 2 to 4 ms means to retain only every other sample of 2 ms data. Alias filtering must be included to avoid possible aliasing. Sometimes called **decimate**. Opposite of **reconstitute**.

subshot: A seismic shot or other energy release; the records from a set of subshots are stacked vertically to

make one profile for input to a processing system. Marine seismic work often employs 2 or 4 subsources per source point.

subsurface coverage: 1. The position under which reflection points from plane horizontal reflectors lie. (Actual reflection points will not lie under the subsurface-coverage position if the reflectors dip.) For a single-end spread, the position extends from the midpoint between the source and the nearest geophone group to the midpoint between the source and the most distant geophone group, therefore of length equal to half the spread length. See Figure D-5. **2.** Sometimes the position of the actual reflection points, allowing for dip. **3.** The multiplicity associated with common-midpoint data.

subsurface stacking chart: See *stacking chart*.

subsurface trace: The locus of subsurface reflecting points as a seismic line is traversed, making allowance for migration perpendicular to the line as well as along the line. Reflection data properly should be plotted along the subsurface trace rather than along the seismic line. Apt to be different for each reflector. See Figure S-26.

subtractive primary colors: See *primary colors*.

subweathering velocity: Velocity immediately below the base of the weathering. Sometimes taken as the velocity of a refraction at the base of the weathering.

summation check: See *check*.

summation method: A method of calculating a weathering correction to seismic arrival times. For adjacent interlocked split-dip profiles with shots just below the low-velocity layer, the correction for each group is half the sum of the first-arrival times at that group from the two interlocking records minus the average high-velocity time between the shotpoints (obtained by subtracting the uphole time from the first-arrival time of the group at one of the shotpoints when the shot is fired at the other shotpoint). See Figure S-22.

superimposed mode: A display in which two presentation modes are superimposed on each other. Usually refers to wiggle trace superimposed on variable density or variable area. See Figure D-14.

supernormal pressure: *Abnormally high pressure* (q.v.).

superposition: 1. The situation where the same end result is obtained by an operation on a whole input as is obtained from the sum of the results of operations on the components of the input If α, β, and γ are operations to be performed on $a(t)$ and $b(t)$, then superposition dictates that

$$\gamma[\alpha a(t) + \beta b(t)] = \alpha \, [\gamma a(t)] + \beta \, [\gamma b(t)].$$

Superposition allows a problem to be broken into a number of component problems which may be easier to solve than the entire problem, and yields the solution for the larger problem. Superposition is a necessary and sufficient condition for linearity; see *linear system*. **2.** *Convolution* (q.v.) involves superposition.

supervisor: 1. Individual overseeing the work of one or more geophysical parties. Immediate supervisor of the party chief. **2.** The function of a computer which keeps tasks in order; **executive**.

suppressed layer: A layer whose resistivity is intermediate between the resistivities of the enclosing layers and which may not be "seen" unless it is very thick.

suppression: See *initial suppression*.

surface anomaly: An anomaly caused by variations at or near the surface of the earth.

surface conductivity: Conduction along the surfaces of certain minerals due to excess ions in the *diffuse layer* (q.v.).

surface-consistent model: A concept used for determining and applying seismic *static corrections* (q.v.) (or amplitude corrections). Time delays (or attenuations) are assigned to source and geophone (and other) locations based on a statistical study. See Sheriff and Geldart, v. 2 (1983, p. 47-49).

surface corrections: Corrections of geophysical measurements for surface anomalies and ground elevation.

surface density: 1. Mass per unit area. **2.** The density used in calculating the *Bouguer correction* (q.v.). **3.** Occasionally refers to the **Green's equivalent layer**, a surface density distribution which has the same potential field outside the closed surface as does a mass distribution over the volume contained by the closed surface. See *Gauss's theorem*.

surface fitting: A method of approximating a data set by a mathematical surface, usually of low order. Used in *residualizing* (q.v.); the mathematical surface represents the **regional**, and the departures of the original surface from the mathematical surface represents the **residual**.

surface impedance: See *impedance*.

surface-ship gravimeter: *Shipboard gravimeter* (q.v.).

surface SH-wave: *Love wave* (q.v.).

surface source: A seismic energy source which is used on the surface of the ground as opposed to one in a borehole.

surface stacking chart: See *stacking chart*.

surface wave: Energy which travels along (or near to) the surface. Motion involved with the wave falls off rapidly with distance from the surface. In seismic exploration usually refers to ground roll, but also includes Rayleigh, Love, hydrodynamic waves, etc. Also called **interface wave** and **long wave**.

surgical mute: A sharp division in offset-time space between elements retained unchanged in magnitude and those deleted entirely. As opposed to a **tapered** or **ramped mute** where the cutoff is gradual. Usually the data **muted** (deleted) are those preceding, including, and immediately following the first breaks, including shallow head waves. Sometimes refers to a slice diagonally cutting across where values have been set equal to zero, as might be done to remove a ground-roll wave train.

survey: To determine the form, extent, position, subsurface characteristics, etc., of an area or prospect by topographical, geologic, or geophysical measurements.

surveying: 1. Determining surface locations. Conventional plane surveying is done using level and chain, stadia, transit, plane table, or other methods. Location networks may be resolved by traversing, triangulation, or trilateration. Surveys over larger areas require corrections for earth curvature; see *spherical excess* and

Legendre theorem. Various radio, sonic, and satellite *positioning methods* (q.v.) are used in marine and airborne work. **2.** Carrying out other types of measurements as a function of location.

susceptibility: See *magnetic susceptibility* and *electric susceptibility*.

SV-wave: *S*-wave energy polarized so that motion is in the vertical plane which also contains the direction of wave propagation. Converted waves are (mostly) *SV*.

S_w: *Water saturation* (q.v.).

swarm: 1. A series of minor earthquakes, none of which may be identified as the main shock, occurring in a limited area and time. **2.** A group of roughly parallel igneous intrusives.

swath method: A type of *three-dimensional (3-D) surveying* (q.v.); see Figure T-3.

S-wave: A body wave in which the particle motion is perpendicular to the direction of propagation. Also called **secondary wave (undae secundae)**, **shear wave**, **transverse wave**, **rotational wave**, **distortional wave**, **equivolumnar wave**, **tangential wave**. *S*-waves are generated by the incidence of *P*-waves on interfaces at other than normal incidence, whereupon they are sometimes called **converted waves** (*SV*-waves). In an isotropic medium the velocity of shear waves V_s is given by

$$V_s = (\mu/\rho)^{1/2} = \{E/[2\rho(1 + \sigma)]\}^{1/2},$$

where μ is the shear modulus, ρ is the density, E is Young's modulus, and σ is Poisson's ratio. *S*-waves have two degrees of freedom and can be polarized in various ways. See *SH-wave* and *SV-wave*. *S*-wave reflection data are often displayed at double the vertical scale of the comparable *P*-wave data to compensate roughly for the differences between *S*-wave and *P*-wave velocities; see Figure S-27.

sweep: 1. A Vibroseis source input. The frequency is varied continuously during a "sweep" period, commonly 7 s or longer. See Figure V-7. **2.** To explore the effect of varying a parameter. **Velocity sweeping** consists of trying various normal moveouts on a set of common-midpoint data to see which stacking velocity emphasizes desired events. **Dip sweeping** is done in the sonograph process; see *sonogram*. **3.** The steady movement of the electron beam across a cathode-ray screen. **4.** A cycle of operations such as a radar antenna making one rotation or a side-scan sonar cycle.

Syledis: A medium-range UHF pulsed time measurement radio-positioning system operating in the 420-450 MHz range. Signal processing using pseudorandom noise coding allows for very accurate time measurements and hence high accuracy. Sercel tradename.

symbolic language: A collection of symbols used in programming to represent operation codes, functions, addresses, etc., with rules of usage.

synchronous data link control: An IBM communications protocol that supports transmission of binary data, multidrop devices, and multiplexing of multiple logical links on one physical line.

synchronous detection: A method of enhancing signal and suppressing noise by synchronizing the detection peri-

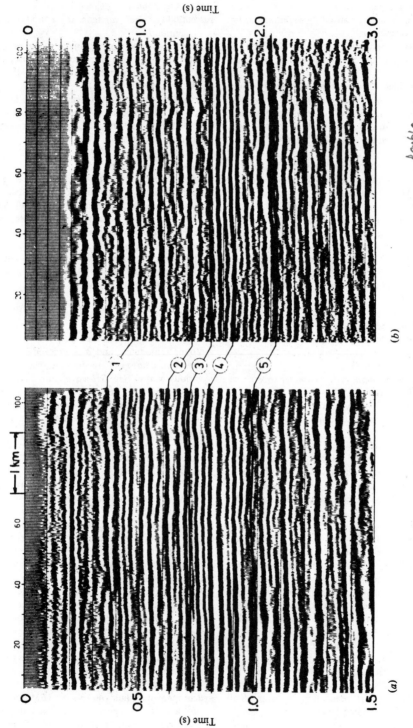

FIG. S-27. *P*-wave and **S-wave** sections compared. (**a**) *P*-wave section; (**b**) *S*-wave section plotted at double the vertical scale used for the *P*-wave section. (Courtesy Compagnie Générale de Géophysique.)

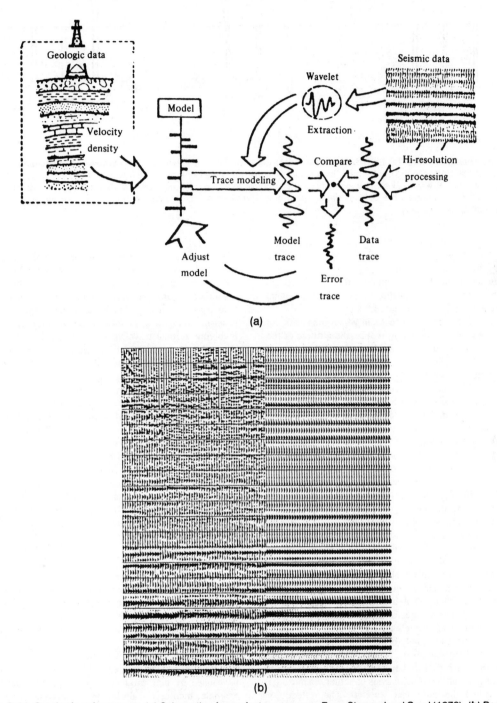

FIG. S-28. **Synthetic seismogram**. (**a**) Schematic of manufacture process. From Stommel and Graul (1978). (**b**) Portion of synthetic seismogram (right half) compared to actual seismic section (left half). Courtesy Seiscom Delta United.

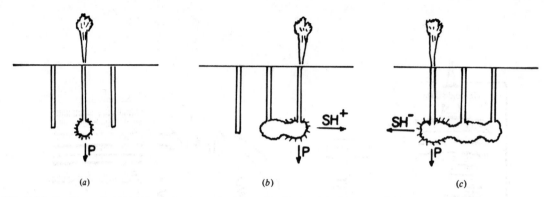

FIG. S-29. Syslap method. (**a**) Explosion in center hole generates mainly *P*-waves; (**b**) because of the asymmetry produced by the explosion in the center hole, the explosion in the right hole generates *SH*-waves as well as *P*-waves; (**c**) the left hole produces *P*-waves plus *SH*-waves of polarity opposite to those from the right hole. Subtracting records from the right and left holes doubles the *SH* contribution and nearly eliminates the *P*-wave contribution. (Courtesy Compagnie Générale de Géophysique.)

od of the voltmeter receiver with the "on" cycle of the current transmitter or reference signal.

synchronous protocol: A communications protocol in which the two stations synchronize to the carrier signal before any data are transmitted.

syncline: A fold in stratified rocks in which the rocks dip toward a central depression, that is, the attitude of the rocks is concave upward; opposite of **anticline**.

synergetic: A combination of data elements such that more information is apparent from the combination than from the elements treated independently. Also spelled **synergistic**. Synergetic Log is a Schlumberger tradename.

synthetic acoustic impedance log: See *seismic log*.

synthetic aperture: The effect of a much larger antenna achieved by summation of readings as the antenna is moved. Usually refers to radar (**SAR**) where the dipole antenna is transported normal to its axis to achieve greater directivity (effectively a smaller aperture).

synthetic fault: A minor fault whose throw is in the same sense as that of the major fault with which it is associated. Opposite of **antithetic fault**.

synthetic seismogram: An artificial seismic reflection record manufactured by assuming that some waveform travels through an assumed model. See Figure S-28. A **one-dimensional synthetic seismogram** is formed by simply convolving an embedded waveform with a reflectivity function (also called a **stickogram** because it is usually plotted as a series of spikes indicating the sign and magnitude of the reflectivity at successive interfaces, the variable usually being two-wave traveltime). The embedded waveform is sometimes an assumed waveform (such as a Ricker wavelet) and sometimes a waveform resulting from analysis of actual seismic data (also called the **equivalent wavelet**). The reflectivity function sometimes involves primary reflections only, sometimes with selected multiples added, sometimes with all multiples added. Sometimes includes simulation of Earth-filtering effects (divergence and other attenuation effects, including frequency-dependent absorption) and instrumental filtering effects. The reflec-

tivity is usually that calculated for normal incidence from velocity and density data, but often velocity changes only are considered because density changes are unknown (or some relationship between density and velocity is assumed). While a one-dimensional synthetic seismogram is a single-channel convolution (in effect, involving vertical travel in the assumed model only), often the model is varied and successive one-dimensional traces are displayed side-by-side to simulate a seismic section. Used to compare with an actual seismogram to aid in identifying events or to predict how variations in the model might show effects on a seismic section. A **two-dimensional synthetic seismogram** allows for wave effects including reflections from dipping reflectors, diffractions, etc. Usually only two-dimensional effects are included but occasionally true three-dimensional effects are included. Often coincident source and receiver are modeled, but sometimes offset-dependent effects are included, sometimes including head waves, surface waves, and other wave modes. The making of a synthetic seismogram is also called **direct modeling**. Also called **theogram**.

synthetic sonic log: See *seismic log*.

Syslap method: A method of generating an *S*-wave record. See Figure S-29. Tradename of CGG.

system: An assemblage united by interactions. For example, the **seismic system** includes the Earth, geophones, amplifiers, seismic wave transmission through the Earth, recorders, processing, and final presentation.

systematic error: An error not attributable to chance alone. Such errors do not average out merely by including more measurements. Systematic errors may be caused by bias produced by the instruments (for example, incorrect scale factor or incorrect zero), by the observer or measuring procedures (for example, not selecting representative samples or altering samples before the measurements), or by the action of factors or physical laws which are not properly allowed for or understood. Much geophysical noise is systematic.

system deconvolution: See *deconvolution*.

T

T: Period; the time between adjacent corresponding points on a periodic wave; the reciprocal of frequency. See Figure W-2.

tab: 1. An indicator of where data begin or end. **2.** A special character which indicates that the next character should be placed at some predetermined position.

tabular body: A body of finite thickness with one edge horizontal but other edges infinitely remote; an **infinite dike.** Implies that a body's width is more than 50 times its thickness. This is one type of model used in potential-field calculations.

tactical characteristics: The characteristics of a ship which determine its maneuverability under various sea conditions.

tadpole plot: A type of plot of dipmeter or drift results; an **arrow plot.** The position of a dot gives the dip angle versus depth and a line segment pointing from the dot gives the direction of dip, using the usual map convention of north being up. See Figure D-11.

Tafel's law: An empirical relationship between overvoltage η and current density J at an anode or cathode:

$$\eta = a - b \log_{10} J,$$

where a and b are experimentally determined constants. This law applies over a greater current-density range than is used in IP field measurements.

Tagg method: A method of interpreting resistivity-sounding data obtained with the Wenner array over a two-layered earth.

tail buoy: A floating marker, usually with a reflector to aid in its location by radar, attached to the end of a seismic streamer.

tail end: The portion of a seismic line behind the source, that is, in the direction from which the source has been moving.

tailing: Lengthening of a waveform, as by adding extra cycles.

tail mute: Deleting data which arrive after some boundary drawn in offset-time space. Used to eliminate ground roll, air waves, or similar slow wave trains.

takeout: A connection point to a multiconductor cable where geophones or geophone flyers can be connected. Takeouts are usually polarized to reduce the likelihood of making the connection backwards.

tamp: To pack material about an explosive in a shothole. Objectives of tamping are to effect better coupling of the explosive energy with the earth and to retard the expulsion of the gaseous and other products of the explosion, thereby improving the conversion of explosive energy to seismic energy and delaying hole-blow effects. Water or mud is usually used, sometimes sand and earth.

tamper: An earth compactor, used as an energy source with *Sosie* (q.v.); **whacker.**

tandem survey: Electromagnetic survey method in which both transmitting and receiving coils are moved simultaneously, maintaining a constant separation between them. Equivalent to **moving-source method.**

tangential stress: Shearing stress; see *stress.*

tangential wave: *S-wave* (q.v.).

tangent projection: See *map projection.*

tape: 1. A *magnetic tape* (q.v.). **2.** A *chain* (q.v.).

tape-guide pins: Pins of nonmagnetic material which help position magnetic tape and prevent tape *skew* (q.v.).

tapered array: A source or geophone array in which elements contribute unequally. Tapered source arrays may be achieved by loading different amounts of explosives in different shotholes of a pattern, by varying the number of source impulses (pops, weight drops, etc.) or their spacing, or by varying the weighting while vertical stacking. Tapered geophone or hydrophone arrays may be achieved by varying the output of the different elements, the spacing of the elements or (more commonly) the number of elements at each location. Array tapering is used to change the shape of array directivity patterns. Generally the attenuation in the reject region is made greater and more nearly constant over the reject region but at the price of widening the major lobe (for the same number of elements and the same overall array size). See Figure D-12c.

tapered sweep: Vibroseis sweep where source varies as

$$A(t) \cos \left[(\omega_0 \pm bt)t \right],$$

$A(t)$ being the time-dependent factor that produces the tapering. Most Vibroseis sweeps are tapered for 100-200 ms at the start and end of the sweep to make it easier on the equipment and to avoid the undesirable effects of a sharp start or stop discontinuity.

tapered window: Window or gate with gradual edges. Compared to an abrupt boxcar window, a tapered window lessens the dependence on high frequencies in order to represent it adequately, simplifying many types of processing. Some tapers in common use are shown in Figure W-11. See also *Gibbs' phenomena.*

taper mix: See *mixing.*

taphrogenic: A term referring to regional block-faulting tectonics. See *tectonic types.*

tap test: A recording made as a geophone is tapped lightly, showing which channel that geophone feeds. Used to check that the spread is properly connected and oriented and also that the geophone is live.

TAR: True-amplitude recovery; a process for removing the effects of variable gain in the field recording and adjusting the amplitude to compensate for spherical divergence and other time-dependent attenuation.

tare: A discontinuity in data indicating an error in measurement or computation rather than an actual jump in the quantity being measured. Sometimes spelled **tear.**

target: 1. The object at which a survey sighting is aimed.

2. A sliding marker on a stadia rod on which one sights while leveling.

Tarrant method: A graphical refraction interpretation method applicable where refractor shape varies; see Sheriff and Geldart, v. 1 (1982, p. 221-222) or Tarrant (1956).

tau: Time to reach a background; see *pulsed neutron capture log*.

tau-gamma mapping: See *tau-p mapping*.

tau-p domain: See Figure T-1.

tau-p mapping: An unstacked seismic record can be described in terms of slope $dt/dx = p$ and intercept time τ, the arrival time obtained by projecting the slope back to $x = 0$, where x = source-geophone distance; see Figure T-1. Common-midpoint gathers can be similarly transformed. The transform process is also called **slant stack.** Filtering can be done on the tau-p map and the filtered result transformed back into a record. The tau-p map has certain interpretational advantages. See Diebold and Stoffa (1981). Similar to **tau-gamma mapping** where gamma is angle of emergence, gamma = γ = $\sin^{-1}(pV_0)$.

Taylor series: A function $f(x)$ can be expressed in terms of the value of the function and its derivatives (f', f'', f''', etc.) at any point $x = b$. In one variable this is

$$f(x) = f(b) + [f'(b)/1!](x - b) + [f''(b)/2!](x - b)^2 + [f'''(b)/3!](x - b)^3 + \cdots,$$

where $f''(b) = (dx'')/(d''t)$, when $x = b$ and ! denotes factorial (e.g., $3! = 3 \cdot 2 \cdot 1 = 6$). This series converges

if

$$\lim_{n \to \infty} \frac{f'(b)\,(x - b)}{nf^{n-1}(b)} \ll 1.$$

Named for Brook Taylor (1685-1731), English mathematician. The **Maclaurin series** is the special case where $b = 0$.

TB: *Time break* (q.v.).

TCA: *Time of closest approach* (q.v.).

Tchebyscheff array: *Chebychev array* (q.v.).

TD: Total depth, the maximim depth reached by a well.

T-D curve: 1. *Time-distance curve* (q.v.). **2.** Plot of *time-depth chart* (q.v.) data.

T^2-D^2: X^2-T^2; see $X^2\,T^2$ *analysis* (q.v.).

T-ΔT analysis: Normal moveout (Δt) as a function of offset (x) and arrival time (t) can be used to yield stacking velocity (V_s):

$$V_s \approx x/(2t\Delta t)^{1/2}.$$

Used to determine V_s and to identify primary and multiple reflections. See *velocity*.

TDEM: Time-domain electromagnetic method; a controlled source method.

tear: See *tare*.

tear fault: A type of strike-slip fault; see Figure F-2.

Techno: The name of an analog magnetic recorder and magnetic tape. Tradename.

tectogene: A downbuckling of the Earth's crust.

tectonic map: A map showing major structural features produced by uplift, downwarp, compression, or fault-

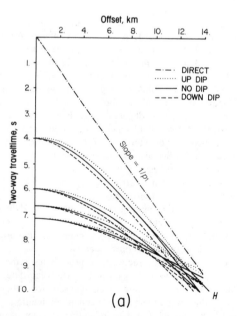

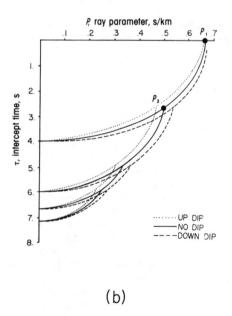

FIG. T-1. **Tau-p mapping. (a)** On-end seismic record is $f(x,t)$ where x = source-geophone distance (offset) and t = arrival time; **(b)** Tau-p transform of (a) is $F(\tau,p)$ where $p = dt/dx = 1/V_{\text{app}}$ and τ = intercept time at $x = 0$. p is the reciprocal of the apparent velocity and also called **apparent slowness.** Hyperbolic reflections transform into ellipses, straight events such as head waves and direct wave into points (the direct wave into p_1, the head wave H into p_2).

ing, with the more significant lineations associated with such features. The term is usually applied to maps covering large areas while maps of smaller areas showing the same features are called **structural maps**.

tectonic types: Four types of broad structural deformations are (a) **orogenic** or mountain-building involving lateral forces, folding and thrusting; (b) **epeirogenic** or relatively gentle warping; (c) **taphrogenic** involving mainly vertical forces and block faulting; and (d) **lineagenic** involving strike-slip faulting.

tectonophysics: See *geophysics*.

tectonosphere: The portion of the earth above the point of isostatic balance to which tectonic activity is confined.

telemetering: The transmission of data over a distance, such as from a point of observation to a recording point. Used to indicate the transmission of data which are digitized near the geophones to the recording unit, either over a wire or by radio. Telemetered time signals can be used for synchronous detection of resistivity and IP signals.

teleprocessing: A data-processing and communications system which permits input/output devices to be remote from the processing devices.

teleseism: An earthquake whose epicenter is over 1000 km away. Earthquakes originating nearer are "local" earthquakes.

televiewer: See *borehole televiewer*.

telluric: Of the earth. Often refers specifically to telluric currents.

telluric current: A natural electrical earth current of very low frequency which extends over a large region. Telluric currents originate in variations of the Earth's magnetic field.

telluric-current method: A method in which orthogonal components of the horizontal electric field associated with currents induced in the earth by natural sources of energy are measured simultaneously at two or more stations. The measurements from one station, which serves as a base, are used to normalize the measurements from the other stations to compensate for variation of the source with time. The normalized measurements, plotted as vectors at each station, may outline an ellipse if signals from several different azimuths are recorded as the source changes. The relative area of the ellipse at each station is (ideally) inversely proportional to the conductance of the sedimentary section above the basement. The orientation of the ellipse yields information about the direction of current flow. Other methods of processing or recording the data such as the triangle or the vectogram methods can be used to obtain the relative ellipse areas.

telluric magnetotelluric method: A reconnaisance magnetotelluric method where the magnetic field at one site is used with the telluric fields measured at neighboring sites.

telluric profiling: A reconnaissance resistivity method in which an array of three in-line electrodes emplaced along the traverse line form two consecutive grounded electric dipoles, with the central electrode in common. Signals from the two dipoles are filtered about a band of high geomagnetic activity (e.g., periods of 20 s), ampli-

fied and recorded to yield amplitude ratio and phase difference across the two dipoles. The array is leap-frogged along the survey line to obtain continuous relative electric-field intensity ratios. When successively multiplied together, these ratios yield a relative amplitude profile of the component of the electric field in the traverse line direction. Exploration depth varies inversely with frequency of the electromagnetic field and usually two or more frequencies are recorded and analyzed. Also known as **in-line tellurics** or **E-field ratio tellurics**. See Beyer (1977).

Tellurometer: An electronic survey instrument for measuring distances with great accuracy. A high-frequency (3×10^9 Hz) radio pulse is transmitted to the "rod" transponder, where it is retransmitted back to the master transmitter which measures the time that has elapsed since the original transmission. Accuracy of a few inches in several miles can be achieved with proper corrections (mainly for moisture content of the air). Line-of-sight limited. Tradename of Tellurometer, Ltd.

Telseis: An analog radiotelemetry system utilizing one VHF radio channel per seismic channel. Fairfield Industries tradename.

TEM: *Transient electromagnetic method* (q.v.).

temperature log: A well log of temperature, often made with a resistance thermometer (thermistor). Used for locating (a) cement behind the casing (because the setting of cement is exothermic and hence raises temperature), (b) intervals which are producing gas (because the expansion of gas as it enters the borehole lowers the temperature), and (c) fluid flows (particularly behind the casing). The differential-temperature log records the difference between two thermometers which are usually about 6 ft apart; this log is especially sensitive to very small changes in temperature gradient.

temperature surveying: Measurements of temperature to locate thermal-energy sources or to investigate groundwater problems, thermal springs, karst cavities, sulfide deposits, dikes, structure (faults) affecting groundwater flow, etc. See *geothermal prospecting*.

template: 1. A pattern. **2.** A sheet of transparent plastic marked with the statics of normal-moveout patterns; used as a guide in picking seismic events. **3.** A transparent overlay for calculating gravity or magnetic effects such as a *dot chart* (q.v.). Templates are used for terrain correction, isostatic correction, or residualizing. See also *graticule* and *zone chart*.

template scan: Filtering with a *matched filter* (q.v.).

temporal frequency: Ordinary *frequency* (q.v.), i.e., cycles per unit time, as opposed to **spatial frequency**, cycles per unit distance.

tensor: A set of quantities that relate different vector fields. A tensor is usually expressed as a matrix. The elastic "constants" relating a stress vector to a strain vector are components of tensors for anisotropic media.

tensor impedance: The impedance obtained from total vector measurements of the electric and magnetic fields. The vector electric field $\mathbf{E}$ and the vector current-density field $\mathbf{J}$ are related by the resistivity tensor ρ in which $\mathbf{E} = \rho\mathbf{J}$; in an isotropic material ρ is a scalar.

tensor magnetotelluric method: A *magnetotelluric method*

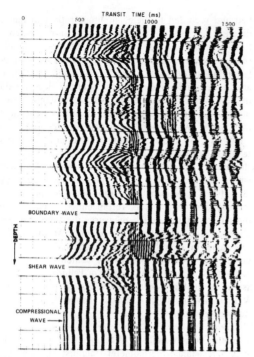

TRANSIT TIME (ms)

BOUNDARY WAVE

SHEAR WAVE

COMPRESSIONAL WAVE

FIG. T-2. **Three-D log**. (Courtesy Birdwell Div., Seismograph Service Corp.)

(q.v.) in which orthogonal measurements of both horizontal magnetic and electric fields are made ($\mathbf{H}_x$, $\mathbf{H}_y$, $\mathbf{E}_x$, $\mathbf{E}_y$) so that the impedance can be described as a complex tensor to account for anisotropy or two-dimensional structure.

tent poling: Sudden deflections of a sonic log resulting from cycle skip or tool sticking.

terminal: An input and/or output device for a computer.

terrain correction: 1. A correction to gravity data required because the surroundings are not all at the same elevation as the meter. Relief in the immediate vicinity of the station may require special surveying (**terrain surveying**), whereas corrections for relief more remote from the station often are made from a topographic map using a terrain-correction template or zone chart. **2.** A correction to seismic data because of the effect of topographic loading on velocity. **3.** A correction to magnetic or electrical data because of terrain effects.

Terrapak: A seismic energy source in which a plate (hammer) is driven against the ground by compressed air. General Dynamics tradename.

terrestial magnetism: See *magnetic field of the Earth* and *geophysics*.

tertiary oil recovery: See *enhanced oil recovery*.

tesla: A unit of magnetic induction, $\mathbf{B}$. 1 tesla = 1 weber/m^2 = 1 newton/amp-m = 10^4 gauss = 10^9 gamma. 1 nanotesla = 1 gamma. Named for Nikola Tesla (1857-1943), American inventor.

thematic mapper: A remote-sensing device which measures the radiation in frequency bands which are sensitive to variations in certain vegetation (for example).

The objective is to map this radiation. A part of the Landsat D satellite.

theodolite: A precision survey instrument; see *transit*.

theogram: *Synthetic seismogram* (q.v.).

theoretical seismogram: *Synthetic seismogram* (q.v.).

thermal conductivity: The heat flow across a surface per unit area per unit time divided by the negative of the rate of change of temperature perpendicular to the surface. Also **heat conductivity**.

thermal-decay time log: See *neutron-lifetime log*. TDT is a Schlumberger tradename.

thermal diffusivity: A quantity which relates to how long it takes for a remote thermal event to have perceptible effect on temperature. The value $k/\rho c_p$ where k = thermal conductivity, ρ = density, and c_p = specific heat at constant pressure. For most rocks it is of the order of 15–60 km²/My.

thermal gradient: The rate of temperature increase within the Earth as a function of depth.

thermal gradient hole: A hole logged by a temperature probe to determine thermal gradient. Usually involves a hole less than 500 ft deep drilled specifically for this purpose.

thermal imaging: Mapping with infrared radiation. See *thermal-infrared*.

thermal-infrared: A remote-sensing method in which an infrared beam outside the natural thermal range is swept back and forth across the earth from an aircraft and the reflected energy recorded. The display often simulates an aerial photograph.

thermal neutron: A neutron whose motion energy corresponds to ambient temperatures; neutrons with mean energies of the order of 0.025 electron volts. See also *neutron log*.

thermal-neutron decay-time log: See *pulsed neutron-capture log* and *neutron-lifetime log*. **TDT** is a Schumberger tradename.

thermal noise: Noise resulting from random thermal energy; **Johnson noise**. The mean-square voltage because of thermal noise in an electrical circuit varies as the absolute temperature, the bandwidth and the resistance.

thermal-relaxation time: See *nuclear-magnetism log*.

thermistor: A device whose electrical resistance varies with temperature. Used with Doppler-sonar units to measure the temperature and hence the velocity of sea water (assuming the salinity).

thermocline: The decrease in water temperature with depth in the ocean. See Figure W-1b.

thermocouple: Two dissimilar conductors welded together at one end. When the junction is heated, a voltage develops across it which is proportional to the temperature difference between the junction and the open ends.

thermoelectric coupling: The ratio of the voltage difference to the temperature difference across a rock.

thermoremanence: See *remanent magnetism*.

thixotropic: A property of gels which allows them to become liquid when agitated. Drilling muds are often thixotropic.

Thomson scattering: Scattering of electromagnetic radiation by free or loosely bound charged particles. The transverse electric field of the radiation accelerates the charged particles which then radiate the energy removed from the primary radiation. A non-relativistic interaction. Named for Sir Joseph John Thomson (1856-1940), English physicist.

Thompson-Haskell method: A frequency-domain method for plane waves in a multilayered half-space, used for both body-wave propagation and surface-wave dispersion problems. See Haskell (1953).

Thornburgh's method: A refraction interpretation method which involves using Huygens' principle to construct wavefronts from reciprocal shotpoints by working backward from the observed arrival times at the surface. The velocities above the refractor must be known for the construction. See *wavefront method* and Thornburgh (1930) or Sheriff and Geldart, v. 1 (1982, p. 223-224).

three-array: A special case of the pole-dipole array in which the three electrodes are equally spaced. See *array (electrical)*.

3-D: Three-dimensional.

three-dimensional dip: The true dip of a reflection or refraction horizon as opposed to the dip component seen in some direction.

three-D log: Three-dimensional velocity log; a display of the seismic wave train received a short distance (3 to 12 ft) from a sonic-wave transmitter. Also called **microseismogram log** and **variable density log**. See also *acoustic wave, full-waveform log, cement-bond log,* and *sonic log*. Three-D log is a Birdwell tradename. See Figures T-2 and C-2.

three-dimensional (3-D) survey: A survey involving collection of seismic data over an area with the objective of determining spatial relations in three dimensions as opposed to their apparent components along lines of survey. Various arrangements are used in such surveys; see Figure T-3. The data from such a survey constitute a volume (Figure T-4) which can be displayed in different ways.

three-layer resistivity models: Models of horizontal electrical layers are given "type" names; see Figure T-5. The curves for such models are called **type curves** and are used in the interpretation of electrical-resistivity observations. Compare *layered earth*.

three-point method: 1. A method for locating a station by taking backsights on three previously located stations. **2.** Determining geographic position by the intersection of bearing lines from three stations. **3.** Determining the strike and dip of a bed from the location and elevation of three points on the bed.

three-point operator: An operator having only three non-zero points, such as a [1/4, 1/2, 1/4] smoother.

threshold: The lower limit that will produce a phenomenon. For example, the IP saturation threshold is the current density above which the IP phenomenon becomes nonlinear.

throw: The vertical component of separation of a rock unit (bed) by a fault. See Figure F-3.

thrust: A reverse fault, especially a low-angle reverse fault where the dip of the fault plane is less than 45 degrees. See Figure F-2.

Thumper: Device for dropping a weight to provide seismic energy. Typically a 3 ton weight is dropped 10 ft. Tradename of Geosource Inc.

TIAC: Texas Instruments automatic computer.

Tiburg rule: A *depth rule* (q.v.) used in magnetic data interpretation: the depth to a pole is 2/3 the horizontal distance at half the maximum amplitude.

tidal correction: A correction to gravity measurements to compensate for the attraction of the sun and moon. Sometimes included as part of the drift correction.

tidal effect: Variations in gravity observations resulting from the attraction of the moon and sun and the distortion of the Earth so produced. Tidal corrections to gravity observations are made by means of tables or are included with drift corrections.

tidal wave: See *tsunami*.

tie: Observations repeated at previously observed points, with the objective of establishing the relationship of two data sets or to check for error.

tie-in: To relate a new station to previously established stations.

tie-line: A survey line which connects other survey lines. Especially such a line which closes a traverse loop.

tie-time: See *time tie*.

tight: 1. Having very low permeability. **2.** Data held confidential.

tight hole: A well about which information is kept secret.

tilt: *Pitch* (q.v).

tilt angle: See *polarization ellipse*.

tilt table: A device on which a meter can be mounted and then tilted by precisely known amounts. Used to test geophones and to calibrate certain kinds of gravimeters.

time: 1. *Record time* (q.v.). **2.** Geologic age; see Appendix L.

time anomaly: Arrival time which is different from that expected.

time-average equation: An empirical equation stating that the transit time $\Delta t = 1/V$ through a rock with matrix velocity V_m and porosity ϕ which is filled with fluid of velocity V_f is approximately

$$\Delta t = 1/V = (1 - \phi)/V_{ma} + \phi/V_f.$$

This relation works well in clean consolidated formations with uniformly distributed pores. In formations containing vugs, the sonic log may not reflect the secondary porosity; and in unconsolidated formations, this relationship may overestimate porosity. In such cases the formula may be empirically modified to give better values. Also called **Wyllie relationship**. Enunciated in Wyllie et al. (1956).

time branch: One of several reflection events produced by the same curved or discontinuous reflector observed at the same point. Coherent reflections may be seen from different portions of the same reflector where synclinal curvature produces *buried-focus effects* (q.v.); see Figures B-8 and D-15. Time branches due to two-dimensional curvature in the plane of the seismic line will migrate into a syncline if the proper velocities are used, but they will not migrate to a point if there is curvature perpendicular to the line.

time break: The mark on a seismic record which indicates the shot instant or the time at which the seismic wave was generated. Abbreviated **TB**. See Figure R-5.

time constant: 1. The time taken for the current in a circuit having a steady emf to reach a definite fraction of its final value after the circuit is closed. The fraction is $(1 - 1/e) = 0.632$. **2.** The time taken for the current to decay to $1/e = 0.368$ of its value after the emf is removed. Also called **decay constant** or **relaxation time**. **3.** AGC time constant. **4.** A time over which readings are averaged to remove statistical fluctuations, as with nuclear-log readings.

time delay: See *delay time, filter correction,* and *Elcord.*

time-depth-chart: A graph or table of reflection time (or sometimes one-way time) against reflector depth for vertically traveling energy. Often written **T-D chart.**

Specific for a particular velocity distribution. Used in converting times to corresponding depths. Compare *time-distance curve.*

time-distance curve: A plot of the arrival time against the shotpoint-to-geophone distance. Also called a **T-X curve.** Used in noise and velocity analysis and in interpreting refracted events (head waves). The slopes of segments of the curve give the reciprocals of the apparent velocities for various refractor beds. See also *normal-traveltime curve, reduced traveltime,* and Figure T-6. Time-distance curves are sometimes composited from measurements made at shorter offsets (see Figure T-7).

time domain: 1. Expression of a variable as a function of time, as opposed to its expression as a function of

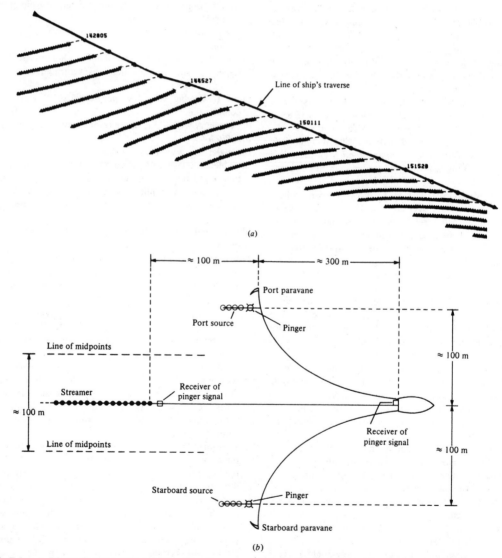

FIG. T-3 a, b. **Three-D surveying.** (a) Use of a crosscurrent to pull streamer off to side; streamer positions are shown at about 15 minute intervals. (b) Use of paravanes to pull marine sources to the side.

frequency (**frequency domain**). Processing can be done using time as the variable, i.e., "in the time domain". For example, convolving involves taking values at successive time intervals, multiplying by appropriate constants, and recombining; this is equivalent to filtering through frequency-selective circuitry. It is also equivalent to Fourier transforming, multiplying the amplitude spectra, and adding the phase spectra ("in the frequency domain"), and then inverse-Fourier transforming. **2.** Time-domain induced polarization is called the *pulse method* (q.v.). **3.** For time-domain electromagnetic methods, see *transient electromagnetic method*.

time-domain sounding: See *electric sounding* and *induced polarization*.

time invariant: Not changing with time. A time-invariant filter has the same action regardless of record time.

time lag: The amount by which arrival times are larger than expected. Indicates that some of the paths from source to detector include a low-speed portion. Delays may also be due to phase shifts in filtering, shot-hole fatigue, etc.

time lead: The amount by which arrival time is smaller than expected indicating that the path from source to detector includes a high-speed segment. See also *lead*.

time line: Line indicating sediments deposited at the same time on a geologic cross-section or correlation diagram. A time line (actually **time surface**) at that time was the surface of the solid Earth.

time of closest approach: The time when a navigation satellite is closest and reaches its maximum elevation angle during a pass. Abbreviated **TCA**.

time sag: A **pull-down** on a seismic section resulting from overlying lateral velocity changes, such as a localized area of low velocity.

time section: A seismic section where the vertical scale is linear in arrival time, i.e., an ordinary seismic section.

time-sequential format: *Multiplexed format* (q.v.).

time series: The series of values of a function sampled at regular time intervals. Sometimes represented as a set of values, as a stickogram, in Z-transform form, etc. A digitized seismic trace is such a series.

time sharing: Simultaneous use of a computing machine on two or more different programs. The operations involve different parts of the computer, and the computer's central processing unit exercises the control to keep the results separate. For example, a computer might be simultaneously reading, searching its memory for certain data, calculating, and writing results as parts of different programs.

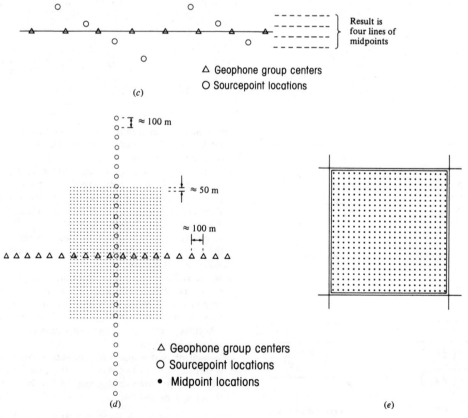

FIG. T-3 c, d, e. **Three-D surveying** (cont). (**c**) Wide-line layout. (**d**) Block layout. (**e**) Seis-loop with geophones and source points around the perimeter of a loop. (From Sheriff and Geldart, v.1, 1982, p. 148-149.)

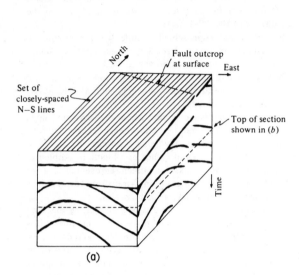

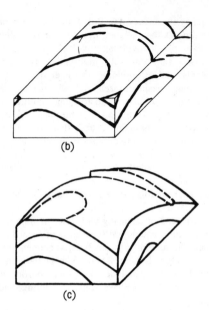

FIG. T-4. **Three-dimensional data** obtained from a set of closely spaced N-S lines. (**a**) Isometric diagram of the volume these data occupy. The easternmost N-S section is shown along with an E-W section made from the southernmost traces on each N-S line. (**b**) The data set with the top portion removed; the top now constitutes a **time-slice map**. (**c**) The data sliced along one reflection constitutes a **horizon-slice map**. (After Sheriff and Geldart, v.1, 1982, p. 170.)

ρ_1

$\rho_2 \ (> \rho_1)$

$\rho_3 \ (> \rho_2)$

A-TYPE SECTION

ρ_1

$\rho_2 \ \left(\begin{smallmatrix} < \rho_1 \\ < \rho_3 \end{smallmatrix} \right)$

ρ_3

H-TYPE SECTION

ρ_1

$\rho_2 \ (< \rho_1)$

$\rho_3 \ (< \rho_1)$

Q-TYPE SECTION

ρ_1

$\rho_2 \ \left(\begin{smallmatrix} > \rho_1 \\ > \rho_3 \end{smallmatrix} \right)$

ρ_3

K-TYPE SECTION

FIG. T-5. **Three-layer** resistivity types.

time signal: A signal indicating an exact instant of time. Such a signal is used to indicate the time of energy release in seismic work.

time-slice map: A display of the seismic measurements corresponding to a single arrival time (or single depth for migrated data) for a grid of data points; a horizontal slice through a volume of 3-D data. See Figures T-4 and T-8. Also called a **Seiscrop section**.

time surface: A surface which one time was the surface of the solid Earth. Reflections parallel time surfaces. Also called **stratal surface**.

time tie: 1. To verify that arrival times are the same for events on different records which possess common raypaths. **2.** To relate data obtained in opposite directions or along intersecting seismic lines. **3.** To relate reflection events to contacts seen in wells where the velocity is known.

time-to-depth conversion: A seismic section commonly has a vertical scale of two-way time (i.e., the time for the signal to go down and come back). For vertically traveling reflected energy:

$$\text{depth} = (\text{average velocity}) \times (\text{two-way time})/2.$$

Average velocity may be obtained from well data,

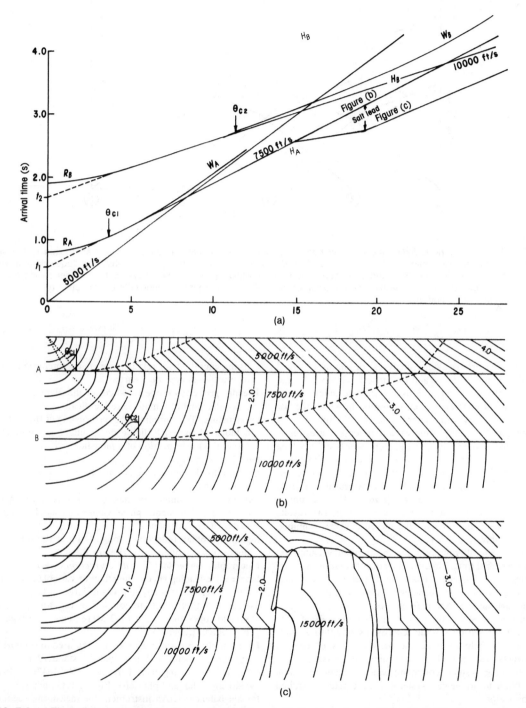

FIG. T-6. (a) **Time-distance curve**. R_A, R_B = reflections from A, B; H_A, H_B = head waves at A, B; W_A, W_B = wide-angle reflections from A, B. t_1, t_2 = intercept times for head waves at A, B. (b) Wavefront diagram showing first arrivals. (c) Wavefront diagram showing an idealized salt dome.

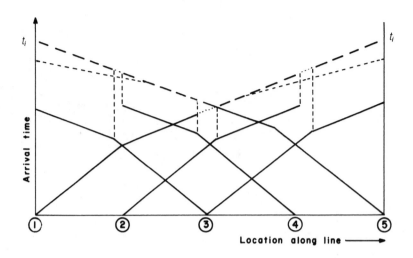

FIG. T-7. Composite **time-distance** curves (dashed lines) constructed from first-arrival curves from shotpoints 1, 2, 3, 4, and 5 (solid lines). Refraction portions from SP 2, 3, and 4 have been displaced upward to show the "equivalent arrival times" as if the profiles from SP 1 and 5 had been longer. Such displaced curves should overlap for confidence and show the same reciprocal time t_i. Actual time-distance curves from SP 1 and 5 might have been different if another refractor had been reached (e.g., dotted lines).

calculated from velocity analyses, or simply assumed (based on experience and the closest velocity information); two-way time is measured from seismic records.

time variable: Describing an operation in which the parameters vary with record time, as in "time-variant filtering". Usually, time-variant processes are implemented by determining parameters over two fairly long portions of a trace at different times in a time-invariant manner, processing the data twice using the two sets of parameters, and then blending the two results together over some time interval (**merge zone**). The blending is usually achieved by varying the mix of the two results in a linear manner over the merge interval; **ramping**. Often abbreviated **TV**.

time-variable filtering: Varying the frequency band-pass with record time. Time-variable deconvolution is sometimes used to compensate for the shift of reflection energy to lower frequencies at late record times.

time-variable gain: See *gain control*.

timing lines: Marks or lines at precise intervals of time such as used on seismic records (usually 0.01 s intervals) to help measure the arrival times of seismic events. See Figure R-5. The timing mechanism in older recording systems commonly was an accurate tuning fork, and in newer systems is a crystal-controlled oscillator.

timing word: A word at the head of a block of data which gives the elapsed time since the shot instant.

tin hat: 1. A hard hat worn as protection to the head. **2.** A hole plug shaped somewhat like a hat.

tipper: A complex quantity whose amplitude is the ratio between the vertical and horizontal magnetic fields perpendicular to apparent strike. Devised by Vozoff

from general tensor relations for magnetotelluric fields over a two-dimensional earth.

Toeplitz property: Property when all the elements on a given diagonal of a matrix are identical.

toe structure: The rumpling or overthrusting at the end of a block of material sliding down under gravitational force; the lower portion of a landslide.

tool: *Sonde* (q.v.).

tomography: Reconstruction of an object from a set of its projections.

Tomoseis: A technique for making *time-slice maps* (q.v) showing instantaneous phase values displayed in color code. CGG tradename.

toplap: An offlap reflection configuration at the top of a depositional sequence; see Figure R-8.

topographic correction: *Terrain correction* (q.v.).

topographic-loading effect: The effect of variable overburden on seismic velocity in areas of large surface relief.

Toran: A set of short- to long-range radio-positioning systems operating in the 1.6-3 MHz range. See Figure P-7. Sercel tradename.

torr: A unit of pressure; the pressure required to support a column of mercury 1 mm high at standard gravity. Named for Evangelista Torricelli (1608-1647), Italian mathematician and physicist. Equals 1.333 mbar.

torsion balance: 1. An instrument for measuring second derivatives of the gravitational potential. In a nonuniform field, the forces on equal masses at opposite ends of a horizontal beam suspended by a very fine torsion wire differ, producing a torque which can be measured by counter balancing it with a known torque. The measured gradients can be integrated to make a gravity map. The curvatures have characteristic patterns associated with gravity anomalies. **2.** A device for measuring other sorts of force fields, such as magnetic or electrical.

tortuosity: The length of the path of a fluid passing through a unit length of rock.

total intensity: Usually refers to the total magnetic intensity as opposed to the components of the intensity in the vertical or horizontal directions.

total reflection: Reflection where the angle of incidence exceeds the critical angle. At and beyond the critical angle the energy is either reflected or converted. So

called because all the energy is reflected in the steady state.

tour: A shift in drilling operations. Pronounced "tower".

tower: *Tour* (q.v.).

towing bridle: Assembly by which the towing ship holds a marine streamer in the proper position amidship and keeps it from rubbing the stern.

township: 1. A unit of area, nominally six miles on a side. It is subdivided into 36 sections, numbered as shown in Figure T-9. **2.** A designation of an east-west row of townships. American townships are specified in rela-

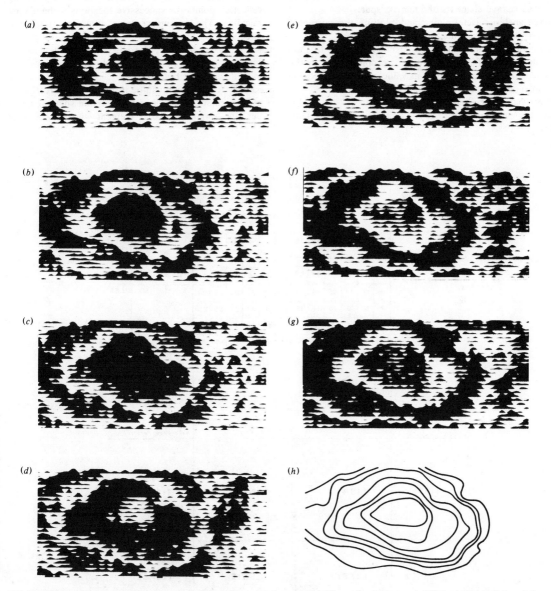

FIG. T-8. **Time-slice maps**. The area is 3.6 by 8.0 km. (**a**) through (**g**): Time slices for $t = 1.580$ to 1.604 s at 4 ms intervals; (**h**) time contour map made by tracing one contour from each of the preceding time-slice maps, starting with the outside of the central area on map (a). (Courtesy GSI.)

tion to standard reference parallels which often are not stated explicitly; T-3-N indicates a township in the third row north of a reference parallel. Canadian townships are counted northward from the 49th parallel. See also *range*. Township is abbreviated **Twp**.

TP: *Turning point* (q.v).

T-phase: A short period (1 s or less) wave which travels through the ocean with the speed of sound in water; it is occasionally identified on the records of earthquakes in which a large part of the path is across deep ocean.

trace: 1. A record of one seismic channel, one electromagnetic channel, etc. See Figure R-5. **2.** A line on one plane representing the intersection of another plane with the first one, such as a fault trace. **3.** The sum of the diagonal elements of a matrix; **spur**.

trace analysis: Determining and plotting the corrected arrival time of events for every trace.

trace equalization: Adjusting a seismic channel so that the amplitudes of adjacent traces are comparable in the sense of having the same rms value over some specified interval or some other criterion.

trace gather: See *gather*.

trace integration: A form of mixing used with the weight-

drop method. Assume 20 input traces are mixed to form 1 output trace, and the last 15 of these input traces plus the next 5 are mixed to form the next output trace; then the "trace integration is 20/5".

trace inversion: Calculating acoustic impedance or velocity from a seismic trace to make a *seismic log* (q.v.).

trace sequential: An arrangement of data on magnetic tape in which one channel (trace) is recorded without interruption, followed sequentially by other channels. As opposed to **time sequential** or **multiplexed format** in which the data for one record time are recorded for all channels, followed by the data for the next record time, etc.

trace, subsurface: A line on a reflecting surface connecting reflection points for successive locations along a seismic line. The subsurface trace allows for migration effects.

track: 1. A *trace* (q.v.). **2.** The data positions which can be read by a single magnetic head. **3.** To follow the movements of an object.

trackball: A ball which can be turned in any direction to move a cursor on a videodisplay so that something can be done about the matrix element whose location is

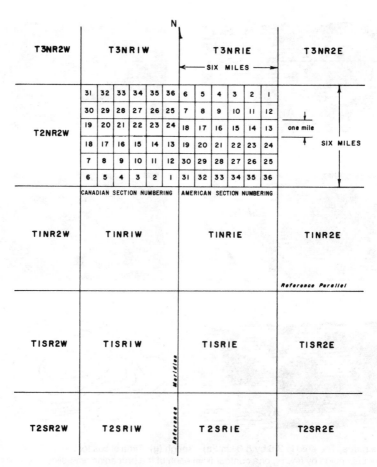

FIG. T-9. **Township-range** location system showing the system for numbering sections.

indicated by the cursor. Function is similar to that of a joystick.

track plotter: A device which continuously displays the position of a ship, operated from navigation (radio-positioning) system signals. Display is often in positioning system coordinates.

traction: 1. Frictional force. **2.** A shearing stress as opposed to a pressure.

train: A series of successive semiperiodic motions, as a "train of waves".

trajectory: The path of a seismic wave. Synonym: *raypath* (q.v.).

transceiver: Device which is both a transmitter and a receiver, such as used in sonar. See *transponder*.

Transcontinental Geophysical Survey: The study of a band 4 degrees wide (about 440 km) centered on latitude 37° N extending across the U.S. and offshore into the Atlantic and Pacific Oceans.

transcord: *Transcribe* (q.v.).

transcribe: To copy information from one storage medium to another, as to make a magnetic tape from a paper seismic record. See also *reformat*.

transcurrent fault: A strike-slip or wrench fault. See Figure F-2.

transducer: 1. A device which converts one form of energy into another. Many types of transducers are reversible; for example, converting electrical energy into acoustical energy and vice versa. The electrodynamic geophone is a reversible transducer; it converts mechanical motion to electrical voltage, or passing a current through the coil causes the coil to move with respect to the case. Other reversible transducers are electrostatic, variable reluctance, magnetostrictive, piezoelectric, etc. Piezoelectric transducers of barium or lead zirconate or titanate are used in many hydrophones and sonar transducers. **2.** Especially a piezoelectric transducer.

transduction factor: Ratio of output to input for a transducer. For digital-grade velocity geophones, it is of the order of 1/4 V/cm/s. For a hydrophone, of the order of 6 V/bar.

transfer characteristics: *Transfer function* (q.v.).

transfer function: The ratio of output to input as a function of frequency. The frequency-domain characteristics of a system (e.g., a filter). The complex function of frequency which changes sinusoidal inputs into outputs. Multiplying the frequency-domain transform of the input by the transfer function yields the frequency-domain transform of the output. The transfer function usually is represented by amplitude-versus-frequency and phase-versus-frequency curves; these contain the same information as the impulse response in the time domain and are convertible into the impulse response through the Fourier transform.

transfer impedance: The complex ratio of a potential difference at one pair of terminals or electrodes to the current at another pair.

transform: To convert information from one representation into another, as with the Fourier transform or Laplace transform.

transformed wave: *Converted wave* (q.v.).

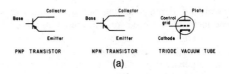

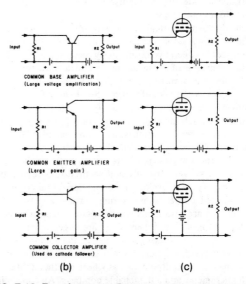

FIG. T-10. **Transistors.** (**a**) Basic transistor and triode representations. (**b**) Schematic amplifier circuits using transistors. (**c**) Schematic amplifier circuits using vacuum-tube triodes.

transform fault: A fault which displaces the rift zone associated with the generation of new crust or which displaces the zone where plates collide. Motion takes place on the portion between the active centers only. See Figures F-2 and P-4.

transform pair: A waveform and its frequency-domain equivalent, or a time-domain operation and its frequency-domain equivalent. See Figures F-17 and F-18. Used also for transforms other than Fourier.

transgression: A landward movement of the shoreline. Opposite of **regression**.

transient: A nonrepetitive pulse of short duration, such as a voltage pulse or seismic pulse.

transient electromagnetic method: TEM. An electromagnetic method in which the waveform of the transmitted signal is a pulse, step function, ramp, or other form which can be considered to be nonperiodic and in which measurements are made after the primary field has stopped changing. Some methods, which may be called **quasi-transient** or **quasi-time-domain methods**, use a train of primary pulses with measurements being made during the off-time between pulses. Principal advantages of transient methods over continuous-wave methods are that the primary field is not present during the measurement of the secondary field and that measurements of the secondary field as a function of time are equivalent to continuous-wave measurements over a wide frequency range. Transient methods are used primarily for depth sounding but quasi-transient airborne methods are used for continuous profiling. See *Input system.*

transient IP method: See *pulse method.*

transient response: Response of a system to a very short transient, ideally an impulse (which would yield *impulse response*, q.v.).

transistor: An electrical device with three or more terminals using a semiconductor for controlling the flow of current between two terminals by means of current flow between one of these terminals and a third terminal. Ideally it is a current-controlled current source which operates in only one direction. See Figure T-10.

transistor/transistor logic: A family of integrated circuit logic in which the multiple inputs on gates are provided by multiple emitter transistors. Abbreviated **TTL.**

transit: 1. A precision instrument for measuring horizontal and vertical angles. It consists of a telescope mounted so as to swivel vertically and secured to a revolvable table carrying a vernier for reading horizontal angles. A graduated circle for measuring vertical angles and a compass are also included. Also called **transit-theodolite.** "Transit" and "theodolite" are largely interchangeable, American usage generally favoring the first, European usage the latter. Usually a transit has an open vernier and a built-in compass whereas a theodolite has a micrometer and a detachable compass. An especially

precise version is sometimes called a **theodolite.** An accuracy of about 10 s of arc can be achieved, although 30-s accuracy is more likely. See Figure T-11. **2.** The passage of a celestial body across a celestial meridian. **3.** The passage of a satellite involved in satellite navigation. "Transit" is the system name of the U.S. Navy Navigation Satellite System.

transit-and-chain surveying: A survey in which directions are determined by transit and distances are measured directly.

transit-and-stadia surveying: A survey in which horizontal and vertical directions are determined by transit and distances are measured by observing a stadia rod through the transit's telescope.

transit-theodolite: See *transit.*

transit time: The traveltime of a wave over 1 ft of distance. See *sonic log.*

translocation method: A positioning technique which uses a nearby fixed station to improve the accuracy of a mobile or other fixed station. **1.** Observation of a satellite transit from a nearby fixed location in order to correct for minor variations in satellite orbit. **2.** Observation of radio-positioning signals at a fixed station for information on sky-wave variations.

transmission coefficient: 1. The ratio of the amplitude of a wave transmitted through an interface to that of the wave incident upon it. This ratio can be greater than

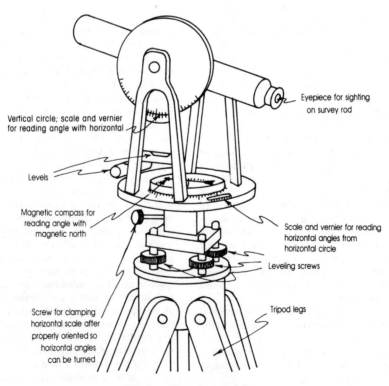

FIG. T-11. **Transit**-theodolite.

one. **2.** A measure of the amplitude of a wave passing through an interface restricted to the case of normal incidence. **3.** Because with reflection we are usually interested in energy which has passed through an interface twice (once going down and once returning upward), a **two-way transmission coefficient** T for normal incidence is often used:

$$T = 1 - R^2 = (4\rho_1 V_1 \rho_2 V_2)/(\rho_1 V_1 + \rho_2 V_2)^2,$$

where R is the reflection coefficient at normal incidence, ρ_1, ρ_2, V_1, and V_2 are the densities and velocities in the upper and lower media. **4.** A ratio of the energy densities (which involves the square of transmission coefficients as defined above).

transmitter: In resistivity and IP surveying, a current waveform generator. Also called a **sender**. In electromagnetic methods, a loop or grounded wire.

transparent: Not evident to the user; something the user does not need to take into account.

Transploder: A marine seismic source employing the vaporization of a wire by passing a large surge of electric current through it. Compare *WASSP*. Prakla-Seismos tradename.

transponder: A device which transmits a signal upon receiving another signal. When the receiver in the device detects the "interrogating signal" it triggers the transmitter which replies with a coded pulse or sequence of pulses. Transponders are used with both electromagnetic and sonar waves. A radar transponder is also called a radar **beacon**, a sonar transponder a **pinger**.

transport: 1. A device or method for moving equipment or personnel, as transport for a geophysical crew. **2.** A device for moving magnetic tape past magnetic heads for reading or writing on the tape.

transpose: The transpose $\mathbf{A}^T$ of a matrix $\mathbf{A}$ is the matrix whose rows are identical to the columns of $\mathbf{A}$. Thus if matrix $\mathbf{A}$ is of size $(m \times n)$, matrix $\mathbf{A}^T$ is of size $(n \times m)$. For conformable matrices,

$$(\mathbf{A}\ \mathbf{B})^T = \mathbf{B}^T\ \mathbf{A}^T.$$

For any matrix, $(\mathbf{A}^T)^T = \mathbf{A}$.

transposed recording: A method of seismic field recording in which a single large geophone group (or a small number of such groups) records energy generated at a succession of positions along the spread line, as opposed to "conventional" shooting in which single or simultaneous shots (or other energy sources) are recorded by a number of geophone groups.

transversely isotropic: Having properties which are the same in two orthogonal directions but different in the third orthogonal direction. See *anisotropy* and Sheriff and Geldart, v. 1 (1982, p. 52-53).

transverse Mercator projection: See *map projection*.

transverse resistance: Resistivity times thickness.

transverse wave: *S-wave* (q.v.).

trap: 1. A waveguide phenomenon; see *channel wave*. **2.** A configuration of rocks which is able to confine fluids (such as oil) which float on other fluids (water). A closed structure in porous formations may be a trap if it has an impermeable cap; an unclosed structure may also be a trap if permeability variations block off the escape route of fluids. Compare *closure*. **3.** A CPU-initiated interrupt which is generated when a predetermined condition, such as an illegal instruction, breakpoint, specified error, or power failure is detected.

trap-door structure: The high area on the upthrown side of each of two intersecting faults.

travel path: The path given by *Fermat's principle* (q.v.). Usually, the least-time path from the source to receiver (geophone).

traveltime: The time between time break and the recording of a seismic event.

traveltime curve: A *time-distance curve* (q.v.).

traverse: 1. A survey line or series of connected survey lines. **2.** A sequence of connected profiles, as a seismic line. **3.** A series of measured distances at measured angles; see Figure P-2. Compare *triangulation* and *trilateration*.

trenching: Electrical *profiling* (q.v.).

trend: 1. That component in a geophysical anomaly map which is relatively smooth, generally caused by regional geologic features. **2.** The systematic smooth component of a function.

trend analysis: 1. The fitting of an analytic surface (the **regional**) to data points as a representation of the "order" in the data, as opposed to the erratic element (the **residual**). The objective may be to analyze trends in the data or to interpolate between data points. The underlying assumption is that the data may be decomposed into a relatively low-order, smooth surface plus more or less random noise. The number of independent parameters in the analytic surface should be appreciably less than (usually less than 10 percent of) the number of data points. Also called **surface fitting. 2.** The two-dimensional Fourier analysis of a surface into spatial frequency components (or wavelengths) so that one can determine preferred orientations and wavelengths.

triad: Three stations (one master and two slaves) which constitute a positioning system, such as loran, raydist, etc. A master and three slaves constitute a **star**.

triangulation: Establishing locations by a system of overlapping triangles where the angles are directly measured but only a few of the sides are directly measured. Sometimes called **method of intersection. First-order triangulation** has an accuracy from one part in 25 000 to one part in 100 000; **second-order**, an accuracy of one part in 10 000; **third-order**, one in 5000. Triangulation is illustrated with a plane table in Figure P-2 (but a plane table is not an accurate way to triangulate). Compare *traverse* and *trilateration*.

trigonal: A grid of equilateral triangles (or regular hexagons) produced by three sets of equally spaced lines positioned at 60 degrees to each other; used in making grid-residual maps. See Figure T-12.

trigonometric identities: Statements which hold for all values of arguments, such as:

$$\sin(x \pm y) = \sin x \cos y \pm \cos x \sin y,$$

$$\cos(x \pm y) = \cos x \cos y \mp \sin x \sin y,$$

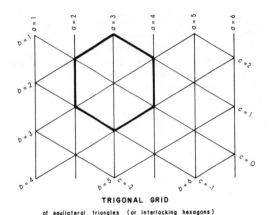

TRIGONAL GRID

of equilateral triangles (or interlocking hexagons)

FIG. T-12. **Trigonal grid** of equilateral triangles (or interlocking hexagons) and a possible grid-identification scheme ($c = a - b$).

$$2 \sin x \sin y = \cos(x - y) - \cos(x + y),$$

$$2 \cos x \cos y = \cos(x - y) + \cos(x + y),$$

$$2 \sin x \cos y = \sin(x - y) + \sin(x + y).$$

trilateration: Establishing locations by a system of overlapping triangles where all sides are directly measured. Compare *triangulation* and *traverse*.

trim: 1. The longitudinal axis of a ship not being horizontal on the average, the bow being raised or depressed. Compare *pitch* and *list*. **2.** To apply statics based on a shallow reflection.

tripartite method: A method of determining the apparent surface velocity and direction of propagation of microseisms or earthquake waves by determining the times at which a wave passes three separated, noncollinear points.

triple junction: A rupture resulting from the tensional forces involved in creating a bulge in the crust of the Earth; three lines of rupture radiate from the point. Often separation continues on two of these lines forming eventually an ocean basin; the third line of rupture is called a **failed arm** or **alaucogen.**

triplets: A method of determining the elevation correction factor (ecf) for gravity data from sets of three readings each. It is assumed that elevation does not correlate with geologic structure. If the difference between the height at a station and the weighted mean of the heights of stations on either side of it is h_i, and the difference between the observed gravity reading at the station and the weighted mean at the neighboring stations is g_i (where the weighting is usually taken as inversely proportional to the distance), then the elevation correction factor k is given by:

$$k = -(\Sigma \, h_i g_i)/\Sigma h_i^2,$$

and the probable error ε in k is

$$\varepsilon = 0.67[(\Sigma g_i^2/\Sigma h_i^2 - k^2)/n]^{1/2}.$$

See Siegert (1942).

tripping: Changing the mode of operation. For example, switching a seismic recording channel from the uphole geophone to the group assigned to the channel or changing from fixed initial gain to AGC or binary gain.

Trisponder: A type of *tellurometer* (q.v.). Del Norte tradename.

troop: *Party* (q.v.).

tropospheric correction: A correction to radio-wave propagation for velocity and refraction variations because of meteorological conditions, principally because of variations in atmospheric moisture. The troposphere is the lower 10-18 km of atmosphere.

tropospheric scatter: The bending of radio waves in the troposphere by scattering instead of by refraction. Responsible for extended-range shoran and other UHF radio positioning beyond the line of sight.

troubleshoot: To look for the cause of a malfunction.

trough: The lowest part of a waveform between successive peaks.

true: The effect if all distortions introduced by the experimental situation were removed, as with "true resistivity" or "true IP effect." See *apparent*.

true bearing: Azimuth with respect to true north.

true dip: Three-dimensional dip; as opposed to the component of dip in some direction.

trumpet log: A *microlaterolog* (q.v.) in which the guard electrodes are concentric about the current electrode so that the current flow is concentrated in a tube which gradually flares out.

truncation error: 1. The error resulting from using only a finite number of terms of a series. The error produced by using only a limited operator length in a convolution, or a finite gate length in a correlation. **2.** The effect of digitizing an analog signal whose corresponding digital value exceeds the maximum value permitted; see *clipped*. Loss of information in digitized data because of truncation of high-order or low-order bits causes errors of different kinds. **3.** In calculating the total mass from a gravity anomaly, the error resulting from the integration not being carried out to the limit of the anomaly.

truth table: A listing which presents all possible input and output states of a logical function. See *Boolean algebra* and *gate*.

TSP: Time at the shotpoint; *up-hole time* (q.v.).

T-spread: A seismic spread in which the shotpoint is offset perpendicular to the center of the spread by an appreciable distance. Also called **broadside.** See Figure S-17.

tsunami: Gravity waves set up by disturbances in the seafloor; **tidal wave.**

TTL: *Transistor/transistor logic* (q.v.).

tube wave: A surface wave in a borehole. See Sheriff and Geldart, v. 1 (1982, p. 51-52).

tumescence: Swelling (increase in elevation) because of increase in a magma chamber underneath.

tuned array: An array of marine seismic sources of different strengths arranged to suppress bubble pulses relative to the initial pulse.

tuned voltmeter: A voltmeter containing a band-pass filter.

tuning fork: A U-shaped bar of hard steel, fused quartz, or other elastic material that vibrates at a definite natural frequency when set in motion. Used as a frequency standard.

Turam method: Electromagnetic-survey method employing an energizing source consisting of a long insulated cable grounded at both ends or a large horizontal loop. The cable is often several kilometers long and energized at 100 to 800 Hz. Measurements are made of the amplitude ratio and phase difference between two receiving coils about 100 ft apart. Usually the plane of the two loops is horizontal. Many profiles may be made using the same source location.

turbidity current: A bottom-flowing current resulting from large amounts of suspended sediment producing a fluid of higher density. Turbidity currents are intermittent but possess considerable erosional power and transport appreciable volumes of sediment.

turkey shoot: A direct comparison of the results of recording with two or more sets of instruments simultaneously under the same field conditions.

turn around: The period of time between submission of data for computer processing and receipt of the results.

turning point: 1. The location of the survey rod in the procedure where transit and survey rod successively leapfrog over each other in traversing along a line. The point on which a foresight is taken from one instrument station in a line of survey and on which a backsight is taken from the next instrument station. Abbreviated **TP. 2.** A point on maximum-depth estimation curves at which limiting depths reach a minimum; an aid in gravity interpretation.

turnkey: 1. A computer console containing a single control, usually a power switch, that can be turned on and off only with a key. **2.** A design and/or installation in which the user receives a complete running system.

turnkey bid: A price for specified work which is all-inclusive, often a fixed price per unit of production (plus cost of consumables and subcontract work).

turtle: A residual structure resulting from the withdrawal of salt surrounding it. Originally a low because of early salt withdrawal and hence involves thicker sediments; the structure inverts as surrounding thinner sediments settle because of later salt withdrawal.

TV: *Time-variant* (q.v.).

TVG: Time-varying gain. See *time-variant*.

T-wave: Tertiary wave; seismic waves from earthquakes characterized by travel within the oceans as ordinary sound waves, which are then converted to *P*-, *S*-, or surface waves for travel on the continents.

two array: *Pole-pole array* (q.v.).

2-D: *Two-dimensional* (q.v.).

two-dimensional (2-D): Having no variation in the direction perpendicular to the plane (usually vertical) which includes the line of measurement, as with a plane perpendicular to the axis of an infinitely long feature. **Infinitely long** means so long that the effects of the ends are negligible.

two-dimensional filtering: Apparent velocity or *f-k* filtering; see *apparent velocity*.

two-dimensional plot: A contour plot of depth-probe or sounding data (delay time, apparent resistivity, metal factor, etc.) as a function of position along a line (often plotted below midpoints) and electrode separation or offset. See *pseudosection* and Figure P-12.

two's complement: The radix complement form for representing negative binary numbers. It can be found by replacing all ones by zeros and all zeros by ones and then adding one. For example, the decimal number 27 is represented as 0011011 and −27 by 1100101. When a number is added to its negative, all registers are empty. There is only one representation for zero: all zeros. Compare *one's complement*.

two-sided: Being defined for both positive and negative values of the argument.

two-way control: *Reverse control* (q.v.).

two-way transmission coefficient: See *transmission coefficient*.

Twp: *Township* (q.v.).

T-X curve: *Time-distance curve* (q.v.).

type curves: Master curves; the computed IP, resistivity, or electromagnetic response plotted against electrode interval, source-receiver spacing, or frequency for various models. Type curves are used for interpreting field data where the conditions of the models appear to hold. Abscissa and ordinate are usually normalized so that the curves are dimensionless and data are usually compared by using transparent overlays. Type-curve derivations often employ the method of images. Models are sometimes given "type" names as in Figure T-5.

U

ULSEL: *Ultra-long-spaced electric log* (q.v.).

ultra-long-spaced electric log: A modified long normal borehole log mounted on a 5000 ft bridle. Abbreviated **ULSEL**. The A to M spacings are 75, 150, 600, and 1000 ft. Differences between the measured resistivities and anticipated resistivities calculated from conventional resistivity logs indicate nearby resistivity anomalies, such as a salt-dome or nearby hydrocarbons. Has been used to sense casing in a blow-out well from an approaching relief well.

uncertainty: The precision with which a measurement or value is known. Often implies a 50-50 chance that any one of a series of measurements would fall within a given range. Does not necessarily imply anything about accuracy, which is comparison with the true value. See *probable error* and *standard deviation*.

unconformity: A surface of erosion or nondeposition that separates younger strata from older rocks. An unconformity is often a good seismic reflector and can be recognized where the layers above and below the unconformity are not parallel (**angular unconformity**).

uncracking: *Unwrapping* (q.v.).

underdamped: See *damping*.

undershooting: Shooting from a shothole on one side of a property into a spread on the opposite side in order to obtain subsurface coverage under the property itself. Used when the surface of the property is inaccessible (such as on a line crossing a river) or to get data beneath some feature whose presence might introduce intolerable uncertainties if measured through it (such as to map beneath a salt dome).

underthrusting: A type of thrust fault in which a lower plate is pushed under relatively passive overlying rock.

Uniboom: Tradename for a profiler source which generates a 300-3000 Hz signal.

unijunction transistor: A transistor made of *n*-type semi-conductive material with a *p*-type alloy region on one side; connections are made to base contacts at either end of the *n*-type material, and also to the *p* region. Primarily used in timing circuits.

union: *Disjunction* (q.v.); see Figure G-1.

Unipulse: An air gun designed to minimize bubble effects. Petty-Ray Geophysical tradename.

unit circle: A circle of unit radius. See *z-transform* and Figure Z-3.

unit impulse: *Impulse* (q.v.) having a value of one.

unit step: A *step function* (q.v.) whose magnitude is one.

Universal Transverse Mercator: A standard rectangular map grid. The projection is onto a cylinder tangent to the earth along a central meridian (i.e., with the cylinder axis perpendicular to the earth's axis). See Figure M-3 and *Mercator projection*. The earth is divided into 60 north-south columns, each 6 degrees of longitude wide. The central meridian is assigned the value of 500 000 m easting and the equator a northing of 0 in the Northern Hemisphere and 10 000 000 m in the Southern Hemisphere. Abbreviated **UTM**.

universe: The complete collection of objects, numbers, functions, etc., with probabilities attached to relevant subcollections; **population**.

unmigrated seismic map: A map showing data posted at midpoints. See also *map migration*.

unwrapping: Determining phase as a continuous function of time or frequency from data where discontinuities of $2n\pi$ (n being an integer) are possible. Also called **uncracking**.

update: To correct a system for deviations or drift which has occurred since it was put into operation or since last updated. Satellites are updated by having new position information injected each day. Doppler-sonar positions are updated with satellite-fix information.

updip: The direction of seismic shooting where reflectors

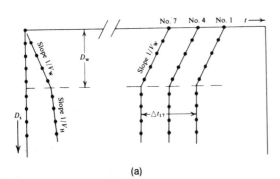

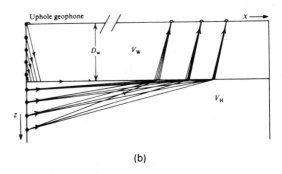

(a)

(b)

FIG. U-1. **Uphole survey.** (**a**) Plots of traveltime versus shot depth for geophones at shotpoint and three offset distances. (**b**) Vertical section showing raypaths. $V_H \cong (x_1 - x_7)/\Delta t_{17} = (x_1 - x_7)/(t_1 - t_7)$.

or refractors dip away from the geophones toward the source point.

uphole geophone: A geophone placed a few feet from a shothole to detect *uphole time* (q.v.). Sometimes called a **bug**.

uphole method: Constructing a shallow wavefront diagram by shooting at several depths and recording on a full surface spread of geophones. See also *Meissner technique*.

uphole seis: *Uphole geophone* (q.v.).

uphole shooting: 1. The successive detonation of a series of charges at varying depths in a shothole in order to determine the velocities of the near-surface formations, the weathering thickness, and (sometimes) the variations of record quality with shot depth. **2.** Sometimes a string of geophones is placed in a hole of the order of 200 ft deep to measure the vertical traveltimes from a nearby shallow shot. See Figure U-1.

uphole stack: The combining of seismic records from shots at different shot depths, after time-shifting based on uphole-time measurements, so that reflection energy is in register. A method of attenuating ghost energy. Sometimes erroneously called "vertical stack", which is a different process.

uphole survey: *Uphole shooting* (q.v.).

uphole time: Time for the first wave from an explosion to reach the surface at or near the shotpoint. Used in determining weathering or near-surface corrections to seismic data. See Figure R-5. Also called **bug time**.

uplap: In-filling of a basin, usually accompanied with greater compactional settling in the center than near the margins so that reflections near the margins dip toward the basin.

UPS: Universal polar stereographic projection, used for latitudes 80-90°; see Figure M-3.

upstairs: In the numerator of an equation.

upsweep: Vibroseis sweep in which the frequency increases with time.

upward continuation: Calculation of the potential field at an elevation higher than that at which the field is known. The continuation involves the application of Green's theorem and is unique if the field is completely known over the lower surface (which is usually true for gravity and magnetic fields) and where all sources above the lower surface are known (usually all are zero). Upward continuation is used to smooth out near-surface effects and to tie aeromagnetic surveys flown at different heights.

utility routine: A standard routine, usually part of a larger software package, which performs service and/or program-maintenance functions, such as file maintenance, file storage and retrieval, media conversions, and production of memory and file printouts.

UTM: *Universal Transverse Mercator projection* (q.v.).

V

VA: Display in *variable-area* (q.v.) form. Also abbreviated VAR.

Vacupulse: Seismic energy source in which a weight is dropped in an evacuated chamber. Geophysical Resources tradename.

vacuum-tube voltmeter: A voltage-measuring device using electronic circuits to obtain high impedance across the measurement probe so that very little current is drawn. Now superseded by solid-state circuitry. Abbreviated VTVM.

validity check: See *check*.

valley: The downward displacement of a single cycle of a seismic trace; a **trough**. Opposite of **peak**.

Vaporchoc: A marine seismic energy source in which a quantity of superheated steam under high pressure is injected into the water. Subsequent condensation of the steam attenuates bubble oscillation. Also called **steam gun**. CGG tradename.

vara: An old Spanish unit of length; about 33 inches or 0.85 m, but the exact length differs from country to country.

variable amplitude recording: Recording *wiggle trace* (q.v.). See Figure D-14.

variable-area: A display method in which the width of a blacked-in area is roughly proportional to the signal strength. Abbreviated VA. See Figure D-14.

variable-density: A display method wherein the photographic density is proportional to signal amplitude. Abbreviated VD. See Figure D-14.

variable-density log: A microseismogram log, *three-D log*, (q.v.) or VDL log; see Figure T-2. Compare *full waveform log*.

variable reluctance geophone: A geophone whose magnetic reluctance is made to vary by mechanically changing the size of an air gap.

variable word length: An attribute of a computer which handles data words containing variable numbers of characters.

variance: See *statistical measures*.

variometer: An instrument for measuring small magnetic variations, used in magnetotelluric work. Consists essentially of a magnet suspended on a torsion fiber; slight rotations of the magnet deflect a light beam reflected by a small mirror attached to the magnet, and the position of the beam is recorded on photographic paper. Used in magnetic observatories to observe diurnal changes in the magnetic field. See also *magnetometer*.

VAX: 1. Variable-area record section; see *variable area*. 2. A computer. Digital Equipment tradename.

V-band: Radar frequencies between 46 and 56 GHz; see Figure E-8.

V-bar: Average velocity; see *velocity*.

VD: *Variable-density* (q.v.).

VDL log: A *three-D log* (q.v.). Schlumberger tradename.

vectogram method: A scheme in telluric surveying in which *x–y* recorders rather than strip-chart recorders are used. The technique yields more accurate results and is faster than triangle or ellipse methods. Exact time-ties are not required to establish simultaneity of electrical vector observations at the station and base. See Yungul (1968).

vector: 1. A quantity having both a magnitude and a sense of direction. Vectors are usually represented by arrows pointing in the appropriate direction whose length is proportional to the vector's magnitude. Vectors are not restricted to three dimensions. **Vector addition** may be done by the **vector-parallelogram method** whereby the tail of the one vector is placed at the head of the other vector; the sum vector then goes from the tail of the second to the head of the first. The **negative of a vector** is represented by the arrow pointing in the opposite direction; subtraction is performed by adding the negative vector. Vectors are often resolved into component vectors in orthogonal directions, the directions being indicated by the unit vectors $\mathbf{i}$, $\mathbf{j}$, and $\mathbf{k}$. The **dot product** (or **inner product**) of two vectors $\mathbf{A}$ and $\mathbf{B}$ is not a vector but a scalar of magnitude $\mathbf{A} \cdot \mathbf{B} = |\mathbf{A}| \, |\mathbf{B}| \cos(\mathbf{A}, \mathbf{B})$, where the cosine is of the angle between their directions. The **cross product** (or **outer product**) for three-dimensional vectors is a vector perpendicular to both $\mathbf{A}$ and $\mathbf{B}$ and of magnitude $\mathbf{A} \times \mathbf{B} = \mathbf{k}|\mathbf{A}| \, |\mathbf{B}| \sin(\mathbf{A}, \mathbf{B})$, where $\mathbf{k}$ is a unit vector perpendicular to the plane of $\mathbf{A}$ and $\mathbf{B}$, pointing in the direction a right-hand screw would advance if turned from $\mathbf{A}$ toward $\mathbf{B}$. For three-dimensional vectors expressed in orthogonal components, these products are:

$$\mathbf{A} = a_1\mathbf{i} + a_2\mathbf{j} + a_3\mathbf{k};$$

$$a_1 = |\mathbf{A}| \cos(\mathbf{A}, \mathbf{i});$$

$$|\mathbf{A}| = \text{magnitude of } \mathbf{A} = (a_1^2 + a_2^2 + a_3^2)^{1/2};$$

$$\mathbf{B} = b_1\mathbf{i} + b_2\mathbf{j} + b_3\mathbf{k};$$

$$\mathbf{A} \pm \mathbf{B} = (a_1 \pm b_1)\mathbf{i} + (a_2 \pm b_2)\mathbf{j} + (a_3 \pm b_3)\mathbf{k};$$

$$\mathbf{A} \cdot \mathbf{B} = a_1b_1 + a_2b_2 + a_3b_3;$$

$$\mathbf{A} \times \mathbf{B} = (a_2b_3 - a_3b_2)\mathbf{i} + (a_3b_1 - a_1b_3)\mathbf{j}$$
$$+ (a_1b_2 - a_2b_1)\mathbf{k}$$

$$= \begin{vmatrix} \mathbf{i} & \mathbf{j} & \mathbf{k} \\ a_1 & a_2 & a_3 \\ b_1 & b_2 & b_3 \end{vmatrix}.$$

With "**vector fields**" (i.e., where a vector $\mathbf{V} = V_1\mathbf{i} + V_2\mathbf{j} + V_3\mathbf{k}$ is associated with every point in space) we define the divergence and curl using the operator del = $\boldsymbol{\nabla}$:

$$\text{div } \mathbf{V} = \mathbf{\nabla} \cdot \mathbf{V} = \frac{\partial V_1}{\partial x} + \frac{\partial V_2}{\partial y} + \frac{\partial V_3}{\partial z}$$

$$\text{curl } \mathbf{V} = \mathbf{\nabla} \times \mathbf{V} = \mathbf{i}\left(\frac{\partial V_3}{\partial y} - \frac{\partial V_2}{\partial z}\right)$$

$$+ \mathbf{j}\left(\frac{\partial V_1}{\partial z} - \frac{\partial V_3}{\partial x}\right) + \mathbf{k}\left(\frac{\partial V_2}{\partial x} - \frac{\partial V_1}{\partial y}\right).$$

Equivalent expressions in cylindrical and spherical coordinates are shown in Figure C-13. For rotating vectors, see *complex notation*. **2.** A sequence of values, such as might represent values of a wavelet at successive discrete time intervals. Thus a wavelet which has the values 0.75, 0.25, −0.50 at three successive time intervals, and is zero at other times, might be considered as a 3-term row vector, [0.75, 0.25, −0.50]. See *matrix*.

vector graphics: Graphic display devices which represent the image as lines drawn between specified points (a set of vectors). Compare *raster graphics*.

vector processor: A computer processor that performs arithmetical and logical operations on data vectors in a pipelined fashion.

vector wave equation: See *wave equation*.

Vela Uniform: A project under the Nuclear Test Detection Office to devise methods of detecting nuclear explosions by the characteristics of the seismic waves which they generate.

vellum: A chemically treated translucent paper used for original drawings in pencil or ink.

velocimeter: 1. An instrument for measuring the velocity of sound in water, used to correct Doppler-sonar data for salinity and temperature variations. An acoustic pulse is transmitted over a fixed distance between transducers in the instrument, amplified, and used to generate the next pulse to be transmitted. The regeneration frequency depends on the traveltime and hence on the velocity of the acoustic wave. This technique is sometimes called **sing around. 2.** A device which measures fluid flow; a *flowmeter* (q.v.).

velocity: 1. A vector quantity which indicates time rate of change of displacement. **2.** Usually refers to the propagation rate of a seismic wave without implying any direction. Velocity is a property of the medium. It is measured (or inferred) from *sonic logs* (q.v.), normal moveout (see *velocity analysis*), in *well shooting* (q.v.) and from refraction *time-distance curves* (q.v.). **3.** **Instantaneous velocity** refers to the speed at any given moment of a wavefront in the direction of the energy propagation (perpendicular to the wavefront for isotropic media); this term is also sometimes used for velocity calculated from reflection amplitude data by inversion (see *seismic log*). **4. Apparent velocity** refers to the apparent speed of a given phase in a particular direction; thus apparent velocity is greater than instantaneous velocity by the secant of the angle between the direction of wave travel and the direction in which the apparent velocity is measured. Apparent velocity is

usually measured in the spread direction. **5. Average velocity** $\overline{V}$ is the ratio of distance along a certain path to the time required for a wave to traverse the path:

$$\overline{V} = \int_0^t V(t)\, dt \bigg/ \int_0^t dt.$$

While it has meaning only with respect to a particular path, a vertical path is often implied, that is, it is given by depth divided by the seismic traveltime to that depth, usually assuming straight-raypath travel. If the section is made of parallel horizontal layers of velocity V_i and thickness z_i so that the traveltime across each layer is $t_i = z_i/V_t$, then the average velocity $\overline{V}$ is

$$\overline{V} = \Sigma\, (V_i t_i)/\Sigma t_i = \Sigma z_i/\Sigma(z_i/V_i).$$

6. Root-mean square (rms) velocity V_{rms} likewise refers to a specific raypath. It is given by

$$V_{rms} = [\Sigma\, (v_i^2 t_i)/\, \Sigma t_i]^{1/2}.$$

Rms velocities are typically a few percent larger than corresponding average velocities. **7. Stacking velocity** or **NMO velocity** is the velocity of a constant homogeneous isotropic layer above a reflector which would give approximately the same offset-dependence (normal moveout) as actually observed. It is the value determined by a *velocity analysis* (q.v.) and is the value used for optimum common-midpoint stacking. In the limit as offset approaches zero, it is the same as rms velocity. Where the velocity layering is not parallel, stacking velocity and rms velocity differ. **8. Interval velocity** (a) is the average velocity over some interval of travel path. It is often calculated from stacking velocities for the interval between reflectors where they are both horizontal. It is also called **Dix velocity** (see Dix, 1955) and is given approximately by

$$V_i = [(V_n^2 t_n - V_{n-1}^2 t_{n-1})/(t_n - t_{n-1})]^{1/2},$$

where V_n is the rms velocity and t_n is the zero-offset arrival time corresponding to the nth reflection. This calculation yields fictitious values if the reflectors are not both horizontal or if there are lateral velocity variations between or above them. **Interval velocity** (b) is also the average velocity over the log span, calculated from sonic logs and (c) calculated from well surveys for the intervals between measurements. For a discussion of the factors affecting seismic velocity, see Sheriff and Geldart, v. 2 (1983, p. 2-10). Usually the apparent speed of a phase (phase velocity) is intended but sometimes the speed of the center of a packet of wave energy (group velocity). See also *group velocity, phase velocity*.

velocity analysis: Calculation of stacking or NMO velocity (see *velocity*) from measurements of normal moveout. In current usage, generally involves common-midpoint data but includes also $T - \Delta T$ *analysis* and $X^2 - T^2$ *analysis* (q.v.). Most analysis schemes assume a normal moveout, measure the coherency at that normal moveout, and then vary the normal moveout in order to maximize the coherency. The stacking velocity value

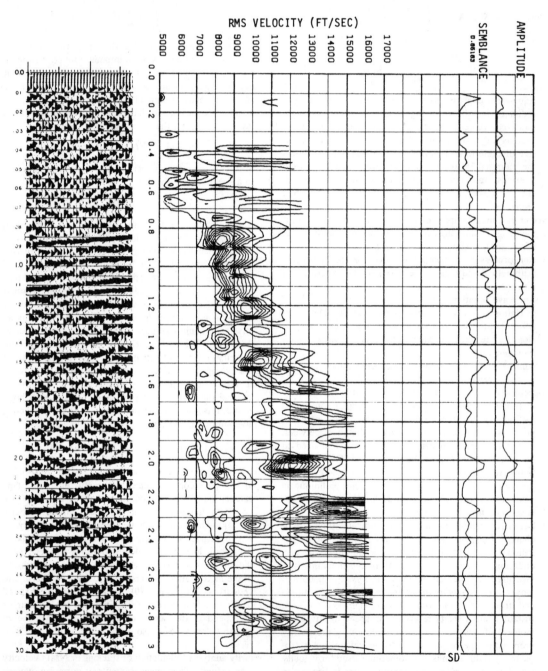

FIG. V-1. **Velocity analysis**. (a) Seismic record section. (b) Velocity analysis of data at the right side of the seismic section. The graph shows semblance as a function of the velocity which normal moveout implies. (c) Graph of the maximum semblance at each arrival time. (d) Graph of the peak amplitude at each arrival time. (Courtesy Seiscom Delta United.)

depends somewhat on the amount of data included in the analysis, that is, on the range of offsets and locations analyzed. Where all reflectors are horizontal and where velocity varies only with depth, the stacking velocity is approximately the rms velocity. See Figure V-1 and Sheriff and Geldart, v. 2 p. 48-52).

velocity anomaly: A fictitious feature which appears because of irregularities in the velocity, usually because of a horizontal variation in velocity which has not been allowed for properly.

velocity contrast: A change in velocity, such as produces reflection or change in wave direction.

velocity correction: A modification made to seismic data based on assumptions with regard to the velocities of the various media through which the seismic signal has passed, made in order to represent as nearly as possible reflectors in their correct relative relationships.

velocity curves: Plots of velocity versus depth.

velocity-depth relationship: See *velocity function.* A number of velocity-depth curves are shown in Sheriff and Geldart, v. 2 (1983, p. 8-9).

velocity discontinuity: An abrupt change of the rate of propagation of seismic waves within the earth, as may occur at an interface between different beds.

velocity filter: Discrimination on the basis of apparent velocity (or dip moveout). Coherent arrivals with certain apparent velocities are attenuated. Also called **apparent velocity filter, fan filter, dip filter,** *f-k* **filter,** and **Pie-Slice filter.** Velocity filtering is different from discrimination based on stacking velocity as accomplished in common-midpoint stacking.

velocity focusing: The bending of seismic rays at curved interfaces which act like lenses in optics, resulting in focusing or defocusing wavefronts and distorting structure and velocities calculated from normal-moveout measurements. The distortion becomes greater as the curvature of the velocity lens increases and as the depth of the feature being studied beneath the velocity lens increases.

velocity function: Seismic velocity expressed as a mathematical function of depth or time. The most common forms of velocity function are (a) linear with depth and (b) linear with arrival time. The use of any functional form is an approximation because lithology and the other factors involved in velocity do not vary systematically and smoothly. The extrapolation of velocity functions can lead to errors. The linear-with-depth velocity function ($V_i = V_0 + kz$) leads to a simple method for migrating data because then both raypaths and wavefronts are arcs of circles; see Figure V-2 and Dix (1952).

velocity gradient: Usually means a lateral (horizontal) change in velocity.

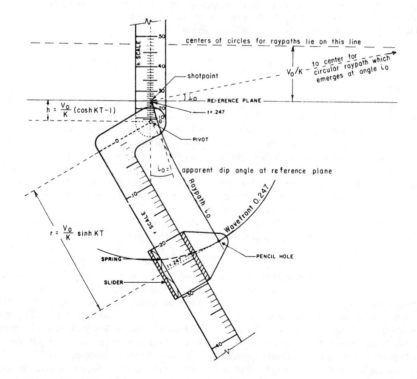

FIG. V-2. **Velocity function** relations for linear increase with depth, $V = V_0 + KZ$. The plotting machine for this function has the *h* and *r* scales calibrated in arrival time *T*. The scales are thus specific for a particular V_0 and *K*. (From Musgrave, 1967.)

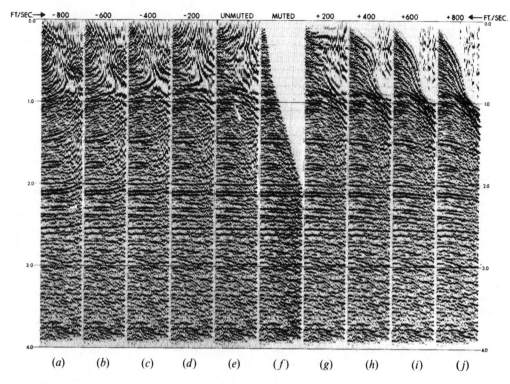

FIG. V-3. **Velocity panel.** Panels (e) and (f) show a common-midpoint gather for the applied stacking velocity V_S; a mute has been applied in (f). Panels (a), (b), (c), (d) show the results where V_s has been decreased by $n = V_s$, $n = 4, 3, 2, 1$. ΔV_S is often 200-500 ft/s. Panels (g), (h), (i), (j) show results for increasing increments of V_S. Comparison of (e) with (f) shows the data to be muted before stack. (From Sheriff and Geldart, v.2, 1983, p. 52.)

velocity-gradient map: A contour map showing the average vertical velocity to a particular horizon. Used to convert seismic reflection-time maps to depth maps.

velocity inversion: A decrease in velocity with depth. Such a situation can result in erroneous refractor depth calculation; see *hidden layers* (q.v.).

velocity layering: The layering formed by contours of velocity values (usually instantaneous velocity, sometimes stacking velocity or average velocity) which in general is different from the layering of bedding. Often implies a series of layers each of which has constant velocity. Involved in ray tracing.

velocity log: *Sonic log* (q.v.).

velocity overlays: Graphs plotted on transparent material which may be placed over seismic data and used as an aid in estimating velocities.

velocity panels: A display of the coherency when various normal moveouts (hence, implying various velocities) are assumed. See Figure V-3 and *velocity analysis*.

velocity profile: Data obtained with an expanding spread designed to record reflections over a large range of offset distance so that velocity can be determined from the time-distance data for reflection events. Also called $X^2 - T^2$ (q.v.).

velocity sag: A *velocity anomaly* (q.v.) involving a time delay.

velocity spectrum: *Velocity panels* (q.v.).

velocity survey: **1.** A series of measurements to determine average velocity as a function of depth, as in *well shooting* (q.v.). **2.** May also refer to running a *sonic log* (q.v.). **3.** Sometimes refers to surface velocity shooting, $X^2 - T^2$ or T-ΔT (q.v.). **4.** See also *vertical seismic profile*.

velocity sweeping: Trying various stacking velocities to see which seems to produce the best results.

velocity wavelet: A wavelet which depicts earth-particle velocity rather than displacement.

Velog: A *seismic log* (q.v.). CGG tradename.

Vening-Meinesz hypothesis: See *isostasy*.

Venn diagram: An illustration of relationships used in logic and in Boolean algebra; see Figure G-1.

vented-gas column: A stream of gas bubbles above a sea floor mound which is caused by gas from depth forcing its way to the sea bottom.

vernal equinox: The point on the celestial sphere occupied by the sun at the time of the vernal equinox (about March 21), which is the reference point from which right ascension and celestial longitude are calculated.

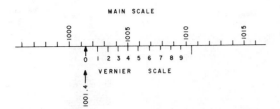

FIG. V-4. **Vernier** principle. A Vernier scale to read to 1/n contains n subdivisions in the space of $(n-1)$ subdivisions on the main scale. The tenths figure is given by the mark on the Vernier scale which lines up with a mark on the main scale (1001.4 in this instance).

Also called the **first point of Aries** and symbolized by τ. One of the two points of intersection of the ecliptic and the celestial equator.

vernier: An auxiliary scale used in conjunction with the main scale of a measuring device to obtain a more precise reading. See Figure V-4.

versine: Versine $\alpha = 1 - \cos \alpha$.

vertical angle: The angle between a direction and the horizontal plane; **attitude**.

vertical closure: See *closure*.

vertical electric sounding: See *electric sounding*. Abbreviated **VES**.

vertical exaggeration: 1. The use of a vertical scale which is larger than the horizontal scale. Exaggeration makes subtle effects more evident but distorts structural relationships. Seismic time sections involve variable vertical exaggeration because the velocity varies with depth. The picking and interpretation of significant features on record sections is greatly affected by vertical exaggeration. See *compressed section* and Figure V-5. **2.** The ratio of vertical scale to horizontal scale; **aspect ratio**.

vertical intensity: The component of the total-intensity field in the vertical direction.

vertical-loop dip-angle method: An electromagnetic-prospecting method in which the transmitter coil or loop is vertical and the receiver coil is in the plane of the transmitter coil. The null or minimum-coupling orientation is observed for the receiver coil. This orientation will be horizontal except near a conducting body. The angle between the plane of the coil and horizontal (dip angle) corresponds to the angle between the major axes of the ellipse of polarization and the horizontal.

vertical parity check: Parity check on a frame of data across tracks of a magnetic tape. Usually the parity track is recorded to make the number of 1 bits in the frame odd. Compare *longitudinal parity check* (q.v.).

vertical profile: See *sounding* and *vertical seismic profiling*.

vertical section: 1. A plot of seismic events directly beneath the point midway between the source and detector locations. Such a section does not represent structural relationships correctly except where reflec-

tors are flat. The vertical scale is usually either vertical time or depth (obtained by multiplying vertical time by average velocity). **2.** A plot of projections of seismic data showing where interfaces attributed to seismic events would intersect a vertical plane. Such a section is a correct structural section to the extent the picking, plotting, velocity, and projecting are correct.

vertical seismic profiling: Measurements of the response of a geophone at various depths in a borehole to shots on the surface. See Figure V-6 and Sheriff and Geldart, v. 1 (1982 p 151-152). Sometimes the location of the surface shots is varied. Where the source point is an appreciable distance from the well head, the result is an **offset VSP**. Abbreviated **VSP**.

vertical stack: 1. Combining of the records from several sources at nearly the same location without correcting for static or offset differences. Used especially with surface sources in which the records from several successive weight drops, vibrations, pops, etc., are combined to give in effect the field record which would have resulted from a much stronger source. **2.** Sometimes used (incorrectly) to mean *uphole stack* (q.v.).

vertical time: 1. The arrival time which would be observed for a given reflection if the travel path had been vertical. The vertical time differs from the observed arrival time

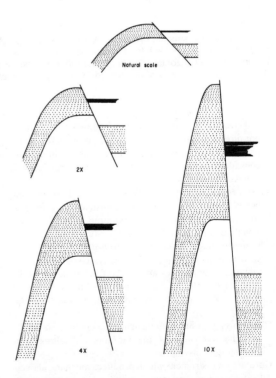

FIG. V-5. **Vertical exaggeration**. Exaggeration allows one to see both vertical detail and horizontal context but severely distorts bed thickness, structural relationship, fault dip, etc.

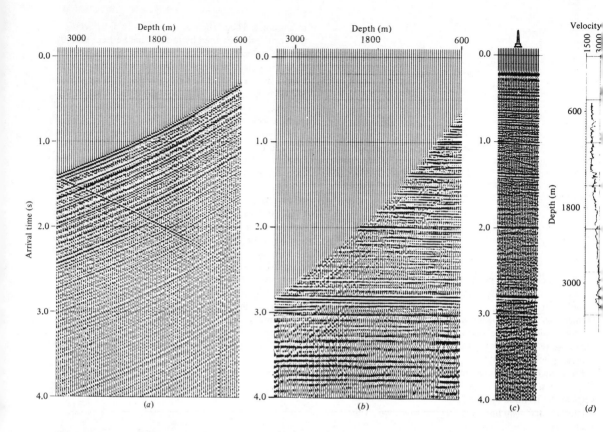

FIG. V-6. **Vertical seismic profile**. (a) Each trace is recorded at a station in a borehole using an air gun source at the surface. (b) Same except each trace has been shifted by the one-way traveltime to the surface, thus aligning reflections (upcoming events) horizontally. A shift in the other direction would align downgoing events horizontally. (c) Portion of a reflection record shot across the well. (d) Sonic log in the well. (From Sheriff and Geldart, v.1, 1982, p. 152.)

if the reflector is dipping so that the reflecting point is not directly under the observation point, if the travel path is longer because geophone and shotpoint are not coincident, or if velocity variations distort raypaths. **2.** *Uphole time* (q.v.).

VES: Vertical *electric sounding* (q.v.), such as Schlumberger sounding, frequency, transient or geometric sounding in electromagnetic exploration, or frequency sounding in magnetotelluric exploration.

vibration monitor: A calibrated recorder of ground and structural acceleration and velocity. Used to measure vibrational amplitudes and modal frequencies of buildings, towers, etc., under ambient conditions, as well as potentially damaging vibrations due to blasting, pile driving, etc.

vibration survey: Study to measure the amount, intensity, and characteristics of the vibrations resulting from blasting, pile driving, etc.

vibrator: An instrument which produces mechanical oscillations used as a seismic source for *Vibroseis* (q.v.).

vibratory plough: A device for burying detonating cord about 50 cm deep for use as a seismic source; a vibrating blade pulled by a tractor.

Vibroseis: A seismic method in which a vibrator is used as an energy source to generate a controlled wave train. A sinusoidal vibration of continuously varying frequency (Figure V-7) is applied during a **sweep period** typically lasting seven seconds or longer. In **upsweeping** the frequency begins low and increases with time, and in **downsweeping** the highest frequencies occur first. The frequency is usually changed linearly with time. The field record which consists of the superposition of many such long wave trains is correlated with the sweep wave train. The correlated record resembles a conventional seismic record such as results from an impulsive source. Conoco tradename.

video display: A display on a cathode-ray screen.

virtual memory: A technique that permits a user to treat secondary (disk) memory as an extension of main processor memory. Blocks of data (**pages**) are transferred between real memory and disk as access to them is or is not needed. Use of the storage is transparent to the user.

viscoelastic: Having a stress-strain relationship which includes terms proportional to both the strain and the rate of change of strain. Leads to attenuation of seismic

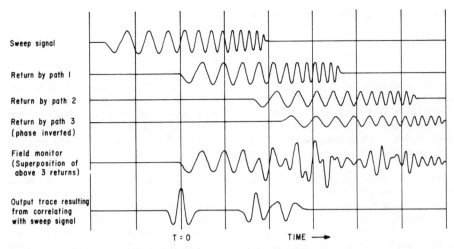

Sweep signal

Return by path 1

Return by path 2

Return by path 3
(phase inverted)

Field monitor
(Superposition of
above 3 returns)

Output trace resulting
from correlating
with sweep signal

T = 0   TIME ⟶

FIG. V-7. **Vibroseis** record. (Courtesy Conoco.)

waves dependent on the square of the frequency. A viscoelastic solid is also call a **Voigt solid**.

viscometer: A device for measuring viscosity.

viscosity: Resistance of a fluid to flow; (stress)/(rate of shear).

viscous magnetization: Remanent magnetization produced by a weak magnetic field over a long period of time. It is generally proportional to the logarithm of the time and parallel to the weak applied field. Viscous magnetization has its origin in thermal energy which is large enough to realign the magnetization direction of domains with rather high energy barriers. The weak field of the Earth acts to bias the direction of these jumps. Abbreviated **VRM**.

VLF: Very low frequency. **1.** Radio transmission at frequencies of 3 to 30 kHz, used for communication with submerged submarines and long-range radio positioning. See Figure E-8. **2.** An electromagnetic prospecting method which uses such transmissions as plane-wave sources. The VLF receiver measures the tilt of the total field by nulling one of two small orthogonal coils in the plane of the primary field. **3.** A radio-location system such as Omega which uses very low frequency energy.

voice grade: A channel with a frequency range from approximately 300 to 3000 Hz.

Voigt solid: See *viscoelastic*.

Voigt waves: *P*-waves in a Voigt or viscoelastic solid.

voltmeter: An electrical instrument used to measure the potential difference between points in a circuit. A voltmeter may respond to average, root-mean-square, or peak voltage values.

volume control: *Gain control* (q.v.).

volume magnetization: Magnetic moment per unit volume.

volume reverberation: *Water-track* (q.v.) mode.

von Schmidt wave: *Head wave* (q.v.).

vortex shedding: A mode of fluid flow involved in pulling a cable through the water. See *fairing*.

vote: See *majority vote*.

VRM: *Viscous magnetization* (q.v.).

VSP: *Vertical seismic profiling* (q.v.).

VTVM: *Vacuum-tube voltmeter* (q.v.).

V₂: Subweathering velocity, often determined from first-break refraction arrivals.

W

walkaway: *Noise analysis (seismic)* (q.v.).

wall resistivity log: A *microresistivity log* (q.v.).

Warburg impedance: Impedance involved in current transfer at an electrode by a *faradaic path* (q.v.); it is a measure of the rate of the faradaic ion-diffusion process. Varies inversely as the square root of the frequency.

Warburg region: The steep part of the resistivity-spectrum curve near the inflection point where the electrode impedance of a rock is dominated by faradaic-path conduction; see *faradaic path*.

washing: Demagnetizing (degaussing) a rock or other material, especially by gradually increasing the demagnetization to remove successively harder portions of the remanent magnetization.

WASSP: *Exploding wire* (q.v.), a marine energy source. Teledyne tradename.

waterbreak: The arrival of energy which travels in the water directly from the source to a waterbreak detector. Used to determine the location of waterbreak detectors in a seismic streamer with respect to the source location and thus the location of the streamer. Filters in the waterbreak circuitry pass components between approximately 500 and 5000 Hz, thus avoiding possible confusion with shallow refractions of lower frequency and background sonic energy of higher frequency; see Figure C-3e. The velocity of seismic (acoustic) waves in water is shown in Figure W-1.

waterbreak detector: A high-frequency detector sensitive to the direct wave carried in the water; see *waterbreak*.

water cut: The volume fraction of water produced from a well.

water gun: A seismic source which propels a slug of water into the water mass, producing an implosive effect.

water injection: A method of drilling which uses air as the principal fluid for removal of cuttings but with enough water added to lubricate the hole and make the hole wall firm enough to prevent excessive caving. Used when drilling dry sand.

water saturation: Fraction of the pore volume filled with formation water. Symbol: S_w. See *Archie's formulas*.

water track: A mode of reflected sonar energy produced by scattering within the first tens of feet below a transducer. When using Doppler-sonar navigation in water deeper than 400 to 1000 ft, the scattered energy dominates the reflection from a deeper ocean bottom. Accuracy drops by about a factor of four compared with "bottom mode." Water-track positioning also is done with respect to the water which may itself be moving, introducing additional systematic error.

water velocity: See Figure W-1.

water wave: A gravity surface wave on water, usually wind-generated. In deep water, wavelength λ depends on wind speed V_w: $\lambda \approx 4\pi^2 g V_w^2$ where g = gravitational acceleration. In shallow water, waves become asymmetric and break when $\lambda \approx h$ where h = water depth.

wave: A disturbance which is propagated through the body or on the surface of a medium without involving net movement of material. Waves are usually characterized by periodicity (Figure W-2). The general expression for a plane wave in rectangular coordinates is

$$f(lx + my + nz - Vt) + g(lx + my + nz + Vt),$$

where f and g are any functions and l,m,n are direction

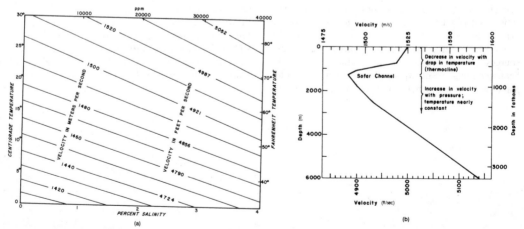

FIG. W-1. **Water velocity**. (**a**) Velocity as a function of temperature and salinity. The speed of P-waves in water is given by $V = 1449.2 + 4.6\,T - 0.055\,T^2 + 0.0003\,T^3 + (1.34 - 0.010\,T)(S - 35) + 0.016\,Z$, where V is in meters/second, T temperature Celsius, S salinity in parts per thousand, and Z depth below the surface in meters. (**b**) Typical velocity versus depth graph showing the low-velocity Sofar channel. (From Ewing et al, 1948.)

cosines for the direction of travel. For a spherical wave, the general expression is

$$(1/r)f(r - Vt) + (1/r)g(r + Vt),$$

where r is the distance from the point source. **Wave amplitude** is usually defined as the maximum displacement from the equilibrium or null position. The **rms amplitude** is the square root of the mean of the squares

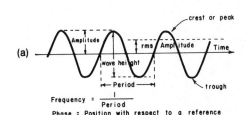

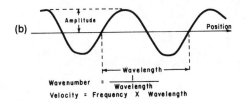

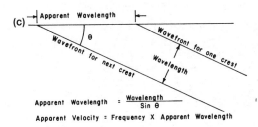

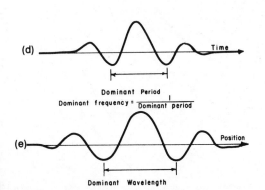

FIG. W-2. **Wave** definitions. For sinusoids: (**a**) How displacement at one point varies with time. (**b**) How a wave looks at different places at a given instant. (**c**) If wavefront approaches at an angle, the apparent wavelength will differ from the true wavelength. For nonperiodic waves: (**d**) Dominant period between principal adjacent troughs (or peaks). (**e**) Dominant wavelength is measured similarly.

of the displacements, which is $\sqrt{2}/2$ times the peak amplitude for sinusoidal waves. The wave **peak** or **crest** is the point at which the displacement is greater (in the positive direction) than at adjacent points. The wave **trough** is the point which is displaced farther than adjacent points in the negative sense. The **wave height** is the difference in displacement between successive peaks and troughs. The **wavelength** (λ) is the distance perpendicular to the wavefront between successive similar points on the wave train. The **wavenumber** is the number of cycles per unit distance, the reciprocal of the wavelength (sometimes defined as 2π/wavelength, which is given the symbol κ). **Body waves** propagate through the body of the medium, and **surface** or **interface waves** propagate along a boundary. Body waves may be either *P-waves* (q.v.) or *S-waves* (q.v.). Wave energy which has traveled partly as a *P*-wave and partly as an *S*-wave is called a **converted wave**. Surface waves (or interface waves) may travel by several modes, the most common of which are *Rayleigh waves* (q.v.) or pseudo-Rayleigh waves. Other surface waves include *Love waves* (q.v.), hydrodynamic waves, coupled waves, and Stoneley waves. A *tube wave* (q.v.) is a surface wave which travels along the surface of a borehole. See also *wave notation*. Wave motion at a point is often described mathematically in terms of harmonic components:

$$f(t) = \Sigma (A_n \cos 2\pi n\nu t + B_n \sin 2\pi n\nu t)$$
$$= \Sigma C_n \cos (2\pi\nu t - \gamma),$$

or in complex notation:

$$f(t) = \Sigma C_n e^{j(2\pi\nu t - \gamma)},$$

where C_n is amplitude, n is an integer, ν is frequency, γ is phase angle, and $j = \sqrt{-1}$.

wave amplitude: The maximum displacement from rest position in an oscillatory motion.

wave attenuation: A decrease in amplitude with distance from the source; see also *absorption* and *divergence*.

wave conductor: *Refractor* (q.v.).

wave equation: An equation which relates the spatial and time dependence of a disturbance which can propagate as a wave. In rectangular coordinates x,y,z, it is

$$\nabla^2\Psi = \frac{\partial^2\Psi}{\partial x^2} + \frac{\partial^2\Psi}{\partial y^2} + \frac{\partial^2\Psi}{\partial z^2} = \frac{1}{V^2}\,\Psi,$$

where Ψ represents wave displacement (pressure, rotation, dilatation, etc.) and V the velocity of the wave. Functions $f(x \pm Vt)$ are solutions to this equation. In spherical coordinates where r is the radius, θ the colatitude, and ϕ the longitude, the wave equation becomes:

$$\frac{1}{V^2}\frac{\partial^2\Psi}{\partial t^2} = \frac{1}{r^2}\left[\frac{\partial}{\partial r}\left(r^2\frac{\partial\Psi}{\partial r}\right) + \frac{1}{\sin\theta}\frac{\partial}{\partial\theta}\left(\sin\theta\frac{\partial\Psi}{\partial\theta}\right)\right.$$
$$\left. + \frac{1}{\sin^2\theta}\left(\frac{\partial^2\Psi}{\partial\theta^2}\right)\right].$$

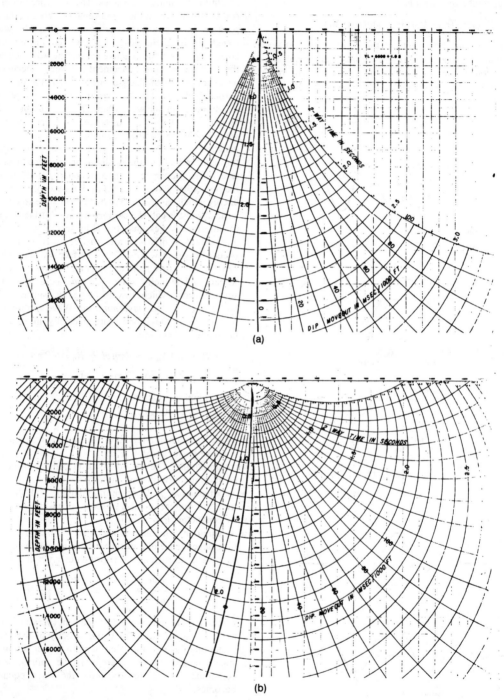

FIG. W-3. **Wavefront charts**. (a) Wavefronts for a particular velocity function become farther apart with depth because of the increase of velocity with depth. Raypaths are orthogonal to wavefronts. (b) Wavefronts where velocity increases from right to left as well as with depth. (Courtesy Chevron Oil Co.)

The foregoing are forms of the **scalar wave equation**. These forms do not provide for the conversion of P-waves to S-waves nor vice versa. The **vector wave equation** is more general; it is

$$[2\mu + \lambda]\, \nabla[\nabla \cdot \Psi] - \mu\nabla x\, [\nabla x\Psi] = \rho\, \partial\, ^2\Psi/\partial t^2,$$

which can be written in component form as

$$\mu\nabla^2\Psi_x + (\mu + \lambda)\frac{\partial}{\partial x}\left(\frac{\partial\Psi_x}{\partial x} + \frac{\partial\Psi_y}{\partial y} + \frac{\partial\Psi_z}{\partial z}\right) = \rho\frac{\partial^2\Psi_x}{\partial t^2}.$$

If div $\Psi = 0$, this gives an S-wave; if curl $\Psi = 0$, a P-wave.

wave equation migration: *Migration* (q.v.) or imaging accomplished by application of the wave equation in one of several ways: solution in the time-domain by a finite-difference method, solution in an integral form (Kirchhoff migration), solution in frequency or wave-number domains or after a two-dimensional transform into frequency-wavenumber domain (frequency-domain migration), or some combination of domains. See Sheriff and Geldart, v. 2 (1983, p. 62-65). Sometimes implies finite-difference solution in the time domain.

waveform: A plot (usually as a function of time) of a quantity involved in wave motion, such as voltage, current, seismic displacement, etc.

wavefront: 1. The surface over which the phase of a traveling wave disturbance is the same. The wavefront moves perpendicular to itself as the disturbance travels in an isotropic medium. A locus of equal traveltime. **2.** The leading edge of a waveform.

wavefront chart: A plot of the location of wavefronts emanating from a point source after various amounts of traveltime. Wavefronts are surfaces rather than curves (see Figure M-7). The shape of wavefronts depends on the velocity distribution. Raypaths corresponding to different values of apparent velocity usually are drawn also on such charts; raypaths are perpendicular to wavefronts for isotropic media; see Figure W-3.

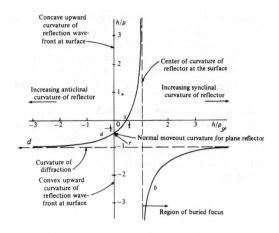

FIG. W-4. **Wavefront curvature**. Normalized curvature of wavefront at the surface (plotted horizontally) as a function of a reflector curvature (plotted vertically) for a point source. "Normalized" means the curvature has been multiplied by the reflector depth. (From Sheriff and Geldart, v.1, 1982, p. 114.)

wavefront curvature: See Figure W-4, *buried-focus effect, normal moveout.*

wavefront healing: The diffraction of energy into shadow zones obscures most of the effect of the shadow-producing obstacle at large enough distances from it; this is called "healing" of the wavefronts. Figure W-5 shows the reflection from a plane containing a hole, the effect of the hole being nearly healed.

wavefront method: A seismic interpretation method (often graphical) which involves reconstructing emerging wavefronts from the arrival times at various geophones from a common shot (or the equivalent). The wavefront for time t is constructed by striking circles about each geophone position of radius $(t_1 - t)V$, where t_1 is the

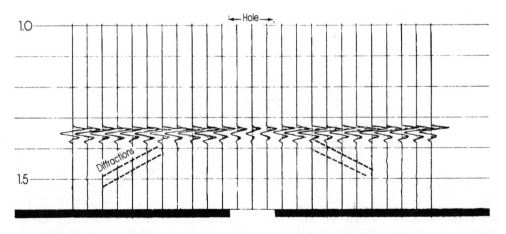

FIG. W-5. **Wavefront healing**. A plane wavefront reflected from a reflector 5000 ft deep containing a slit three trace-intervals wide shows how the diffraction process "heals" the wavefront. (Courtesy Chevron Oil Co.)

traveltime observed by that geophone and V is the velocity of the upper layer; see Figure W-6. Similar wavefronts can be constructed from other shotpoints or from the reversed profile of a refraction interpretation. The solution which locates the reflector or refractor must satisfy the observed arrival times. See Rockwell (1967).

waveguide: 1. An arrangement which constrains wave travel to within a unit by repeated reflection at the boundaries. Natural waveguides have lower velocity than adjacent beds. Wave travel in a waveguide is called **normal-mode propagation** and the waves are called **channel waves.** See Sheriff and Geldart, v. 1

(1982, p. 70-73) and Figure C-3. **2.** A man-made device through which high-frequency electromagnetic waves can be transmitted.

wave impedance: 1. The ratio of orthogonal components of electric-field to magnetic-field intensities. See *impedance*. **2.** The complex ratio of particle velocity (or pressure) to displacement in a wave as a function of frequency.

wavelength: The distance between successive similar points on two adjacent cycles of a monochromatic wave, measured perpendicular to the wavefront. Symbol λ:

$$\lambda = V/\nu = 2\pi/\kappa = 1/\text{wavenumber},$$

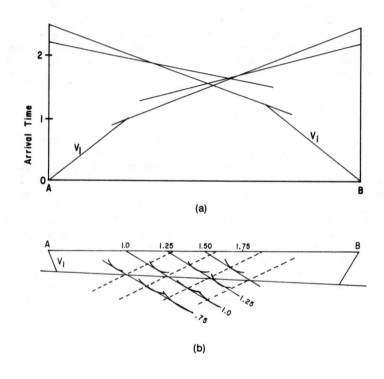

(a)

(b)

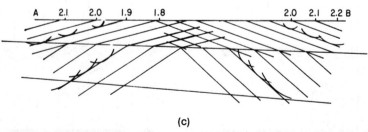

(c)

FIG. W-6. **Wavefront methods.** (a) Reversed time-distance curves showing both primary and secondary head waves. (b) Arcs of radii $(t_n - t)V_1$ about geophone positions (where t_n = arrival time at geophone n) locate wavefront at time t. Symmetry of critical angles shows refractor dip and reconciling traveltimes locates refractor. (c) Extension to second refractor; new arcs are drawn about points where wavefronts impinge on first refractor to locate wavefronts in second layer.

where V = wave velocity, and ν = frequency. See Figure W-2. **Dominant wavelength** refers to the wavelength of the dominant frequency component.

wavelet: A seismic pulse usually consisting of only a few cycles. An *embedded wavelet* (q.v.) or **basic wavelet** is the time-domain reflection shape from a single positive reflector at normal incidence. See also *Ricker wavelet*.

wavelet equalization: A form of wavelet processing used with the objective of making the equivalent or *embedded wavelet* (q.v.) the same on different traces.

wavelet extraction: Wavelet processing used to determine the shape of the *embedded wavelet* (q.v.).

wavelet processing: Deconvolution processing which attempts to determine the basic wavelet shape or to control or manipulate the shape of the basic wavelet. The objective usually is to achieve some specified wavelet shape. The specified wavelet is often (but not necessarily) zero-phase and short in length. See Sheriff and Geldart, v. 2 (1983, p. 44-45 and 75).

wave notation: 1. Waves are conventionally identified using letters to designate the nature of the wave over various legs of their travel paths. Arrivals of seismic waves from earthquakes (see *earthquake seismology*) are identified in this way; see Figure E-2.

P = P-wave in the crust or mantle,
S = S-wave in the crust or mantle,
K = P-wave in the core,
I = P-wave in the inner core,
J = S-wave in the inner core,
$G = Q = L_Q$ = Love surface wave,
$R = L_R$ = Rayleigh surface wave,
L_g = guided wave in the continental crust, (which disappears if any of the travel path is through oceanic crust), and
W_2 = surface wave which has traveled around the earth the long way.

Repetition of P and S indicates reflection at the surface of the earth (as in *PP, PS, SS*). Reflection at the outer surface of the core is indicated by c (e.g., *PcP, PcS*), at the inner surface of the core by *KK* (e.g., *PKKP, PKKS*). $P' = PKP$, pP = P-wave ghost (for a deep earthquake). A Jeffreys-Bullen chart (Figure J-1) shows the normal arrival time for various types of waves. See also *T-wave, H-wave*, and *C-wave*. **2.** Guided electromagnetic waves are classified as TE, TM, or TEM to indicate that the electric or magnetic field or both have transverse components only.

wavenumber: 1. The number of waves per unit distance perpendicular to a wavefront, that is, the reciprocal of the wavelength. It is thus equal to $\kappa/2\pi$, that is, wavenumber is to κ as frequency is to angular frequency. Some authors define κ as the wavenumber. **2.** Spatial frequency, the number of wave cycles per unit of distance in a given direction (direction of the spread); **apparent wavenumber.** Specifically, the reciprocal of the apparent wavelength λ_{app} along the spread direction.

$$1/\lambda_{app} = \nu/V_{app} = \kappa_{app}/2\pi,$$

where ν = frequency and V_{app} is apparent velocity. If a wavefront makes the angle θ with the given direction,

$$\kappa = 2\pi\nu \sin \theta/V,$$

where V is the actual velocity of the wavefront. Hence a wavenumber of zero indicates a wavefront striking the line of detectors simultaneously. See *f-k plot*. **3.** See *propagation constant*.

wavenumber filtering: Filtering of certain wavenumbers such as performed by spatial sampling and mixing. See Figure D-12.

wavenumber-time domain: The result of Fourier-transforming in the spatial direction a space-time array (such as a seismic section).

waveshape kit: A modification of an air gun so that air bleeds into the expanding bubble to increase the pressure in the bubble during the collapse, thereby decreasing the sharpness of the collapse and simplifying the seismic waveform.

waveshape stabilization: Shaping frequency spectra so as to make the waveshape for adjacent traces more nearly the same. Sometimes accomplished by *cross-equalization* (q.v.).

wave slowness: The reciprocal of wave velocity, often regarded as a vector.

wave spreading: See *divergence*.

wave surface: A *wavefront* (q.v.).

wave test: Walkaway or *noise analysis (seismic)* (q.v.).

wavetilt: The ratio of the horizontal to vertical components of the magnetic or electric field. In some cases only the modulus is considered (see *polarization ellipse*). In an electromagnetic method using VLF and higher-frequency radio waves, the apparent resistivity of the earth is calculated from measurements of the quadrature component of the electric wavetilt; for a horizontally layered earth, the results are equivalent to a magnetotelluric measurement made at the same frequency.

wave train: A wave having several cycles.

wave-vector filtering: *Wavenumber filtering* (q.v.).

wave velocity: The speed with which a wave advances. If the medium is dispersive (i.e., if the speed depends on frequency), individual wave crests move with a velocity different from that with which the energy moves; the former is **phase velocity**, the latter **group velocity**.

weathered layer: See *weathering*.

weathering: A near-surface low-velocity layer, usually the portion where air rather than water fills the pore spaces of rocks and unconsolidated earth. Seismic weathering is usually different from **geologic weathering** (the result of rock decomposition). The term **LVL (low-velocity layer)** is often used for seismic weathering. Frequently the base of the weathering is the water table. Sometimes the weathering velocity is gradational, sometimes it is sharply layered. Weathering velocities are typically 500 to 800 m/s (although weathering velocity may be 150 m/s for the first few cm) compared to subweathering velocities of 1500 m/s or greater. Weathering thickness is calculated from uphole-survey data and from refraction first breaks. See *weathering correction*.

OBJECTIVE OF DRILLING	INITIAL CLASSIFICATION WHEN DRILLING IS STARTED	FINAL CLASSIFICATION AFTER COMPLETION OR ABANDONMENT	
		SUCCESSFUL ● ✿ ✿	UNSUCCESSFUL ◇
Drilling for a new field on a structure or in an environment never before productive	1. NEW-FIELD WILDCAT	NEW-FIELD DISCOVERY WILDCAT	DRY NEW-FIELD WILDCAT
Drilling for a new pool on a structure or in a geological environment already productive — NEW POOL TESTS — Drilling outside limits of a proved area of pool	2. NEW-POOL (PAY) WILDCAT	NEW-POOL DISCOVERY WILDCAT (Sometimes an extension well)	DRY NEW-POOL WILDCAT
Drilling inside limits of proved area of pool — For a new pool below deepest proven pool	3. DEEPER POOL (PAY) TEST	NEW-POOL DISCOVERY WELLS (Sometimes extension wells) — DEEPER POOL DISCOVERY WELL	DRY NEW-POOL TESTS — DRY DEEPER POOL TEST
For a new pool above deepest proven pool	4. SHALLOWER POOL (PAY) TEST	SHALLOWER POOL DISCOVERY WELL	DRY SHALLOWER POOL TEST
Drilling for long extension of a partly developed pool	5. OUTPOST or EXTENSION TEST	EXTENSION WELL (Sometimes a new-pool discovery well)	DRY OUTPOST OR DRY EXTENSION TEST
Drilling to exploit or develop a hydrocarbon accumulation discovered by previous drilling	6. DEVELOPMENT WELL	DEVELOPMENT WELL	DRY DEVELOPMENT WELL

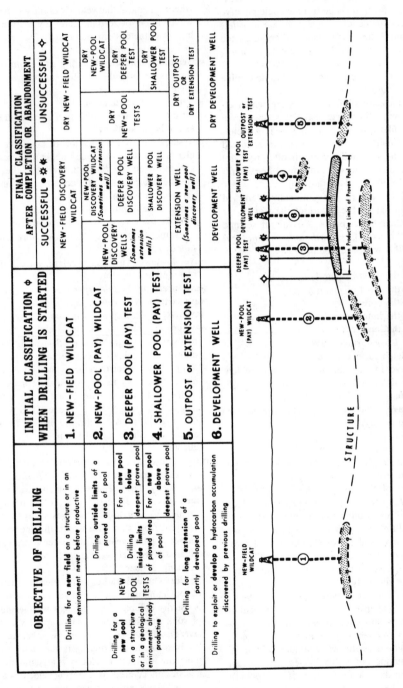

FIG. W-7. AAPG and AGI **well classification.** (Courtesy AAPG.)

Well log types

Cuttings samples (lithology, "shows," rock types)

Cores (lithology, porosity, permeability, grain density, formation factor, saturation exponent, capillary-pressure curves, initial-residual oil saturation, acoustic velocity)

Acoustic logs (velocity or $\triangle t$, amplitude, waveform displays, variable intensity, cement-bond logs, borehole scanner, shear-wave velocity)

Radioactivity logs (gamma ray, neutron, density, chlorine, nuclear magnetism, neutron lifetime, spectral)

SP-resistivity logs (self-potential, electrical, focused resistivity, flushed-zone resistivity, induction, dipmeter)

Borehole fluid logs (hydrocarbon component concentrations in drilling fluid)

Productivity and fluid testing (wireline formation tests, drill stem tests, production tests)

Gravity measurement (borehole gravimeter)

Well log applications

1. Estimates of hydrocarbon volumes in place.
2. Estimates of recoverable hydrocarbon volume.
3. Rock typing; lithology determination; parameters in log response equations; initial versus residual oil saturation relations.
4. Identification of geologic environments.
5. Water flood feasibility.
6. Location of fluid contacts.
7. Reservoir "quality" mapping.
8. Water salinity determination.
9. Determination of reservoir fluid pressure.
10. Fracture detection.
11. Determination of reservoir engineering parameters.
12. Prediction of interzone fluid communication in annulus.
13. Determination of porosity and pore size distribution.
14. Monitoring of fluid movement in reservoir.

FIG. W-8. Types of **well-log** measurements and applications.

weathering correction: A correction of seismic reflection or refraction times to remove the delay in the weathering or low-velocity layer (LVL). The simplest correction is based on uphole times from shots in the sub-weathering layer (see *uphole shooting*). Correction methods based on first-break times include the *ABC method,* the *Blondeau method,* the *summation method,* and the *first-break intercept-time method* (see Figure S-22 and separate entries). Automatic statics-correction programs (see Sheriff and Geldart, v. 2, 1983, p. 47-49), based on maximizing the coherence of reflection events are usually an early stage in digital processing. See also *double-layer weathering.*

weathering map: A map showing the thickness (occasionally velocity) of the weathering or LVL.

weathering shot: A special shot to give weathering data. Where pattern shots or geophone patterns are used or where surface sources are used, the first-break quality may be too poor for good weathering corrections and a separate single-hole weathering shot may be recorded into a single geophone of each group. See also *uphole shooting.* Sometimes called a **poop shot** or **short shot.**

weber: The unit of magnetic flux in the SI system, being one joule/ampere. The analogous unit in the cgs system is the maxwell = one erg/abampere. 1 weber = 108 maxwells. Named for Wilhelm Edward Weber (1804-1891), German physicist.

weight-drop: Use of a dropped weight as a seismic source; thumper.

weighted array: Pattern in which geophones or seismic sources are distributed along a line (or over an area) so that the contributions of various parts of the line (or area) are unequal. Sometimes achieved by varying the geometric distribution of geophones and/or shotpoints or weightdrops, by varying the outputs of the different geophone elements or varying the charge size in different holes, or by varying the geophone/source spacing. See also *tapered array.*

weighted average: The sum of a set of values x_i multiplied by weighting values w_i, normalized by the sum of the weights:

$$\Sigma(w_i \, x_i)/\Sigma w_i.$$

Weiss's theory of magnetism: A ferromagnetic material is made up of small regions or **domains** magnetized to saturation (i.e., spins aligned cooperatively) despite the tendency of thermal agitation to disorient the spins (as in the case in paramagnetism). A weak external field can orient the domains with the field direction, and if strong enough, can align the domains irreversibly and so make a permanent magnet.

well classification: The AAPG-API terminology for wells is shown in Figure W-7.

well log: A record of one or more physical measurements as a function of depth in a borehole. Also called **borehole log.** Distinction is sometimes made between a log as an entire record (which may contain curves showing several measurements) and the individual curves themselves, which are also called logs. **1. Wireline logs** are recorded by means of sondes carrying sensors which are lowered into the hole by a cable.

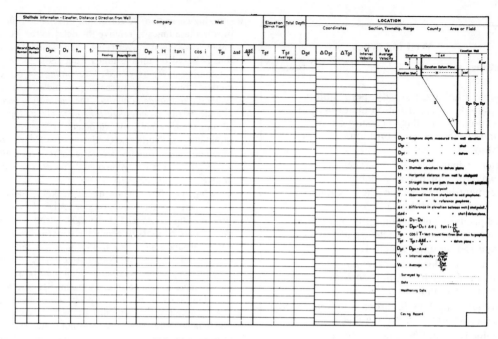

FIG. W-9. **Well-shooting** computation form.

Examples include logs which record electrical measurements (SP, resistivity, etc.), acoustic measurements (sonic, 3-D, etc.), nuclear measurements (natural radioactivity, neutron logs, etc.), and miscellaneous measurements (hole size, temperature, etc.). **2.** Other types of well logs are made of data collected at the surface; examples are core logs, *mud logs* (q.v.), drilling-time logs, etc. **3.** Still other logs show quantities calculated from other measurements; examples are movable-oil plots, synergetic logs, etc. Pickett (1970) lists types and applications of well-log measurements, as shown in Figure W-8.

well shooting: A method of determining the average velocity as a function of depth by lowering a geophone into a hole and recording energy from shots fired from surface shotholes. Often run in addition to a sonic log to supply a reference time at the base of the casing and to check the integrated time. A standard computation form is shown in Figure W-9.

well tie: Running a seismic line by a well so that seismic events may be correlated with subsurface information. The planning of well ties sometimes fails to consider the migration of seismic data where dip is present, resulting in failure to achieve the desired well tie.

well-velocity surveys: *Well shooting* (q.v.).

Wenner electrode array: Electrode arrangement used in resistivity surveying, consisting of four equally spaced collinear electrodes, with the outer two serving as current sources and the inner two as potential-measuring points. See Figure A-12.

Wentworth scale: A scale of particle size. See Figure W-10.

Werner filtering: A method of inverting magnetic measurements to regularly spaced magnetic data such as obtained from aeromagnetic surveys. The method assumes that anomalies are produced by dikes with infinite strike and depth which are perpendicular to the line of measurement (or else produced by thin magnetic sheets). The anomaly produced by a dike can be expressed in terms of four unknowns so that in a noise-free environment, values at four successive points suffice for a solution. Usually two sometimes three) additional unknowns are added to allow for interference and the solution is found for each successive 6 (or 7) points on an overlapping basis. The results are then filterd to remove erratic solutions. Sometimes the solution is done both for adjacent points and alternate points and for both a dike and a thin-sheet model; sometimes incorporating not only total field but also horizontal and vertical derivative measurements. The form of the solution is expressed as a convolution, hence calling the operation "filtering." See Hartman et al. (1971, p. 891-918).

westing: See *departure*.

wet auger: See *auger*.

whacker: An earth impactor used as a seismic source for shallow-penetration studies. Used with Mini-Sosie method.

whetstones: A set of computational operations used to measure and compare computer performance.

white: Containing all frequencies in equal proportion.

whiten: To adjust the amplitudes of all frequency components within a certain band-pass to the same level. A method of deconvolving.

white noise: Random energy containing all frequency components in equal proportions within the bandwidth but with random phases.

white-noise level: The amount of white noise added to data undergoing analysis for inverse-filter design. An inverse filter tends to build up frequencies in which meaningful data are virtually missing, an undesirable consequence of which is that noise at such frequencies is magnified. The addition of white noise (or what is equivalent, superimposing an impulse or biasing the amplitude-frequency response curve) for filter-design purposes limits the extent to which this can occur. (White noise is not added to the final data, only to the inverse filter design.) See Sheriff and Geldart, v. 2 (1983, p. 42-43).

whole-body excitation: *Mise-à-la-masse method* (q.v.).

wide-angle reflection: Reflection where the angle of incidence is near or greater than the critical angle. Reflection coefficients may have large values near the critical

Sieve size opening in mm	U.S. Standard sieve mesh number	Phi (ϕ) units	Wentworth class
256		−8	Boulder
64		−6	Cobble
4.0	5	−2	Pebble
3.36	6	−1.75	
2.83	7	−1.50	Granule
2.38	8	−1.25	
2.00	10	−1.00	
1.68	12	−0.75	
1.41	14	−0.50	Very coarse sand
1.19	16	−0.25	
1.00	18	0	
0.841	20	0.25	
0.707	25	0.50	Coarse sand
0.595	30	0.75	
0.500	35	+1.00	
0.420	40	1.25	
0.354	45	1.50	Medium sand
0.297	50	1.75	
0.250	60	+2.00	
0.210	70	2.25	
0.177	80	2.50	Fine sand
0.149	100	2.75	
0.125	120	+3.00	
0.105	140	3.25	
0.088	170	3.50	Very fine sand
0.074	200	3.75	
0.0625	230	+4.00	
0.0526	270	4.25	
0.0442	325	4.50	Coarse silt
0.0372	400	4.75	
(0.031)		+5.00	
(0.0156)		+6	Medium silt
(0.0078)		+7	Fine silt
(0.0039)		+8	Very fine silt
			Clay (smaller than 0.0039 mm)

FIG. W-10. **Wentworth scale;** $\phi = \log_2 \text{mm} = -(\log_{10} \text{mm}/\log_{10} 2)$, where mm = grain size in mm.

angle so that reflection energy is exceptionally strong. See Figures C-15 and Z-1.

wide-band stack: A stack which does not produce appreciable frequency discrimination. See also *optimum wide-band*.

Wide-line profiling: A technique for obtaining, processing and displaying three-dimensional data. See Figure T-3. Tradename of Compagnie Generale de Geophysique.

Widess limit: See *resolvable limit*.

width: 1. The width of an anomaly usually is measured between either half amplitude or inflection points. See also *half-width*. **2.** The width of a pulse is the width of a *boxcar* (q.v.) with the same peak height and containing the same area.

Wiener filter: A causal filter which will transform an input into a desired output as nearly as possible, subject to certain constraints. "As nearly as possible" implies in a least-squares sense, that is, the sum of the square of differences between the filter output and the desired result is minimized. The filter will optimize (in a least-squares sense) standout of a signal S (which is a function of frequency ν) in the presence of noise N (also a function of frequency). The filter is given by the *normal equations* (q.v.). Each frequency is passed proportional to

$$[S(\nu)]^2/[S(\nu)^2 + N(\nu)^2].$$

If the desired output is specified, the Wiener filter will give the output for an actual input which will come closest to the desired output. Also called a **least-squares filter**. See *Wiener-Hopf equations* and Sheriff and Geldart, v. 2 (1983, p. 41, 151, 190-191, 195).

Wiener-Hopf equations: 1. The Wiener-Hopf equation of the first kind is an integral equation in the unknown $f(t)$:

$$\phi_{xz}(\tau) = \int_0^\infty f(t)\phi_{xx}(\tau - t)\, dt, \ \tau > 0.$$

This equation is the necessary and sufficient condition for minimizing the mean-square error between a desired output $z(t)$ and the actual output $y(t)$ which results from passing an input $x(t)$ through a causal filter with an impulse response $f(t)$. $\phi_{xx}(\tau)$ is the autocorrelation of x, and ϕ_{xz} is the crosscorrelation between z and x. When digital processing is involved, this equation becomes the normal set of linear simultaneous equations (**normal equations**). **2.** The Wiener-Hopf equation of the second kind, which applies to a nonstationary input, involves a time-varying filter $f(t,\delta)$ and time-varying correlation functions:

$$\phi_{xz}(t,\tau) = \int_0^\infty f(t, \delta)\phi_{xx}(\delta, \tau)\, d\delta.$$

See *Wiener filter* and Lee (1960).

Wiener-Levinson algorithm: See *Levinson algorithm*.

wiggle trace: A graph of amplitude against time, as on a conventional seismic recording with mirror galvanometers. Also called "squiggle" recording. See Figure D-14.

wild: Having very large and often unpredictable amplitudes, e.g., a noisy seismic channel at high gain.

wildcat well: An exploratory well in an area where no oil or gas has been found in commercial quantities. See Figure W-7.

wind noise: 1. Random noise attributed principally to ground unrest caused by the wind moving plants, trees, etc., and shaking their roots. **2.** Seismic background noise (in the absence of a shot) regardless of the source, i.e., ambient noise. **3.** Noise voltage induced in a suspended wire of an IP, resistivity, or telluric-current measuring system, caused by oscillations of the wires in the Earth's magnetic field.

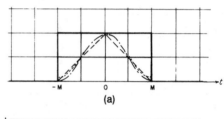

(a)

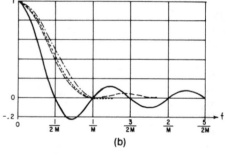

(b)

—————— Boxcar

— — — — Triangular

.... Hamming

—·——·— Hanning

FIG. W-11. **Windows** in time and frequency domains.

Boxcar (rectangular window)
$y=1$,	$\|t\| \leqslant M$;
$y=0$	$\|t\| > M$;

Triangular (Bartlett) window
$y=1-t/M$,	$\|t\| \leqslant M$;
$y=0$	$\|t\| \geqslant M$;

Hamming window
$y=0.54+0.46\cos(\pi t/M)$,	$\|t\| \leqslant M$;
$y=0$	$\|t\| > M$;

Hanning window
$y=0.5+0.5\cos(\pi t/M)$,	$\|t\| \leqslant M$;
$y=0$	$\|t\| > M$;

Daniell window
$y=\mathrm{sinc}(\pi t/M)$	$\|t\| \leqslant M$;
$y=0$	$\|t\| > M$.

(a) Window shapes in time-domain. **(b)** Spectral shapes. The effective width of the boxcar is greater than that for the other window shapes and hence its central lobe is sharper.

window: 1. A portion of a seismic record free from certain disturbances; that is, where certain important noise trains are absent. **2.** The portion of a data set chosen for consideration, such as for designing operators to be used for autocorrelation or frequency analysis. Also called *gate* (q.v.). See also *window carpentry*. The **equivalent width** of a window is the width of a boxcar with the same peak amplitude which contains the same energy.

window carpentry: Abrupt changes such as at the edges of a boxcar window produce undesirable ringing and overshoot effects. Window carpentry concerns designing the boundaries of windows to minimize undesirable effects. The values within a window are weighted according to some scheme. Window weightings in common use are shown in Figure W-11. Note the tradeoff between narrowness of the major lobe and low sidelobe energy.

window pair: A window and its Fourier transform, such as shown in Figure W-11.

wind scale: Wind force is often given in terms of a Beaufort number. See Figure B-1.

wink technique: Rapid alternation of two displays which are nearly alike, so that the places of difference appear to jitter whereas places which are alike remain stationary.

wipe-out zone: 1. A region without internal reflections, possibly representing gassy sediments or some other type of zone without internal contrasts (such as a mud-filled channel). **2.** A region from which reflections cannot be obtained because of raypath disturbances or excessive attenuation in a shallower region.

wireline corer: See *corer*.

wireline log: A *well log* (q.v.) recorded while being withdrawn by a sonde which has been lowered into the borehole by a cable.

witness marker: A marked location (such as a blaze on a tree) which helps locate a survey point whose location is known with respect to the witness marker.

WKBJ solutions: A method attributed to Wentzel, Kramers, Brillouin, and Jeffreys for finding approximate solutions to the equation

$$d^2\phi/dx^2 + \omega^2 s^2 \phi = 0,$$

where ω is large and positive and s^2 is a monotonically increasing function of x. See Aki and Richards (1980, p. 416-418).

word: A group of characters occupying one storage location in a computer. This unit is treated by the computer as an entity.

Worden: A type of gravimeter. Lightweight and insulated by a thermos bottle, the meter is very portable. See Figure G-5.

World Data Centers: Centers for the collection, exchange, and distribution of data from various geophysical disciplines, e.g., solid-earth geophysics, solar-terrestrial geophysics, oceanography, glaciology, meteorology, tsunamis. They were originally established for the International Geophysical Year but are being continued under the auspices of the International Council of Scientific Unions (ICSU). World Data Center A is in Boulder, Colorado, World Data Center B is in Moscow, World Data Center C is split with parts in Japan and parts in the Western Europe.

wow: Variations in the speed of a magnetic tape or a camera, evidenced by an irregular timing-line pattern. Often periodic and of low frequency.

wrap around: 1. Aliasing in the *f-k domain* (q.v.). **2.** The effect produced when a digital memory element (usually a register) is incremented (decremented) past its maximum (minimum) value. For example, a 4-bit register can contain any value from 0 to 15. When it contains 15, incrementing it results in a value of zero.

Wulff net: See *sterographic projection*.

WWSS: World-wide standarized seismographs, a world-wide earthquake monitoring network initiated and monitored by the U.S. Geological Survey.

WWV: The U.S. Bureau of Standards radio station which broadcasts time and frequency standards. WWV (Ft. Collins, Colorado) and WWVH (Maui, Hawaii) broadcast continuously on 2.5, 5, 10, 15, 20, and 25 MHz (the last two only by WWV). WWV is off the air for 4 minutes commencing at 45 minutes 15 sec after each hour and WWVH for 4 minutes commencing at 15 min 15 sec after each hour. Each second is marked by a signal or tick. A voice announcement is given every minute, for example, "National Bureau of Standards, WWV, Fort Collins, Colorado. At the tone, 17 hours, 16 minutes, Coordinated Universal time." WWVB broadcasts a binary-coded-decimal version of WWV.

W_x: *Weathering* (q.v.).

Wyllie relationship: The *time-average equation* (q.v.).

Wyrobek method: A refraction interpretation method based on applying delay and intercept times to continuous refraction profiling, even where the profiles are not reversed. See Wyrobek (1956), or Sheriff and Geldart, v. 1 (1982, p. 196, 222-223).

X

x: **1.** The distance from the source to a particular geophone group; **offset. 2.** Cross; an **x-spread** is a cross-spread; see Figure S-17.

x-**band:** See *radar* and Figure E-8.

x-**hole:** Crosshole; see *crosshole method*.

xmit: To transmit.

xo: Subscript used with log terms to indicate values appropriate to the flushed zone adjacent to the borehole.

XR: Extended range; see *extended-range shoran*.

x-**spread:** *Cross-spread* (q.v.).

X^2-T^2 **analysis:** A method of determining stacking velocity V_s and the depth of a reflector z from the arrival time versus offset relationship:

$$V_s^2 t^2 = 4z^2 + x^2.$$

The square of the source-to-geophone distance or offset (x^2) is plotted against the square of the reflection time (t^2); the slope gives the inverse of the velocity squared and the depth can be obtained from the intercept. Applies only to a constant velocity material. Because of the variation of velocity with depth, these curves are not perfectly straight lines. For horizontal velocity layering and horizontal reflectors, the stacking velocity V_s is given by the slope at the origin. See also *velocity* and Sheriff and Geldart, v. 2 (1983, p. 18-19).

X-Y **reader:** A device for converting the positions of points on a map or graph to digital coordinates. **Coordinatograph**.

Y

yardstick: Standard of performance for evaluating devices or processes.

yaw: Rotational motion of a ship or aircraft about a vertical axis. A steady heading at an angle to the course (such as to compensate for cross wind or cross sea) is called **crab**. Compare *pitch* and *roll*.

Young's modulus: See *elastic constant*.

yo-yo: 1. A method of marine seismic shooting in which the seismic cable or streamer being towed by the recording boat is alternately released so that it floats freely in the water during recordings, and then reeled in between recordings to catch up with the recording boat which is traveling steadily ahead all the while. **2.** Depth oscillation of a logging sonde because of unequal drag as it is pulled up the hole.

Z

Z/A: The ratio of atomic number to atomic weight and, hence, proportional to the ratio of electron density (as measured by the density log) to mass density. **Apparent density** equals true density where $Z/A = 1/2$, which is closely approximated for many minerals (e.g., quartz, calcite, anhydrite, dolomite). The apparent density is larger than the true density when $Z>1/2$ (e.g., gypsum, oil, 110 percent for water), smaller when $Z<1/2$ (e.g., salt).

Zeeman effect: A splitting of spectral lines in the radiation emitted by atoms or molecules in a magnetic field. Named for Pieter Zeeman (1865-1943), Dutch physicist.

Zener diode: A silicon diode in which the breakdown voltage in the reverse direction (Zener voltage) is used for voltage stabilization or voltage reference.

zenith: A point directly overhead. **Zenith distance** is the angle between zenith and a body.

zero: A *root* (q.v.) of an equation.

zero crossing: Where a seismic trace crosses the zero-deflection axis, the phase of a semiperiodic signal is zero.

zero frequency: Alternating-current phenomena extrapolated to zero frequency in the frequency domain.

zero-frequency component: DC shift.

zero-frequency seismology: Study of long-term displacements, strains, and tilts.

zero-lag correlation: The value of an autocorrelation $\phi_{xx}(0)$ or crosscorrelation $\phi_{zx}(0)$ for zero time shift; a measure of the mean power or crosspower.

zero-length spring: A spring whose effective length, as

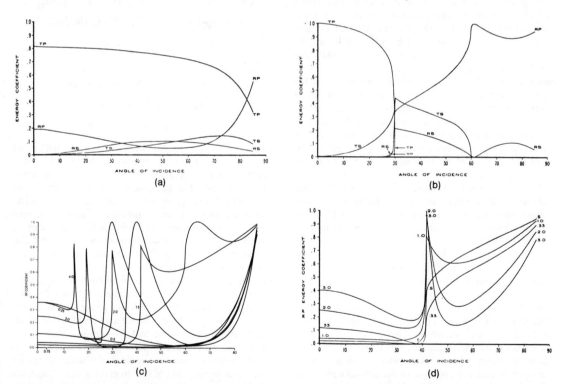

FIG. Z-1. **Zoeppritz equations** yield the amplitude of the waves which result from oblique incidence on an interface. (**a**) Energy ratio of reflected P-waves RP, reflected S-waves RS, transmitted P-waves TP, and transmitted S-waves TS for a velocity ratio of 0.5, density ratio of 0.8, Poisson's ratio of 0.3 in the upper medium and 0.25 in the lower. Similar to (a) except for velocity ratio of 2.0 and density ratio of 0.5. (**c**) Energy ratio for reflected P-waves as a function of P-wave velocity ratio; no density contrast and Poisson's ratios are 0.25. (**d**) Energy ratio for reflected P-waves as a function of density ratio; P-wave velocity contrast 1.5, Poisson's ratios 0.25. (From Tooley et al., 1965.)

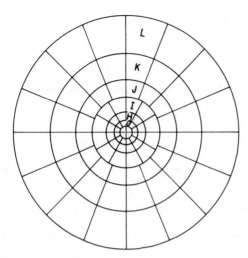

FIG. Z-2. **Zone chart** for terrain corrections. (From Nettleton, 1940, p. 146.)

measured from its fixed point of support, is zero when the external forces acting upon it are zero. The stress-strain relationship between its points of attachment projected back to zero length has zero strain. Gravimeters designed with a zero-length spring are linear and have extreme sensitivity combined with stability and reduced sensitivity to leveling error.

zero-phase: 1. A filter for which the phase shift is zero for all frequencies. Zero-phase filters are anticipatory and hence are not physically realizable, i.e., half of the energy arrives before the time reference so that one gets output before the input arrives. If the input to a zero-phase filter is symmetric, then the output will also be symmetric. Zero-phase filtering can be approximated by using a linear-phase filter, a mixed-phase filter which shifts component frequencies proportional to their frequency, and then delaying the time reference. A zero-phase filter produces no phase distortion. see Figure P-1 and *phase characteristics*. **2.** A wavelet symmetric about zero time.

zero time: The reference time for a seismic trace, with respect to which arrival times are measured.

zeta potential: Potential drop across the diffuse layer in an electrolyte which consists of a group of relatively mobile ions at the interface between a solid and a liquid.

Zietz-Andreasen method: A magnetic interpretation method; see Zietz and Andreasen (1967).

Zoeppritz's equations: Equations which express the partition of energy when a plane wave impinges on an acoustic impedance contrast. In the general case for an interface between two solids when the incident angle is not zero, four waves are generated: reflected P-wave and S-wave and transmitted P-wave and S-wave. The partition of energy among these is found from four boundary conditions which require continuity of normal and tangential displacement and stress at the boundary.

Using the symbols given in Figure S-11, Snell's law states:

$$(\sin \theta_{P1})/V_{P1} = (\sin \theta_{S1})/V_{S1} = (\sin \theta_{P2})/V_{P2}$$
$$= (\sin \theta_{S2})/V_{S2},$$

which defines all angles. For an incident plane P-wave of unity amplitude, continuity of tangential displacement (x-direction) requires that:

$$(1 + A) \sin \theta_{P1} + B \cos \theta_{S1}$$
$$= C \sin \theta_{P2} - D \cos \theta_{S2},$$

where we take the P-wave displacement to be positive in the direction of propagation and the S-wave displacement to be positive to the right of the direction of propagation. A,B,C,D are the amplitudes respectively of the reflected P-wave, reflected S-wave, transmitted P-wave, and transmitted S-wave. Continuity of normal displacement (z-direction) requires that:

$$(1 - A) \cos \theta_{P1} + B \sin \theta_{S1}$$
$$= C \cos \theta_{P2} + D \sin \theta_{S2}.$$

Continuity of normal and tangential stress yields (after some manipulation) the two equations:

$$(-1 + A) \sin 2\theta_{p1} + (V_{P1}/V_{S1}) B \cos 2\theta_{S1}$$
$$= - (\rho_2 V_{S2}^2 V_{P1}/\rho_1 V_{S1}^2 V_{P2}) C \sin 2\theta_{P2}$$
$$+ (\rho_2 V_{S2} V_{P1}/\rho_1 V_{S1}^2) D \cos \theta_{S2},$$

and

$$(1 + A) \cos 2\theta_{S1} - (V_{S1}/V_{P1}) B \sin 2\theta_{S1}$$
$$= (\rho_2 V_{P2}/\rho_1 V_{P1}) C \cos 2\theta_{S2}$$
$$+ (\rho_2 V_{S2}/\rho_1 V_{P1}) D \sin 2\theta_{S2}.$$

Figure Z-1 shows the variation of amplitude with angle for several sets of parameters. Beyond the critical angles for P- and S-waves, the respective refracted waves vanish. The increase in reflection energy near the critical angle is sometimes referred to as the **wide-angle phenomenon** and is sometimes exploited in seismic surveying. The same relationships in terms of potentials are called **Knott's equations**; see Sheriff and Geldart, v. 1 (1982, p. 65-66). These equations do not give head-wave amplitude.

zone chart: A template for making terrain corrections or isostatic corrections. The zone chart (Figure Z-2) is laid over a topographic map with its center at the station being corrected. The difference in mean absolute elevation between each zone and the station's elevation is tabulated without regard for sign (because the correction is always positive regardless of whether zones are higher or lower than the station elevation) to determine the terrain correction. With a zone chart for isostatic correction, the mean elevation above sea level in each zone is used in calculating the correction.

z-plane: A representation of the z-transform polynomial associated with a waveform in sampled form; see *z-transform* and Figure Z-3.

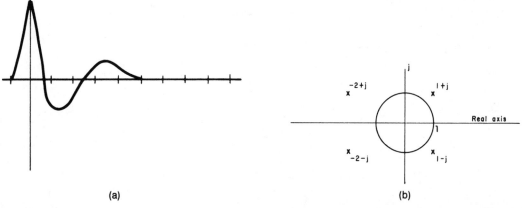

(a)   (b)

FIG. Z-3. **z-plane**. (**a**) The wavelet $[10, -2, -1, 1, 1]$ has the **z**-transform $10 - 2z - z^2 + z^3 + z^4$, which may be factored $(2 + j + z)(2 - j + z)(-1 + j + z)(-1 - j + z)$, which has the roots $(-2-j)$, $(-2+j)$, $(1-j)$, $(1+j)$. (**b**) A plot of these roots in the **z**-plane is shown. Since all roots lie outside a circle of radius 1 (the **unit circle**), the wavelet is minimum-phase.

ZSR: Zero source-receiver distance; zero offset.

z-transform: A transform useful for representing time series and calculating the effects of various operations. If the sample values of a wavelet at successive times are:

$$x_t = [x_0, x_1, x_2, x_3, \cdots, x_n],$$

then the z-transform of the wavelet is

$$x(z) = x_0 + x_1 z + x_2 z^2 + x_3 z^3 + \cdots + x_n z^n.$$

The z-transform may be thought of as

$$z = e^{-i\omega t},$$

where ω = angular frequency; this allows one to relate it to the Fourier transform. The z-transform technique is an easy way of converting (by inspection) from the time domain into a form which can be treated (in some ways) as in the frequency domain. The convolution operation can be accomplished by merely multiplying the z-transforms of the waveforms being convolved. The inverse of a filter can be found by finding the reciprocal of the filter's z-transform. Used in digital seismic processing. The z-transform polynomial can be factored and expressed as the product of couplets of the form:

$$f(z) = (z - a)(z - b)(z - c) \cdots, (z - n).$$

The values for which this expression vanishes, i.e., $z = a$, $z = b$, etc. (a, b, etc. may be complex) are called **zeros** or **roots** of the expression. Values greater than unity are said to "lie outside the unit circle;" see Figure Z-3. If all the roots lie outside the unit circle, the function is **minimum phase**; if all are inside, it is **maximum phase**. Values for which an expression becomes infinite [such as r in $1/(z - r)$] are called **poles** or **singularities**. Filters are sometimes designed in the z-plane. See Robinson and Treitel (1964) and Sheriff and Geldart, v. 2 (1983, p. 178-180). Sometimes the opposite convention is used, i.e., successive sample values are multiplied by successively higher negative powers of z. Use of this convention reverses the criteria for minimum and maximum phase with respect to the unit circle.

Appendix A: SI units

"SI" is an abbreviation for Le Système International d'Unités, an international system of units adopted by many national and international authorities, associations, professional societies, and agencies. SI is closely related to but not identical with the former cgs, mks, and mksa systems of metric units. Official information, development history, and more detail on SI can be found in Bureau of Standards Special Publication 330 (1974) and in the SEG Subcommittee on Metrification (1981) publication.

SI is based on seven **base units** listed in Table 1 and two **supplementary units** (the last two).

SI **derived** units are formed by combining the base and supplementary units. Some of the derived units are listed in Table 2.

Table 3 shows how larger or smaller units can be made by adding prefixes. When prefixes are used, the first syllable is accented. Note that k and M stand for 10^3 and 10^6, whereas M and MM (or m and mm) are sometimes used in the oil industry for designating thousands and millions (of gas volumes). Prefixes are raised to the power of the unit employed; for example, km^3 means cubic kilometers, not thousands of cubic meters. Prefixes are not compounded (GW rather than kMW).

SEG allows the forms in Table 4 in addition to those in Tables 1-3.

Table 5 relates cgs electromagnetic and electrostatic units to SI units. Figure M-1 relates cgs and SI magnetic units.

Rules about writing units

Symbols are written in Roman (not italics) type. They are never pluralized.

Unit names, including prefixes, are not capitalized except at the beginning of a sentence or in titles. Unit names are pluralized in the usual manner, as 100 meters, 70 henries, except for lux, hertz, and siemens. Fractional values require the singular form.

Periods are not used after symbols, that is, symbols are not abbreviations.

Symbols are lower case except when named for a person (exception: L for liter).

A space separates a numerical value and the unit symbol (except for °C); thus, 10 m, 0.112 s, 1.5 g/cm^3, 20°C. A hyphen separates value and symbol when used as an adjective; thus, 35-mm film. No space separates a prefix and the symbol; thus, ms for milliseconds, kW for kilowatt.

The symbols "/" or "." are used to indicate the compounding of symbols (for example, km/s or N.m for kilometers per second and Newton-meter), but / and . are not used when units are written out. Where symbols are compounded, parentheses should be used to avoid ambiguity, as W/(m.k). "P" is not acceptable as an abbreviation for "per". "Per" should not be compounded; thus, "meters per second squared", not "meters per second per second". Use × rather than . for products of numbers; thus 6.2×5, not 6.2.5. A space should be used on each side of symbols for multiplication, addition, subtraction, convolution ($\times$, $+$, $-$, $*$) and for the division symbol $\div$ but not for /.

Numbers with many decimal places should be grouped by threes separated by a space rather than by a comma (which Europeans read as a decimal point); thus, 4 720 525 or 0.528 75. For numbers smaller than one, a zero should be shown in the units place. A space is not necessary for four-digit numbers.

Table 1. SI base and supplementary units

Quantity	SI unit, symbol
Length	meter or metre, m
Mass	kilogram, kg
Time	second, s
Electric current	ampere, A
Thermodynamic temperature	kelvin, K
Amount of substance	mole, mol
Luminous intensity	candela, cd
Plane angle	radian, rad
Solid angle	steradian, sr

Note that the kilogram is not a unit of force (weight). The word "weight" is often ambiguous and its use should be avoided. The temperature unit kelvin is not "degree kelvin."

"Squared" or "cubed" should follow unit names except for areas and volumes; thus, meter per second squared, square meter, watt per cubic meter.

The spellings "metre" and "litre" are preferred but "meter" and "liter" are the official U.S. forms of spelling. The use of liter as a cubic decimeter is discouraged.

Table 2. SI derived units

Quantity	Derived unit, symbol	Quantity	Derived unit, symbol
Absorbed dose	gray, Gy = J/kg	Luminous flux	lumen, lm = cd.sr
Acceleration	meters per second squared, m/s^2	Magnetizing force	ampere per meter, A/m
Activity (of radionuclides)	becquerel, Bq = l/s	Magnetic flux	weber, Wb = V.s
Angular acceleration	radian per second squared, rad/s^2	Magnetic flux density	tesla, $T = Wb/m^2$
		Potential difference	volt, V = W/A
Angular velocity	radian per second, rad/s	Power	watt, W = J/s
Area	square meter, m^2	Pressure	pascal, $Pa = N/m^2$
Density	kilogram per cubic meter, kg/m^3	Quantity of electricity	coulomb, C = A.s
		Quantity of heat	joule, J = N.m
Electric capacitance	farad, F = A. s/V = C/V	Radiant flux	watt, W = J/s
		Radiant intensity	watt per steradian, W/sr
Electric charge	coulomb, C = A/s	Specific heat capacity	joule per kilogram kelvin, J/kg.K
Electrical conductance	siemen, S = A/V		
Electric field strength	volt per meter, V/m	Stress	pascal, $Pa = N/m^2$
Electric inductance	henry, H = V.s/A = Wb/A	Thermal conductivity	watt per meter kelvin, W/m.K
Electric potential	volt, V = W/A	Torque	newton meter (not joule)
Electric resistance	ohm, Ω = V/A		
Electromotive force	volt, V = W/A	Velocity	meter per second, m/s
Energy	joule, J = N.m	Viscosity, dynamic	pascal second, Pa.s
Entropy	joule per kelvin, J/K	Viscosity, kinematic	square meter per second, m^2/s
Force	newton, $N = kg. m/s^2$		
Frequency	hertz, Hz = l/s	Voltage	volt, V = W/A
Illiminance	lux, $lx = lm/m^2$	Volume	cubic meter, m^3
Luminance	candela per square meter, cd/m^2	Wavenumber	per meter, l/m
		Work	joule, J = N.m

Table 3. SI prefixes

Multiplier	Prefix, symbol
10^{18}	exa, E
10^{15}	peta, P
10^{12}	tera, T
10^{9}	giga, G
10^{6}	mega, M
10^{3}	kilo, k
10^{2}	hecto, h
10	deka, da
10^{-1}	deci, d
10^{-2}	centi, c
10^{-3}	milli, m
10^{-6}	micro, μ
10^{-9}	nano, n
10^{-12}	pico, p
10^{-15}	femto, f
10^{-18}	atto, a

When prefixes are used, the first syllable is accented.

Table 4. Additional units allowed by SEG

Quantity	Unit and equivalence
Acceleration	milligal, mGal = 10^{-5} m/s^2
Angular velocity	revolutions per minute, revolutions per second
Area	hectare, ha = 10^4 m^2
Calorific value	kilowatt hours per kilogram
Energy	kilowatt hour, kw.h = (1/3600)J
Energy unit	electron volt, eV
Length	centimeter, cm = 10^{-2} m
Magnetic flux density	gamma = nT
Mass	tonne = 10^3 kg
Plane angle	degree = 0/017 453 29 rad
Pressure	bar = 100 kPa
Temperature	degree Celsius, °C = Kt 273.15
Time	minute, min = 60 s
	hour, h = 3600 s
	day, d
	year, a
Volume	liter or litre, L = dm^3
	hectare meter, ha.m = 10^4 m^3
Yield	liter/tonne

Note that "degree" in "degree Celsius" is lower case. "Centigrade" is now obsolete. The symbol ° to indicate degree is not used when temperature is expressed in kelvin.

Table 5. SI equivalents of cgs units

Quantity	SI unit	cgs-emu	cgs-esu
Length	meter	= 10^2 centimeter	
Mass	kilogram	= 10^3 gram	
Force	newton	= 10^5 dyne	
Energy (work)	joule	= 10^7 erg	
Current	ampere	= 10^{-1} abampere	= 3×10^9 statampere
Charge	coulomb	= 10^{-1} abcoulomb	= 3×10^9 statcoulomb
Electrical potential	volt	= 10^8 abvolt	= (1/300) statvolt
Resistance	ohm	= 10^9 abohm	= $(9 \times 10^{11})^{-1}$ statohm
Capacitance	farad	= 10^{-9} abfarad	= 9×10^{11} statfarad
Magnetic flux density	tesla	= 10^4 gauss	
Magnetic flux	weber	= 10^8 maxwell	
Magnetizing force	ampere turn/m	= $4\pi \times 10^3$ oersted	
Inductance	henry	= 10^9 abhenry	

For some of the above units, magnitude depends on the speed of light in vacuum, here taken as 3×10^5 km/s.

Appendix B: Greek alphabet

A	α	alpha	I	ι	iota	P	ρ	rho
B	β	beta	K	κ	kappa	Σ	σ	sigma
Γ	γ	gamma	Λ	λ	lambda	T	τ	tau
Δ	δ	delta	M	μ	mu	Υ	υ	upsilon
E	ϵ	epsilon	N	ν	nu	Φ	ϕ	phi
Z	ζ	zeta	Ξ	ξ	xi	X	χ	chi
H	η	eta	O	o	omicron	Ψ	ψ	psi
Θ	θ	theta	Π	π	pi	Ω	ω	omega

Appendix C: Symbols used in geophysical exploration

The following symbols are recommended for publications of the SEG. Additional symbols used in well logging and mathematics are listed in Appendices D and E.

a	Apparent (as a subscript); velocity gradient with depth.
A	Area.
AMT	Audiomagnetotelluric.
BA	Electric vector potential.
b/B	Magnetic induction in time/frequency domain.
c	Velocity (of light).
C	Capacitance $= Q/V$.
d/D	Electric displacement in time/frequency domain.
d_s	Depth of shot.
E	Young's modulus; elevation; voltage (EMF).
e/E	Electric field strength in time/frequency domain.
EM	Electromagnetic.
F	Magnetic vector potential.
f, F	Fair (reliable but with less accuracy than desirable); frequency.
FE	Frequency effect.
FEM	Frequency-domain electromagnetic.
g	Acceleration of gravity.
G	Conductance.
G, **G**	Scalar/tensor Green's function.
h	Thickness.
h/H	Magnetic field strength in time/frequency domain.
I	Current.
I	Intensity of magnetization (a vector).
i, j	$(-1)^{1/2}$.
i, j, k	Unit vector in the x-, y-, z-direction.
IP	Induced electric polarization.
J	Free charge current density (a vector).
k	Bulk modulus; susceptibility.
K_e	Relative dielectric permittivity $[k_e = 1 \, (4\pi\epsilon_0)]$.
K_m	Relative magnetic permeability $[k_m = (\mu_0/4\pi)]$.
*L	Inductance; area over a decay curve; length.
m	Volume chargeability.
M	Mutual inductance, integral chargeability.
m, M	Magnetic polarization in time/frequency domain.
MF	Metal factor.
MIP	Magnetic induced electric polarization.
MMR	Magnetometric resistivity.
MT	Magnetotelluric.
n	Unit normal vector.
n	Index of refraction.
p	Raypath parameter; pressure.
P	P-wave; polarization; dipole moment/volume; poor (probably reliable with poor accuracy); pressure.
PFE	Percent frequency effect.
P,P	Magnetic polarization in time/frequency domain.

q   Charge.
Q   Heat-flow rate.
r   Position vector.
r   Radial distance.
R   Resistance; reflectivity (reflection coefficient); radius.
s   Laplace transform variable.
S   S-wave; admittance = $1/Z$; surface.
t   Time; traveltime.
T   Period; temperature; age.
TEM   Time-domain electromagnetic.
$\mathbf{u}_x$, $\mathbf{u}_y$, $\mathbf{u}_z$ Unit vectors in Cartesian coordinates (or i, j, k).
$\mathbf{u}_\rho$, $\mathbf{u}_\theta$, $\mathbf{u}_z$ Unit vectors in cylindrical coordinates.
$\mathbf{u}_r$, $\mathbf{u}_\theta$, $\mathbf{u}_\phi$ Unit vectors in spherical coordinates.
*U   Group velocity; magnetic scalar potential.
*v, V   Velocity; volume
V   Voltage; electric or gravity scalar potential.
vp, VP   Very poor (in both reliability and accuracy).
w_x Weathering.
x   Offset distance; distance.
X   Reactance; $X_C = 1/(2\pi\nu C)$; $X_L = 2\pi\nu L$.
$\hat{y}$ Admittivity.
y   Admittance.
z   Depth
$\hat{z}$ Impedivity.
z^n Time delay of n units.
Z   Impedance.

α alpha P-wave velocity; proportional to; attenuation factor (seis-
                        mology), propagation constant (electromagnetic), phase
                        constant; α-particles.
β beta S-wave velocity, β-particles, attenuation constant (electro-
                        magnetic).
γ gamma   Skewness; gyromagnetic ratio; unit of magnetic field
                        strength; phase angle; gravitational constant; gamma ray.
δ delta   Impulse; logarithmic decrement; skin depth; depth of pene-
                        tration.
$\tan \delta$, $\tan \delta_m$,   Dielectric, magnetic, electromagnetic loss tangent.
  $\tan \delta_{em}$
Δ Delta   Difference; dilatation; skin depth or attenuation length.
ϵ epsilon   Permittivity; eccentricity.
η eta   Overvoltage; absorption coefficient.
θ theta   Angle; induction number.
θ_c, Θ Theta   Critical angle.
κ kappa 2π (wavenumber).
λ lambda   Wavelength; coefficient of anisotropy; Lamé's constant.
μ mu   Magnetic permeability = B/H; micro; Lamé's shear modu-
                        lus; damping factor, attenuation constant (radiometric);
                        viscosity.

ν	nu	Frequency; ν_0 = natural frequency.
ξ	xi	Dip.
ρ	rho	Density; electric resistivity; radius of curvature; radial distance, charge density.
σ	sigma	Poisson's ratio; electrical conductivity; standard deviation; stress.
τ	tau	Time delay; damping factor.
ϕ	phi	Porosity; flux; correlation function; latitude.
χ	chi	Magnetic susceptibility.
ψ	psi	Wave function.
ω	omega	Angular frequency.
Ω	Omega	Ohm.
∇	Del.	Gradient operator (a vector).
$*$		Convolution operator.

Appendix D: Additional symbols used in well logging

a	Activity; air requirement; air; apparent; atmospheric
A	Area, areal
b	Reciprocal of formation-volume-factor; bank; bubble point; bulk; burned
B	Formation volume factor = volume at reservoir conditions/volume at standard conditions; turbulence
c	Capillary; contact; conversion; critical; electrochemical; compressibility
cf	Flowing casing (pressure)
cs	Static casing (pressure)
C	Coefficient of back-pressure curve; concentration; specific heat; water-drive constant
C_1	Methane
C_2	Ethane
d	Diameter; depletion; dew-point; differential separation; dip; displaced; drainage
D	Deliverability; depth; diffusion coefficient; dimensionless; displacement
e	Influx rate; cumulative influx; effective; external boundary conditions
ext	Extrapolated (subscript)
E	Efficiency; experimental
f	Coefficient of friction; fraction; fugacity; flash separation; fluid; formation; front
F	Factor; force; formation resistivity factor = R_0/R_w; free; fuel
g, G	Gas
h	Hole; hydrocarbon; thermal heat
H	Enthalpy
i	Injection rate; initial; injected; invaded zone
I	Injectivity index; resistivity index = R_t/R_0; invasion
j	$(-1)^{1/2}$
J	Productivity index
k	Absolute permeability; electrokinetic
K	Spontaneous electromotive force; equilibrium ratio
L	Moles of liquid phase; liquid
lim	Limiting value
m	Mud; fuel; fuel consumption; mass; slope; ratio of initial reservoir free gas volume/initial reservoir oil volume; porosity exponent
ma	Solid matrix (subscript)
mc	Mud cake (subscript)
mf	Mud filtrate (subscript)
max	Maximum (subscript)
min	Minimum (subscript)
M	Mobility ratio; molecular weight; molal
n	Net; total moles; back-pressure exponent; saturation exponent
N	Dimensionless; initial oil in place
N_2	Nitrogen

o		Oil (subscript) (except with resistivity); 100 percent water saturated (resistivity)
O_2		Oxygen
p		Cumulative produced; particle; pore; production time; mean
pc		Pseudocritical
P		Pattern
q		Flow rate
r		Radial distance; resistance; reduced; relative; residual
R		Recovery; reservoir; gas-oil ratio; universal gas constant
Re		Reynolds Number
s		Skin; segregation; solid; solution; specific; stabilization; surrounding formation; swept region
sb		Solution at bubble-point conditions
sc		Standard conditions
sw		Solution in water
S		Saturation
t		True; total; gross
tf		Flowing tubing (pressure)
ts		Static tubing (pressure)
u		Flux; flow rate/area; unburned
v		Specific volume; vaporization
V		Volumetric; moles of vapor phase
w		Water; well conditions; mass flow rate
wa		Apparent wellbore (subscript)
wf		Flowing bottom-hole (pressure); well flowing conditions
wg		Wet gas; water in gas cap
ws		Static bottom-hole; well static conditions
W		Initial water in place; work
x		Mole fraction in liquid phase
xo		Flushed zone (subscript)
y		Mole fraction in vapor phase
z		Mole fraction in mixture; compressibility factor $=pV/nRT$
Z		Elevation with respect to datum
α	alpha	Angle; SP reduction factor
β	beta	Thermal cubic expansion coefficient
γ	gamma	Specific gravity
δ	delta	Displacement ratio
ϵ	epsilon	Hydraulic diffusivity
η	eta	Kinematic viscosity
Θ	Theta	Acoustic transit time per unit length
λ	lambda	Mobility; wavelength
σ	sigma	Surface tension; interfacial tension
τ	tau	Tortuosity
ϕ	phi	Fluid potential
Ψ	Psi	Stream function

Appendix E: Mathematical symbols

Bold face $\begin{cases} \text{Vectors } \mathbf{V} \\ \text{Matrices } \underset{\sim}{\mathbf{A}} \end{cases}$

Superscripts $\begin{cases} \text{Powers as } a^n, \sin^n x = (\sin x)^n \ (n\text{-positive}) \\ \text{Inverse functions as } \sin^{-1} x = \text{arc sin } x \\ \text{Order of differentiation as } x' = dx/dt; \ x'' = d^2x/dt^2 \end{cases}$

Subscripts $\qquad$ Position in a sequence, set, or matrix as $\left\| \begin{matrix} a_{11} & a_{12} \\ a_{11} & a_{22} \end{matrix} \right\|$

Parentheses $\begin{cases} \text{Aggregation as } (a + b) \\ \text{Argument of a function } f(t) \end{cases}$

Vertical bars $\begin{cases} \text{Absolute value, modulus, magnitude of vector } \mathbf{V} \\ \text{Determinant } \begin{vmatrix} a_{11} & a_{12} \\ a_{21} & a_{22} \end{vmatrix} \\ \text{Evaluated at } f(x)|_a = f(a) \end{cases}$

Double bars $\qquad$ Matrix $\|a_{ij}\| = \left\| \begin{matrix} a_{11} & a_{12} \\ a_{11} & a_{21} \end{matrix} \right\|$

Brackets	Ordered set $[x, y, z]$ or $[g_1, g_2, g_3, \cdots]$
Superscript line	Mean; complex conjugate
$=$	Equal to
$\neq$	Not equal to
$\approx$	Approximately equal to
$>$	Greater than
$<$	Smaller than
$\geq$	Greater than or equal to
$\leq$	Smaller than or equal to
$+$	Plus
$-$	Minus
$\pm$	Plus or minus
$\cup$	Union
$\cap$	Intersection
$\supset$	Contains; implies
$\rightarrow$	Approaches

$\leftrightarrow$	Transforms to (either way)
/	Division, as $1/2 = \frac{1}{2}$
*	$\begin{cases} \text{Convolved with} \\ \text{As superscript: complex conjugate} \end{cases}$
$\cdots$	And so forth as $a_0, a_1, \ldots, a_n$
!	Factorial, as $3! = 3 \cdot 2 \cdot 1 = 6$
$\int_b^a$	Integral from a to b
∞	Infinity
$\triangle$	Difference; sizeable increment
∇	Del (a vector): ∇^2 = Laplacian
δ	Very small increment
∂	Partial derivative, $\dfrac{\partial f(x, y)}{\partial x}$
Σ	Sum, as $\displaystyle\sum_{i=1}^{3} a_i = a_1 + a_2 + a_3$; $\sum$ = sum of all appropriate elements.
Π	Product, as $\displaystyle\prod_{i=1}^{3} a_i = a_1 a_2 a_3$
$\phi_{xy}(\tau)$	Correlation of x with y as function of time shift τ
σ	Standard deviation
ω	Angular velocity (frequency)
abs	Absolute
arc	Inverse, as arc $\sin x = \sin^{-1} x$ = angle whose sine is x
arg	Argument of
av	Average
cis θ	Cos $\theta + i \sin \theta$
d	Differential
det	Determinant
div	Divergence
erf	Error function, as $\text{erf}(x) = \dfrac{2}{\sqrt{\pi}} \displaystyle\int_0^x e^{-v^2} dv$
exp	Exponential function, as $\exp[x] = e^x$
Im	Imaginary part of
lim	Limit
ln	Natural logarithm $= \log_e$ (log to the base e)
max	Maximum
P[E]	Probability of E
P [$E \cap F$]	Probability of both E and F
P[E/F]	Probability of E given F
Re	Real part of
rms	Root-mean-square
sgn	Sign of

Cylinder functions of order ν and argument x
General $Z_\nu(x)$
Bessel function of first kind $J_\nu(x)$
Bessel function of second kind $N_\nu(x)$
Modified Bessel function of second kind $K_\nu(x)$
Hankel function of first kind $H_\nu^{(1)}(x)$
Hankel function of second kind $H_\nu^{(2)}(x)$
Spherical cylinder function of order ν and argument x

$$z_\nu(x) = \left(\frac{\pi}{2x}\right)^{1/2} Z_{\nu+1/2}(x)$$

$j_\nu(x)$
$n_\nu(x)$
$i_\nu(x)$
$k_\nu(x)$
$h_\nu^{(1)}(x)$
$k_\nu^{(2)}(x)$

Gamma function $\Gamma(x)$
Struve function $H_\nu(x)$
Modified Struve function $L_\nu(x)$
Error function erf (x)
Complementary error function $\text{erf}c(x)$
Legendre functions $P_\nu(x)$
Associated Legendre functions $P_\nu^m(x)$

Fourier transform pair $f(t) = \dfrac{1}{2\pi} \displaystyle\int_{-\infty}^{\infty} F(\omega)\, e^{i\omega t} d\omega$

$$F(\omega) = \int_{-\infty}^{\infty} f(t)e^{-i\omega t}\, dt$$

Laplace transform $F(s) = L\{f(t)\} = \displaystyle\int_{0}^{\infty} f(t)e^{-st}\, dt$

Hilbert transform pair $\text{Im}\{f(\omega)\} = \dfrac{1}{\pi} \displaystyle\int_{\infty}^{\infty} \dfrac{\text{Re}(\alpha)}{\omega - \alpha}\, d\alpha$

$$\text{Re}\{f(\omega)\} = \frac{-1}{\pi} \int_{\infty}^{\infty} \frac{\text{Im}(\alpha)}{\omega - \alpha}\, d\alpha$$

Hankel transform pair $f_m(\rho) = \displaystyle\int_{0}^{\infty} F_m(\lambda)J_m(\lambda\rho)\lambda d\lambda$

$$F_m(\lambda) = \int_{0}^{\infty} f_m(\rho)J_m(\lambda\rho)\rho d\rho$$

For definitions of functions and related matters, refer to the following:

Abramowitz, M., and Stegun, I. A., 1972, Handbook of mathematical functions: Dover Publications, Inc.

Appendix F: Map and rock symbols

A. Well Symbols

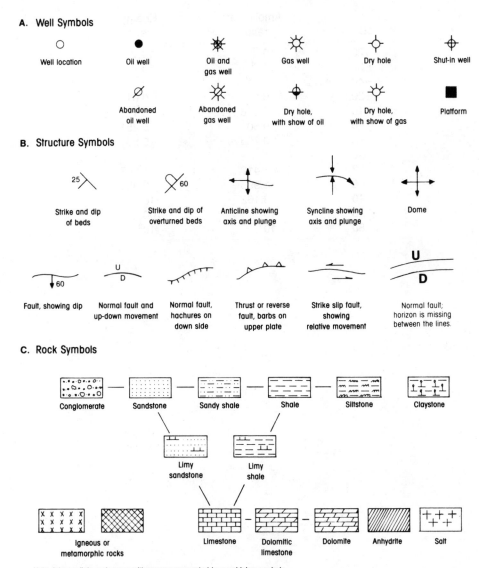

Well location	Oil well	Oil and gas well	Gas well	Dry hole	Shut-in well
	Abandoned oil well	Abandoned gas well	Dry hole, with show of oil	Dry hole, with show of gas	Platform

B. Structure Symbols

Strike and dip of beds

Strike and dip of overturned beds

Anticline showing axis and plunge

Syncline showing axis and plunge

Dome

Fault, showing dip

Normal fault and up-down movement

Normal fault, hachures on down side

Thrust or reverse fault, barbs on upper plate

Strike slip fault, showing relative movement

Normal fault; horizon is missing between the lines.

C. Rock Symbols

Conglomerate — Sandstone — Sandy shale — Shale — Siltstone — Claystone

Limy sandstone

Limy shale

Igneous or metamorphic rocks

Limestone — Dolomitic limestone — Dolomite — Anhydrite — Salt

Note: Intermediate rock compositions are represented by combining symbols.

Appendix G: Decibel conversion

dB	Amplitude ratio	Energy ratio
−120	10^{-6}	10^{-12}
−80	10^{-4}	10^{-8}
−40	0.01	10^{-4}
−20	0.1	0.01
−10	0.316	0.1
−6	0.501	0.251
−3	0.708	0.501
0	1	1
+3	1.413	1.997
+6	1.995	3.980
+10	3.162	10
+20	10	100
+80	10^4	10^8

Appendix H: Names for commercial borehole logging services

		Dresser atlas	Gearhart	Schlumberger	Welex
Resistivity/Conductivity	1.	Electrolog	—	Electrical Log ES	Electric Log EL
	2.	Induction Electrolog, IEL	Induction Electric Log, IEL	Induction Electric Survey log, IES	Induction Electric Log, IEL
	3.	Dual Induction/ Focused Log, DIFL	Dual Induction/ Laterolog, DIL	Dual Induction Laterolog-8, DIL	Dual Induction Guard Log, DIGL
		—	—	Dual Induction Spherically Focused Log, DIL-SFL	
	4.	Dual Laterolog, DLL	Dual Laterolog, DLL	Dual Laterolog, DLL	Guard Log, GL Dual Guard, DG
Microresistivity	5.	Minilog, ML	Micro Electric Log, MEL	Microlog, ML	Contact Log, CONC
	6.	Micro-Laterolog, MLL	Miro Laterolog MLL	Microlaterolog, MLL	FoRxo Logs, FORXO
		—	Microspherically Focused Log, MSFL	Micro Spherically Focused Log, MSFL	—
		—	—	—	FoRxo, DGF
		Proximity	—	Proximity Log	—
Dipmeter	7.	Diplog	Dipmeter	High Resolution Dipmeter, HDT	Resistivity Dip Log
		—	—	Dual Dipmeter	—
		Strata Dip (Computed)	GEO	Stratigraphic High Resolution Dip-meter, SHDT	Stratigraphic Plot (Computed)
		Dip FracLog	Fracture Ident. Log	Fracture Ident. Log, FIL	Fracture Profile Log
Acoustic	8.	Acoustilog	—	Sonic Log	Acoustic Velocity Log, AVL
		BHC Acoustilog	Borehole Compensated Sonic Log, BCS	Borehole Compensated Sonic Log, BHCS	Compensated Acoustic Velocity Log, CAVL
		Long-Spaced B.H.C. Acoustilog	—	Long Spaced Sonic Log, LSS	Long Spaced Velocity Log, LSV
	9.	Acoustic Fraclog With Compensated Amplitudes	Sonic Frac Log	Amplitude Log	Fracture Finder Micro-Seismogram Log, FFMSG
	10.	—	Seismic Spectrum	—	—
		—	—	Cement Evaluation Log, CET	Micro-Seismogram MSGB
		Acoustic Cement Bond Log	Cement Bond Log, CBL	Cement Bond Log, CBL	Acoustic Cement Bond Log, ACBL
		3D Acoustic Variable Density	—	Variable Density Log, VDL	—

	Dresser Atlas	Gearhart	Schlumberger	Welex
	11. Densilog	—	Formation Density Log, FD	Density Log, DL
	Compensated Densilog	Compensated Density Log, CDL	Compensated Formation Density Log, FDC	Compensated Density Log, CDL
	—	—	Litho-Density Log LDL	—
	12. Gamma Ray-Neutron Log, GRN	Gamma Ray Neutron Log, GNL	—	Gamma Ray Neutron Log, GRN
	Compensated Neutron Log, CNL	Compensated Neutron/ Gamma Ray, CNS/ GR	Compensated Neutron Log, CNL	—
Radioactive	—	—	Dual Porosity Compensated Neutron (Epithermal/ Thermal), DNL	Dual Spaced Neutron Log, DSN
	Sidewall, Epithermal Neutron Log, SWN	Sidewall Neutron Porosity Log, SNL	Sidewall Neutron Porosity Log, SNP	Sidewall Neutron Log SWN
	13. Gamma Ray Log, GR Spectralog	Gamma Ray Log, GR Natural Gamma Ray Spectral Log, NGS	Gamma Ray Log, GR Natural Gamma Ray Spectrometry, NGS	Gamma Ray Log, GR Compensated Spectral Natural Gamma Log CSNG
	— Carbon/Oxygen Log	— —	Induced Gamma Ray Spectrometry, GST	— —
	14. Dual Detector Neutron Lifetime Log, DNLL	—	Thermal Decay Time Log, TDT	Thermal Multigate Delay Log, TMD
	15. Formation Tester, FT	Selective Formation Tester, SFT	Repeat Formation Tester, RFT	Multiset Tester, MST
	16. Corgun	Sidewall Coring Tool, SCT	Sidewall Core Sample Taker, CST	Sidewall Coring, SWC
Other	17. —	—	Nuclear Magnetism Log, NML	—
	18. —	—	Electromagnetic Propagation Tool, EPT	—

Appendix I: Abbreviations for geophysical contractors and organizations

Geophysical contractors

CDP Computer Data Processors
CGG Compagnie Générale de Géophysique
GAI Geophysical Associates (EG&G)
GMX Gravity Meter Exploration (EG&G)
GRC Geophysical Research Corp.
GSI Geophysical Service Inc.
GUS Globe Universal Services Inc.

ONI Offshore Navigation Inc.
SEI Seismic Engineering
SIE Southwestern Industrial Electronics Co.
SPC Satellite Positioning Corp.
SSC Seismograph Service Corp.
SSL Seismograph Service Ltd.
TI Texas Instruments Inc.

Organizations

Am. Assoc. for Advance. Sci.	American Association for the Advancement of Science
Am. Assoc. Petr. Geol.	American Association of Petroleum Geologists, publishers of the *Bulletin* of the AAPG
Assoc. Eng. Geol.	Association of Engineering Geologists
Am. Geol. Inst.	American Geological Institute
Am. Geophys. Union	American Geophysical Union
Am. Inst. Min., Metallurg., Petr. Eng.	American Institute of Mining, Metallurgical, and Petroleum Engineers
Am. Inst. Phys.	American Institute of Physics
Am. Inst. Prof. Geol.	American Institute of Professional Geologists
Am. Math. Soc.	American Mathematical Society
Am. Petr. Inst.	American Petroleum Institute
Can. Soc. Expl. Geophys.	Canadian Society of Exploration Geophysicists
Eur. Assoc. Expl. Geophys.	European Association of Exploration Geophysicists, publishers of *Geophysical Prospecting* and *First Break*.
Geol. Soc. Am.	Geological Society of America
Int. Assoc. Geophys. Contr.	International Association of Geophysical Contractors
Inst. Electr. Electron. Eng.	Institute of Electrical and Electronics Engineers Inc.
Int. Union Geod. Geophys.	International Union of Geodesy and Geophysics
Can. Expl. Geophys. Soc.	Canadian Exploration Geophysical Society
Org. Petr. Export. Countries	Organization of Petroleum Exporting Countries
Offshore Tech. Conf.	Offshore Technology Conference
Soc. Expl. Geophys.	Society of Exploration Geophysicists, publishers of the journals GEOPHYSICS and THE LEADING EDGE.
Soc. Econ. Geol.	Society of Economic Geologists
Soc. Econ. Paleont. Mineral.	Society of Economic Paleontologists and Mineralogists
Soc. Indep. Prof. Earth Sci.	Society of Independent Professional Earth Scientists
Soc. Petr. Eng., Am. Inst. Min., Metall. Petr. Eng.	Society of Petroleum Engineers of AIME
U.S. Geol. Surv.	United States Geological Survey

Appendix J: Conversion of units

Each of the fractions below has the value 1 (i.e., numerator and denominator are equal).

Circular measure

$$\frac{0.01745 \text{ radians}}{\text{degree}} \qquad \frac{57.30 \text{ degrees}}{\text{radian}} \qquad \frac{9.55 \text{ rev/minute}}{\text{rad/s}}$$

Linear measure

$$\frac{0.3048 \text{ m}}{\text{ft}} \qquad \frac{3.281 \text{ ft}}{\text{m}} \qquad \frac{1.609 \text{ km}}{\text{statute mile}} \qquad \frac{0.6214 \text{ statute mile}}{\text{km}} \qquad \frac{10^{10} \text{ Angström}}{\text{m}}$$

$$\frac{9.46 \text{ m}}{10^{-15} \text{ light year}} \qquad \frac{10^{6} \text{ micron}}{\text{m}} \qquad \frac{66 \text{ ft}}{\text{chain}} \qquad \frac{6 \text{ ft}}{\text{fathom}} \qquad \frac{1.1516 \text{ statute mile}}{\text{nautical mile}}$$

$$\frac{1.852 \text{ km}}{\text{nautical mile}} \qquad \frac{100 \text{ link}}{\text{chain}} \qquad \frac{60 \text{ nautical miles}}{\text{degree at equator}} \qquad \frac{1 \text{ league}}{3 \text{ statute miles}}$$

Velocity

$$\frac{1.689 \text{ ft/s}}{\text{knot}} \qquad \frac{1.15157 \text{ mile/hour}}{\text{knot}} \qquad \frac{0.5148 \text{ m/s}}{\text{knot}} \qquad \frac{1.852 \text{ km/hour}}{\text{knot}}$$

Area

$$\frac{10^{28} \text{ barn}}{\text{m}^2} \qquad \frac{640 \text{ acres}}{\text{mile}^2} \qquad \frac{1 \text{ section}}{\text{mile}^2} \qquad \frac{2.471 \text{ acres}}{\text{hectare}} \qquad \frac{2.590 \text{ km}^2}{\text{mile}^2} \qquad \frac{100 \text{ hectares}}{\text{km}^2}$$

Volume

$$\frac{3.785 \text{ liters}}{\text{U.S. gallon}} \qquad \frac{4.546 \text{ liters}}{\text{British gallon}} \qquad \frac{7.4805 \text{ U.S. gallons}}{\text{ft}^3} \qquad \frac{0.15899 \text{ m}^3}{\text{bbl}}$$

$$\frac{1 \text{ acre ft}}{1233.5 \text{ m}^3} \qquad \frac{7758 \text{ bbl}}{\text{acre ft}} \qquad \frac{5.61 \text{ bbl}}{\text{ft}^3} \qquad \frac{1 \text{ bbl}}{0.159 \text{ m}^3} \qquad \frac{42 \text{ U.S. gallons}}{\text{bbl}}$$

Mass

$$\frac{2.2046 \text{ lb}}{\text{kg}} \qquad \frac{0.4536 \text{ kg}}{\text{lb}} \qquad \frac{1.120 \text{ short ton}}{\text{long ton}} \qquad \frac{1.102 \text{ short ton}}{\text{metric tonne}} \qquad \frac{0.9842 \text{ long ton}}{\text{metric tonne}}$$

Pressure

$$\frac{1.01325 \text{ pascal}}{10^{-5} \text{ atmosphere}} \quad \frac{1 \text{ bar}}{10^5 \text{ pascal}} \quad \frac{29.92 \text{ inches of Hg}}{\text{atmosphere}} \quad \frac{14.223 \text{ lb/inch}^2}{10^4 \text{ kg/m}^2} \quad \frac{1 \text{ cm of Hg}}{1333 \text{ pascal}}$$

$$\frac{14.7 \text{ psi}}{\text{atmosphere}} \quad \frac{1 \text{ newton/m}^2}{\text{pascal}} \quad \frac{0.06895 \text{ bar}}{\text{lb/inch}^2} \quad \frac{703.07 \text{ kg/m}^2}{\text{lb/inch}^2} \quad \frac{0.1333 \text{ kPa}}{\text{torr}} \quad \frac{16.018 \text{ kg/m}^3}{\text{lb/ft}^3}$$

Work (Energy)

$$\frac{1055 \text{ joules}}{\text{Btu}} \quad \frac{4186 \text{ joules}}{\text{kilocalorie}} \quad \frac{3600 \text{ joules}}{\text{watt hour}} \quad \frac{1.6020 \text{ joule}}{10^{19} \text{ electron volt}} \quad \frac{0.2930 \text{ watt/hour}}{\text{Btu}}$$

$$\frac{10^7 \text{ erg}}{\text{joule}}$$

Power

$$\frac{745.7 \text{ watts}}{\text{horsepower}} \quad \frac{0.001341 \text{ horsepower}}{\text{watt}}$$

Other

$$\frac{3.7 \times 10^{10} \text{ becquerel}}{\text{curie}} \quad \frac{\text{abamp}}{10 \text{ amp}} \quad \frac{10^8 \text{ abvolt}}{\text{volt}} \quad \frac{299.79 \text{ volt}}{\text{statvolt}} \quad \frac{2,9979 \times 10^9 \text{ statamp}}{\text{amp}}$$

$$\frac{\text{mPa.s}}{\text{centipoise}} \quad \frac{\text{langley}}{\text{joule/m}^2} \quad \frac{16.018 \text{ kg/m}^3}{\text{pound/ft}^3} \quad \frac{41.86 \text{ mW/m}^2}{10^{-2}\text{cal/m}^2\text{s} = 1 \text{ HFU}} \quad \frac{1 \text{ neper}}{8.686 \text{ dB}} \quad \frac{10^5 \text{ dyne}}{\text{newton}}$$

Appendix K: Numerical Constants

π	$= 3.1415927$
e	$= 2.71828$
Velocity light	$c = 2.997925 \times 10^8$ m/s
Solar day	$= 86,400$ s
Sidereal day	$= 86,164$ s
Gravitational constant	$= 6.670 \times 10^{-11}$ newton m^2/kg^2
Planck constant	$h = 6.626 \times 10^{-34}$ joules
Boltzmann constant	$k = 1.3805 \times 10^{-23}$ joule/degree
Avogadro's number	$= 6.0226 \times 10^{23}$/mole
Gas constant	$R = 8.314$ joules/mole degree
Volume of gas at STP	$= 22.414$ liters/mole
Absolute zero	0 K $= -273.15°$C
Faraday	$= 96494$ coulombs/gm equivalent
Electronic charge	$= 1.6021 \times 10^{-19}$ coulomb
Rest mass of electron	$= 9.109 \times 10^{-31}$ kg
Proton mass	$= 1.672 \times 10^{-27}$ kg
Standard gravity	$= 9.81274$ m/s$^2 = 32.1937$ ft/s^2
Equatorial gravity	$= 9.78032$ m/s^2
Mass of the Earth	$= 5.983 \times 10^{24}$ kg
Mean density of the Eearth	$= 5.522$ g/cm^3
Equatorial radius	$= 6,378,388$ m $= 3963.34$ miles
Polar radius	$= 6,356,912$ m $= 3949.99$ miles
Flattening	$= 1/298.25$
Surface area of the Earth	$= 5.10 \times 10^{14}$ m^2
Mean height of continents	$= 840$ m
Mean depth of oceans	$= 3800$ m
Mean radius of Earth's core	$= 3.47 \times 10^6$ m
Magnetic field of Earth	$\sim \frac{1}{2} \times 10^{-4}$ tesla
Near-surface temperature gradient	$= 30$ K/km
Mean heat flow at Earth's surface	$= 50$ mW/m^2

Appendix L: Geologic time scale

Era	Period (system)	Epoch (series)	Age (stage)	Beginning of epoch, millions of years B.P.
Phanerozoic Cenozoic	Quaternary	Holocene		0.005
		Pleistocene		2.5
	Tertiary	Pliocene	Upper	
			Lower	7
		Miocene	Upper	
			Middle	
			Lower	26
		Oligocene	Upper	
			Middle	
			Lower	38
		Eocene	Jacksonian	
			Claibornian	
			Wilcoxian	54
		Paleocene	Midwayan	65
Mesozoic	Cretaceous	Late (upper)	Maestrichtian	
			Campanian	
			Santonian	
			Coniacian	
			Turonian	
			Cenomanian	
		Early (lower)	Albian	
			Aptian	
			Barremian	
			Hauterivian	
			Valanginian	
			Berriasian	136
	Jurassic	Late (upper)	Purbeckian	
			Portlandian	
			Kimmeridgian	
			Oxfordian	
		Middle	Callovian	
			Bathonian	
			Bajocian	
		Early (lower)	Toarcian	
			Pliensbachian	
			Sinemurian	
			Hettangian	190

Era	Period (system)	Epoch (series)	Age (stage)	Beginning of epoch, millions of years B.P.
	Triassic	Late (upper)	Rhaetian Norian Carnian	
		Middle	Ladinian Anisian	
		Early (lower)	Scythian	225
Paleozoic	Permian	Ochoaian Guadalupian Leonardian Wolfcampian		280
	Carboniferous	Pennsylvanian[1]	Virgilian Missourian Des Moinesian Atokan Morrowian	325
		Mississippian[1]	Chesterian Meramecan Osagian Kinderhookian	345
	Devonian	Bradfordian Chautauquan Senecan Erian Ulsterian	Fammenian Frasnian Givetian Eifelian Coblenzian Gedinnian	395
	Silurian	Cayugan Niagaran Albion		430
	Ordovician	Cincinnatian Champlainian Canadian	Ashgillian[2] Caradocian[2] Llandellian[2] Skiddavian[2] Tremadocian[2]	500
	Cambrian	Late (upper) Middle Early (lower)		570
Precambrian Proterozoic				2000
Archeozoic				4600

[1] Pennsylvanian and Mississippian are periods rather than epochs. Their subdivisions shown are epochs.

[2] The Ordovician subdivisions are the British names of epochs.

The terms in parentheses refer to the time-rock scale (as compared with the time scale).

Appendix M: Instructions to authors

Revised September, 1984

by the Editors of GEOPHYSICS

EDITORIAL POLICIES

Introduction

Conformity to these instructions is a prerequisite for publication. Manuscripts submitted to GEOPHYSICS in a form other than described in these instructions will be returned to the author

Submit an **original** and **four** copies of each technical paper and short note, complete with **original** and **four** copies of illustrations, list of references, and list of figure captions to the Editor. Regular manuscripts require an abstract, short notes (less than ~4 printed pages) do not.

Submit only an **original** and **one** copy of a discussion.

All communications should be sent to the **Editor** whose address is on the masthead.

All contributions submitted in English are considered, regardless of the status of the author's membership in the Society. Technical contributions are accepted with the understanding that they have not been accepted for publication nor published elsewhere, and are not currently under consideration by another journal nor will be submitted to another journal while under consideration for GEOPHYSICS.

Translations of papers published in or submitted to another journal or printed in a foreign language will not be accepted unless specifically solicited by the Editor. Some translations will, from time to time, be solicited.

Authors will usually be notified of the status of their paper within 8 to 12 weeks.

Type of material

Regular technical papers, short notes, discussions, letters to the Editor, reviews of geophysical or other pertinent literature, and tutorial or review papers are welcome. Case histories are especially solicited. In general, the subject matter should relate to petroleum, mining, geothermal, or engineering geophysics, in land, air, or marine surveys, although the relation need not be direct. For example, any manuscript on fundamental scientific principles basic to geophysical exploration methods is most welcome including papers on exploration of the deep crust and upper mantle. GEOPHYSICS should have broad appeal, ranging from pragmatic field studies to the most sophisticated theory.

Review and editing

With the exception of reviews (which should be sent to the Chairman of the Reviews Committee), all technical manuscripts and short notes are processed by the Editor. Usually, four copies of each submitted manuscript are sent to an appropriate Associate Editor for technical review. The Associate Editor will seek evaluation from three reviewers qualified to judge the technical value of the paper. He will collate the reviews received, synthesize their contents, and make a recommendation to the Editor concerning the acceptability of the manuscript. Most commonly this calls for revision of the manuscript. After submission, the revised manuscript normally is sent only to an Associate Editor for his approval of the revisions relative to the recommendations of the reviewers. The final decision on a paper's disposition is made by the Editor.

Discussions are sent to the senior author of the manuscript under discussion for a reply. Returned replies are sent to the author of the discussion so that he may, if he wishes, revise his discussion. If no reply is received, the discussion will be printed without a reply.

Accepted manuscripts are edited by the Editor and edited and styled for the printer by a copy editor at SEG Headquarters. An effort is made to improve the effectiveness of the communication between the authors and their readers and, in particular, to eliminate ambiguities. When extensive editing is required, with the consequent danger of altered meanings, the edited manuscript is returned to the author for approval before type is set.

Galley proofs are sent to the author, the Editor, and the copy editor for review. **Alterations must be kept to a minimum.**

An order blank for reprints is sent to authors with the galley proofs.

For a discussion of the review and editorial process, see the editorial by S. H. Ward in GEOPHYSICS, **49**, 1137-1138.

WRITING PAPERS

Before beginning to write your paper, organize your material carefully. Include all the data necessary to support your conclusions, but exclude those which are redundant or unnecessary.

In writing your paper, prepare a first draft that includes all the data, arguments, and conclusions which you had planned to cover. Then edit your manuscript carefully. From the reader's point of view, is the text clear, are the figures thoroughly integrated with the text? Go through this process at least twice, having a new draft typed each time.

When you are satisfied, test your success on a colleague, preferably one who is not well acquainted with the subject matter. Be prepared for criticism. If one reader doesn't understand parts of your text, others will have the same problems. Remember, you are thoroughly acquainted with your subject, your reader is not.

For details of style and usage, such as capitalization, punctuation, etc., the University of Chicago's *A Manual of Style* is the official guide for GEOPHYSICS.

The *Encyclopedic Dictionary of Exploration Geophysics* (revised edition), by R. E. Sheriff, is the standard for terms peculiar to geophysical technology.

SEG's manual *The SI Metric System of Units and SEG Tentative Metric Standard* contains the preferred units and abbreviations for units. Other systems of units will be phased out in the near future.

Choose the active voice more often than the passive, for the passive usually requires more words and sometimes obscures the meaning. Use the first person, not the third; do not use first person plural when singular is appropriate.

MANUSCRIPTS

Format

Five copies of each manuscript, complete with illustrations, list of references, and list of figure captions are required. Manuscripts are to be typed **double-spaced on one side only.** Use white, medium-weight stock, 8½ × 11 inch paper. The typewritten line should not exceed 5½ or 6 inches in length, and ample margins should be left at the top and bottom. Do **not** use single-space for footnotes, quoted matter, references, and so on. Do not forget to number the pages.

When an addition is to be made in the text, type the addition and splice it into its proper place. Brief inserts or corrections of a word or two may be written above the line with the position indicated by a caret.

Neatly print by hand in ink all equations except simple ones that a typewriter or word processor can handle. Give particular attention to proper placement of subscripts and superscripts. Equations should be double-spaced and set off from the text.

Illustrations must be submitted in $8\frac{1}{2} \times 11$ inch format. An **original** set (or good-quality photographic prints) of each illustration and **four** extra sets of prints, such as Xerox copies, are required. High-quality computer-drawn illustrations are acceptable if they measure $8\frac{1}{2} \times 11$ inches.

A list of references and a list of figure and table captions are to be typed on separate pages and attached at the end of the manuscript.

Title

Make the title of your paper as brief as possible. The first word should be a significant one, suitable both for classifying and indexing the paper. The right running head (**RRH**) of GEOPHYSICS can accommodate only **38** characters, including spaces. If your title is longer than 38 characters, please provide a 38-character **RRH.**

If your paper was given orally at a meeting, include a footnote specifying the name of the organization that held the meeting, where held, and the date of presentation.

Abstract

Every technical paper must be accompanied by an abstract of approximately 200 to 300 words. It should **emphasize the conclusions** reached in the paper and should not be simply a list, in sentence form, of the topics covered. An abstract is not published with a short note.

Figure captions and text references

Each illustration should be given a figure number (not a plate number) and referred to by that number in the text. Figure numbers are to be arabic numerals.

Each figure must have a caption or title;

include a list of illustrations at the end of the manuscript. Each caption should be explicit enough that the significance of the illustration can be understood by the reader without reference to the text. List the figure number in the form "FIG. 3." **In text references, "Figure" should be spelled out with the first letter capitalized.**

Headings

In the manuscript, principal headings should be typed at the center of the page in capital letters. Headings of the next lower rank should be typed in small letters (first word of heading and proper nouns are capitalized) from the left side, without indentation, and underlined. The following text should begin on the next line. For headings of lower rank, indent, underline, place a period and dash after the heading, and follow with text on the same line. If still lower rank headings are necessary, introduce them by a lower case letter i.e., (a), indent, underline, and follow with a period and dash. Follow with text on the same line.

Preparation of illustrations

All line drawings should be made with black drawing ink on white paper (preferably tracing paper) or mylar. Computer-drawn illustrations are acceptable. Explanatory data, such as contour interval, etc., should be added to the caption. The author's last name should be written in pencil on the margins for identification (or on back). Please indicate your preference of whether an illustration should be set as one column or two columns in width when printed. Please conserve pages or the Editor will not be able to accommodate your preference.

Foldouts or color illustrations are encouraged, but will be billed to the author at cost. Contact the SEG Publications Office for cost estimates.

Lettering should be done with some formal lettering device. Typed lettering is **unacceptable**.

Tables

Each table should be accompanied by a title. It is satisfactory to type the table titles on the same sheet as the figure captions. Tables must be drafted or computer-drawn so that they may be reproduced photographically, like illustrations.

Bibliographic references

All references should be grouped alphabetically by author at the end of the paper under the heading **REFERENCES**. For a given author referenced more than once, use a chronological listing with suffix, a, b, etc., to distinguish references of the same year.

References **not** cited in the text should be grouped separately under the heading **REFERENCES FOR GENERAL READING**.

In the text, literature citations should be referred to with the year of publication in parentheses after the author's name, e.g., Nettleton (1940). If the author is not referred to by name in the text itself, his name and the year should be inserted in parentheses at the point where the reference applies (Nettleton, 1940).

If there is more than one reference to the same author at a given point in the text, list in chronological order with a comma between years. When more than one author is referenced at a given point in the text, separate references by semicolons. If a specific page is referenced, include it in the parenthetical insertion following the year.

In the list of references, the following order and punctuation should be observed for

Papers from journals:

Author(s)   —Last name(s) first, then initial(s).

Comma

Year of publication

Comma

Title   —Capitalize only first words of title and proper nouns (except in German where all nouns are capitalized).

No quotation marks unless they are actually a part of the title.

Colon

Journal name   —Abbreviations follow the American National Standards Institute guides

Comma

Volume number   —Arabic numerals in bold face, i.e., **39**.

Comma

Page number   —Where possible, use beginning and ending pages, using Arabic numerals. Do not use "p." before page numbers.

Period

For reference to books:

Author(s)   —As above.

Comma

Year of Publication

Comma

Title   —Capitalize as above.

Colon

Publisher   —Firm name.

Period

In a reference to a paper within a book, list the Editor before the book title, in the same way the author is listed before the paper title. Between the paper and the book titles, set the word *in* in *italics,* e.g.,

Baker, D. W., and Carter, N. L., 1972, Seismic velocity anisotropy calculated for ultramafic minerals and aggregates, *in* Heard, H. C., Borg, I. V., Carter, N. L., and Raleigh, C. B., Eds., Flow and fracture of rocks: Geophys. Mono. 16, Am. Geophys. Union, 157-166.

EXAMPLES OF STYLE

Examples of style now being used are as follows:

Style of terms

pseudosection	wave field	radio positioning (hyphen when adjective)	wave test
traveltime	wave stack		waveform
		radio location (hyphen when adjective)	wavefront
			waveguide
		baseline	wavelength
		crossover	wavenumber

modeling   waveshape
semiinfinite   wavetilt
half-space   whole space
free-space   dike
wide band   bandwidth
  (hyphen when adjective) Chebychev
Earth (used as a planet)

Style in text

- The abbreviations et al., i.e., and e.g., are set with periods and commas.
- Mathematical symbols will serve as verbs. Equations are punctuated as sentences.
- Do not use italics for foreign and Latin words which have become common in English text. Examples are ''a priori,'' ''et al.''
- Refer to a special phrase/term in quotes the first time it appears **only**. Reserve italics in text for emphasis.
- No hyphens in words formed with prefixes.
  Examples are nonlinear, semimajor axis, antisymmetric.
  Exceptions are words with quasi-, as in quasi-static.
- The text papers with single authors should use the singular pronoun ''I'' rather than ''we.''
- Do not use split infinitives.
- Do not end a sentence with a preposition.
- Do not begin a sentence with a conjunction.
- Do not put a hyphen between an adverb and the word it modifies, e.g., horizontally layered.
- Do not use ''wrt'' in text. Instead, use ''with respect to.''
- In the text, the names of institutions should not be abbreviated, e.g., Geophysical Service Incorporated, not Geophys. Serv. Inc.
- Contractions such as it's, we've, I've, can't, won't, etc., are acceptable if they conserve space.
- If an oral presentation is referenced, the author, year, and title are listed as above. The colon after the title is followed by a reference to the meeting, e.g., 40th Annual International SEG Meeting, followed by the date of the presentation and the city in which the meeting was held (November 10, New Orleans, La.).

Footnotes

Footnotes should be avoided unless essential and, under any conditions, held to a minimum. All footnotes introduced in the text of a paper should be numbered consecutively from the beginning to the end of the manuscript. In the manuscript, footnotes **must** be inserted at the bottom of the page to which they refer.

Style of abbreviations and units

pers. comm. for personal communication
s for seconds
rms for root-mean-square
CDP for common depth point
CMP for common midpoint
$\Omega \cdot$m for ohm-meters
S/m for mho/m or Siemens/m
Hz as unit, hertz as word
m/s for meters per second (not ms^{-1})
10 000 for 10,000
times sign instead of dot for multiplication
space between number and unit
mGal, not mgal
10 m, not 10m
Note: all of the above conform with the SI metric standards.

Mathematical material

One of the most complicated and expensive operations in publishing GEOPHYSICS is typesetting mathematical formulas. You can help in reducing these costs by writing equations in their simplest form. Often a complicated expression can be simplified if various terms are assigned symbols which are defined individually. The paper by Katsube and Collett (1973, 76-105) furnishes some good examples.

In the text, the shilling fraction using the solidus (/) should be used rather than the built-up fraction (two decks).

Fractional exponents should be used instead of radicals wherever feasible. Radicals are preferred, however, for simple square roots, e.g., $\sqrt{2}$ rather than $2^{1/2}$.

Subscripts and superscripts, where there is any doubt as to whether they will be clear to the typesetter, should be indicated by carets and inverted carets, for example:

$$q_{\hat{ij}}; \ p^{\check{2}}$$

It will be helpful to the printer if handwritten Greek letters and other unusual symbols are labeled where they are introduced for the first time. For example, a "kappa" will not be set up as a "k" if there is a marginal notation of "lower case kappa" connected with the character in question by a penciled line. A list of all symbols used in the manuscript on a separate sheet of paper is very helpful.

All equations should be punctuated where their position in the sentence calls for punctuation. An equation should be followed by a comma where good English usage requires a comma at that position in the sentence.

The printer is instructed to set all mathematical symbols and all isolated letters in the text in italic type, if there are no markings to the contrary. Italics are proper in the case of all symbols for scalar quantities. Symbols signifying vector quantities, on the other hand, will be set in roman boldface. A **wavy line** (i.e., tilde) should be drawn under such symbols and "roman boldface" written in the margin at the point where the vectors are first introduced. The use of the bar over the symbol to indicate a vector causes confusion, since the bar is also used to denote averages, for example. A double tilde (e.g., $\underset{\sim}{A}$) should be written under matrices and tensors. Matrices and tensors will be **set** in GEOPHYSICS as **boldface** letters with a tilde $\underset{\sim}{A}$. For those authors possessing boldface word-processing capability, set vectors **A** and matrices and tensors as $\underset{\sim}{A}$.

With the object of standardizing space and time coordinates, we suggest the following. Lower case letters x, y, z should be used for Cartesian space coordinates. Corresponding axes are designated by x-axis, y-axis, and z-axis. The time coordinate is designated by t. For representing traveltime and stepout time of seismic waves, t and δt should be used in preference to T and ΔT. Figures should be lettered consistently with these conventions.

When referring to equations by number in the text, put the numbers in parentheses, e.g., "As shown in equation (10), . . ."

If a paper contains an appendix, any equation in the appendix should be numbered with the prefix A- (such as A-1).

Summations do not always require specific limits.

Appendix N: Publishing houses

Abelard-Schuman, Ltd. (See Thomas Y. Crowell Co.)

Abingdon Press, 201 8th Ave. S., Nashville, TN 37203.

Harry N. Abrams, Inc., 110 E. 59th St., New York, NY 10022.

Academic Press Inc., 111 5th Ave., New York, NY 10003.

Addison-Wesley Pub. Co., Reading, MA 01867.

AHM Publishing Corp., 3110 N. Arlington Heights Rd., Arlington Heights, IL 60004.

Albion Publishing Co., 1736 Stockton St., San Francisco, CA 94133.

Aldine Publishing Co., 200 Saw Mill River Rd., Hawthorne, NY 10532.

Allyn and Bacon, Inc., College Div., Rockleigh, NJ 07647.

American Accounting Assn., 5717 Bessie Drive, Sarasota, FL 33583.

American Bible Society, 1865 Broadway, New York, NY 10023.

American Inst. of Certified Public Accountants, 1211 Ave., of the Americas, New York, NY 10036.

American Law Institute, 4025 Chestnut St., Philadelphia, PA 19104.

American Library Assn., 50 E. Huron St., Chicago, IL 60611.

American Map Co., Inc., 1926 Broadway, New York, NY 10023.

American Psychological Assn., 1200 17th St., N.W., Washington, DC 20036.

American Sciences Press, Inc., P.O. Box 21161, Columbus, Ohio 43221-0161.

American Technical Publishers, Inc., 12235 S. Laramie Ave., Alsip, NY 60658.

American Veterinary Publications, Inc., Drawer KK, Santa Barbara, CA 93102.

Apollo Editions, 10 East 53rd St., New York, NY 10022.

Architectural Book Pub. Co., Inc., 10 E. 40th St., New York, NY 10016.

Arco Pub. Co., Inc., 219 Park Ave. S., New York, NY 10003.

Association Press. (See Follett Publishing Co.)

Atheneum Publishers, 597 Fifth Ave., New York, NY 10017.

Augsburg Publ. House, 426 S. 5th St., Box 1209, Minneapolis, MN 55440.

The AVI Pub. Co., 250 Post Rd., E., P.O. Box 831, Westport, CT 06881.

Avon Books (Div. of The Hearst Corp.)., 959 Eighth Ave., New York, NY 10019.

Baker Book House, P.O. Box 6287, Grand Rapids, MI 49506.

Ballantine Books, Inc., 201 E. 50th St., New York, NY 10022.

Ballinger Publishing Co., 17 Dunster St., Cambridge, MA 02138.

Banks-Baldwin Law Pub. Co., University Center, P.O. Box 1974, Cleveland, OH 44106.

Bantam Books, Inc., 666 Fifth Ave., New York, NY 10019.

A.S. Barnes & Co., 11175 Flintkote Ave., San Diego, CA 92121.

Barnes & Noble Publications (See Harper & Row).

Barron's Educational Series, Inc., 113 Crossways Park Dr., Woodbury, NY 11797.

Basic Books, Inc., 10 E. 53rd St., New York, NY 10022.

The Beacon Press, 25 Beacon St., Boston, MA 02108.

Matthew Bender & Co., Inc., 235 E. 45th St., New York, NY 10017.

Benjamin/Cummings Publishing Co., Inc., 2725 Sandhill Rd., Menlo Park, CA 94025.

Bennett Publishing Co., 809 W. Detweiller Dr., Peoria, IL 61615.

Berkley/Jove Publishing Corp., 200 Madison Ave., New York, NY 10016.

The Bethany Press, 2640 Pine Blvd., PO Box 179, St. Louis, MO 63166.

Binford & Mort Publishers, 2536 S.E. 11th Ave., Portland, OR 97202.

The Bobbs-Merrill Co., 4300 W. 62nd St., Indianapolis, IN 46268.

Boosey & Hawkes, Inc., P.O. Box 130, Oceanside, NY 11572.

Boston Music Co., 116 Boylston St., Boston, MA 02116.

R.R. Bowker Co., 1180 Ave. of the Americas, New York, NY 10036.

Boyd & Fraser Publishing Co., 3627 Sacramento St., San Francisco, CA 94118.

Charles T. Branford Co., Box 41, Newton Centre, MA 02159.

George Braziller, Inc., 1 Park Ave., New York, NY 10016.

Brigham Young Univ. Press, 206 Univ. Press Bldg., Provo, UT 84602.

Broadman Press, 127 9th Ave., N., Nashville, TN 37234.

The Brookings Institution, 1775 Mass. Ave. N.W., Washington, DC 20036.

Brooks/Cole Publishing Co., (See Wadsworth Publishing Co., Inc.)

Wm. C. Brown Company, 2460 Kerper Blvd., Dubuque, IA 52001.

Brown Univ. Press, (See University Press of New England).

Burgess Publ. Co., 7108 Ohms Ln., Minneapolis, MN 55435.

Business Publications, Inc. (See Richard D. Irwin).

Callaghan & Co., 3201 Old Glenview Rd., Wilmette, Il 60091.

Cambridge Univ. Press, 32 E. 57th St., New York, NY 10022.

The Catholic Univ. of America Press, Inc., 620

Michigan Ave. N.E., Washington, DC 20064.

The Caxton Printers, Ltd., 308-320 Main St., Caldwell, ID 83605.

Chandler Publishing Co., (See Harper & Row Publishers, Inc.)

Chelsea Publishing Co., 432 Park Ave. S., New York, NY 10016.

Chemical Publishing Co., 155 W. 19th St., New York, NY 10011.

Chilton Book Co., Radnor, PA 19089.

Christopher Publishing House, 1405 Hanover St., West Hanover, MA 02339.

The Citadel Press, 120 Enterprise Ave., Secaucus, NJ 07094.

Clark Boardman Co., Ltd., 435 Hudson St., New York, NY 10014.

Cliffs Notes, 1701 "P" St. P.O. Box 80728, Lincoln, NE 68501.

Collier Books (See The Macmillan Co.).

William Collins Publishers, Inc., 2080 W. 117th St., Cleveland, OH 44111.

Columbia Univ. Press, 562 W. 113th St., New York, NY 10025.

Commerce Clearing House, Inc., 4025 W. Peterson Ave., Chicago, IL 60646.

Comm. For Economic Development, 477 Madison Ave., New York, NY 10022.

CompuSoft Publishing, 1050 Pioneer Way, Ste. E, El Cajon, CA 92020.

Concordia Publishing House, 3558 S. Jefferson Ave., St. Louis, MO 63118.

Congressional Quarterly, Inc., 1414 22nd St., N.W., Washington, DC 20037.

Contemporary Books, Inc. (formerly Henry Regnery Co.), 180 N. Michigan Ave., Chicago, IL 60601.

Cornell Maritime Press Inc., P.O. Box 456, Centreville, MD 21617.

Cornell University Press (Also Comstock Pub. Assoc.), Box 250 124 Roberts Pl., Ithaca, NY 14850.

George F. Cram Co., Inc., 301 S. LaSalle St., P.O. Box 426, Indianapolis, IN 46206.

Criterion Books (See Harper & Row).

Thomas Y. Crowell Co., 10 East 53rd St., New York, NY 10022.

Crown Publishers, Inc., 1 Park Ave., New York, NY 10016.

Dance Horizons, Inc., 1801 E. 26th St., Brooklyn, NY 11229.

Data-Guide, Inc., 154-01 Barclay Ave., Flushing, NY 11355.

Davis Publications, Inc., 50 Portland St. Worcester, MA 01608.

F.A. Davis Co., 1915 Arch St., Philadelphia, PA 19103.

Delacorte Press, 1 Dag Hammerskjold Plaza, 245 E. 47th St., New York, NY 10017.

Dell Pub. Co., Inc., 1 Dag Hammarskjold Plaza, 245 E. 47th St., New York, NY 10017.

Delmar Publishers, Inc., 50 Wolf Rd., Albany, NY 12205.

Dennis & Co., Inc., 251 Main St., Buffalo, NY 14203.

Denoyer-Geppert Co., 5235 Ravenswood Ave., Chicago, IL 60640.

The Devin-Adair Co., 143 Sound Beach Ave., Old Greenwich, CT 06870.

DeVorss & Co., Inc., P.O. Box 550, Marina Del Rey, CA 90291.

The Dial Press, 1 Dag Hammarskjold Plaza, 245 E. 47th St., New York, NY 10017.

D.C. Divry, Inc., 293 Seventh Ave., New York, NY 10001.

Dodd, Mead & Co., Inc., 79 Madison Ave., New York, NY 10016.

Dorrance & Co., Cricket Terrace Center, Ardmore, PA 19003.

Dorsey Press. (See Richard D. Irwin).

Doubleday & Co., Inc., 501 Franklin Ave., Garden City, NY 11530.

Dover Publications, Inc., 180 Varick St., New York, NY 10014.

Dryden Press, 901 N. Elm, Hinsdale, IL 60521.

Duke University Press, College Station, Box 6697, Durham, NC 27708.

Dushkin Publishing Group, Inc., Sluice Dock, Guilford, CT 06437.

E.P. Dutton & Co., Inc., 2 Park Ave. New York, NY 10016.

Duxbury Press (See Wadsworth Inc.).

Wm. B. Eerdmans Publishing Co., 255 Jefferson Ave., S.E., Grand Rapids, MI 49503.

Eliot Books, Inc., 34 Exchange Place, Jersey City, NJ 07302.

Elot Publishing Co., Inc., P.O. Box 8218, Long Beach, CA 90808.

Encyclopaedia Britannica, Inc., 425 N. Mich. Ave., Chicago, IL 60611.

Engineering Press, Inc., P.O. Box 1, San Jose, CA 95103.

Engineering Technology, Inc., 503 E. Main St., Mohomet, IL 61853.

Expression Co., Publishers, P.O. Box 153, Londonderry, NH 03053.

Fairchild Book & Visuals, 7 E. 12th St., New York, NY 10003.

Family Serv. Assoc. of Amer., 44 E. 23rd, New York, NY 10010.

Farrar, Straus & Giroux, Inc., 19 Union Sq. W., New York, NY 10003.

Fawcett Books Group, CBS Publications, 1515 Broadway, New York, NY 10036.

Fearon-Pitman Publishers, Inc. (See Pitman Learning).

Frederick Fell, Inc., 386 Park Ave. S., New York, NY 10016.

Field Museum of Natural History, Roosevelt Rd. & Lake Shore Dr., Chicago, IL 60605.

Follett Pub. Co., 1010 W. Washington Blvd., Chicago, IL 60607.

Fordham University Press, Univ. Box L, Bronx, NY 10458.

Foreign Policy Ass'n, Inc., 205 Lexington Ave., New York, NY 10016.

Forest Press, Inc., 85 Watervliet Ave., Albany, NY 12206.

Fortress Press, 2900 Queen Lane, Philadelphia, PA 19129.

Four Continent Book Corporation, 149 Fifth Ave., New York, NY 10010.

Free Press (See The Macmillan Co.).
W.H. Freeman & Co., 660 Market St., San Francisco, CA 94104.
French & Spanish Book Corp., 115 Fifth Ave., New York, NY 10003.
French Book Guild, 11-03 46th Ave., Long Island City, NY 11101.
Samuel French, Inc., 25 W. 45th St., New York, NY 10036.
Friendship Press, 475 Riverside Dr., New York, NY 10015.
Funk & Wagnalls, 10 East 53rd St., New York, NY 10022.

The Geological Society of America, Inc., 3300 Penrose Pl., P.O. Box 9140 Boulder, CO 80301.
Glencoe Publishing Co., Inc., 17337 Ventura Blvd., Encino, CA 91316.
Ginn Custom Publishing, 191 Spring St., Lexington, MA 02173.
Golden Press (See Western Publishing Co.).
Goodheart-Willcox, 123 W. Taft Dr., S. Holland, IL 60473.
Goodyear Publishing Co., 1640 Fifth St., Santa Monica, CA 90401.
Gordon & Breach, Science Publishers, Inc., 1 Park Ave., New York, NY 10016.
Warren H. Green, Inc., 8356 Olive Blvd., St. Louis, MO 63132.
Gregg Div., McGraw-Hill Book Co., 1221 Ave. of The Americas, New York, NY 10020.
Grid Publishing, Inc., 2950 N. High St., Columbus, OH 43202.
Grosset & Dunlap, Inc., 51 Madison Ave., New York, NY 10010.
Grove Press, Inc., 196 W. Houston St., New York, NY 10014.
Grune & Stratton, Inc., 111 Fifth Ave., New York, NY 10003.

Hafner Press (See Macmillan Publishing Co., Inc.).
Hammond, Inc., 515 Valley St., Maplewood, NJ 07040.
Harcourt Brace Jovanovich, Inc., 757 Third Ave., New York, NY 10017.
Harper & Row Publishers, Inc., 10 E. 53rd St., New York, NY 10022.
Harvard University Press, 79 Garden St., Cambridge, MA 02138.
Hastings House, Publishers, Inc., 10 E. 40th St., New York, NY 10016.
Hawthorn Books, Inc. (See E.P. Dutton).
Hayden Book Co., Inc., 50 Essex St., Rochelle Park, NJ 07662.
D.C. Heath and Co., 125 Spring St., Lexington, MA 02173.
Hebrew Pub. Co., 80 Fifth Ave., New York, NY 10011.
W.S. Heinman, Imported Books, 1966 Broadway, New York, NY 10023.
Herald Press, 616 Walnut Ave., Scottdale, PA 15683.
Hill & Wang, Inc. (See Farrar, Straus & Giroux.).
Holden-Day, Inc., 500 Sansome St., San Francisco, CA 94111.
A.J. Holman Co., Box 956, E. Washington Sq., Philadelphia, PA 19107.

Holt, Rinehart & Winston, 383 Madison Ave., New York, NY 10017.
Horizon Press, Inc., 156 Fifth Ave., New York, NY 10010.
Thomas Horton and Daughters, 26662 S. Newtown Dr., Sun Lakes, AZ 85224.
Houghton Mifflin Co., 1 Beacon St., Boston, MA 02107.
Humanities Press, Inc., 171 First Ave., Atlantic Highlands, NJ 07716.
Hunter Publishing Co., P.O. Box 5867, 2505 Empire Dr., Winston-Salem, NC 27103.

Indiana University Press, 10th & Morton St., Bloomington, In 47401.
The Industrial Press, 200 Madison Ave., New York, NY 10157.
Institute of Modern Languages, Inc., 2622 Pittman Dr., Silver Springs, MD 20910.
The Instructor Publications, Inc., Instructor Park, Dansville, NY 14437.
International City Management Ass'n, 1140 Conn. Ave., N.W., Washington, DC 20036.
International Music Co., 545 Fifth Ave., New York, NY 10017.
International Publishers Co., Inc., 381 Park Ave. S., New York, NY 10016.
International Universities Press, 315 Fifth Ave., New York, NY 10016.
The Interstate Printers and Publishers, Inc., 19-27 N. Jackson St., Danville, IL 61832.
Iowa State University Press, S. State Ave., Ames, IA 50010.
Irvington Publishers, Inc., 551 Fifth Ave., New York, NY 10017.
Richard D. Irwin, Inc., 1818 Ridge Rd., Homewood, IL 60430.

The Johns Hopkins University Press, Baltimore, MD 21218.
Johnson Reprint Corp., 111 Fifth Ave., New York, NY 10003.
Marshall Jones Co., Francestown, NH 03043.
Judson Press, Valley Forge, PA 19481.

Kendall/Hunt Publishing Co., 2460 Kerper Blvd., Dubuque, IA 52001.
P.J. Kenedy & Sons (See Macmillan Publishing Co.).
Kent State Univ. Press, Kent, OH 44242.
Alfred A. Knopf, Inc. (See Random House).
John Knox Press, 341 Ponce de Leon Ave., N.E., Atlanta, GA 30365.

Lane Publishing Co., Menlo Park, CA 94025.
Lange Medical Publications, Drawer "L", Los Altos, CA 94022.
Lansford Publishing Co., P.O. Box 8711, San Jose, CA 95155.
Larousse & Co., Inc., 572 Fifth Ave., New York, NY 10036.
Lea & Febiger, 600 S. Washington Sq., Philadelphia, PA 19106.
Leswing Press, P.O. Box 3577, San Rafael, CA 94912.
Lingual House, P.O. Box 3537, Tucson, AZ 85722.
Lippincott/Harper Health Professions Publications,

E. Washington Sq., Philadelphia, PA 19105.

Little, Brown & Co., Inc., 34 Beacon St., Boston, MA 02106.

Littlefield, Adams & Co., 81 Adams Dr., Totowa, NJ 07512.

Liveright Pub. Corp., 500 Fifth Ave., New York, NY 10036.

Louisiana State University Press, Louisiana State Univ., Baton Rouge, LA 70803.

Loyola University Press, 3441 N. Ashland Ave., Chicago, IL 60657.

Mack Pub. Co., 20th & Northampton St., Easton, PA 18042.

The Macmillan Publishing Co., Inc., 866 Third Ave., New York, NY 10022.

Mayfield Publishing Co., 285 Hamilton Ave., Palo Alto, CA 94301.

McCormick-Mathers Pub. Co., Inc., Div. of Litton Ind., 7625 Empire Dr., Florence, KY 41042.

McGraw-Hill Book Co., (Div. of McGraw-Hill, Inc.), 1221 Ave. of the Americas, New York, NY 10020.

David McKay Co., Inc., 2 Park Ave., New York, NY 10016.

McKnight Pub. Co., P.O. Box 2854, Bloomington, IL 61701.

Merck & Co., Inc., Rahway, NJ 07065.

G. & C. Merriam Co., 47 Federal St., Springfield, MA 01101.

Chas. E. Merrill Pub. Co., 1300 Alum Creek Dr., Columbus, OH 43216.

Michigan State University Press, 1405 S. Harrison Rd., 25 Manly Miles Bldg., E. Lansing, MI 48824.

M.I.T. Press, 28 Carleton St., Cambridge, MA 02142.

Modern Language Ass'n., 62 Fifth Ave., New York, NY 10011.

Modern Library, Inc. (See Random House).

Monarch Press (A Simon & Schuster Div. of Gulf & Western Corp.), Simon & Schuster Bldg., 1230 Ave. of the Americas, New York, NY 10020.

Moody Press, 2101 West Howard, Chicago, IL 60645.

William Morrow & Co., 105 Madison Ave., New York, NY 10016.

The C.V. Mosby Co., 11830 Westline Industrial Dr., St. Louis, MO 63141.

National Council of Teachers of English, 1111 Kenyon Rd., Urbana, IL 61801.

National Education Ass'n., 1201 16th St., N.W., Washington, DC 20036.

National Geographic Society, Washington, DC 20036.

National League for Nursing, Inc., 10 Columbus Circle, New York, NY 10019.

National Retail Merchants Ass'n., 100 W. 31st St., New York, NY 10001.

National Textbook Co., 8259 Niles Center Rd., Skokie, IL 60076.

Thomas Nelson Publishers, 407 Seventh Ave. S., Nashville, TN 37203.

The New American Library, Inc., 1633 Broadway, New York, NY 10019.

New Directions Publishing Corp., 80 Eighth Ave., New York, NY 10011.

New York Graphic Society, 41 Mt. Vernon St., Boston, MA 02106.

New York University Press, Washington Sq., New York, NY 10003.

Newbury House, Publishers, 54 Warehouse Ln., Rowley, MA 01969.

Nitty Gritty Productions, P.O. Box 5457, Concord, CA 94520.

Noonday Press, Inc. (See Farrar, Straus & Giroux).

Northern Illinois Univ. Press, DeKalb, IL 60115.

Northwestern University Press, 1735 Benson, Evanston, IL 60201.

W.W. Norton & Co., Inc., 500 Fifth Ave., New York, NY 10036.

Oceana Publications, Inc., Dobbs Ferry, NY 10522.

The Odyssey Press (See Bobbs-Merrill Co.).

Ohio State University Press, 2070 Neil Ave., Columbus, OH 43210.

Ohio Univ. Press-Swallow Press, Athens, OH 45701.

Open Court Pub. Co., P.O. Box 599, LaSalle, IL 61301.

Oxford Book Co., Inc., 11 Park Pl., New York, NY 10007.

Oxford University Press, Inc., 16-00 Pollitt Dr., Fair Lawn, NJ 07410.

Paulist Press, 545 Island Rd., Ramsey, NJ 07446.

F.E. Peacock Publishers, Inc., 115 N. Prospect, Itasca, IL 60143.

Pegasus (See Bobbs-Merrill Co.).

Penguin Books, 625 Madison Ave., New York, NY 10022.

Pergamon Press, Inc., Maxwell House, Fairview Park, Elmsford, NY 10523.

Peter Pauper Press, 135 W. 50 St., New York, NY 10020.

Petersen Publishing Co., 6725 Sunset Blvd., Los Angeles, CA 90028.

Philosophical Library, 200 W. 57th St., New York, NY 10019.

Pocket Books, Simon & Schuster Bldg., 1230 Ave. of The Americas, New York, NY 10020.

Praeger Publishers, 383 Madison Ave., New York, NY 10017.

Prentice-Hall, Inc., Englewood Cliffs, NJ 07632.

Princeton University Press, Princeton, NJ 08540.

Prindle, Webber & Schmidt, 20 Providence St., Statler Office Bldg., Boston, MA 02116.

Pruett Publishing Co., 3235 Prairie Ave., Boulder, CO 80302.

The Psychological Corp., 757 Third Ave., New York, NY 10017.

Public Affairs Pamphlets, 381 Park Ave. S., New York, NY 10016.

Public Affairs Press, 419 New Jersey Ave., S.E., Washington, DC 20003.

G.P. Putnam's Sons, 200 Madison Ave., New York, NY 10016.

Quandrangle (See Harper & Row).

Quick Fox, Inc., 33 W. 60th St., New York, NY 10023.

Rand McNally & Co., P.O. Box 7600, Chicago, IL 60680.

Random House, Inc., 201 E. 50th St., New York, NY 10022.

Regents Press of Kansas, 303 Carruth-O'Leary, Lawrence, KS 66045.

Fleming H. Revell Co., Old Tappan, NJ 07675.

Rizzoli International Publications, Inc., 712 Fifth Ave., New York, NY 10019.

The Ronald Press Co. (See John Wiley & Sons.).

The H.M. Rowe Co., 624 N. Gilmore St., Baltimore, MD 21217.

Rutgers Univ. Press, 30 College Ave., New Brunswick, NJ 08903.

St. Martin's Press, Inc., 175 Fifth Ave., New York, NY 10010.

Albert Saifer: Publisher, P.O. Box 239 W.O.B., W. Orange, NJ 07052.

Howard W. Sams & Co., Inc. (See Bobbs-Merrill Co.).

W.B. Saunders Co. W. Washington Sq., Philadelphia, PA 19105.

Schaum Outline Series (See McGraw-Hill Book Co.).

Schenkman Publishing Co., Inc., 3 Mt. Auburn Pl., Cambridge, MA 02138.

Schirmer Music NY, 40 W. 62nd St., New York, NY 10023.

Schocken Books, Inc., 200 Madison Ave., New York, NY 10016.

Schoenhof's Foreign Books, Inc., 1280 Mass. Ave., Cambridge, MA 02138.

Science Research Associates, Inc., College Division, 1540 Page Mill Rd., Palo Alto, CA 94304.

Scott, Foresman & Co., 1900 E. Lake Ave., Glenview, IL 60025.

Scribner Book Companies, Inc., 597 Fifth Ave., New York, NY 10017.

Seabury Press, Inc., 815 Second Ave., New York, NY 10017.

Sheed Andrews & McMeel, Inc., (See Andrews & McMeel, Inc.).

Sheridan House, Inc., 175 Orawaupum St., White Plains, NY 10606.

Silver Burdett Co., 250 James St., Morristown, NJ 07960.

Simmons-Boardman Pub. Corp., 1809 Capitol Ave., Omaha, NE 68102.

Simon & Schuster, 1230 Ave. of the Americas, New York, NY 10020.

Sinauer Associates, Inc., Publishers, Sunderland, MA 01375.

Peter Smith Publisher, Inc., 6 Lexington Ave., Gloucester, MA 01931.

Southern Illinois Univ. Press, P.O. Box 3697, Carbondale, IL 62901.

Southern Methodist University Press, Southern Methodist University, Dallas TX 75275.

South-Western Pub. Co., 5101 Madison Rd., Cincinnati, OH 45227.

Sportshelf & Soccer Assoc., P.O. Box 634, New Rochelle, NY 10802.

Springer Pub. Co., Inc., 200 Park Ave. S., New York, NY 10003.

Springer-Verlag New York, Inc., 175 Fifth Ave., New York, NY 10010.

Stackpole Books, Cameron & Kelker Sts., P.O. Box 1831, Harrisburg, PA 17105.

Stanford University Press, Stanford, CA 94305.

State Univ. of New York Press, State University Plaza, Albany, NY 12246.

Sterling Pub. Co., Inc., 2 Park Ave., New York, NY 10016.

Sunburst Books (See Farrar, Straus and Giroux, Inc.).

The Superintendent of Documents, U.S. Government Printing Office, Washington, DC 20402.

Swallow Press, (See Ohio University Press).

Syracuse University Press, 1011 E. Water St., Syracuse, NY 13210.

Taplinger Publishing Co., Inc., 132 W. 22nd St., New York, NY 10011.

Teachers College Press, Teachers College, Columbia Univ., 1234 Amsterdam Ave., New York, NY 10027.

Texas Christian University Press, Box 30783, Ft. Worth, TX 76129.

Theatre Arts Books, 153 Waverly Place, New York, NY 10014.

Paul Theobald & Co., 5 N. Wabash Ave., Chicago, IL 60602.

Charles C. Thomas, Publisher, 301-327 E. Lawrence Ave., Springfield, IL 62717.

Thomas Law Book Co., 1909 Washington Ave., St. Louis, MO 63103.

Time-Life Books, 777 Duke St., Alexandria, VA 22314.

Tudor Publishing Co. (See Amiel Book Distributors Corp.).

Charles E. Tuttle Co., Inc., 29 S. Main St., Rutland, VT 05701.

Frederick Ungar Pub. Co., Inc. 250 Park Ave. S., New York, NY 10003.

Union of American Hebrew Congregations, 838 Fifth Ave., New York, NY 10021.

United Church Press (See Seabury Press)

United Nations, New York, NY 10017.

U.S. Naval Institute, Annapolis, MD 21402.

Universe Books, 381 Park Ave. S., New York, NY 10016.

University Books, Inc., 120 Enterprise Ave., Secaucus, NJ 07094.

Univ. of Alabama Press, P.O. Box 2877, University, AL 35486.

Univ. of Arizona Press, P.O. Box 3398, Tucson, AZ 85722.

Univ. of British Columbia Press, 303-6344 Memorial Rd., Univ. of British Columbia, Vancouver, BC V6T 1W5.

Univ. of California Press, 2223 Fulton St., Berkeley, CA 94720.

Univ. of Chicago Press, 5801 Ellis Ave., Chicago, IL 60637.

Univ. of Delaware Press, 4 Cornwall Dr., East Brunswick, NY 08816.

Univ. of Georgia Press, U. of Ga., Athens, GA 30602.

Univ. of Illinois Press, Box 5081, Station A, Champaign, IL 61820.

Univ. of Iowa Press, Graphic Services Bldg., Iowa City, IA 52242.

Univ. of Massachusetts Press, Box 429, Amherst, MA 01002.

Univ. of Miami Press, P.O. Box 4836, Hampden Station, Baltimore, MD 21211.

Univ. of Michigan Press, P.O. Box 1104, Ann Arbor, MI 48106.

Univ. of Minnesota Press, 2037 University Ave. S.E., Minneapolis, MN 55455.

Univ. of Missouri Press, P.O. Box 7088, Columbia, MO 65205.

Univ. of Nebraska Press, 901 N. 17th St., Lincoln, NE 68588.

The Univ. of New Mexico Press, Albuquerque, NM 87131.

Univ. of North Carolina Press, Box 2288, Chapel Hill, NC 27514.

Univ. of Notre Dame Press, Notre Dame, IN 46556.

Univ. of Oklahoma Press, 1005 Asp Ave., Norman, OK 73019.

Univ. of Pennsylvania Press, 3933 Walnut St., Philadelphia, Pa 19104.

Univ. of Pittsburgh Press, 127 N. Bellefield Ave., Pittsburgh, PA 15260.

Univ. of South Carolina Press, Columbia, SC 29208.

Univ. of Tennessee Press, 293 Communications Bldg., Knoxville, TN 37916.

Univ. of Texas Press, P.O. Box 7819, Austin, TX 78712.

Univ. of Toronto Press, Front Campus, Univ. of Toronto, Toronto, Ont., M5S 1A6.

Univ. of Utah Press, 101 USB, Salt Lake City, UT 84112.

Univ. of Washington Press, Seattle, WA 98195.

Univ. of Wisconsin Press, 114 N. Murray St., Madison, WI 53715.

Univ. Park Press, 300 N. Charles St., Baltimore, MD 21201.

Univ. Press of Hawaii, 2840 Kolowalu St., Honolulu, HI 96822.

Univ. Press of Kentucky, Lexington, KY 40506.

Univ. Press of New England, Box 979, 3 Lebanon St., Hanover, NH 03755.

Univ. Press of Virginia, Box 3608 University Sta., Charlottesville, VA 22903.

Univ. Press of Washington, D.C., 212 Univ. Press Bldg., Dellbrook Campus, C.A.S., Riverton, VA 22651.

Univ. Presses of Florida, 15 N.W. 15th St., Gainesville, FL 32603.

Urban Institute Press, 2100 M St., N.W., Washington, DC 20037.

D. Van Nostrand Co., 135 W. 50th St., New York, NY 10020.

Vanderbilt University Press, Nashville, TN 37203.

Vanguard Press, 424 Madison Ave., New York, NY 10017.

S.F. Vanni, 30 W. 12th St., New York, NY 10011.

Vantage Press, Inc., 516 W. 34th St., New York, NY 10001.

The Viking Press, Inc., 625 Madison Ave., New York, NY 10022.

Wadsworth, Inc., Belmont, CA 94002.

Frank R. Walker Co., 5030 N. Harlem Ave., Chicago, IL 60656.

Frederick Warne & Co., Inc., 2 Park Ave., New York, NY 10016.

Watson-Guptill Publications, 1515 Broadway, New York, NY 10036.

Franklin Watts, Inc., 730 Fifth Ave., New York, NY 10019.

Wayne State University Press, 5959 Woodward Ave., Detroit, MI 48202.

Wesleyan Univ. Press, 136 S. Broadway, Irvington, NY 10533.

Western Pub. Co., Inc., 1220 Mound Ave., Racine, WI 53404.

The Westminster Press, 925 Chestnut St., Phila., PA 19107.

John Wiley & Sons, Inc., 605 Third Ave., New York, NY 10016.

The Williams & Wilkins Co., 428 E. Preston St., Baltimore, MD 21202.

Willis Music Co., 7380 Industrial Rd., Florence, KY 41042.

Wittenborn Art Books, Inc., 1018 Madison Ave., New York, NY 10021.

World Book Co. (See Harcourt Brace Jovanovich).

Worth Publishers, Inc., 444 Park Ave. S., New York, NY 10016.

The Writer, Inc., 8 Arlington St., Boston, MA 02116.

Writer's Digest Books, 9933 Alliance Rd., Cincinnati, OH 45242.

Xerox Individualized Publishing (See Ginn Custom Publishing).

Yale University Press, 92 A Yale Station, New Haven, CT 06520.

Year Book Medical Pub., Inc., 35 E. Wacker Dr., Chicago, IL 60601.

Zaner-Bloser Co., 612 N. Park St., Columbus, OH 43215.

Zondervan Pub. House, 1415 Lake Dr. S.E., Grand Rapids, MI 49506.

References

Abramowitz, M., and Stegun, I. A., 1972, Handbook of mathematical functions: Dover Publications, Inc.

Adachi, R., 1954, On the proof of fundamental formula concerning refraction method of geophysical prospecting and some remarks: Kumamoto J. Sci., ser. A, 2, 18-23.

Aki, K., and Richards, P. G., 1980, Quantitative seismology: Theory and methods: W. H. Freeman and Co.

Anstey, N. A., 1964, Correlation techniques: A review: Geophys. Prosp., 12, 355-382.

Backus, M. M., 1959, Water reverberations: Their nature and elimination: Geophysics, 24, 233-261.

Balch, A. H., and Smolka, F. R., 1970, Plane and spherical Voigt waves: Geophysics, 35, 745-761.

Barry, K. M., 1967, Delay time and its application to refraction profile interpretation, in Seismic refraction prospecting: Musgrave, A. W., Ed., Soc. of Expl. Geophys., 348-361.

Barthelmes, A. J., 1946, Application of continuous profiling to refraction shooting: Geophysics, 11, 24-42.

Bath, M., 1966, Earthquake energy and magnitude: Phys. Chem. Earth, 7, 115-165.

Baumgarte, J. von, 1955, Konstruktive Darstellung von seismischen Horizonten: Geophys. Prosp., 3, 126-162.

Bayless, J. W., and Brigham, E. O., 1970, Application of the Kalman filter: Geophysics, 35, 2-23.

Beyer, J. H., 1977, Telluric and dc resistivity techniques applied to the geophysical investigation of Basin and Range geothermal systems, Part 1: The E-field radio telluric method: Ph.D. thesis, Univ. of California, Berkeley; Lawrence Berkeley Lab rep. LBL-6325 1/3.

Bureau of Standards, 1974, The international system of units (SI): Spec. pub. 330, catalog no. C13.12;330/3, U.S. Government Printing Office.

Clarke, 1974, Josephson junction detectors: Science, 184, 1235-1242.

Clay, C. S., and Medwin, H., 1977, Acoustical oceanography: John Wiley & Sons, Inc.

Cooley, J. W., and Tukey, J. W., 1965, An algorithm for the machine calculation of complex Fourier series: Math. of Comput., 19, 297-301.

Diebold, J. B., and Stoffa, P. L., 1981, The traveltime equation, tau-p mapping, and inversion of common-midpoint data: Geophysics, 46, 238-254.

Dix, C. H., 1955, Seismic velocities from surface measurements: Geophysics, 20, 68-86.

————— 1981, Seismic prospecting for oil: Revised ed., Int. Human Resources Dev. Corp.

Dobrin, M. B., Ingalls, A. L., and Long, J. A., 1965, Velocity and frequency filtering of seismic data using laser light: Geophysics, 30, 1144-1178.

Edwards, R. N., Lee, H., and Nabighian, M. N., 1978, On the theory of magnetometric resistivity methods: Geophysics, 43, 1176-1203.

Evenden, B. S., Stone, D. R., and Anstey, N. A., 1971, Seismic prospecting instruments, v. 1: Gebruder-Borntraeger.

Ewing, M., Worzel, J., and Pekeris, C. L., 1948, Geol. Soc. of Am. Memoir 27.

Faust, L. Y., 1951, Seismic velocity as a function of depth and geologic time: Geophysics, 16, 192-206.

————— 1953, A velocity function including lithologic variation: Geophysics, 18, 271-288.

Flinn, E. A., Robinson, E. A., and Treitel, S., 1967, MIT Geophysical Analysis Group reports: Spec. issue, Geophysics, 32, 411-521.

Fuller, B. D., 1967, Two-dimensional frequency analysis and design of grid operators, in Mining geophysics, v. 2: Soc. of Expl. Geophys., 658-708.

Gamble, T. D., Goubau, W. M., and Clarke, J., 1979, Error analysis for remote reference magnetotellurics: Geophysics, 44, 959-968.

Gardner, G. H. F., Gardner, L. W., and Gregory, A. R., 1974, Formation velocity and density—the diagnostic basics for stratigraphic traps: Geophysics, 39, 770-780.

Gardner, L. W., 1939, An areal plan of mapping subsurface structure by refraction shooting: Geophysics, 4, 247-259.

————— 1949, Seismograph determination of saltdome boundary using well detector deep on dome flank: Geophysics, 14, 29-38.

Garland, G. D., 1979, Introduction to geophysics: W. B. Saunders Co.

Gassmann, F., 1951, Elastic waves through a packing of spheres: Geophysics, 16, 673-685.

Geertsma, J., 1961, Velocity-log interpretation: The effect of rock bulk compressibility: Soc. Petrol. Eng. J., 1, 235-248.

Glicken, M., 1962, Eötvös corrections for a moving gravity meter: Geophysics, 27, 531-533.

Goupillaud, P. L, 1961, An approach to inverse filtering of near-surface effects from seismic records: Geophysics, 26, 754-760.

Grossling, B., 1969, Color mimicry in geology and geophysics: Geophysics, 34, 249-254.

Gutenberg, B., 1959, Physics of the earth's interior: Academic Press Inc.

Hagedoorn, J. G., 1954, A process of seismic reflection interpretation: Geophys. Prosp., 2, 85-127.

————— 1959, The plus-minus method of interpreting seismic refraction sections: Geophys. Prosp., 7, 158-182.

Hales, F. W., 1958, An accurate graphical method for interpreting seismic refraction lines: Geophys. Prosp., **6**, 285-314.

Hammer, S., 1940, Terrain corrections for gravimeter stations: Geophysics, **4**, 184-194.

Hartman, R. R., Teskey, D. J., and Friedberg, J. L., 1971, A system for rapid digital aeromagnetic interpretation: Geophysics, **36**, 891-918.

Haskell, N. A., 1953, The dispersion of surface waves on multi-layered media: Bull., Seismol. Soc. Am. **43**, 17.

Heiskanen, W., and Meinesz, V., 1958, The earth and its gravity field: McGraw-Hill Book Co.

Hood, P., 1964, The Koenigsberger ratio and the dipping dike equation: Geophys. Prosp., **12**, 440-456.

Hoover, G. M., 1972, Acoustical holography using digital processing: Geophysics, **37**, 1-19.

Howell, B. F., 1959, Introduction to geophysics: McGraw-Hill Book Co.

Isaacs, B., Oliver, J., and Sykes, L. R., 1968, Seismology and the new global tectonics: J. Geophys. Res., **73**, 5855-5899.

Kaufman, H., 1953, Velocity functions in seismic prospecting: Geophysics, **18**, 289-297.

Koefoed, O., 1965, Direct methods of interpreting resistivity observations: Geophys. Prosp., **13**, 568-591.

―――― 1968, Application of the kernel function in interpreting geoelectric resistivity measurements: Gebruder-Borntraeger.

Kramer, F. S., Peterson, R. A., and Walter, W. C., Eds., 1968, Seismic energy sources 1968 handbook: Bendix United Geophysical.

Lee, Y., 1969, Statistical theory of communication: John Wiley & Sons, Inc.

Levinson, N. W., 1949 (HB) 1964 (PB), 1960, Extrapolation, interpolation, and smoothing of stationary time series: Appendix B in Wiener, MIT Technology Press.

Lindseth, R. O., 1970, Recent advances in digital processing of geophysical data: Computer Data Processors.

―――― 1979, Synthetic sonic logs—A process for stratigraphic interpretation: Geophysics, **44**, 3-26.

Love, A. E. H., 1944, A treatise on the mathematical theory of elasticity: Dover Publ. Inc.

Maillet, R., 1947, The fundamental equations of electrical prospecting: Geophysics, **12**, 529-556.

Meissner, R., 1965, P- and SV-waves from uphole shooting: Geophys. Prosp., **13**, 433-459.

―――― 1966, Interpretation of wide-angle measurements in the Bavarian Molasse basin: Geophys. Prosp., **14**, 7-16.

Morse, P. M., and Feshbach, H., 1967, Methods of theoretical physics, Part 1: McGraw-Hill Book Co.

Musgrave, A. W., and Bratton, R. H., 1967, Practical application of Blondeau weathering solution, in Seismic refraction prospecting: Soc. of Expl. Geophys., 231-246.

Musgrave, A. W., Woolley, W. C., and Gray, H., 1967, Outlining of salt and shale masses by refraction methods, in Seismic refraction prospecting: Soc. of Expl. Geophys., 426-458.

Muskat, M., and Meres, M. W., 1940, Reflection and transmission coefficients for plane waves in elastic media: Geophysics, **5**, 115-148.

Naudy, H., 1970, Une methode d'analysise fine des profils aeromagnetiques: Geophys. Prosp., **18**, 56-63.

Nettleton, L. L., 1940, Geophysical prospecting for oil: McGraw-Hill Book Co.

Nourbehecht, B., 1963 Irreversible thermodynamic effects in inhomogeneous media and their application in certain geoelectric problems: Ph.D. thesis, Mass. Inst. of Tech.

Palmer, D., 1980, The generalized reciprocal method of seismic refraction interpretation: Soc. of Expl. Geophys.

Parasnis, D. S., 1961, Magnetism: Hutchinson.

Paul, M. K., 1968, Notes on "Direct interpretation of resistivity profiles for Wenner electrode configuration" by O. Koefoed: Geophys. Prosp., **16**, 159-162.

Pekeris, C. L., 1940, Direct method of interpretation in resistivity prospecting: Geophysics, **5**, 31-42.

Peters, L. J., 1949, The direct approach to magnetic interpretation and its practical applications: Geophysics, **14**, 290-319.

Pickett, G. R., 1970, Applications for borehole geophysics in geophysical exploration: Geophysics, **35**, 81-92.

Poley, J. P., and van Stevennick, J., 1970, Geothermal prospecting: Delineation of shallow salt domes and surface faults: Geophys. Prosp., **18**, 666-700.

Postic, A., Fourmann, J., and Claerbout, J., 1980, Parsimonious deconvolution: Presented at the 50th Annual International SEG Meeting, Houston.

Poulter, T. C., 1950, The Poulter seismic method of geophysical exploration: Geophysics, **15**, 181-207.

Ricker, N., 1953, The form and laws of propagation of seismic wavelets: Geophysics, **18**, 10-40.

Rieber, F., 1936, Visual presentation of elastic wave patterns under various structural conditions: Geophysics, **1**, 196-218.

Robinson, E. A., and Treitel, S., 1964, Principles of digital filtering: Geophysics, **29**, 395-404.

Rockwell, D. W., 1967, A general wavefront method, in Seismic refraction prospecting: Musgrave A. W., Ed., Soc. of Expl. Geophys., 363-415.

Sangree, J. B., and Widmier, J. M., 1979, Interpretation of depositional facies from seismic data: Geophysics, **44**, 131-160.

Schneider, W. A., Larner, K. L., Burg, J. P., and Backus, M. M., 1964, A new data processing technique for the elimination of ghost arrivals on reflection seismograms: Geophysics, **29**, 783-805.

Schneider, W. A., Prince, E. R., and Giles, B. F., 1965, A new data processing technique for multiple attenuation exploiting differential normal moveout: Geophysics, **30**, 348-362.

SEG Technical Standards Committee, 1980, Digital tape standards: Soc. of Expl. Geophys.

Seigel, H. O., and Wait, J. R., 1959, Overvoltage research and geophysical applications: Pergamon Press, Inc.

Seismological tables, 1940: London, British Assoc. for Advancement of Science.

Sheriff, R. E., 1978, A first course in geophysical exploration and interpretation: Int. Human Resources Dev. Corp.

———— 1980, Seismic stratigraphy: Int. Human Resources Dev. Corp.

———— 1982, Structural interpretation of seismic data: Am. Assoc. of Petr. Geol. education course note ser. 23.

Sheriff, R. E., and Geldart, L. P., 1982, Exploration seismology, v. 1: Cambridge University Press.

———— 1983, Exploration seismology, v. 2: Cambridge University Press.

Siegert, A. J. F., 1942, Determination of the Bouguer correction constant: Geophysics, **7**, 29-34.

Slotnick, M. M., 1950, A graphical method for the interpretation of refraction profile data: Geophysics, **15**, 163-180.

Smith, B. D., and Ward, S. H., 1974, Short note on the computation of polarization ellipse parameters: Geophysics, **39**, 867-869.

Stacey, F. D., 1969, Physics of the earth: John Wiley & Sons, Inc.

Sumner, J. S., 1976, Principles in induced polarization for geophysical exploration: Elsevier.

Taner, M. T., Koehler, F., and Sheriff, R. E., 1979, Complex trace analysis: Geophysics, **44**, 1041-1063.

Tarrant, L. H., 1956, A rapid method of determining the form of a seismic refractor from line profile results: Geophys. Prosp., **4**, 131-139.

Telford, W. M., Geldart, L. P., Sheriff, R. E., and Keys, D. A., 1976, Applied geophysics: Cambridge University Press.

Thornburgh, H. R., 1930, Wave front diagrams in seismic interpretation: Bull., Am. Assoc. Petr. Geol., **14**, 185-200.

Toksöz, M. N., and Johnston, D. H., 1982, Seismic wave attenuation: Soc. of Expl. Geophys.

Tooley, R. D., Spencer, T. W., and Sagoci, H. F., 1965, Reflection and transmission of plane compressional waves: Geophysics, **30**, 552-570.

Treitel, S., and Robinson, E. A., 1969, Optimum digital filters for signal-to-noise ratio enhancement: Geophys. Prosp., **17**, 248-293.

Trorey, A. W., 1961, The information content of a Rieber sonogram: Geophysics, **26**, 761-764.

———— 1962, Theoretical seismograms with frequency and depth dependent absorption: Geophysics, **27**, 766-785.

Vail, P. R., Mitchum, R. M., and Thompson, S., 1977, Changes of sea level from coastal onlap, *in* Seismic stratigraphy—Applications to hydrocarbon exploration: Payton, C. E., Ed., Am. Assoc. of Petr. Geol. Mem. 26, 63-81.

Vozoff, K., 1956, Numerical resistivity analysis: Horizontal layers: Geophysics, **23**, 536-556.

———— 1972, The magnetotelluric method in the exploration of sedimentary basins: Geophysics, **37**, 98-141.

Ward, S. H., O'Brien, D. P., Parry, J. R., and McKnight, B. K., 1968, AFMAG interpretation: Geophysics, **33**, 621-644.

Weinstock, H., and Overton, W. C., 1981, SQUID applications to geophysics: Soc. of Expl. Geophys.

White, J. E., 1964, Motion product seismograms: Geophysics, **29**, 288-299.

———— 1969, SEG presidential address: Geophysics, **34**, 1.

Woollard, G. P., 1979, The new gravity system—changes in international gravity base values and anomaly values: Geophysics, **44**, 1352-1366.

Wyllie, M. R. J., Gregory, A. R., and Gardner, L. W., 1956, Elastic wave velocities in heterogeneous and porous media: Geophysics, **21**, 41-70.

Wyrobek, S. M., 1956, Application of delay and intercept times in the interpretation of multilayer refraction time-distance curves: Geophys. Prosp., **4**, 112-130.

Yungul, S., 1968, Measurement of telluric relative-ellipse area by means of vectograms: Geophysics, **33**, 127-131.

Zemanek, J., Glenn, E. E., Norton, L. J., and Caldwell, R. L., 1970, Formation evaluation by inspection with the borehole televiewer: Geophysics, **35**, 254-269.

Zietz, I., and Andreasen, G. E., 1967, Remanent magnetization and aeromagnetic interpretation *in* Mining geophysics: Soc. of Expl. Geophys., 569-590.